UNIT 1

INTRODUCTION

Welcome to Collins New GCSE Maths for Edexcel Modular Foundation Book 1. The first part of this book covers all the Core content you need for your Unit 1 and Unit 2 exams. The second part covers the content that is specific for Unit 1.

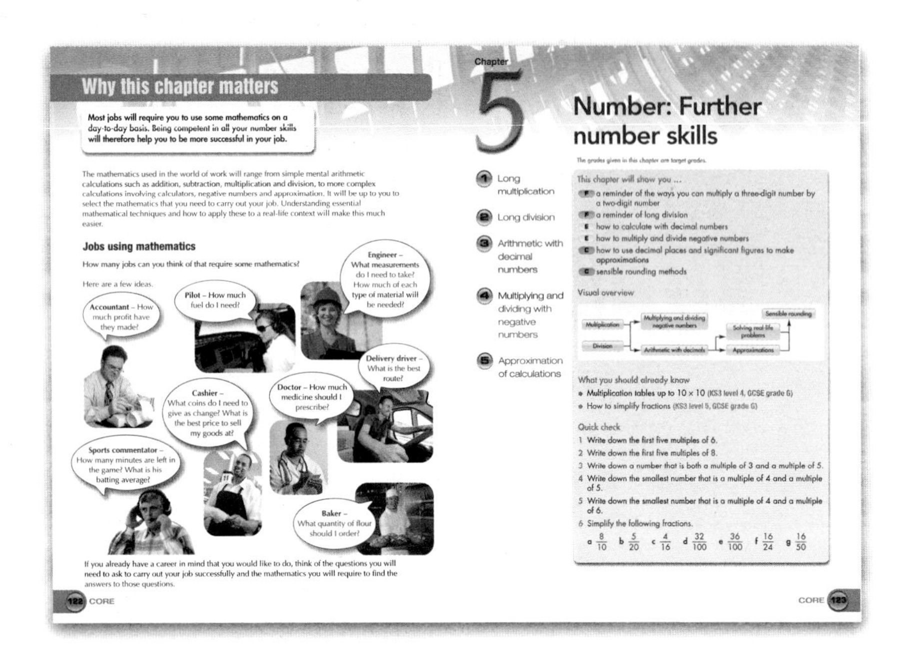

Why this chapter matters

Most jobs will require you to use some mathematics on a day-to-day basis. Being competent in all your number skills will therefore help you to be more successful in your job.

The mathematics used in the world of work will range from simple mental arithmetic calculations such as addition, subtraction, multiplication and division, to more complex calculations involving calculators, negative numbers and approximation. It will be up to you to select the mathematics that you need to carry out your job. Understanding essential mathematical techniques and how to apply these to a real-life context will make this much easier.

Jobs using mathematics

How many jobs can you think of that require some mathematics?

Here are a few ideas.

If you already have a career in mind that you would like to do, think of the questions you will need to ask to carry out your job successfully and the mathematics you will require to find the answers to those questions.

122 CORE

Chapter 5

Number: Further number skills

1 Long multiplication
2 Long division
3 Arithmetic with decimal numbers
4 Multiplying and dividing with negative numbers
5 Approximation of calculations

The grades given in this chapter are target grades.

This chapter will show you ...

- F a reminder of the ways you can multiply a three-digit number by a two-digit number
- F a reminder of long division
- E how to calculate with decimal numbers
- E how to multiply and divide negative numbers
- C how to use decimal places and significant figures to make approximations
- C sensible rounding methods

Visual overview

What you should already know

- Multiplication tables up to 10×10 (KS3 level 4, GCSE grade G)
- How to simplify fractions (KS3 level 5, GCSE grade G)

Quick check

1 Write down the first five multiples of 6.
2 Write down the first five multiples of 8.
3 Write down a number that is both a multiple of 3 and a multiple of 5.
4 Write down the smallest number that is a multiple of 4 and a multiple of 5.
5 Write down the smallest number that is a multiple of 4 and a multiple of 6.
6 Simplify the following fractions.

a $\frac{8}{10}$ b $\frac{5}{20}$ c $\frac{4}{16}$ d $\frac{32}{100}$ e $\frac{36}{100}$ f $\frac{16}{24}$ g $\frac{16}{50}$

CORE 123

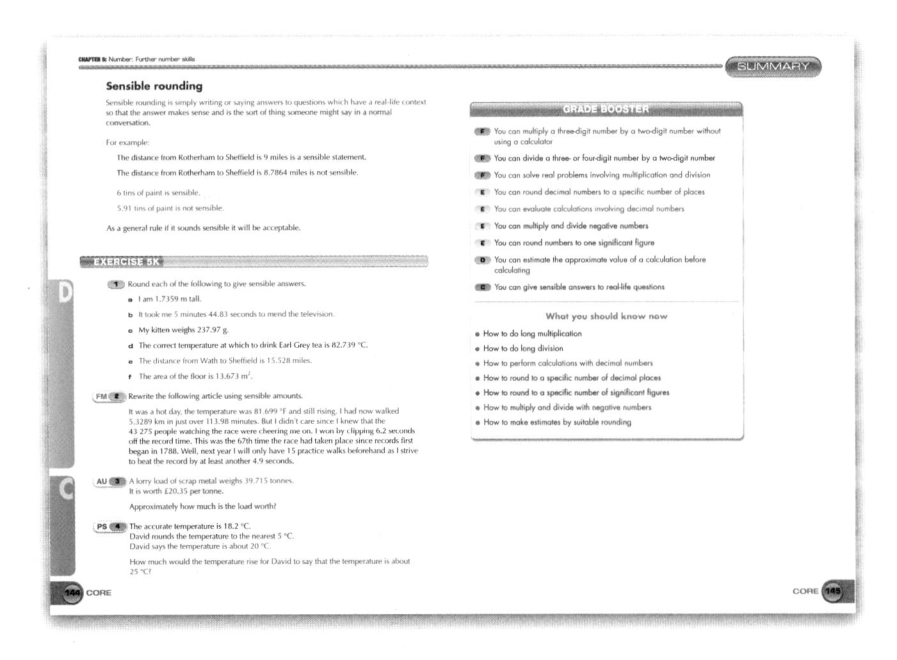

CHAPTER 5: Number: Further number skills

Sensible rounding

Sensible rounding is simply writing or saying answers to questions which have a real-life context so that the answer makes sense and is the sort of thing someone might say in a normal conversation.

For example:

The distance from Rotherham to Sheffield is 9 miles is a sensible statement.

The distance from Rotherham to Sheffield is 8.7864 miles is not sensible.

6 tins of paint is sensible.

5.91 tins of paint is not sensible.

As a general rule if it sounds sensible it will be acceptable.

EXERCISE 5K

1 Round each of the following to give sensible answers.

a I am 1.7359 m tall.
b It took me 5 minutes 44.83 seconds to mend the television.
c My kitten weighs 237.97 g.
d The correct temperature at which to drink Earl Grey tea is 82.739 °C.
e The distance from Wath to Sheffield is 15.528 miles.
f The area of the floor is 13.673 m^2.

FM 2 Rewrite the following article using sensible amounts.

It was a hot day, the temperature was 81.699 °F and still rising. I had now walked 5.3289 km in just over 113.98 minutes. But I didn't care since I knew that the 43 275 people watching the race were cheering me on. I won by clipping 6.2 seconds off the record time. This was the 67th time the race had taken place since records first began in 1788. Well, next year I will only have 15 practice walks beforehand as I strive to beat the record by at least another 4.9 seconds.

AU 3 A lorry load of scrap metal weighs 39.715 tonnes.
It is worth £20.35 per tonne.

Approximately how much is the load worth?

PS 4 The accurate temperature is 18.2 °C.
David rounds the temperature to the nearest 5 °C.
David says the temperature is about 20 °C.

How much would the temperature rise for David to say that the temperature is about 25 °C?

144 CORE

SUMMARY

GRADE BOOSTER

- F You can multiply a three-digit number by a two-digit number without using a calculator
- F You can divide a three- or four-digit number by a two-digit number
- F You can solve real problems involving multiplication and division
- E You can round decimal numbers to a specific number of places
- E You can evaluate calculations involving decimal numbers
- E You can multiply and divide negative numbers
- E You can round numbers to one significant figure
- D You can estimate the approximate value of a calculation before calculating
- C You can give sensible answers to real-life questions

What you should know now

- How to do long multiplication
- How to do long division
- How to perform calculations with decimal numbers
- How to round to a specific number of decimal places
- How to round to a specific number of significant figures
- How to multiply and divide with negative numbers
- How to make estimates by suitable rounding

CORE 145

Why this chapter matters

Find out why each chapter is important through the history of maths, seeing how maths links to other subjects and cultures, and how maths is related to real life.

Chapter overviews

Look ahead to see what maths you will be doing and how you can build on what you already know.

Colour-coded grades

Know what target grade you are working at and track your progress with the colour-coded grade panels at the side of the page.

Use of calculators

Questions when you must or could use your calculator are marked with an icon. Explanations involving calculators are based on the *Casio fx-83ES*.

Grade booster

Review what you have learnt and how to get to the next grade with the Grade booster at the end of each chapter.

Worked examples

Understand the topic before you start the exercise by reading the examples in blue boxes. These take you through questions step by step.

Collins

Student Book, **Foundation 1**

Delivering the Edexcel Specification

NEW GCSE MATHS
Edexcel Modular

Fully supports the 2010 GCSE Specification

Brian Speed • Keith Gordon • Kevin Evans • Trevor Senior

CONTENTS

CORE

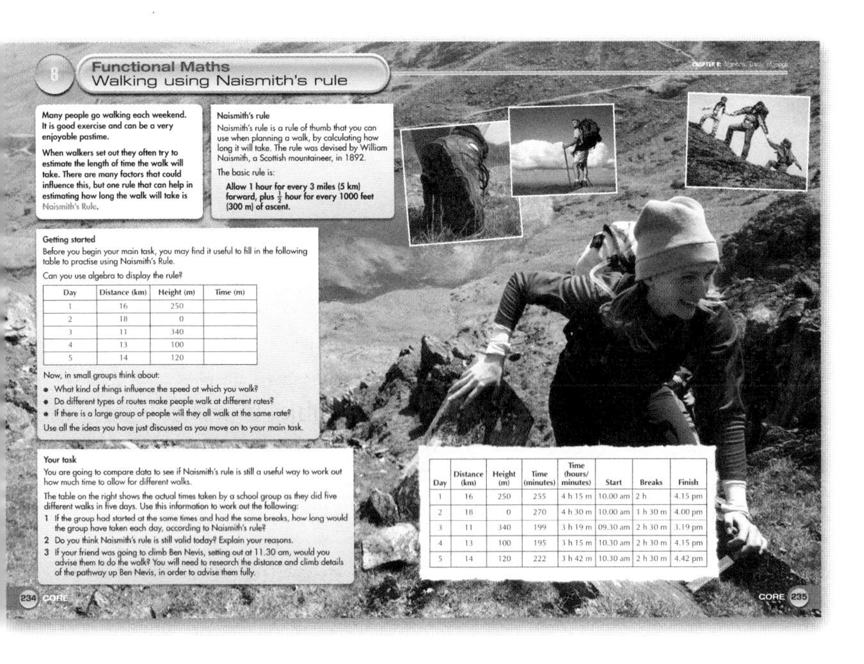

8 Functional Maths
Walking using Naismith's rule

Many people go walking each weekend. It is good exercise and can be a very enjoyable pastime.

When walkers set out they often try to estimate the length of time the walk will take. There are many factors that could influence this, but one rule that can help in estimating how long the walk will take is Naismith's Rule.

Naismith's rule

Naismith's rule is a rule of thumb that you can use when planning a walk, by calculating how long it will take. The rule was devised by William Naismith, a Scottish mountaineer, in 1892.

The basic rule is:

Allow 1 hour for every 3 miles (5 km) forward, plus $\frac{1}{2}$ hour for every 1000 feet (300 m) of ascent.

Getting started

Before you begin your main task, you may find it useful to fill in the following table to practise using Naismith's Rule.

Can you use algebra to display the rule?

Day	Distance (km)	Height (m)	Time (m)
1	16	250	
2	18	0	
3	11	340	
4	13	100	
5	14	120	

Now, in small groups think about:

- What kind of things influence the speed at which you walk?
- Do different types of routes make people walk at different rates?
- If there is a large group of people will they all walk at the same rate?

Use all the ideas you have just discussed as you move on to your main task.

Your task

You are going to compare data to see if Naismith's rule is still a useful way to work out how much time to allow for different walks.

The table on the right shows the actual times taken by a school group as they did five different walks in five days. Use this information to work out the following:

1 If the group had started at the same times and had the same breaks, how long would the group have taken each day, according to Naismith's rule?
2 Do you think Naismith's rule is still valid today? Explain your reasons.
3 If your friend was going to climb Ben Nevis, setting out at 11.30 am, would you advise them to do the walk? You will need to research the distance and climb details of the pathway up Ben Nevis, in order to advise them fully.

Day	Distance (km)	Height (m)	Time (minutes)	Time (hours/minutes)	Start	Breaks	Finish
1	16	250	255	4 h 15 m	10.00 am	2 h	4.15 pm
2	18	0	270	4 h 30 m	10.00 am	1 h 30 m	4.00 pm
3	11	340	199	3 h 19 m	09.30 am	2 h 30 m	3.19 pm
4	13	100	195	3 h 15 m	10.30 am	2 h 30 m	4.15 pm
5	14	120	222	3 h 42 m	10.30 am	2 h 30 m	4.42 pm

Functional maths

Practise functional maths skills to see how people use maths in everyday life. Look out for practice questions marked **FM**.

There are also extra functional maths and problem-solving activities at the end of every chapter to build and apply your skills.

New Assessment Objectives

Practise new parts of the curriculum (Assessment Objectives AO2 and AO3) with questions that assess your understanding marked **AU** and questions that test if you can solve problems marked **PS**. You will also practise some questions that involve several steps and where you have to choose which method to use; these also test AO2. There are also plenty of straightforward questions (AO1) that test if you can do the maths.

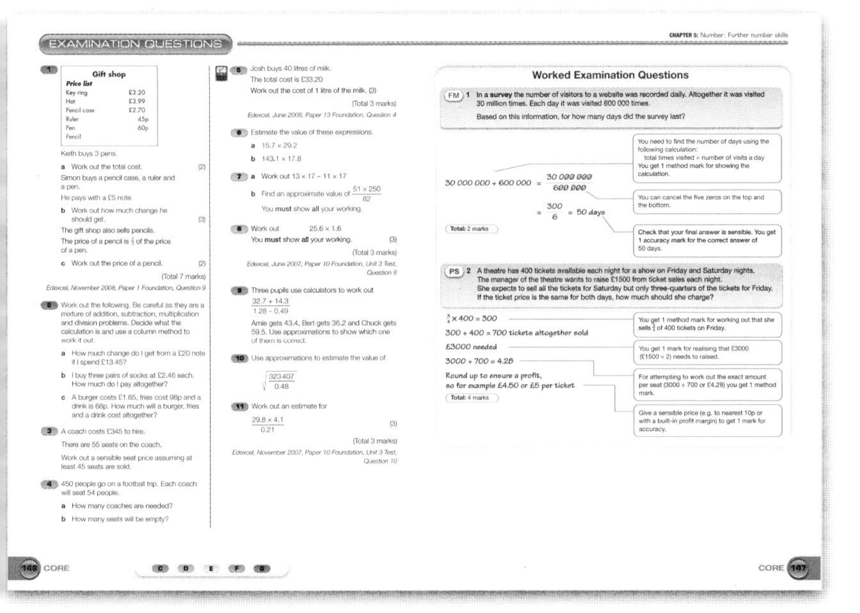

EXAMINATION QUESTIONS

1

Gift shop

Price list

Key ring	£3.20
Hat	£3.99
Pencil case	£2.70
Ruler	45p
Pen	60p
Pencil	

Keith buys 3 pens.

a Work out the total cost. (2)

Simon buys a pencil case, a ruler and a pen.

He pays with a £5 note.

b Work out how much change he should get. (3)

The gift shop also sells pencils.

The price of a pencil is [illegible] of the price of a pen.

c Work out the price of a pencil. (2)

(Total 7 marks)

Edexcel, November 2008, Paper 1 Foundation, Question 9

2 Work out the following. Be careful as they are a mixture of addition, subtraction, multiplication and division problems. Decide what the calculation is and use a column method to work it out.

a How much change do I get from a £20 note if I spend £13.45?

b I buy three pairs of socks at £2.46 each. How much do I pay altogether?

c A burger costs £1.65, fries cost 98p and a drink is 68p. How much will a burger, fries and a drink cost altogether?

3 A coach costs £345 to hire.

There are 55 seats on the coach.

Work out a sensible seat price assuming at least 45 seats are sold.

4 450 people go on a football trip. Each coach will seat 54 people.

a How many coaches are needed?

b How many seats will be empty?

5 Josh buys 40 litres of milk.

The total cost is £33.20

Work out the cost of 1 litre of the milk. (3)

(Total 3 marks)

Edexcel, June 2008, Paper 13 Foundation, Question 4

6 Estimate the value of these expressions.

a 15.7×29.2

b 143.1×17.8

7 a Work out $13 \times 17 - 11 \times 17$

b Find an approximate value of $\frac{51 \times 250}{82}$

You **must** show **all** your working.

8 Work out 25.6×1.6

You **must** show **all** your working. (3)

(Total 3 marks)

Edexcel, June 2007, Paper 10 Foundation, Unit 3 Test, Question 8

9 Three pupils use calculators to work out

$$\frac{32.7 + 14.3}{1.28 - 0.49}$$

Arnie gets 43.4, Bert gets 36.2 and Chuck gets 59.5. Use approximations to show which one of them is correct.

10 Use approximations to estimate the value of

$$\sqrt{\frac{323\,407}{0.48}}$$

11 Work out an estimate for

$$\frac{29.8 \times 4.1}{0.21}$$ (3)

(Total 3 marks)

Edexcel, November 2007, Paper 10 Foundation, Unit 3 Test, Question 10

Worked Examination Questions

FM 1 In a **survey** the number of visitors to a website was recorded daily. Altogether it was visited 30 million times. Each day it was visited 600 000 times.

Based on this information, for how many days did the survey last?

$$30\,000\,000 \div 600\,000 = \frac{30\,000\,000}{600\,000}$$

$$= \frac{300}{6} = 50 \text{ days}$$

You need to find the number of days using the following calculation:
total times visited ÷ number of visits a day
You get 1 method mark for showing the calculation.

You can cancel the five zeros on the top and the bottom.

Check that your final answer is sensible. You get 1 accuracy mark for the correct answer of 50 days.

Total: 2 marks

PS 2 A theatre has 400 tickets available each night for a show on Friday and Saturday nights. The manager of the theatre wants to raise £1500 from ticket sales each night. She expects to sell all the tickets for Saturday but only three-quarters of the tickets for Friday. If the ticket price is the same for both days, how much should she charge?

$\frac{3}{4} \times 400 = 300$

$300 + 400 = 700$ tickets altogether sold

£3000 needed

$3000 \div 700 = 4.28$

Round up to ensure a profit, so for example £4.50 or £5 per ticket

You get 1 method mark for working out that she sells $\frac{3}{4}$ of 400 tickets on Friday.

You get 1 mark for realising that £3000 (£1500 × 2) needs to raised.

For attempting to work out the exact amount per seat (3000 ÷ 700 or £4.28) you get 1 method mark.

Give a sensible price (e.g. to nearest 10p or with a built-in profit margin) to get 1 mark for accuracy.

Total: 4 marks

Exam practice

Prepare for your exams with past exam questions and detailed worked exam questions with examiner comments to help you score maximum marks.

Quality of Written Communication (QWC)

Practise using accurate mathematical vocabulary and writing logical answers to questions to ensure you get your QWC (Quality of Written Communication) marks in the exams. The Glossary and worked exam questions will help you with this.

Why this chapter matters

Thousands of years ago, many different civilisations developed different number systems. Most of these ways of counting were based on the number 10 (decimal numbers), simply because humans have 10 fingers and 10 toes.

The Egyptian number system (from about 3000BC) used different symbols to represent 1, 10, 100, 1000 and so on.

So the number 23 would be written: ∩∩ |||

Decimal number	Egyptian symbol	
1 =	𓏺	staff
10 =	𓎆	heel bone
100 =	𓍢	piece of rope
1000 =	𓆼	flower
10 000 =	𓂭	pointing finger
100 000 =	𓆐	tadpole
1 000 000 =	𓁨	man

The Chinese number system used sticks.

|	||	|||	||||	|||||	⊤	⊤|	⊤||	⊤|||
1	2	3	4	5	6	7	8	9

—	=	≡	≣	5 lines	⊥	⊥ with 2 lines	⊥ with 3 lines	⊥ with 4 lines
10	20	30	40	50	60	70	80	90

So, the number 23 would be written: = |||

The Roman number system is still used today, often on clocks, and is based on the numbers five and ten.

I	V	X	L	C	D	M
1	5	10	50	100	500	1000

So, the number 23 would be written: XXIII

The system we use today is called the **Hindu-Arabic system** and uses the symbols 0, 1, 2, 3, 4, 5, 6, 7, 8 and 9. This has been widely used since about 900AD.

This system provides an almost universal 'language' of maths that has allowed us to make sense of the world around us and communicate ideas to others, even when their spoken language may differ from ours.

Here you can see the Roman numerals for 1–12 on a clock face.

Roman numerals also appear on Big Ben.

Chapter 1

Number: Basic number

The grades given in this chapter are target grades.

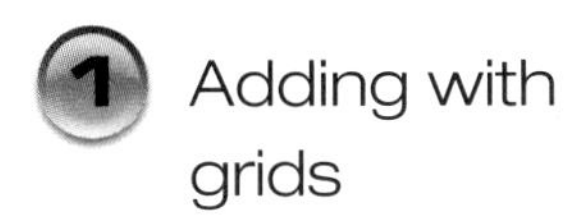

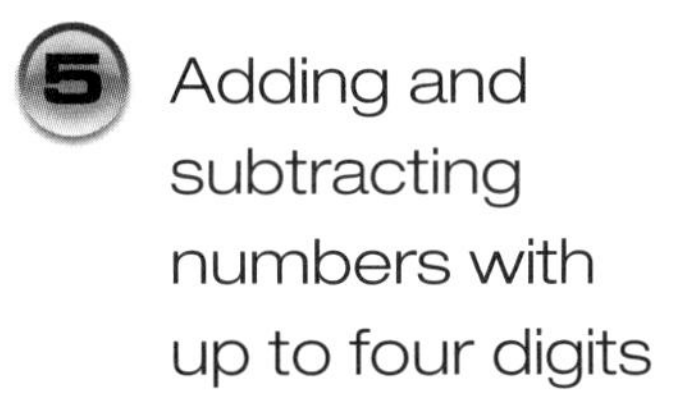

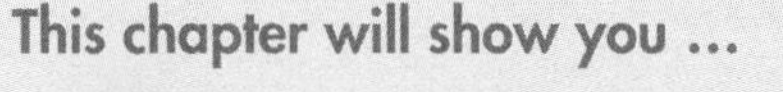

This chapter will show you …

to G F how to use basic number skills without a calculator

Visual overview

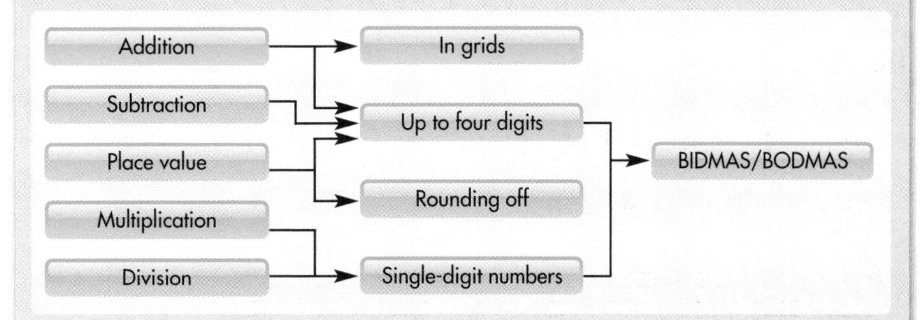

What you should already know

- Multiplication tables up to 10 × 10 **(KS3 level 4)**
- Addition and subtraction of numbers less than 20 **(KS3 level 2)**
- Simple multiplication and division **(KS3 level 3)**
- How to multiply numbers by 10 and 100 **(KS3 level 3)**

Quick check

How quickly can you complete these?

1 4 × 6	2 3 × 7	3 5 × 8	4 9 × 2
5 6 × 7	6 13 + 14	7 15 + 15	8 18 – 12
9 19 – 7	10 11 – 6	11 50 ÷ 5	12 48 ÷ 6
13 35 ÷ 7	14 42 ÷ 6	15 36 ÷ 9	16 8 × 10
17 9 × 100	18 3 × 10	19 14 × 100	20 17 × 10

1.1 Adding with grids

This section will show you how to:

- add and subtract single-digit numbers in a grid
- use row and column totals to find missing numbers in a grid

Key words
add
column
grid
row

ACTIVITY

Adding with grids

You need a set of cards marked 0 to 9.

0 1 2 3 4 5 6 7 8 9

Shuffle the cards and lay them out in a three by three **grid**. You will have one card left over.

3	5	0
7	6	4
8	2	9

Copy your grid onto a piece of paper. Then **add** up the numbers in each **row** and each **column** and write down their totals. Finally, find the grand total and write it in the box at the bottom right.

3	5	0	8
7	6	4	17
8	2	9	19
18	13	13	44

Look out for things that help. For example:

- in the first column, 3 + 7 make 10 and 10 + 8 = 18
- in the last column, 9 + 4 = 9 + 1 + 3 = 10 + 3 = 13

Reshuffle the cards, lay them out again and copy the new grid. Copy the new grid again on a fresh sheet of paper, leaving out some of the numbers.

4	5	8
0	2	6
9	1	7

4	5	8	17
0	2	6	8
9	1	7	17
13	8	21	42

4	□	8	17
□	2	□	8
9	□	7	□
□	8	21	42

Pass this last grid to a friend to work out the missing numbers. You can make it quite hard because you are using only the numbers from 0 to 9. Remember: once a number has been used, it *cannot* be used again in that grid.

Example Find the numbers missing from this grid.

□	□	9	17
□	2	□	11
8	□	□	□
19	3	17	□

Clues The two numbers missing from the second column must add up to 1, so they must be 0 and 1. The two numbers missing from the first column add up to 11, so they could be 7 and 4 or 6 and 5. Now, 6 or 5 won't work with 0 or 1 to give 17 across the top row. That means it has to be:

7	1	9	17
4	2	□	11
8	0	□	□
19	3	17	□

giving

7	1	9	17
4	2	5	11
8	0	3	11
19	3	17	39

as the answer.

You can use your cards to try out your ideas.

EXERCISE 1A

1 Find the row and column totals for each of these grids.

a

1	3	7	□
9	2	8	□
6	5	4	□
□	□	□	□

b

0	6	7	□
8	1	4	□
9	5	3	□
□	□	□	□

c

0	8	7	□
1	6	2	□
9	3	4	□
□	□	□	□

d

2	4	6	□
3	5	7	□
8	9	1	□
□	□	□	□

e

5	9	3	□
6	1	8	□
2	7	4	□
□	□	□	□

f

0	8	3	□
7	2	4	□
1	6	5	□
□	□	□	□

g

9	4	8	□
7	0	5	□
1	6	3	□
□	□	□	□

h

0	8	6	□
7	1	4	□
5	9	2	□
□	□	□	□

i

1	8	7	□
6	2	5	□
0	9	3	□
□	□	□	□

FM **2** **a** Adam has two £10 notes, one £5 note and eight £1 coins in his pocket. Does he have enough money to buy a shirt costing £32?

b Belinda puts some plants into pots. There are six roses, eight sunflowers and nine lilies. How many plants altogether does she put into pots?

G

G

AU 3 Here is a list of numbers.

4 5 7 8 11 12

a From the list write down **two** numbers that add up to 18.

b From the list work out the largest total that can be made using **three** numbers.

c From the list work out the largest **even** total that can be made using **three** numbers.

PS 4 What number could be being described?

'It is bigger than three'

'It is less than 10'

'It is an even number'

F

5 Find the numbers missing from each of these grids. Remember: the numbers missing from each grid must be chosen from 0 to 9 without any repeats.

a

1	7	□	16
□	3	6	9
5	□	2	11
6	14	16	36

b

1	□	3	6
□	5	4	15
7	8	□	24
14	15	16	35

c

9	3	□	18
4	□	5	9
□	2	8	11
14	5	19	38

d

□	□	□	16
2	□	4	13
8	5	0	13
19	13	10	42

e

2	□	6	17
□	1	□	□
5	□	8	13
11	□	17	38

f

1	□	□	16
□	2	4	12
□	9	3	□
12	□	15	□

g

0	2	□	3
9	□	□	□
□	4	5	17
17	□	13	42

h

□	□	3	4
□	7	4	□
9	6	□	20
18	□	12	□

i

□	□	4	10
□	2	□	□
8	□	□	15
15	□	□	36

PS 6 Find **three different** numbers that add up to 36.

PS 7 Find **two different odd** numbers that add up to 24.

1.2 Multiplication tables check

This section will show you how to:
- recall and use your knowledge of multiplication tables

Key words
multiplication
multiplication tables
sign
times table

ACTIVITY

Special table facts

You need a sheet of squared paper.

Start by writing in the easy multiplication tables (or **times tables**). These are the 1 ×, 2 ×, 5 ×, 9 × and 10 × tables.

Now draw up a 10 by 10 table square before you go any further. (Time yourself doing this and see if you can get faster.)

Once you have a complete tables square, shade out all the **multiplications** that you already know. You should be left with something like the square on the right.

×	1	2	3	4	5	6	7	8	9	10
1										
2										
3			9	12		18	21	24		
4			12	16		24	28	32		
5										
6			18	24		36	42	48		
7			21	28		42	49	56		
8			24	32		48	56	64		
9										
10										

Now cross out **one** of each pair that have the same answer, such as 3 × 4 and 4 × 3. This leaves you with:

×	1	2	3	4	5	6	7	8	9	10
1										
2										
3			9							
4			12	16						
5										
6			18	24		36				
7			21	28		42	49			
8			24	32		48	56	64		
9										
10										

Now there are just 15 table facts. Do learn them.

The rest are easy tables, so you should know all of them. But keep practising!

EXERCISE 1B

G

1 Write down the answer to each of the following without looking at the multiplication square.

a 4×5	**b** 7×3	**c** 6×4	**d** 3×5	**e** 8×2
f 3×4	**g** 5×2	**h** 6×7	**i** 3×8	**j** 9×2
k 5×6	**l** 4×7	**m** 3×6	**n** 8×7	**o** 5×5
p 5×9	**q** 3×9	**r** 6×5	**s** 7×7	**t** 4×6
u 6×6	**v** 7×5	**w** 4×8	**x** 4×9	**y** 6×8

FM **z** Matt works for 6 hours each day and is paid £8 an hour.

He is saving for a bike that costs £75.

Will he have enough after two days?

Show how you worked out your answer.

2 Write down the answer to each of the following without looking at the multiplication square.

a $10 \div 2$	**b** $28 \div 7$	**c** $36 \div 6$	**d** $30 \div 5$	**e** $15 \div 3$
f $20 \div 5$	**g** $21 \div 3$	**h** $24 \div 4$	**i** $16 \div 8$	**j** $12 \div 4$
k $42 \div 6$	**l** $24 \div 3$	**m** $18 \div 2$	**n** $25 \div 5$	**o** $48 \div 6$
p $36 \div 4$	**q** $32 \div 8$	**r** $35 \div 5$	**s** $49 \div 7$	**t** $27 \div 3$
u $45 \div 9$	**v** $16 \div 4$	**w** $40 \div 8$	**x** $63 \div 9$	**y** $54 \div 9$

FM **z** Viki works for 7 hours and is paid £42. She wants to save £60 to buy two tickets to the ballet.

How many hours will she need to work in order to save enough?

3 Write down the answer to each of the following. Look carefully at the **signs**, because they are a mixture of ×, +, – and ÷ .

a $5 + 7$	**b** $20 - 5$	**c** 3×7	**d** $5 + 8$	**e** $24 \div 3$
f $15 - 8$	**g** $6 + 8$	**h** $27 \div 9$	**i** 6×5	**j** $36 \div 6$
k 7×5	**l** $15 \div 3$	**m** $24 - 8$	**n** $28 \div 4$	**o** $7 + 9$
p $9 + 6$	**q** $36 - 9$	**r** $30 \div 5$	**s** $8 + 7$	**t** 4×6
u 8×5	**v** $42 \div 7$	**w** $8 + 9$	**x** 9×8	**y** $54 - 8$

FM **z** Ahmed works for 3 hours and is paid £11 an hour. Ben works for 4 hours and is paid £8 an hour. Who is paid more?

G

AU **4** Here are four single-digit number cards.

5	8	3	6

The cards are used for making calculations. Complete the following.

a [3] [5] + [...] [...] = 121

b [8] [3] − [...] [...] = 27

c [...] [...] [8] + [...] = 364

5 Write down the answer to each of the following.

a 3×10	**b** 5×10	**c** 8×10	**d** 10×10	**e** 12×10
f 18×10	**g** 24×10	**h** 4×100	**i** 7×100	**j** 9×100
k 10×100	**l** 14×100	**m** 24×100	**n** 72×100	**o** 100×100
p $20 \div 10$	**q** $70 \div 10$	**r** $90 \div 10$	**s** $170 \div 10$	**t** $300 \div 10$
u $300 \div 100$	**v** $800 \div 100$	**w** $1200 \div 100$	**x** $2900 \div 100$	**y** $5000 \div 100$

F

PS **6** **a** Two numbers when multiplied together give an answer of 24. One of the numbers is odd and greater than one. What is the other number?

b One whole number is divided by another whole number and gives an answer of 90. One of the numbers is 10. What is the other number?

PS **7** Consecutive numbers are numbers that are next to each other, for example 2 and 3.

Phil says that when two consecutive numbers are multiplied together the answer is always even.

Show, with two different examples, that this is true.

PS **8** Jenna says that when you multiply three consecutive numbers together you always get a number in the six times table, for example $2 \times 3 \times 4 = 24 = 4 \times 6$

Show, with two different examples, that this is true.

1.3 Order of operations and BIDMAS/BODMAS

This section will show you how to:

- work out the answers to a problem with a number of different mathematical operations

Key words
brackets
operation
order

Suppose you have to work out the answer to $4 + 5 \times 2$. You may say the answer is 18, but the correct answer is 14.

There is an **order** of **operations** which you *must* follow when working out calculations like this. The × is always done *before* the +.

In $4 + 5 \times 2$ this gives $4 + 10 = 14$.

Now suppose you have to work out the answer to $(3 + 2) \times (9 - 5)$. The correct answer is 20.

You have probably realised that the parts in the **brackets** have to be done *first*, giving $5 \times 4 = 20$.

So, how do you work out a problem such as $9 \div 3 + 4 \times 2$?

To answer questions like this, you *must* follow the BIDMAS (or BODMAS) rule. This tells you the order in which you *must* do the operations.

B	Brackets	**B**	Brackets
I	Indices (Powers)	**O**	pOwers
D	Division	**D**	Division
M	Multiplication	**M**	Multiplication
A	Addition	**A**	Addition
S	Subtraction	**S**	Subtraction

For example, to work out $9 \div 3 + 4 \times 2$:

First divide:	$9 \div 3 = 3$	giving	$3 + 4 \times 2$
Then multiply:	$4 \times 2 = 8$	giving	$3 + 8$
Then add:	$3 + 8 = 11$		

And to work out $60 - 5 \times 3^2 + (4 \times 2)$:

First, work out the brackets:	$(4 \times 2) = 8$	giving	$60 - 5 \times 3^2 + 8$
Then the index (power):	$3^2 = 9$	giving	$60 - 5 \times 9 + 8$
Then multiply:	$5 \times 9 = 45$	giving	$60 - 45 + 8$
Then add:	$60 + 8 = 68$	giving	$68 - 45$
Finally, subtract:	$68 - 45 = 23$		

ACTIVITY

Dice with BIDMAS/BODMAS

You need a sheet of squared paper and three dice.

Draw a five by five grid and write the numbers from 1 to 25 in the spaces.

The numbers can be in *any order*.

14	13	18	7	24
15	1	16	17	6
23	8	2	12	5
3	22	4	10	19
25	21	9	20	11

Now throw three dice. Record the score on each one.

Use these numbers to make up a number problem.

You must use all three numbers, and you must not put them together to make a number (such as making 136 with the three dice shown above). For example, with 1, 3 and 6 you could make:

$1 + 3 + 6 = 10$ $3 \times 6 + 1 = 19$ $(1 + 3) \times 6 = 24$

$6 \div 3 + 1 = 3$ $6 + 3 - 1 = 8$ $6 \div (3 \times 1) = 2$

and so on. The answer to the problem must be from 1 to 25. Remember to use **BIDMAS/BODMAS**.

You have to make only one problem with each set of numbers.

When you have made a problem, cross the answer off the grid and throw the dice again. Make up a problem with the next three numbers and cross that answer off the grid. Throw the dice again and so on.

The first person to make a line of five numbers across, down or diagonally is the winner.

You must write down each problem and its answer so that they can be checked.

Just put a line through each number on the grid, as you use it. Do not cross it out so that it cannot be read, otherwise your problem and its answer cannot be checked.

This might be a typical game.

~~14~~	13	~~18~~	7	~~24~~
15	~~1~~	16	17	6
23	8	~~2~~	12	5
3	22	4	~~10~~	19
25	21	9	20	~~11~~

First set (1, 3, 6)	$6 \times 3 \times 1 = 18$
Second set (2, 4, 4)	$4 \times 4 - 2 = 14$
Third set (3, 5, 1)	$(3 - 1) \times 5 = 10$
Fourth set (3, 3, 4)	$(3 + 3) \times 4 = 24$
Fifth set (1, 2, 6)	$6 \times 2 - 1 = 11$
Sixth set (5, 4, 6)	$(6 + 4) \div 5 = 2$
Seventh set (4, 4, 2)	$2 - (4 \div 4) = 1$

EXERCISE 1C

G

1 Work out each of these.

a $2 \times 3 + 5 =$ **b** $6 \div 3 + 4 =$ **c** $5 + 7 - 2 =$

d $4 \times 6 \div 2 =$ **e** $2 \times 8 - 5 =$ **f** $3 \times 4 + 1 =$

g $3 \times 4 - 1 =$ **h** $3 \times 4 \div 1 =$ **i** $12 \div 2 + 6 =$

j $12 \div 6 + 2 =$ **k** $3 + 5 \times 2 =$ **l** $12 - 3 \times 3 =$

2 Work out each of the following. Remember: first work out the bracket.

a $2 \times (3 + 5) =$ **b** $6 \div (2 + 1) =$ **c** $(5 + 7) - 2 =$

d $5 + (7 - 2) =$ **e** $3 \times (4 \div 2) =$ **f** $3 \times (4 + 2) =$

g $2 \times (8 - 5) =$ **h** $3 \times (4 + 1) =$ **i** $3 \times (4 - 1) =$

j $3 \times (4 \div 1) =$ **k** $12 \div (2 + 2) =$ **l** $(12 \div 2) + 2 =$

3 Copy each of these and put a loop round the part that you do first. Then work out the answer. The first one has been done for you.

a $\boxed{3 \times 3} - 2 = 7$ **b** $3 + 2 \times 4 =$ **c** $9 \div 3 - 2 =$

d $9 - 4 \div 2 =$ **e** $5 \times 2 + 3 =$ **f** $5 + 2 \times 3 =$

g $10 \div 5 - 2 =$ **h** $10 - 4 \div 2 =$ **i** $4 \times 6 - 7 =$

j $7 + 4 \times 6 =$ **k** $6 \div 3 + 7 =$ **l** $7 + 6 \div 2 =$

4 Work out each of these.

a $6 \times 6 + 2 =$ **b** $6 \times (6 + 2) =$ **c** $6 \div 6 + 2 =$

d $12 \div (4 + 2) =$ **e** $12 \div 4 + 2 =$ **f** $2 \times (3 + 4) =$

g $2 \times 3 + 4 =$ **h** $2 \times (4 - 3) =$ **i** $2 \times 4 - 3 =$

j $17 + 5 - 3 =$ **k** $17 - 5 + 3 =$ **l** $17 - 5 \times 3 =$

m $3 \times 5 + 5 =$ **n** $6 \times 2 + 7 =$ **o** $6 \times (2 + 7) =$

p $12 \div 3 + 3 =$ **q** $12 \div (3 + 3) =$ **r** $14 - 7 \times 1 =$

s $(14 - 7) \times 1 =$ **t** $2 + 6 \times 6 =$ **u** $(2 + 5) \times 6 =$

v $12 - 6 \div 3 =$ **w** $(12 - 6) \div 3 =$ **x** $15 - (5 \times 1) =$

y $(15 - 5) \times 1 =$ **z** $8 \times 9 \div 3 =$

5 Copy each of these and then put in brackets where necessary to make each answer true.

a $3 \times 4 + 1 = 15$ **b** $6 \div 2 + 1 = 4$ **c** $6 \div 2 + 1 = 2$

d $4 + 4 \div 4 = 5$ **e** $4 + 4 \div 4 = 2$ **f** $16 - 4 \div 3 = 4$

g $3 \times 4 + 1 = 13$ **h** $16 - 6 \div 3 = 14$ **i** $20 - 10 \div 2 = 5$

j $20 - 10 \div 2 = 15$ **k** $3 \times 5 + 5 = 30$ **l** $6 \times 4 + 2 = 36$

m $15 - 5 \times 2 = 20$ **n** $4 \times 7 - 2 = 20$ **o** $12 \div 3 + 3 = 2$

p $12 \div 3 + 3 = 7$ **q** $24 \div 8 - 2 = 1$ **r** $24 \div 8 - 2 = 4$

6 Three dice are thrown. They give scores of three, one and four.

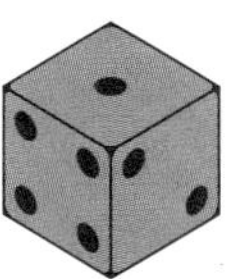

A class makes the following questions with the numbers. Work them out.

a $3 + 4 + 1 =$ **b** $3 + 4 - 1 =$ **c** $4 + 3 - 1 =$

d $4 \times 3 + 1 =$ **e** $4 \times 3 - 1 =$ **f** $(4 - 1) \times 3 =$

g $4 \times 3 \times 1 =$ **h** $(3 - 1) \times 4 =$ **i** $(4 + 1) \times 3 =$

j $4 \times (3 + 1) =$ **k** $1 \times (4 - 3) =$ **l** $4 + 1 \times 3 =$

AU 7 Jack says that $5 + 6 \times 7$ is equal to 77.

Is he correct?

Explain your answer.

AU 8 This is Micha's homework.

Copy the questions where she has made mistakes and work out the correct answers.

a $2 + 3 \times 4 =$ 20 **b** $8 - 4 \div 4 =$ 7 **c** $6 + 3 \times 2 =$ 12

d $7 - 1 \times 5 =$ 30 **e** $2 \times 7 + 2 =$ 16 **f** $9 - 3 \times 3 =$ 18

9 Three different dice give scores of 2, 3, 5. Add ÷, ×, + or – signs to make each calculation work.

a 2 3 5 = 11 **b** 2 3 5 = 16 **c** 2 3 5 = 17

d 5 3 2 = 4 **e** 5 3 2 = 13 **f** 5 3 2 = 30

AU 10 Which is smaller

$4 + 5 \times 3$ or $(4 + 5) \times 3$?

Show your working.

PS 11 Here is a list of numbers, some signs and one pair of brackets.

2 5 6 18 – × = ( )

Use **all** of them to make a correct calculation.

PS 12 Here is a list of numbers, some signs and one pair of brackets.

3 4 5 8 – ÷ = ( )

Use **all** of them to make a correct calculation.

FM 13 Jeremy has a piece of pipe that is 10 m long.

He wants to use his calculator to work out how much pipe will be left when he cuts off three pieces, each of length 1.5 m.

Which calculations would give him the correct answer?

$10 - 3 \times 1.5$ $10 - 1.5 + 1.5 + 1.5$ $10 - 1.5 - 1.5 - 1.5$

1.4 Rounding

This section will show you how to:
- round a number

Key words
approximation
rounded down
rounded up

You use rounded information all the time. Look at the examples on the right. All of these statements use rounded information. Each actual figure is either above or below the **approximation** shown here. But if the rounding is done correctly, you can find out what the maximum and the minimum figures really are. For example, if you know that the number of matches in the packet is rounded to the nearest 10, and it states that there are 30 matches in a packet:

- the smallest figure **rounded up** to 30 is 25, and
- the largest figure **rounded down** to 30 is 34 (because 35 would be rounded up to 40).

So, there could actually be from 25 to 34 matches in the packet.

What about the number of runners in the marathon? If you know that the number 23 000 is rounded to the nearest 1000:

- the smallest figure rounded up to 23 000 is 22 500, and
- the largest figure rounded down to 23 000 is 23 499 (because 23 500 would be rounded up to 24 000).

So, there could actually be from 22 500 to 23 499 people in the marathon.

EXERCISE 1D

G

1 Round each of these numbers to the nearest 10.

a 24	**b** 57	**c** 78	**d** 54	**e** 96
f 21	**g** 88	**h** 66	**i** 14	**j** 26
k 29	**l** 51	**m** 77	**n** 49	**o** 94
p 35	**q** 65	**r** 15	**s** 102	**t** 107

2 Round each of these numbers to the nearest 100.

a 240	**b** 570	**c** 780	**d** 504	**e** 967
f 112	**g** 645	**h** 358	**i** 998	**j** 1050
k 299	**l** 511	**m** 777	**n** 512	**o** 940
p 350	**q** 650	**r** 750	**s** 1020	**t** 1070

3 On the shelf of a sweetshop there are three jars like the ones below.

Jar 2

Jar 3

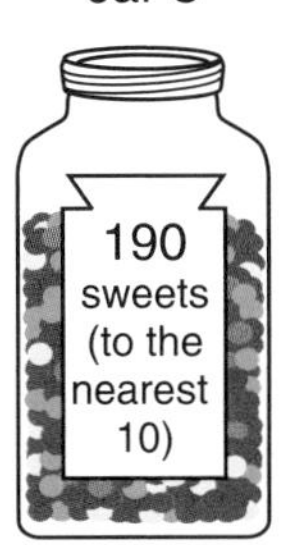

Look at each of the numbers below and write down which jar it could be describing. (For example, 76 sweets could be in jar 1.)

a 78 sweets	**b** 119 sweets	**c** 84 sweets	**d** 75 sweets
e 186 sweets	**f** 122 sweets	**g** 194 sweets	**h** 115 sweets
i 81 sweets	**j** 79 sweets	**k** 192 sweets	**l** 124 sweets

m Which of these numbers of sweets *could not* be in jar 1: 74, 84, 81, 76?

n Which of these numbers of sweets *could not* be in jar 2: 124, 126, 120, 115?

o Which of these numbers of sweets *could not* be in jar 3: 194, 184, 191, 189?

4 Round each of these numbers to the nearest 1000.

a 2400	**b** 5700	**c** 7806	**d** 5040	**e** 9670
f 1120	**g** 6450	**h** 3499	**i** 9098	**j** 1500
k 2990	**l** 5110	**m** 7777	**n** 5020	**o** 9400
p 3500	**q** 6500	**r** 7500	**s** 1020	**t** 1770

5 Round each of these numbers to the nearest 10.

a 234	**b** 567	**c** 718	**d** 524	**e** 906
f 231	**g** 878	**h** 626	**i** 114	**j** 296
k 279	**l** 541	**m** 767	**n** 501	**o** 942
p 375	**q** 625	**r** 345	**s** 1012	**t** 1074

G

AU 6

Welcome to Elsecar	Welcome to Hoyland	Welcome to Jump
Population 800 (to the nearest 100)	Population 1200 (to the nearest 100)	Population 600 (to the nearest 100)

Which of these sentences could be true and which must be false?

a There are 789 people living in Elsecar. **b** There are 1278 people living in Hoyland.

c There are 550 people living in Jump. **d** There are 843 people living in Elsecar.

e There are 1205 people living in Hoyland. **f** There are 650 people living in Jump.

FM 7 A sign maker is asked to create a sign similar to those shown in question 6 for Swinton, which has a population of 1385.

Make a diagram of the sign he should paint.

FM 8 These were the numbers of spectators in the crowds at nine Premier Division games on a weekend in May 2005.

Match	Spectators
Aston Villa v Man City	39 645
Blackburn v Fulham	18 991
Chelsea v Charlton	42 065
C. Palace v Southampton	26 066
Everton v Newcastle	40 438
Man.Utd v West Brom	67 827
Middlesbrough v Tottenham	34 766
Norwich v Birmingham	25 477
Portsmouth v Bolton	20 188

a Which match had the largest crowd?

b Which had the smallest crowd?

c Round all the numbers to the nearest 1000.

d Round all the numbers to the nearest 100.

9 Give these cooking times to the nearest 5 minutes.

a 34 min **b** 57 min **c** 14 min **d** 51 min **e** 8 min

f 13 min **g** 44 min **h** 32.5 min **i** 3 min **j** 50 s

PS 10 Matthew and Viki are playing a game with whole numbers.

a What is the smallest number Matthew could be thinking of?

I am thinking of a number. Rounded to the nearest 10 it is 380.

I am thinking of a different number. Rounded to the nearest 100 it is 400.

F

b Is Viki's number smaller than Matthew's? How many possible answers are there?

E

PS AU 11 The number of adults attending a comedy show is 80 to the nearest 10.

The number of children attending is 50 to the nearest 10.

Katie says that 130 adults and children attended the comedy show.

Give an example to show that she may **not** be correct.

1.5 Adding and subtracting numbers with up to four digits

This section will show you how to:
- add and subtract numbers with more than one digit

Key words
addition
column
digit
subtract

Addition

There are three things to remember when you are adding two whole numbers.

- The answer will always be larger than the bigger number.
- Always add the units **column** first.
- When the total of the **digits** in a column is more than nine, you have to carry a digit into the next column on the left, as shown in Example 2. It is important to write down the carried digit, otherwise you may forget to include it in the **addition**.

EXAMPLE 1

Add: **a** 167 + 25 **b** 2296 + 1173

a
```
  167
+  25
  ---
  192
   1
```

b
```
  2296
+ 1173
  ----
  3469
   1
```

Subtraction

These are four things to remember when you are subtracting one whole number from another.

- The bigger number must always be written down first.
- The answer will always be smaller than the bigger number.
- Always **subtract** the units column first.
- When you have to take a bigger digit from a smaller digit in a column, you must 'borrow' a 10 by taking one from the column to the left and putting it with the smaller digit, as shown in Example 3.

EXAMPLE 2

Subtract: **a** 874 − 215 **b** 300 − 163

a
```
  8 ⁶7̸ ¹4
−  2 1 5
  ------
   6 5 9
```

b
```
  ²3̸ ⁹0̸ ¹0
−  1 6 3
  ------
   1 3 7
```

EXERCISE 1E

G

1 Copy and work out each of these additions.

a $\begin{array}{r} 365 \\ +\ 348 \\ \hline \end{array}$ **b** $\begin{array}{r} 95 \\ +\ 56 \\ \hline \end{array}$ **c** $\begin{array}{r} 4872 \\ +\ 1509 \\ \hline \end{array}$ **d** $\begin{array}{r} 317 \\ 416 \\ +\ 235 \\ \hline \end{array}$ **e** $\begin{array}{r} 287 \\ +\ 335 \\ \hline \end{array}$

f $\begin{array}{r} 483 \\ +\ 832 \\ \hline \end{array}$ **g** $\begin{array}{r} 4676 \\ +\ 3584 \\ \hline \end{array}$ **h** $\begin{array}{r} 438 \\ 147 \\ +\ 233 \\ \hline \end{array}$ **i** $\begin{array}{r} 175 \\ +\ 276 \\ \hline \end{array}$ **j** $\begin{array}{r} 562 \\ 93 \\ +\ 197 \\ \hline \end{array}$

2 Copy and complete each of these additions.

a 128 + 518 **b** 563 + 85 + 178 **c** 3086 + 58 + 674
d 347 + 408 **e** 85 + 1852 + 659 **f** 759 + 43 + 89
g 257 + 93 **h** 605 + 26 + 2135 **i** 56 + 8407 + 395
j 89 + 752 **k** 6143 + 557 + 131 **l** 2593 + 45 + 4378
m 719 + 284 **n** 545 + 3838 + 67 **o** 5213 + 658 + 4073

3 Copy and complete each of these subtractions.

a $\begin{array}{r} 637 \\ -\ 187 \\ \hline \end{array}$ **b** $\begin{array}{r} 908 \\ -\ 345 \\ \hline \end{array}$ **c** $\begin{array}{r} 954 \\ -\ 472 \\ \hline \end{array}$ **d** $\begin{array}{r} 572 \\ -\ 158 \\ \hline \end{array}$ **e** $\begin{array}{r} 732 \\ -\ 447 \\ \hline \end{array}$

f $\begin{array}{r} 673 \\ -\ 187 \\ \hline \end{array}$ **g** $\begin{array}{r} 602 \\ -\ 358 \\ \hline \end{array}$ **h** $\begin{array}{r} 638 \\ -\ 354 \\ \hline \end{array}$ **i** $\begin{array}{r} 650 \\ -\ 317 \\ \hline \end{array}$ **j** $\begin{array}{r} 580 \\ -\ 364 \\ \hline \end{array}$

k $\begin{array}{r} 6254 \\ -\ 3362 \\ \hline \end{array}$ **l** $\begin{array}{r} 8043 \\ -\ 3626 \\ \hline \end{array}$ **m** $\begin{array}{r} 8432 \\ -\ 4665 \\ \hline \end{array}$ **n** $\begin{array}{r} 8034 \\ -\ 3947 \\ \hline \end{array}$ **o** $\begin{array}{r} 5375 \\ -\ 3547 \\ \hline \end{array}$

4 Copy and complete each of these subtractions.

a 354 – 226 **b** 285 – 256 **c** 663 – 329
d 506 – 328 **e** 654 – 377 **f** 733 – 448
g 592 – 257 **h** 753 – 354 **i** 6705 – 2673
j 8021 – 3256 **k** 7002 – 3207 **l** 8700 – 3263

FM 5 The distance from Cardiff to London is 152 miles.
The distance from London to Edinburgh is 406 miles.

a How far is it to travel from London to Cardiff and then from Cardiff to Edinburgh?

b How much further is it to travel from London to Edinburgh than from Cardiff to London?

AU 6 Jon is checking the addition of two numbers.

His answer is 843.

One of the numbers is 591.

What should the other number be?

7 Copy each of these additions and fill in the missing digits.

a $\begin{array}{r} 5\ 3 \\ +\,2\ \square \\ \hline \square\ 9 \end{array}$ **b** $\begin{array}{r} \square\ 7 \\ +\,3\ \square \\ \hline 8\ 4 \end{array}$ **c** $\begin{array}{r} 4\ 5 \\ +\,\square\ \square \\ \hline 9\ 3 \end{array}$ **d** $\begin{array}{r} 4\ \square\ 7 \\ +\,\square\ 5\ \square \\ \hline 9\ 3\ 6 \end{array}$

e $\begin{array}{r} \square\ 1\ 8 \\ +\,2\ 5\ \square \\ \hline 8\ \square\ 7 \end{array}$ **f** $\begin{array}{r} 5\ 4\ \square \\ +\,\square\ \square\ 6 \\ \hline 8\ 2\ 2 \end{array}$ **g** $\begin{array}{r} 4\ 6\ 9 \\ +\,\square\ \square\ \square \\ \hline 7\ 3\ 5 \end{array}$ **h** $\begin{array}{r} \square\ \square\ \square \\ +\,3\ 4\ 8 \\ \hline 8\ 0\ 7 \end{array}$

i $\begin{array}{r} \square\ 4\ \square \\ +\,3\ 3\ 7 \\ \hline 7\ \square\ 5 \end{array}$ **j** $\begin{array}{r} 3\ 5\ 7\ 8 \\ +\,\square\ \square\ \square\ \square \\ \hline 8\ 0\ 7\ 6 \end{array}$

AU 8 Lisa is checking the subtraction: 614 – 258.

Explain how you know that her answer of 444 is incorrect without working out the whole calculation.

9 Copy each of these subtractions and fill in the missing digits.

a $\begin{array}{r} 7\ 4 \\ -\,2\ \square \\ \hline \square\ 1 \end{array}$ **b** $\begin{array}{r} \square\ 7 \\ -\,3\ \square \\ \hline 5\ 4 \end{array}$ **c** $\begin{array}{r} 8\ 5 \\ -\,\square\ \square \\ \hline 2\ 7 \end{array}$ **d** $\begin{array}{r} 6\ 7\ \square \\ -\,\square\ \square\ 3 \\ \hline 1\ 3\ 5 \end{array}$

e $\begin{array}{r} \square\ 1\ 4 \\ -\,2\ 5\ \square \\ \hline 3\ \square\ 7 \end{array}$ **f** $\begin{array}{r} 5\ 4\ \square \\ -\,\square\ \square\ 6 \\ \hline 3\ 2\ 5 \end{array}$ **g** $\begin{array}{r} 4\ 6\ 2 \\ -\,\square\ \square\ \square \\ \hline 1\ 8\ 5 \end{array}$ **h** $\begin{array}{r} \square\ \square\ \square \\ -\,2\ 4\ 7 \\ \hline 3\ 0\ 9 \end{array}$

i $\begin{array}{r} \square\ 4\ \square \\ -\,5\ 5\ 8 \\ \hline 2\ \square\ 5 \end{array}$ **j** $\begin{array}{r} 8\ 0\ 7\ 6 \\ -\,\square\ \square\ \square\ \square \\ \hline 6\ 1\ 8\ 7 \end{array}$

PS 10 A two-digit number is subtracted from a three-digit number.

The answer is 154.

Work out one pair of possible values for the numbers.

1.6 Multiplying and dividing by single-digit numbers

This section will show you how to:
- multiply and divide by a single-digit number

Key words
division
multiplication

Multiplication

There are two things to remember when you are multiplying two whole numbers.

- The bigger number must always be written down first.
- The answer will always be larger than the bigger number.

EXAMPLE 3

a Multiply 231 by 4.

$$\begin{array}{r} 213 \\ \times \quad 4 \\ \hline 852 \\ \hline {\scriptstyle 1} \end{array}$$

b Multiply 543 by 6.

$$\begin{array}{r} 543 \\ \times \quad 6 \\ \hline 3258 \\ \hline {\scriptstyle 2\,1} \end{array}$$

Note that in part **a** the first multiplication, 3 × 4, gives 12. So, you need to carry a digit into the next column on the left, as in the case of addition. Similar actions are carried out in part **b**.

Division

There are two things to remember when you are dividing one whole number by another whole number:

- The answer will always be smaller than the bigger number.
- Division starts at the *left-hand side*.

EXAMPLE 4

a Divide 417 by 3.

417 ÷ 3 is set out as:

$$\begin{array}{r} 1\;3\;9 \\ 3\overline{\smash{)}\,4^{1}1^{2}7} \end{array}$$

b Divide 508 by 4.

508 ÷ 4 is set out as:

$$\begin{array}{r} 1\;2\;7 \\ 4\overline{\smash{)}\,5^{1}0^{2}8} \end{array}$$

This is how the division was done in part **a**:

- First, divide 3 into 4 to get 1 and remainder 1. Note where to put the 1 and the remainder 1.
- Then, divide 3 into 11 to get 3 and remainder 2. Note where to put the 3 and the remainder 2.
- Finally, divide 3 into 27 to get 9 with no remainder, giving the answer 139.

A similar process takes place in part **b**.

EXERCISE 1F

1 Copy and work out each of the following multiplications.

a $\begin{array}{r} 14 \\ \times\ 4 \\ \hline \end{array}$ **b** $\begin{array}{r} 13 \\ \times\ 5 \\ \hline \end{array}$ **c** $\begin{array}{r} 17 \\ \times\ 3 \\ \hline \end{array}$ **d** $\begin{array}{r} 19 \\ \times\ 2 \\ \hline \end{array}$ **e** $\begin{array}{r} 18 \\ \times\ 6 \\ \hline \end{array}$

f $\begin{array}{r} 23 \\ \times\ 5 \\ \hline \end{array}$ **g** $\begin{array}{r} 34 \\ \times\ 6 \\ \hline \end{array}$ **h** $\begin{array}{r} 42 \\ \times\ 7 \\ \hline \end{array}$ **i** $\begin{array}{r} 53 \\ \times\ 4 \\ \hline \end{array}$ **j** $\begin{array}{r} 85 \\ \times\ 5 \\ \hline \end{array}$

k $\begin{array}{r} 50 \\ \times\ 3 \\ \hline \end{array}$ **l** $\begin{array}{r} 200 \\ \times\ 4 \\ \hline \end{array}$ **m** $\begin{array}{r} 320 \\ \times\ 3 \\ \hline \end{array}$ **n** $\begin{array}{r} 340 \\ \times\ 4 \\ \hline \end{array}$ **o** $\begin{array}{r} 253 \\ \times\ 6 \\ \hline \end{array}$

2 Calculate each of the following multiplications by setting the work out in columns.

a 42 × 7 **b** 74 × 5 **c** 48 × 6

d 208 × 4 **e** 309 × 7 **f** 630 × 4

g 548 × 3 **h** 643 × 5 **i** 8 × 375

j 6 × 442 **k** 7 × 528 **l** 235 × 8

m 6043 × 9 **n** 5 × 4387 **o** 9 × 5432

3 Calculate each of the following divisions.

a 438 ÷ 2 **b** 634 ÷ 2 **c** 945 ÷ 3

d 636 ÷ 6 **e** 297 ÷ 3 **f** 847 ÷ 7

g 756 ÷ 3 **h** 846 ÷ 6 **i** 576 ÷ 4

j 344 ÷ 4 **k** 441 ÷ 7 **l** 5818 ÷ 2

m 3744 ÷ 9 **n** 2008 ÷ 8 **o** 7704 ÷ 6

AU 4 Dean, Sean and Andy are doing a charity cycle ride from Huddersfield to Southend. The distance from Huddersfield to Southend is 235 miles.

a How many miles do all three travel altogether?

b Dean is sponsored £6 per mile.
Sean is sponsored £5 per mile.
Andy is sponsored £4 per mile.

How much money do they raise altogether?

G

F

F

FM **5** The 235-mile charity cycle ride in question 4 is planned to take five days.

a What is the least number of miles each cyclist has to cover each day?

b The cyclists travel 60 miles on the first day and 50 miles on the second day.

Fill in a copy of the plan for the remaining days so they complete the ride on time. Each day the number of miles covered has to be fewer than for the previous day.

	Mileage
Day 3	
Day 4	
Day 5	

6 By doing a suitable multiplication, answer each of these questions.

a How many days are there in 17 weeks?

b How many hours are there in four days?

c Eggs are packed in boxes of six. How many eggs are there in 24 boxes?

d Joe bought five boxes of matches. Each box contained 42 matches. How many matches did Joe buy altogether?

e A box of Tulip Sweets holds 35 sweets. How many sweets are there in six boxes?

7 By doing a suitable division, answer each of these questions.

a How many weeks are there in 91 days?

b How long will it take me to save £111, if I save £3 a week?

c A rope, 215 m long, is cut into five equal pieces. How long is each piece?

d Granny has a bottle of 144 tablets. How many days will they last if she takes four each day?

e I share a box of 360 sweets among eight children. How many sweets will each child get?

PS **8** Here is part of the 38 times table.

×1	×2	×3	×4	×5
38	76	114	152	190

Show how you can use this table to work out

a 9×38 **b** 52×38 **c** 105×38.

Letter sets

Find the next letters in these sequences.

a **O, T, T, F, F, ...**

b **T, F, S, E, T, ...**

Valued letters

In the three additions below, each letter stands for a single numeral. But a letter may not necessarily stand for the same numeral when it is used in more than one sum.

a

```
  ONE
+ ONE
-----
  TWO
```

b

```
   TWO
+  TWO
------
  FOUR
```

c

```
  FOUR
+ FIVE
------
  NINE
```

Write down each addition in numbers.

Four fours

Write number sentences to give answers from 1 to 10, using only fours and any number of the operations +, –, × and ÷ . For example:

$1 = (4 + 4) \div (4 + 4)$

$2 = (4 \times 4) \div (4 + 4)$

Heinz 57

Pick any number in the grid on the right. Circle the number and cross out all the other numbers in the row and column containing the number you have chosen. Now circle another number that is not crossed out and cross out all the other numbers in the row and column containing this number. Repeat until you have five numbers circled. Add these numbers together. What do you get? Now do it again but start with a different number.

19	8	11	25	7
12	1	4	18	0
16	5	8	22	4
21	10	13	27	9
14	3	6	20	2

Magic squares

This is a magic square. Add the numbers in any row, column or diagonal. The answer is *always* 15.

8	1	6
3	5	7
4	9	2

Now try to complete this magic square using every number from 1 to 16.

1		14	
	6		9
8		11	
	3		16

Hints

Letter sets Think about numbers.

Valued letters **a** Try E = 2, N = 3 **b** Try O = 7, U = 3 **c** Try N = 5, O = 9

There are other answers to each sum.

GRADE BOOSTER

- **G** You can add columns and rows in grids
- **G** You know the multiplication tables up to 10 × 10
- **G** You can use BIDMAS/BODMAS to work out calculations in the correct order
- **G** You can round numbers to the nearest 10 and 100
- **G** You can add and subtract numbers with up to four digits
- **G** You can multiply numbers by a single-digit number
- **F** You can solve problems involving multiplication and division by a single-digit number

What you should know now

- How to use BIDMAS/BODMAS
- How to round to the nearest 10, 100, 1000
- How to solve simple problems in context, using the four operations of addition, subtraction, multiplication and division

1 Work out

i $3 \times 3 - 5$ (1)

ii $20 \div (12 - 2)$ (1)

iii $7 + 8 \div 4$ (1)

(Total 3 marks)

Edexcel, May 2008, Paper 1 Foundation, Question 10

2 Look at the numbers in the cloud.

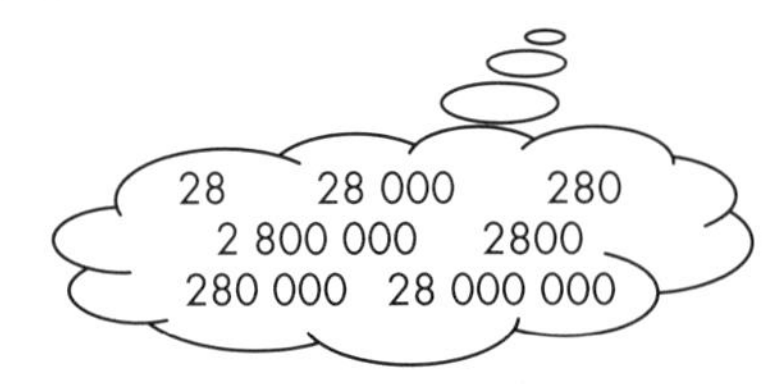

a Write down the number from the cloud which is

i twenty eight million

ii two thousand eight hundred.

b What number should go in the boxes to make the calculation correct?

i $28 \times \square = 2800$

ii $2\,800\,000 \div \square = 280\,000$

3 Mount Everest is 8848 metres high.

Mount Snowdon is 1085 metres high.

How much higher is Mount Everest than Mount Snowdon?

Give your answer to the nearest 10 metres.

4 It costs £7 per person to visit a show.

a 215 people attend the show on Monday.

How much do they pay altogether?

b On Tuesday the takings were £1372.

How many fewer people attended on Tuesday than on Monday?

5 The 2009 population of Plaistow is given as 7800 to the nearest hundred.

a What is the lowest number that the population could be?

b What is the largest number that the population could be?

6 Murray and Harry both worked out $2 + 4 \times 7$.
Murray calculated this to be 42.
Harry worked this out to be 30.
Explain why they both got different answers.

7 Work out the following. Be careful as they could involve addition, subtraction, multiplication or division. Decide what the calculation is and use a column method to work it out.

a Trays of pansies contain 12 plants each. How many plants will I get in 8 trays?

b There are 192 pupils in year 7. They are in 6 forms. How many pupils are in each form?

c A school term consists of 42 days. If a normal school week is 5 days.

i How many full weeks will there be in the term?

ii How many odd dates will there be?

d A machine produces 120 bolts every minute.

i How many bolts will be produced by the machine in 9 minutes?

ii The bolts are packed in bags of 8. How many bags will it take to pack 120 bolts?

8 **a** There are 7 days in a week.

i How many days are there in 15 weeks?

ii How many weeks are there in 161 days?

b Bulbs are sold in packs of 6.

i How many bulbs are there in 12 packs?

ii How many packs make up 186 bulbs?

9 A teacher asked her pupils to work out the following calculation without a calculator

$2 \times 3^2 + 6$

a Alice got an answer of 42. Billy got an answer of 30. Chas got an answer of 24. Explain why Chas was correct

b Put brackets into these calculations to make them true.

i $2 \times 3^2 + 6 = 42$

ii $2 \times 3^2 + 6 = 30$

10 The following are two pupils' attempts at working out $3 + 5^2 - 2$

Adam $3 + 5^2 - 2 = 3 + 10 - 2 = 13 - 2 = 11$

Bekki $3 + 5^2 - 2 = 8^2 - 2 = 64 - 2 = 62$

a Each pupil has made one mistake. Explain what this is for each of them.

b Work out the correct answer to $3 + 5^2 - 2$.

Worked Examination Questions

1 Here are four number cards, showing the number 2745.

2	7	4	5

Using all four cards, write down:

a the largest possible number

b the smallest possible number

c the missing numbers from this problem. ☐ 7 × 2 = ☐ ☐

a 7542

1 mark

Start with the largest number as the thousands digits, use the next largest as the hundreds digit and so on.
This is worth 1 mark.

b 2457

1 mark

Start with the smallest number as the thousands digits, use the next smallest as the hundreds digit and so on. Note the answer is the reverse of the answer to part **a**.
This is worth 1 mark.

c 27 × 2 = 54

2 marks

There are three numbers left, 2, 5, 4. The 2 must go into the first box and then you can work out that 2 × 27 is 54.
You get 1 mark for identifying the units digit of the answer as 4 and 2 marks if the whole answer is correct.

Total: 4 marks

2 Write the number 7845 to:

a the nearest 10

b the nearest 100.

a 7850

A halfway value such as 45 rounds up to 50.
This is worth 1 mark.

b 7800

7845 rounds down to 7800. Do not be tempted to round the answer to part (i) up to 7900.
This is worth 1 mark.

Total: 2 marks

Worked Examination Questions

FM **3** I want to buy a burger, a portion of chips and a bottle of cola for my lunch. I have the following coins in my pocket: £1, 50p, 50p, 20p, 20p, 10p, 2p, 2p, 1p.

a Do I have enough money?

b What about if I replace the chips with beans?

Price List	
Burger	£1.20
Chips	90p
Beans	50p
Cola	60p

a Total cost for a burger, chips and a cola = £2.70.

Total of coins in my pocket = £2.55

No, as £2.70 > £2.55

This is a question in which you could be assessed on your quality of written communication, so set out your answer clearly.

Show the total cost of the meal and the total of the coins in your pocket. This is worth 2 marks.

It is important to show a clear conclusion. Do not just say 'No'. Use the numbers to back up your answer. This is worth 1 mark.

3 marks

b Beans are 40p cheaper and £2.30 < £2.55.

So, I could afford this.

Explain clearly why you could afford this meal. This is worth 1 mark.

1 mark

Total: 4 marks

PS **4** Jack has £5 to spend on plants. Plants cost 85p each.

How much change will he receive if he buys the maximum number of plants he can afford?

$5 \div 0.85$

You get 1 mark for showing a method to work out the maximum he could buy. You could also use the calculation 85 + 85 + 85, etc but this is not as easy as dividing.

5 plants, or £4.25

You get 1 mark for realising that the most he could buy is five or showing £4.25.

75p change

1 mark is available for the answer.

Total: 5 marks

1 Functional Maths
Activity holiday in Wales

You are planning to go on an activity holiday in Wales with a group of friends. You are arriving late on Sunday evening and will be staying all week until the end of Friday.
You have chosen the area that you are staying in because there are lots of activities available and you want to make sure that everyone in the group can find something that they will enjoy on the holiday.

Your task

Work in groups of three or four.

Decide what the different interests in your group are and work together to make a timetable of activities that addresses all of these interests. You will need to make sure that everyone has the time to do all of the things that they would like to do.

You must then work out how much the holiday is going to cost for each of you.

Getting started

Use the following points to get you started:

- On your own, decide the activities you would like to do. You don't have to do everything.
- As a group, think of the questions you are going to ask as you do this task to find out the information that you will need.
- Think about how you might keep track of all the information that you need.
- Decide on how you will work out the cost for each member of your group.

Windsurfing Half-day £59 Full day £79	**Paragliding** Half-day £99 for 1 person £189 for 2 people	**Quadbikes** £21 per hour Race Event (2 hours) £100 for three or four people	**Horse riding** Half-day (Mondays, Wednesdays and Fridays) £32
Fishing Trip Half-day £18 per person	**Coast jumping** 2 hours £85 per group of up to 4 people	**Kayaking** Half-day £29 Full Day £49	**Raft racing** Half-day £60 per team (minimum 2 people)
Diving 3 hours £38 per person	**Spa** 2 hours £24 for 1 person £40 for 2 people	**Shopping Trip** Half-day £60 budget	**Steam train up mountain** 2 hours £5.60 per person (10% discount for groups of 4)

Why this chapter matters

The word 'fraction' comes from the Latin word *fractus*, meaning broken. Just think of when you fracture an arm or leg – you have cracked (or broken) it into parts.

We use fractions to break things – from measurements to shapes – into parts.

Chapter 1 (page 6) showed how there used to be several different number systems, developed by many different ancient civilisations. Each of these civilisations developed their own way of expressing fractions. Most of these fraction systems died out, but one – the Arabic system – directly led to the fractions that we use today.

The Babylonians – the first fractions

Fractions can be traced back to the Babylonians, in around 1800BC. This civilisation in Mesopotamia (modern day Iraq) was the first to develop a sensible way to represent a fraction.

Their fractions were based on the number 60 as was its whole number system. However, its number symbols could not accurately represent the fractions and the fractions had no symbol to show that they were fractions rather than standard numbers. This made their system of fractions very complicated.

Egyptian fractions

The Egyptians (around 1000BC) were known to use fractions but generally these were unit fractions (fractions with a numerator of 1), for example, $\frac{1}{2}, \frac{1}{3}, \frac{1}{4}$.

The Egyptians used this symbol to represent the "one":

Then, using the number symbols shown in Chapter 1, they made their fractions:

$= \frac{1}{3}$

They also had special symbols for $\frac{1}{2}$, $\frac{2}{3}$, and $\frac{3}{4}$.

$= \frac{1}{2}$ | $= \frac{2}{3}$ | $= \frac{3}{4}$

Greek fractions

The Greeks also used unit fractions but the way they wrote them led to confusion. They had symbols for numbers, for example, the number 2 was β (beta) and to make the fraction $\frac{1}{2}$ they added a dash so $\frac{1}{2} = \beta'$.

Number	Symbol	Fraction	Symbol
2	β (beta)	$\frac{1}{2}$	β′
3	γ (gamma)	$\frac{1}{3}$	γ′
4	δ (delta)	$\frac{1}{4}$	δ′

Fractions in India

In India, around 500AD a system called *brahmi* was devised using symbols for the numbers. Fractions were written as a symbol above a symbol but without a line as used now.

Brahmi symbols

1	2	3	4	5	6	7	8	9
—	=	≡	+	h	ɥ	?	ς	ɂ

So $\frac{5}{9}$ was written as:

h
ɂ

Arabic fractions

The Arabs, probably around 1200AD, built on the number system – including the fractions – developed by the Indians. It was in Arabia that the line was first introduced into fractions – sometimes drawn horizontally and sometimes slanting – leading to the clear fractions that we use today.

$\frac{3}{4}$ $\frac{1}{8}$ $\frac{1}{2}$ $\frac{2}{3}$ $\frac{1}{4}$ $\frac{5}{10}$

Chapter

Number: Fractions

The grades given in this chapter are target grades.

1 Recognise a fraction of a shape

2 Recognise equivalent fractions, using diagrams

3 Equivalent fractions and simplifying fractions by cancelling

4 Finding a fraction of a quantity

5 One quantity as a fraction of another

6 Arithmetic with fractions

This chapter will show you …

G to C how to add, subtract, multiply, divide and order simple fractions

G how to use shapes and diagrams to recognise fractions

G how to cancel fractions

F how to calculate a fraction of a quantity

Visual overview

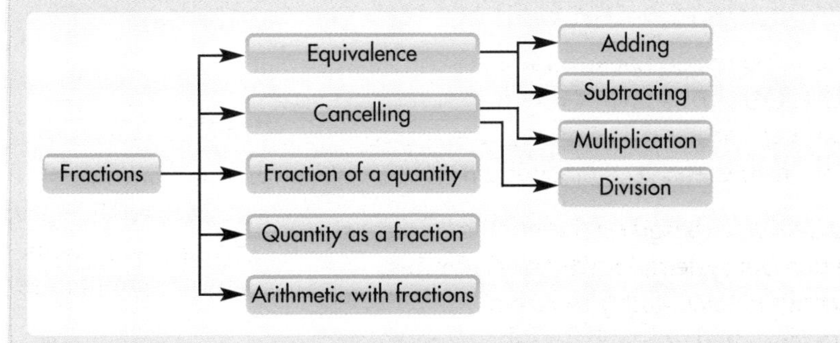

What you should already know

- Multiplication tables up to 10 × 10 **(KS3 level 4, GCSE grade G)**
- What a fraction is **(KS3 level 3, GCSE grade G)**

Quick check

How quickly can you calculate these?

1 2 × 4	2 5 × 3	3 5 × 2	4 6 × 3
5 2 × 7	6 4 × 5	7 3 × 8	8 4 × 6
9 9 × 2	10 3 × 7	11 half of 10	12 half of 12
13 half of 16	14 half of 8	15 half of 20	16 a third of 9
17 a third of 15	18 a quarter of 12	19 a fifth of 10	20 a fifth of 20

2.1 Recognise a fraction of a shape

This section will show you how to:
- recognise what fraction of a shape has been shaded
- shade a given simple fraction of a shape

Key words
fraction
shape

A **fraction** is a part of a whole. The top number is called the **numerator**. The bottom number is called the **denominator**. So, for example, $\frac{3}{4}$ means you divide a whole into four portions and take three of them.

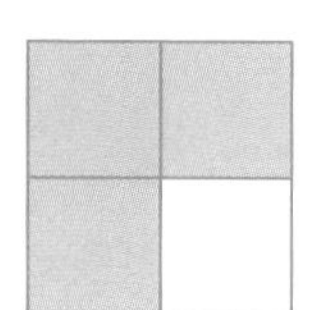

It really does help if you know the multiplication tables up to 10 × 10. They will be tested in the non-calculator paper, so you need to be confident about tables and numbers.

EXERCISE 2A

G

1 What **fraction** is shaded in each **shape** in these diagrams?

a 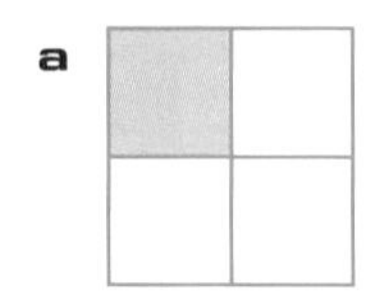

b

c

d 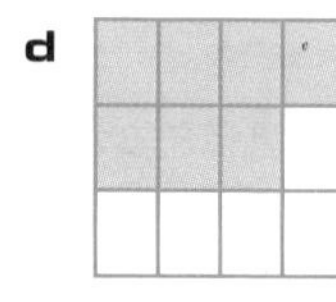

e

f

g

h 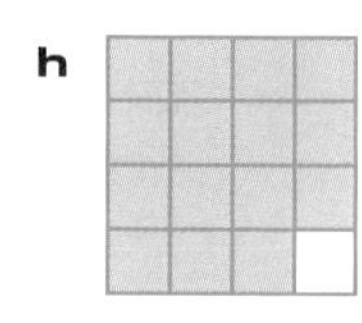

i

j

k 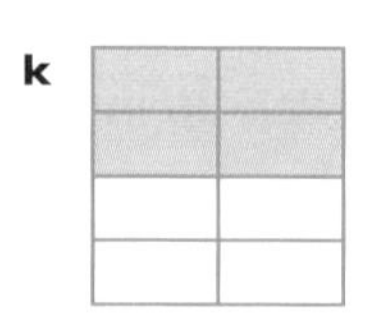

l 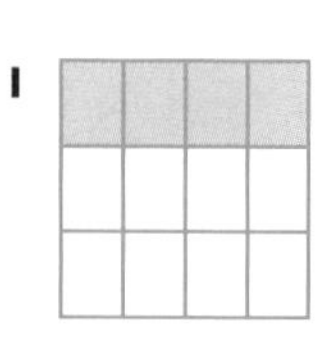

m

n 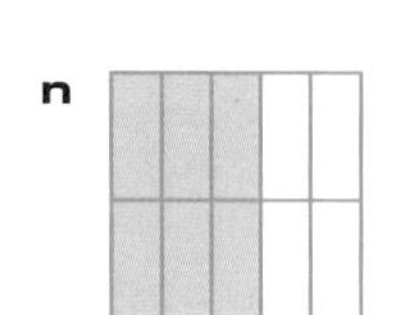

o 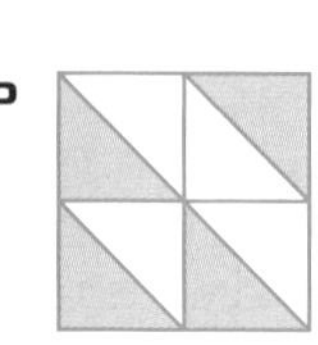

p 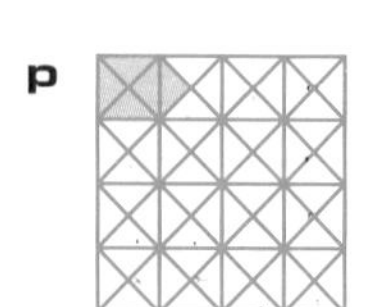

G

2 Draw diagrams as in question 1 to show these fractions.

a $\frac{3}{4}$ **b** $\frac{2}{3}$ **c** $\frac{1}{5}$ **d** $\frac{5}{8}$ **e** $\frac{1}{6}$ **f** $\frac{8}{9}$

g $\frac{1}{9}$ **h** $\frac{1}{10}$ **i** $\frac{4}{5}$ **j** $\frac{2}{7}$ **k** $\frac{3}{8}$ **l** $\frac{5}{6}$

3 A farmer decides to divide up a field into two parts so that one part is twice as big as the other part.

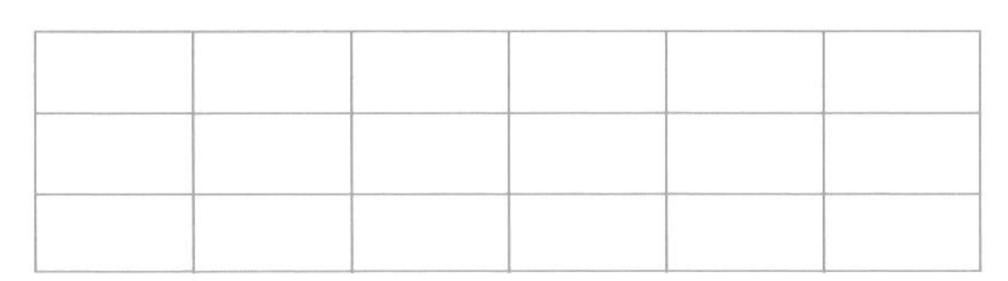

Make two copies of the grid and show two different ways that she can do this.

Shade the smaller area in each case.

F

PS 4 Look again at the diagrams in question 1.

For each question, decide which diagram, if any, has the greater proportion shaded.

a a and g **b** b and i **c** b and l **d** e and j

e k and o **f** e and m **g** g and k **h** e and i

AU 5 Three fractions are shown on the grids below.

A

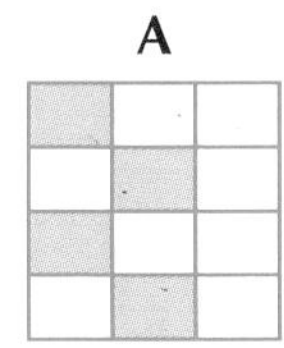

B

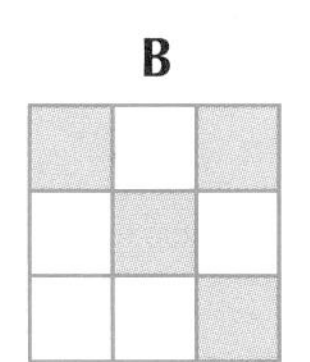

C

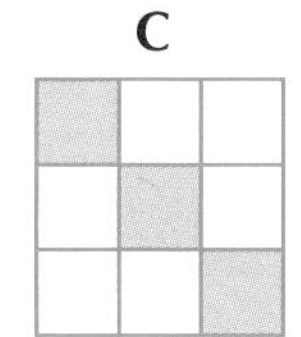

The fraction in grid A is the odd one out because it does not have a denominator of 9.

Give reasons as to why the fractions in grid B and grid C could be the odd one out.

FM 6 A field is divided up into six allotments, numbered 1 to 6.

1	2
3	4
5	6

Each allotment is the same shape and size.

The owners want to sell potatoes at market and decide that they need to plant half the field with potatoes to meet demand.

Potatoes are planted in half of allotment 4 and all of allotments 3 and 6. There are no potatoes in allotments 1 and 5.

The owner of allotment 2 agrees to plant some potatoes so that half the field will be potatoes.

What fraction of her allotment needs to be potatoes?

2.2 Recognise equivalent fractions, using diagrams

This section will show you how to:
- recognise equivalent fractions, using diagrams

Key words
equivalent
equivalent fractions

ACTIVITY

Making eighths

You need lots of squared paper and a pair of scissors.

Draw three rectangles, each 4 cm by 2 cm, on squared paper.

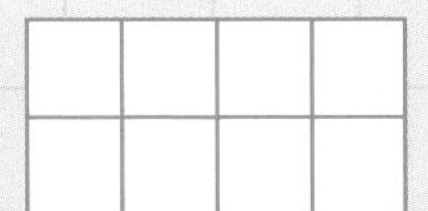

Each small square is called an *eighth* or $\frac{1}{8}$.

Cut one of the rectangles into halves, another into quarters and the last one into eighths.

$\frac{1}{2}$ $\frac{1}{2}$ $\frac{1}{4}$ $\frac{1}{4}$ $\frac{1}{4}$ $\frac{1}{4}$ $\frac{1}{8}$ $\frac{1}{8}$ $\frac{1}{8}$ $\frac{1}{8}$ $\frac{1}{8}$ $\frac{1}{8}$ $\frac{1}{8}$ $\frac{1}{8}$

You can see that the strip equal to one half takes up 4 squares, so:

$$\frac{1}{2} = \frac{4}{8}$$

These are called **equivalent fractions**.

1 Use the strips to write down the following fractions as eighths.

a $\frac{1}{4}$ **b** $\frac{3}{4}$

2 Use the strips to work out the following problems. Leave your answers as eighths.

a $\frac{1}{4} + \frac{3}{8}$ **b** $\frac{3}{4} + \frac{1}{8}$ **c** $\frac{3}{8} + \frac{1}{2}$

d $\frac{1}{4} + \frac{1}{2}$ **e** $\frac{1}{8} + \frac{1}{8}$ **f** $\frac{1}{8} + \frac{1}{4}$

g $\frac{3}{8} + \frac{3}{4}$ **h** $\frac{3}{4} + \frac{1}{2}$

Making twenty-fourths

You need lots of squared paper and a pair of scissors.

Draw four rectangles, each 6 cm by 4 cm, on squared paper.

Each small square is called a *twenty-fourth* or $\frac{1}{24}$. Cut one of the rectangles into quarters, another into sixths, another into thirds and the remaining one into eighths.

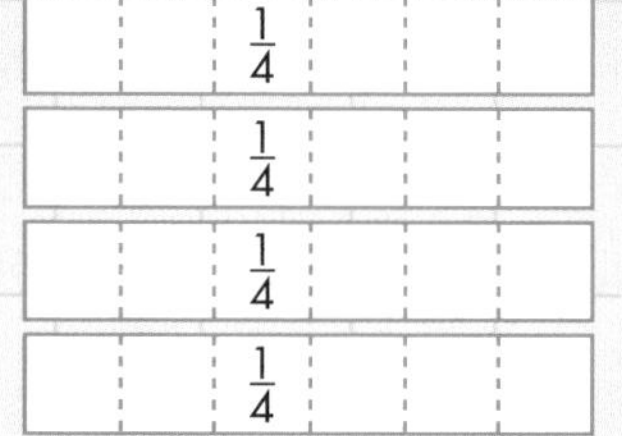

You can see that the strip equal to a quarter takes up six squares, so:

$$\frac{1}{4} = \frac{6}{24}$$

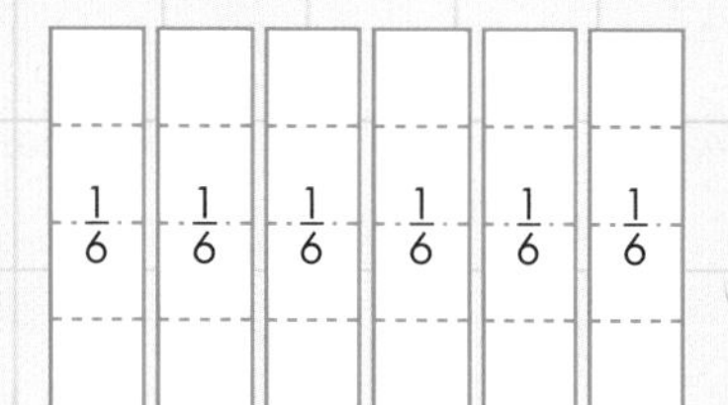

This is another example of equivalent fractions.

This idea is used to add fractions together. For example:

$$\frac{1}{4} + \frac{1}{6}$$

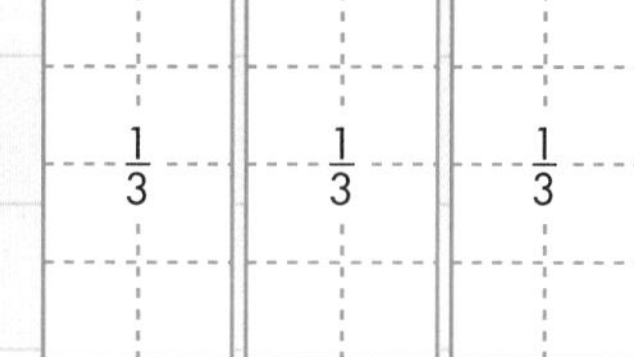

can be changed into:

$$\frac{6}{24} + \frac{4}{24} = \frac{10}{24}$$

EXERCISE 2B

1 Use the strips from the activity to write down each of these fractions as twenty-fourths.

a $\frac{1}{6}$ **b** $\frac{1}{3}$ **c** $\frac{1}{8}$ **d** $\frac{2}{3}$ **e** $\frac{5}{6}$

f $\frac{3}{4}$ **g** $\frac{3}{8}$ **h** $\frac{5}{8}$ **i** $\frac{7}{8}$ **j** $\frac{1}{2}$

2 Use the strips from the activity to write down the answer to each of the following problems. Write each answer in twenty-fourths.

a $\frac{1}{3}+\frac{1}{8}$ **b** $\frac{1}{8}+\frac{1}{4}$ **c** $\frac{1}{6}+\frac{1}{8}$ **d** $\frac{2}{3}+\frac{1}{8}$ **e** $\frac{5}{8}+\frac{1}{3}$

f $\frac{1}{8}+\frac{5}{6}$ **g** $\frac{1}{2}+\frac{3}{8}$ **h** $\frac{1}{6}+\frac{3}{4}$ **i** $\frac{5}{8}+\frac{1}{6}$ **j** $\frac{1}{3}+\frac{5}{8}$

3 Draw three rectangles, each 5 cm by 4 cm. Cut one into quarters, another into fifths and the last into tenths.

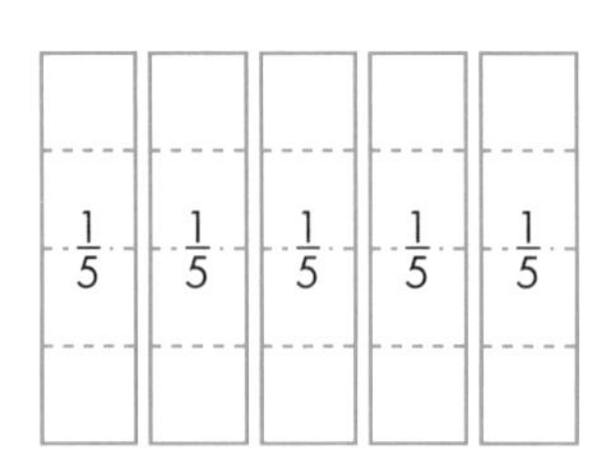

Use the strips to find the equivalent fraction, in twentieths, to each of the following.

a $\frac{1}{4}$ **b** $\frac{1}{5}$ **c** $\frac{3}{4}$ **d** $\frac{4}{5}$ **e** $\frac{1}{10}$

f $\frac{1}{2}$ **g** $\frac{3}{5}$ **h** $\frac{2}{5}$ **i** $\frac{7}{10}$ **j** $\frac{3}{10}$

4 Use the strips from question 3 to write down the answer to each of the following.

a $\frac{1}{4}+\frac{1}{5}$ **b** $\frac{3}{5}+\frac{1}{10}$ **c** $\frac{3}{10}+\frac{1}{4}$ **d** $\frac{3}{4}+\frac{1}{5}$ **e** $\frac{7}{10}+\frac{1}{4}$

PS 5 Use the strips from the activity to decide which fraction in each set of three, if any, is the odd one out.

a $\frac{6}{24}, \frac{2}{6}, \frac{2}{8}$ **b** $\frac{6}{8}, \frac{2}{3}, \frac{3}{4}$ **c** $\frac{1}{2}, \frac{4}{8}, \frac{3}{6}$ **d** $\frac{2}{3}, \frac{4}{6}, \frac{6}{8}$

AU **6** Here is a multiplication table.

×	2	3	4	5
2	4	6	8	10
5	10	15	20	25

The table can be used to pick out equivalent fractions.

For example, $\frac{2}{5} = \frac{6}{15}$.

a Use the table to write down three more fractions that are equivalent to $\frac{2}{5}$.

b Extend this table to write down a different fraction that is equivalent to $\frac{2}{5}$.

AU **7** **a** Complete this multiplication table.

×	2	3	4	5
3				
4				

b Use this table to write down four fractions that are equivalent to $\frac{3}{4}$.

8 You will need lots of squared paper and a pair of scissors.
Draw three rectangles on squared paper, each 6 cm by 2 cm.

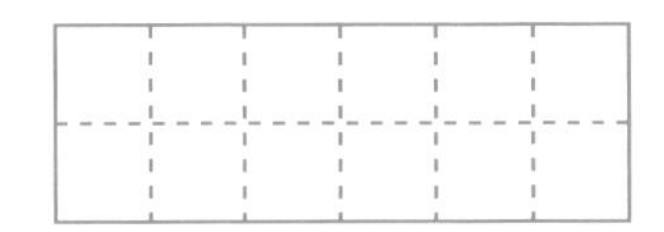

a Write down the fraction of a rectangle that each small square represents.

b Use your own choice of strips to write down the answer to each of the following.

i $\frac{1}{2} + \frac{1}{12}$ **ii** $\frac{5}{12} + \frac{1}{3}$ **iii** $\frac{1}{12} + \frac{5}{6}$ **iv** $\frac{2}{3} + \frac{1}{12}$ **v** $\frac{1}{3} + \frac{1}{4}$

G

F

2.3 Equivalent fractions and simplifying fractions by cancelling

This section will show you how to:
- create equivalent fractions
- simplify fractions by cancelling

Key words
cancel
denominator
lowest terms
numerator
simplest form
simplify

Equivalent fractions are two or more fractions that represent the same part of a whole.

EXAMPLE 1

Complete the following.

a $\frac{3}{4} \longrightarrow \frac{\times 4}{\times 4} = \frac{\square}{16}$ b $\frac{2}{5} = \frac{\square}{15}$

a Multiplying the **numerator** by 4 gives 12. This means $\frac{12}{16}$ is an equivalent fraction to $\frac{3}{4}$.

b To change the **denominator** from 5 to 15, you multiply by 3. Do the same thing to the numerator, which gives $2 \times 3 = 6$. So, $\frac{2}{5} = \frac{6}{15}$.

The fraction $\frac{3}{4}$, in Example 1a, is in its **lowest terms** or **simplest form**.

This means that the only number that is a factor of both the numerator and denominator is 1.

EXAMPLE 2

Write these fractions in their simplest forms.

a $\frac{15}{35}$ b $\frac{24}{54}$

a Here is one reason why you need to know the multiplication tables. What is the biggest number that has both 15 and 35 in its multiplication table? You should know that this is the five times table. So, divide both the numerator and denominator by 5.

$$\frac{15}{35} = \frac{15 \div 5}{35 \div 5} = \frac{3}{7}$$

You can say that you have '**cancelled** by five'.

b The biggest number that has both 24 and 54 in its multiplication table is 6. So, divide both the numerator and denominator by 6.

$$\frac{24}{54} = \frac{24 \div 6}{54 \div 6} = \frac{4}{9}$$

Here, you have 'cancelled by six'.

You have **simplified** the fractions to write them in their lowest terms.

EXAMPLE 3

Put the following fractions in order, with the smallest first.

$\frac{5}{6}, \frac{2}{3}, \frac{3}{4}$

First write each fraction with the same denominator by using equivalent fractions.

$\frac{5}{6} = \frac{10}{\mathbf{12}}$

$\frac{2}{3} = \frac{4}{6} = \frac{6}{9} = \frac{8}{\mathbf{12}}$

$\frac{3}{4} = \frac{6}{8} = \frac{9}{\mathbf{12}}$

This shows that $\frac{5}{6} = \frac{10}{12}$, $\frac{2}{3} = \frac{8}{12}$ and $\frac{3}{4} = \frac{9}{12}$.

In order, the fractions are:

$\frac{2}{3}, \frac{3}{4}, \frac{5}{6}$

EXERCISE 2C

1 Copy and complete the following.

a $\frac{2}{5} \rightarrow \frac{\times 4}{\times 4} = \frac{\square}{20}$ **b** $\frac{1}{4} \rightarrow \frac{\times 3}{\times 3} = \frac{\square}{12}$ **c** $\frac{3}{8} \rightarrow \frac{\times 5}{\times 5} = \frac{\square}{40}$

d $\frac{4}{5} \rightarrow \frac{\times 3}{\times 3} = \frac{\square}{15}$ **e** $\frac{5}{6} \rightarrow \frac{\times 3}{\times 3} = \frac{\square}{18}$ **f** $\frac{3}{7} \rightarrow \frac{\times 4}{\times 4} = \frac{\square}{28}$

g $\frac{3}{10} \rightarrow \frac{\times \square}{\times 2} = \frac{\square}{20}$ **h** $\frac{1}{3} \rightarrow \frac{\times \square}{\times \square} = \frac{\square}{9}$ **i** $\frac{3}{5} \rightarrow \frac{\times \square}{\times \square} = \frac{\square}{20}$

j $\frac{2}{3} \rightarrow \frac{\times \square}{\times \square} = \frac{\square}{18}$ **k** $\frac{3}{4} \rightarrow \frac{\times \square}{\times \square} = \frac{\square}{12}$ **l** $\frac{5}{8} \rightarrow \frac{\times \square}{\times \square} = \frac{\square}{40}$

m $\frac{7}{10} \rightarrow \frac{\times \square}{\times \square} = \frac{\square}{20}$ **n** $\frac{1}{6} \rightarrow \frac{\times \square}{\times \square} = \frac{4}{\square}$ **o** $\frac{3}{8} \rightarrow \frac{\times \square}{\times \square} = \frac{15}{\square}$

G

G

2 Copy and complete the following.

a $\frac{1}{2} = \frac{2}{\square} = \frac{3}{\square} = \frac{\square}{8} = \frac{\square}{10} = \frac{6}{\square}$

b $\frac{1}{3} = \frac{2}{\square} = \frac{3}{\square} = \frac{\square}{12} = \frac{\square}{15} = \frac{6}{\square}$

c $\frac{3}{4} = \frac{6}{\square} = \frac{9}{\square} = \frac{\square}{16} = \frac{\square}{20} = \frac{18}{\square}$

d $\frac{2}{5} = \frac{4}{\square} = \frac{6}{\square} = \frac{\square}{20} = \frac{\square}{25} = \frac{12}{\square}$

e $\frac{3}{7} = \frac{6}{\square} = \frac{9}{\square} = \frac{\square}{28} = \frac{\square}{35} = \frac{18}{\square}$

3 Copy and complete the following.

a $\frac{10}{15} = \frac{10 \div 5}{15 \div 5} = \frac{\square}{\square}$

b $\frac{12}{15} = \frac{12 \div 3}{15 \div 3} = \frac{\square}{\square}$

c $\frac{20}{28} = \frac{20 \div 4}{28 \div 4} = \frac{\square}{\square}$

d $\frac{12}{18} = \frac{12 \div \square}{\square \div \square} = \frac{\square}{\square}$

e $\frac{15}{25} = \frac{15 \div 5}{\square \div \square} = \frac{\square}{\square}$

f $\frac{21}{30} = \frac{21 \div \square}{\square \div \square} = \frac{\square}{\square}$

FM 4 A shop manager is working out how much space to use for different items on a shelf.

The shelf has six equal sections.

He puts baked beans on three sections, tomatoes on two sections and spaghetti on one section.

a Baked beans are in boxes of 48 tins.
Each section of shelf holds 500 tins.
How many boxes will be needed to fill the baked bean sections?

b What fraction of the shelf has baked beans on it?
Give your answer in its simplest form.

c What fraction of the shelf does not have spaghetti on it?
Give your answer in its simplest form.

5 Cancel each of these fractions to its simplest form.

a $\frac{4}{6}$ **b** $\frac{5}{15}$ **c** $\frac{12}{18}$ **d** $\frac{6}{8}$ **e** $\frac{3}{9}$

f $\frac{5}{10}$ **g** $\frac{14}{16}$ **h** $\frac{28}{35}$ **i** $\frac{10}{20}$ **j** $\frac{4}{16}$

k $\frac{12}{15}$ **l** $\frac{15}{21}$ **m** $\frac{25}{35}$ **n** $\frac{14}{21}$ **o** $\frac{8}{20}$

p $\frac{10}{25}$ **q** $\frac{7}{21}$ **r** $\frac{42}{60}$ **s** $\frac{50}{200}$ **t** $\frac{18}{12}$

u $\frac{6}{9}$ **v** $\frac{18}{27}$ **w** $\frac{36}{48}$ **x** $\frac{21}{14}$ **y** $\frac{42}{12}$

6 Put the fractions in each set in order, with the smallest first.

a $\frac{1}{2}, \frac{5}{6}, \frac{2}{3}$

b $\frac{3}{4}, \frac{1}{2}, \frac{5}{8}$

c $\frac{7}{10}, \frac{2}{5}, \frac{1}{2}$

d $\frac{2}{3}, \frac{3}{4}, \frac{7}{12}$

e $\frac{1}{6}, \frac{1}{3}, \frac{1}{4}$

f $\frac{9}{10}, \frac{3}{4}, \frac{4}{5}$

g $\frac{4}{5}, \frac{7}{10}, \frac{5}{6}$

h $\frac{1}{3}, \frac{2}{5}, \frac{3}{10}$

AU 7 Here are four unit fractions.

$\frac{1}{2}$ $\frac{1}{3}$ $\frac{1}{4}$ $\frac{1}{5}$

a Which two of these fractions have a sum of $\frac{7}{12}$?

Show clearly how you work out your answer.

b Which fraction is the biggest?
Explain your answer.

PS 8 **a** Use the fact that $2 \times 2 \times 2 \times 2 \times 2 = 32$ to cancel the fraction $\frac{64}{320}$ to its lowest terms.

b Use the fact that $7 \times 11 \times 13 = 1001$ to cancel the fraction $\frac{13}{1001}$ to its simplest form.

PS 9 What fraction of each of these grids is shaded?

a

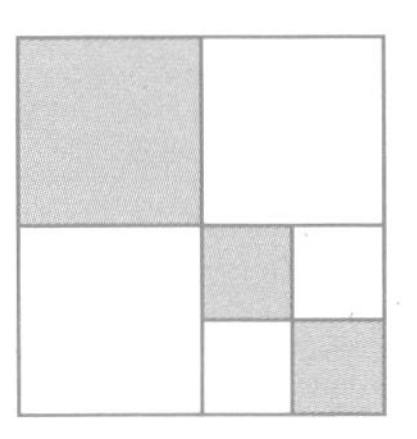

b

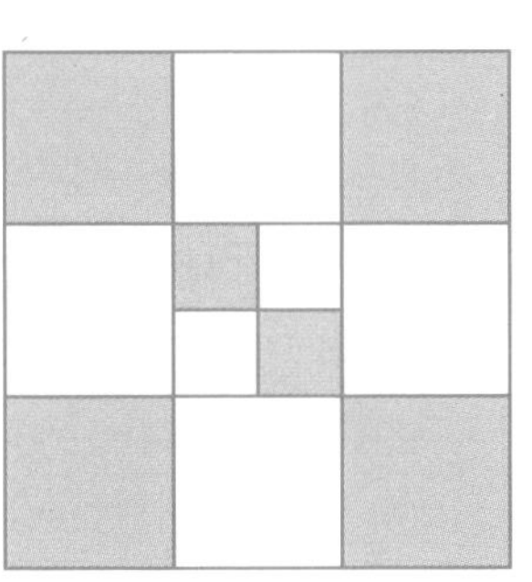

c

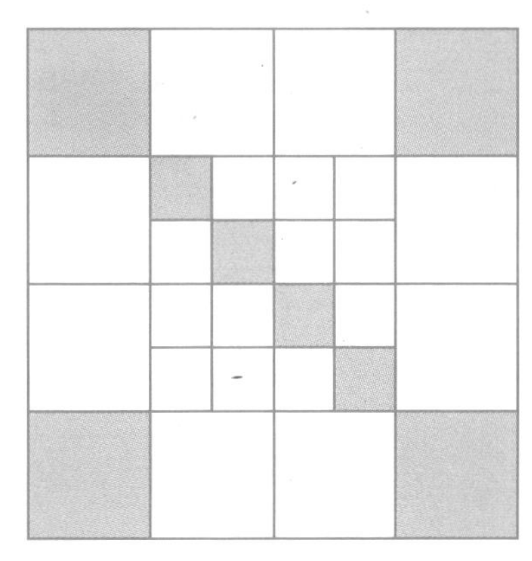

2.4 Finding a fraction of a quantity

This section will show you how to:
- find a fraction of a given quantity

Key words
fraction
quantity

To do this, you simply multiply the **fraction** by the **quantity**, for example, $\frac{1}{2}$ of 30 is the same as $\frac{1}{2} \times 30$.

Remember: In mathematics 'of' is interpreted as ×.

For example, two lots of three is the same as 2×3.

EXAMPLE 4

Find $\frac{3}{4}$ of £196.

First, find $\frac{1}{4}$ by dividing by 4. Then find $\frac{3}{4}$ by multiplying your answer by 3:

$196 \div 4 = 49$ then $49 \times 3 = 147$

The answer is £147.

Of course, you can use your calculator to do this problem by either:

- pressing the sequence: 1 9 6 ÷ 4 × 3 =
- or using the [fraction] key: [fraction] 3 4 × 1 9 6 =

EXERCISE 2D

F

1 Calculate each of these.

a $\frac{3}{5}$ of 30 **b** $\frac{2}{7}$ of 35

c $\frac{3}{8}$ of 48 **d** $\frac{7}{10}$ of 40

e $\frac{5}{6}$ of 18 **f** $\frac{3}{4}$ of 24

g $\frac{4}{5}$ of 60 **h** $\frac{5}{8}$ of 72

F

2 Calculate each of these quantities.

a $\frac{3}{4}$ of £2400 **b** $\frac{2}{5}$ of 320 grams

c $\frac{5}{8}$ of 256 kilograms **d** $\frac{2}{3}$ of £174

e $\frac{5}{6}$ of 78 litres **f** $\frac{3}{4}$ of 120 minutes

g $\frac{4}{5}$ of 365 days **h** $\frac{7}{8}$ of 24 hours

i $\frac{3}{4}$ of 1 day **j** $\frac{5}{9}$ of 4266 miles

3 In each case, find out which is the larger number.

a $\frac{2}{5}$ of 60 or $\frac{5}{8}$ of 40 **b** $\frac{3}{4}$ of 280 or $\frac{7}{10}$ of 290

c $\frac{2}{3}$ of 78 or $\frac{4}{5}$ of 70 **d** $\frac{5}{6}$ of 72 or $\frac{11}{12}$ of 60

e $\frac{4}{9}$ of 126 or $\frac{3}{5}$ of 95 **f** $\frac{3}{4}$ of 340 or $\frac{2}{3}$ of 381

4 A director receives $\frac{2}{15}$ of his firm's profits. The firm made a profit of £45 600 in one year. How much did the director receive?

5 A woman left £84 000 in her will.

She left $\frac{3}{8}$ of the money to charity.

How much did she leave to charity?

AU 6 In the season 2008/2009, the attendance at Huddersfield Town versus Leeds United was 20 928. Of this crowd, $\frac{3}{8}$ were female. How many were male?

7 Two-thirds of a person's weight is water. Paul weighs 78 kg. How much of his body weight is water?

8 **a** Information from the first census in Singapore suggests that then $\frac{2}{25}$ of the population were Indian. The total population was 10 700. How many people were Indian?

b By 1990 the population of Singapore had grown to 3 002 800. Only $\frac{1}{16}$ of this population were Indian. How many Indians were living in Singapore in 1990?

F

9 Mark normally earns £500 a week. One week he is given a bonus of $\frac{1}{10}$ of his wage.

a Find $\frac{1}{10}$ of £500.

b How much does he earn altogether for this week?

10 The contents of a standard box of cereals weigh 720 g. A new larger box holds $\frac{1}{4}$ more than the standard box.

a Find $\frac{1}{4}$ of 720 g.

b How much do the contents of the new box of cereals weigh?

FM 11 The price of a new TV costing £360 is reduced by $\frac{1}{3}$ in a sale.

a Find $\frac{1}{3}$ of £360.

b How much does the TV cost in the sale?

E

FM 12 A car is advertised at Lion Autos at £9000 including extras but with a special offer of one-fifth off this price.

The same car is advertised at Tiger Motors for £6000 but the extras add one-quarter to this price.

Which garage is the cheaper?

D

FM 13 A jar of instant coffee normally contains 200 g and costs £2.

There are two special offers on a jar of coffee.

Offer A: $\frac{1}{4}$ extra for the same price.

Offer B: Same weight for $\frac{3}{4}$ of the original price.

Which offer is the best value?

2.5 One quantity as a fraction of another

This section will show you how to:
- express one quantity as a fraction of another

Key words
fraction
quantity

You may often be asked to give one amount or **quantity** as a **fraction** of another.

EXAMPLE 5

Write £5 as a fraction of £20.

As a fraction this is written $\frac{5}{20}$. This cancels to $\frac{1}{4}$.

So £5 is one-quarter of £20.

EXERCISE 2E

D

1 In each of the following, write the first quantity as a fraction of the second.

a 2 cm, 6 cm

b 4 kg, 20 kg

c £8, £20

d 5 hours, 24 hours

e 12 days, 30 days

f 50p, £3

g 4 days, 2 weeks

h 40 minutes, 2 hours

2 In a form of 30 pupils, 18 are boys. What fraction of the form are boys?

3 During March, it rained on 12 days. For what fraction of the month did it rain?

4 Reka wins £120 in a competition and puts £50 into her bank account. What fraction of her winnings does she keep to spend?

FM **5** Jon earns £90 and saves £30 of it.

Matt earns £100 and saves £35 of it.

Who is saving the greater proportion of their earnings?

AU **6** In two tests Isobel gets 13 out of 20 and 16 out of 25. Which is the better mark?

Explain your answer.

2.6 Arithmetic with fractions

This section will show you how to:
- convert a decimal number to a fraction
- convert a fraction to a decimal
- add and subtract fractions with different denominators
- multiply a mixed number by a fraction
- divide one fraction by another fraction

Key words
decimal
denominator
fraction
mixed number
numerator

Changing a decimal into a fraction

A **decimal** can be changed into a **fraction** by using a place-value table.

For example, $0.32 = \frac{32}{100}$

Units	.	Tenths	Hundredths	Thousandths
0	.	3	2	

EXAMPLE 6

Express 0.32 as a fraction.

$$0.32 = \frac{32}{100}$$

This cancels to $\frac{8}{25}$

So, $0.32 = \frac{8}{25}$

Changing a fraction into a decimal

You can change a fraction into a decimal by dividing the **numerator** by the **denominator**. Example 7 shows how this can be done without a calculator.

EXAMPLE 7

Express $\frac{3}{8}$ as a decimal.

$\frac{3}{8}$ means $3 \div 8$. This is a division calculation:

$$8 \overline{\smash{)}\,3.{}^{3}0{}^{6}0{}^{4}0} \quad = 0.375$$

So, $\frac{3}{8} = 0.375$

Notice that extra zeros have been put at the end to be able to complete the division.

EXERCISE 2F

1 Change each of these decimals to fractions, cancelling where possible.

a 0.7 **b** 0.4 **c** 0.5 **d** 0.03 **e** 0.06

f 0.13 **g** 0.25 **h** 0.38 **i** 0.55 **j** 0.64

2 Change each of these fractions to decimals. Where necessary, give your answer correct to three decimal places.

a $\frac{1}{2}$ **b** $\frac{3}{4}$ **c** $\frac{3}{5}$ **d** $\frac{9}{10}$ **e** $\frac{1}{3}$

f $\frac{5}{8}$ **g** $\frac{2}{3}$ **h** $\frac{7}{20}$ **i** $\frac{7}{11}$ **j** $\frac{4}{9}$

3 Put each of the following sets of numbers in order, with the smallest first.

a $0.6, 0.3, \frac{1}{2}$ **b** $\frac{2}{5}, 0.8, 0.3$

c $0.35, \frac{1}{4}, 0.15$ **d** $\frac{7}{10}, 0.72, 0.71$

e $0.8, \frac{3}{4}, 0.7$ **f** $0.08, 0.1, \frac{1}{20}$

g $0.55, \frac{1}{2}, 0.4$ **h** $1\frac{1}{4}, 1.2, 1.23$

HINTS AND TIPS

Convert the fractions to decimals first.

FM 4 Two shops are advertising the same T-shirts.

Which shop has the best offer?
Give a reason for your answer.

PS 5 Which is bigger $\frac{7}{8}$ or 0.87?

Show your working.

PS 6 Which is smaller $\frac{2}{3}$ or 0.7?

Show your working.

D

PS 7 Complete this statement with a decimal.

$\frac{1}{4}$ of 50 is the same as ...... × 100

PS 8 Complete this statement with a fraction.

$\frac{2}{3}$ of 60 is the same as ...... of 80

! This topic will be assessed in Unit 2.

Addition and subtraction of fractions

Fractions can only be added or subtracted after you have converted them to equivalent fractions with the same denominator.

EXAMPLE 8

i $\frac{2}{3} + \frac{1}{5}$

Note you can change both fractions to equivalent fractions with a denominator of 15.

This then becomes:

$$\frac{2 \times 5}{3 \times 5} + \frac{1 \times 3}{5 \times 3} = \frac{10}{15} + \frac{3}{15} = \frac{13}{15}$$

ii $2\frac{3}{4} - 1\frac{5}{6}$

Split the calculation into $\left(2 + \frac{3}{4}\right) - \left(1 + \frac{5}{6}\right)$.

This then becomes:

$$2 - 1 + \frac{3}{4} - \frac{5}{6}$$

Note you can change both fractions to equivalent fractions with a denominator of 12.

$$= 1 + \frac{9}{12} + \frac{10}{12} = 1 - \frac{1}{12}$$

$$= \frac{11}{12}$$

EXERCISE 2G

1 Evaluate the following.

a $\frac{1}{3} + \frac{1}{5}$ **b** $\frac{1}{3} + \frac{1}{4}$ **c** $\frac{1}{5} + \frac{1}{10}$

d $\frac{2}{3} + \frac{1}{4}$ **e** $\frac{3}{4} + \frac{1}{8}$ **f** $\frac{1}{3} + \frac{1}{6}$

g $\frac{1}{2} - \frac{1}{3}$ **h** $\frac{1}{4} - \frac{1}{5}$ **i** $\frac{1}{5} - \frac{1}{10}$

j $\frac{7}{8} - \frac{3}{4}$ **k** $\frac{5}{6} - \frac{3}{4}$ **l** $\frac{5}{6} - \frac{1}{2}$

m $\frac{5}{12} - \frac{1}{4}$ **n** $\frac{1}{3} + \frac{4}{9}$ **o** $\frac{1}{4} + \frac{3}{8}$

p $\frac{7}{8} - \frac{1}{2}$ **q** $\frac{3}{5} - \frac{8}{15}$ **r** $\frac{11}{12} + \frac{5}{8}$

s $\frac{7}{16} + \frac{3}{10}$ **t** $\frac{4}{9} - \frac{2}{21}$ **u** $\frac{5}{6} - \frac{4}{27}$

2 Evaluate the following.

a $2\frac{1}{7} + 1\frac{3}{14}$ **b** $6\frac{3}{10} + 1\frac{4}{5} + 2\frac{1}{2}$ **c** $3\frac{1}{2} - 1\frac{1}{3}$

d $1\frac{7}{18} + 2\frac{3}{10}$ **e** $3\frac{2}{6} + 1\frac{9}{20}$ **f** $1\frac{1}{8} - \frac{5}{9}$

g $1\frac{3}{16} - \frac{7}{12}$ **h** $\frac{5}{6} + \frac{7}{16} + \frac{5}{8}$ **i** $\frac{7}{10} + \frac{3}{8} + \frac{5}{6}$

j $1\frac{1}{3} + \frac{7}{10} - \frac{4}{15}$ **k** $\frac{5}{14} + 1\frac{3}{7} - \frac{5}{12}$

3 In a class of children, three-quarters are Chinese, one-fifth are Malay and the rest are Indian. What fraction of the class are Indian?

4 **a** In a class election, half of the people voted for Aminah, one-third voted for Janet and the rest voted for Peter. What fraction of the class voted for Peter?

b One of the following is the number of people in the class.

25 28 30 32

How many people are in the class?

D

C

C

FM 5 A group of people travelled from Hope to Castletown. One-twentieth of them decided to walk, one-twelfth went by car and all the rest went by bus. What fraction went by bus?

FM 6 A one-litre flask filled with milk is used to fill two glasses, one of capacity half a litre and the other of capacity one-sixth of a litre. What fraction of a litre will remain in the flask?

FM 7 Katie spent three-eighths of her income on rent, and two-fifths of what was left on food. What fraction of her income was left after buying her food?

PS 8 Mick says that $1\frac{1}{3} + 2\frac{1}{4} = 3\frac{2}{7}$.

He is incorrect.

What is the mistake that he has made?

Work out the correct answer.

Multiplication of fractions

Remember:

- To multiply two fractions, multiply the numerators (top numbers) and multiply the denominators (bottom numbers) and cancel if possible.
- When multiplying a **mixed number**, change the mixed number to an improper fraction before you start multiplying.

EXAMPLE 9

Work out

$$1\frac{3}{4} \times \frac{2}{5}$$

Change the mixed number to an improper fraction.

$$1\frac{3}{4} \text{ to } \frac{7}{4}$$

The problem is now:

$$\frac{7}{4} \times \frac{2}{5}$$

So, $\frac{7}{4} \times \frac{2}{5} = \frac{14}{20}$

This cancels to $\frac{7}{10}$.

EXAMPLE 10

A boy had 930 stamps in his collection. $\frac{2}{15}$ of them were British stamps. How many British stamps did he have?

The problem is:

$$\frac{2}{15} \times 930$$

First, calculate $\frac{1}{15}$ of 930.

$$\frac{1}{15} \times 930 = 930 \div 15 = 62$$

So, $\frac{2}{15}$ of $930 = 2 \times 62 = 124$

He had 124 British stamps.

EXERCISE 2H

1 Evaluate the following, leaving each answer in its simplest form.

a $\frac{1}{2} \times \frac{1}{3}$ **b** $\frac{1}{4} \times \frac{2}{5}$ **c** $\frac{3}{4} \times \frac{1}{2}$ **d** $\frac{3}{7} \times \frac{1}{2}$

e $\frac{2}{3} \times \frac{4}{5}$ **f** $\frac{1}{3} \times \frac{3}{5}$ **g** $\frac{1}{3} \times \frac{6}{7}$ **h** $\frac{3}{4} \times \frac{2}{5}$

i $\frac{2}{3} \times \frac{3}{4}$ **j** $\frac{1}{2} \times \frac{4}{5}$

2 Evaluate the following, leaving each answer in its simplest form.

a $\frac{5}{16} \times \frac{3}{10}$ **b** $\frac{9}{10} \times \frac{5}{12}$ **c** $\frac{14}{15} \times \frac{3}{8}$ **d** $\frac{8}{9} \times \frac{6}{15}$

e $\frac{6}{7} \times \frac{21}{30}$ **f** $\frac{9}{14} \times \frac{35}{36}$

3 I walked two-thirds of the way along Pungol Road which is 4.5 km long. How far have I walked?

4 One-quarter of Alan's stamp collection was given to him by his sister. Unfortunately two-thirds of these were torn. What fraction of his collection was given to him by his sister and were not torn?

5 Bilal eats one-quarter of a cake, and then half of what is left. How much cake is left uneaten?

6 A merchant buys 28 crates, each containing three-quarters of a tonne of waste metal. What is the total weight of waste metal?

E

D

D

C

7 Because of illness, on one day two-fifths of a school was absent. If the school had 650 pupils on the register, how many were absent that day?

8 To increase sales, a shop reduced the price of a car stereo radio by $\frac{2}{5}$. If the original price was £85, what was the new price?

9 Two-fifths of a class were boys. If the class contained 30 children, how many were girls?

10 Evaluate the following, giving each answer as a mixed number where possible.

a $1\frac{1}{4} \times \frac{1}{3}$ **b** $1\frac{2}{3} \times 1\frac{1}{4}$ **c** $2\frac{1}{2} \times 2\frac{1}{2}$ **d** $1\frac{3}{4} \times 1\frac{2}{3}$

e $3\frac{1}{4} \times 1\frac{1}{5}$ **f** $1\frac{1}{4} \times 2\frac{2}{3}$ **g** $2\frac{1}{2} \times 5$ **h** $7\frac{1}{2} \times 4$

11 Which is larger, $\frac{3}{4}$ of $2\frac{1}{2}$ or $\frac{2}{5}$ of $6\frac{1}{2}$?

PS 12 After James spent $\frac{2}{5}$ of his pocket money on magazines, and $\frac{1}{4}$ of his pocket money at a football match, he had £1.75 left. How much pocket money did he have in the beginning?

13 Which is the biggest: half of 96, one-third of 141, two-fifths of 120, or three-quarters of 68?

PS 14 At a burger-eating competition, Liam ate 34 burgers in 20 minutes while Ahmed ate 26 burgers in 20 minutes. Assuming they ate the burgers at a steady rate, how long after the start of the competition would they have consumed a total of 21 burgers between them?

PS 15 If £5.20 is two-thirds of three-quarters of a sum of money, what is the sum?

PS 16 Emily lost $\frac{3}{4}$ of her money in the market, but then found $\frac{3}{5}$ of what she had lost. She now had £21 altogether. How much did she start with?

Dividing fractions

Look at the problem $3 \div \frac{3}{4}$. This is like asking, 'How many $\frac{3}{4}$s are there in 3?'

Look at the diagram.

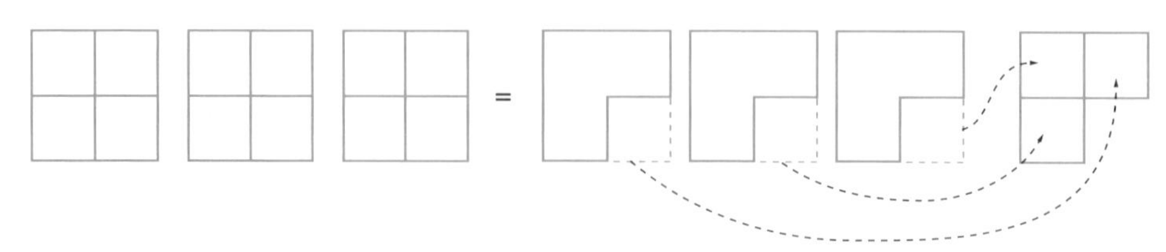

Each of the three whole shapes is divided into quarters. What is the total number of quarters divided by 3?

Can you see that you could fit the four shapes on the right-hand side of the = sign into the three shapes on the left-hand side?

i.e. $3 \div \frac{3}{4} = 4$

or $3 \div \frac{3}{4} = 3 \times \frac{4}{3} = \frac{3 \times 4}{3} = \frac{12}{3} = 4$

So, to divide by a fraction, you turn the fraction upside down (finding its reciprocal), and then multiply.

EXERCISE 2I

1 Evaluate the following, giving your answer as a mixed number where possible.

a $\frac{1}{4} \div \frac{1}{3}$ **b** $\frac{2}{5} \div \frac{2}{7}$ **c** $\frac{4}{5} \div \frac{3}{4}$ **d** $\frac{3}{7} \div \frac{2}{5}$

e $5 \div 1\frac{1}{4}$ **f** $6 \div 1\frac{1}{2}$ **g** $7\frac{1}{2} \div 1\frac{1}{2}$ **h** $3 \div 1\frac{3}{4}$

i $1\frac{5}{12} \div 3\frac{3}{16}$ **j** $3\frac{3}{5} \div 2\frac{1}{4}$

2 A grain merchant has only thirteen and a half tonnes in stock. He has several customers who are all ordering three-quarters of a tonne. How many customers can he supply?

3 For a party, Zahar made twelve and a half litres of lemonade. His glasses could each hold $\frac{5}{16}$ of a litre. How many of the glasses could he fill from the twelve and a half litres of lemonade?

4 How many strips of ribbon, each three and a half centimetres long, can I cut from a roll of ribbon that is fifty-two and a half centimetres long?

5 Joe's stride is three-quarters of a metre long. How many strides does he take to walk the length of a bus twelve metres long?

6 Evaluate the following, giving your answers as a mixed number where possible.

a $2\frac{2}{9} \times 2\frac{1}{10} \times \frac{16}{35}$ **b** $3\frac{1}{5} \times 2\frac{1}{2} \times 4\frac{3}{4}$

c $1\frac{1}{4} \times 1\frac{2}{7} \times 1\frac{1}{6}$ **d** $\frac{18}{25} \times \frac{15}{16} \div 2\frac{2}{5}$

e $\left(\frac{2}{5} \times \frac{2}{5}\right) \times \left(\frac{5}{6} \times \frac{5}{6}\right) \times \left(\frac{3}{4} \times \frac{3}{4}\right)$ **f** $\left(\frac{4}{5} \times \frac{4}{5}\right) \div \left(1\frac{1}{4} \times 1\frac{1}{4}\right)$

C

B

SUMMARY

GRADE BOOSTER

- **G** You can state the fraction of a shape that is shaded
- **G** You can shade in a fraction of a shape
- **G** You know how to identify equivalent fractions
- **G** You can simplify fractions by cancelling (when possible)
- **G** You can put simple fractions into order of size
- **F** You can find a fraction of a quantity
- **E** You can add, subtract and multiply more difficult fractions
- **E** You can compare two fractions of quantities
- **D** You can write a quantity as a fraction of another quantity
- **C** You can divide a fraction by a fraction

What you should know now

- How to recognise and draw fractions of shapes
- How to work out equivalent fractions
- How to calculate a fraction of a quantity
- How to add, subtract, multiply and divide with fractions
- How to solve simple practical problems using fractions

1 **a** What fraction of the shape below is shaded?

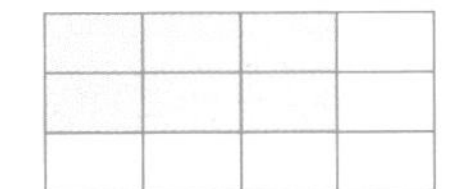

Give your answer as a fraction in its simplest form.

b Copy out and shade $\frac{3}{4}$ of the shape below.

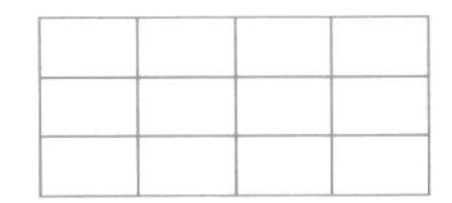

2 Put the following fractions into order, smallest first.

$\frac{3}{4}$ $\frac{1}{2}$ $\frac{2}{5}$ $\frac{1}{3}$

3 Work out

$\frac{7}{10} - \frac{2}{5}$

4 Alison travels by car to her meetings. Alison's company pays her 32p for each mile she travels.

One day Alison writes down the distance readings from her car.

Start of the day: 2430 miles

End of the day: 2658 miles

a Work out how much the company pays Alison for her day's travel. (2)

The next day Alison travelled a total of 145 miles.

She travelled $\frac{2}{5}$ of this distance in the morning.

b How many miles did she travel during the rest of the day? (2)

(Total 4 marks)

Edexcel, Question 9, Paper 2 Foundation, June 2005

5 **a** Here are some fractions.

$\frac{2}{4}$ $\frac{4}{8}$ $\frac{2}{5}$ $\frac{7}{14}$

Which one of these fractions is not equal to $\frac{1}{2}$? (1)

Give a reason for your answer. (1)

b Work out $\frac{3}{4}$ of 20 (2)

(Total 4 marks)

Edexcel, May 2008, Paper 1 Foundation, Question 11

6 Here are two fractions $\frac{4}{5}$ and $\frac{3}{4}$

Explain which is the larger fraction.

You may use the grids to help with your explanation. (2)

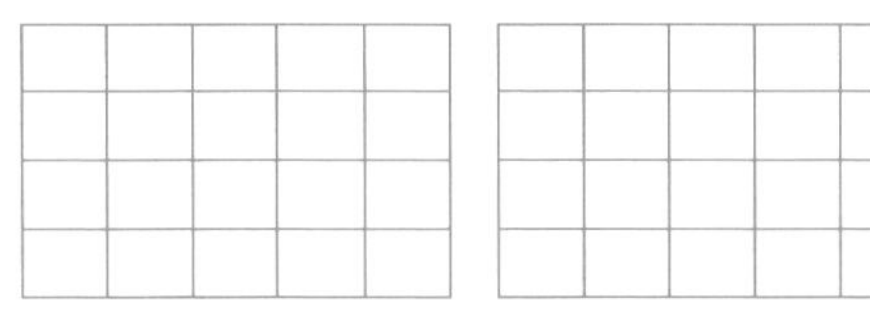

(Total 2 marks)

Edexcel, June 2007, Paper 10 Foundation, Question 5

7 A box contains 200 tissues.

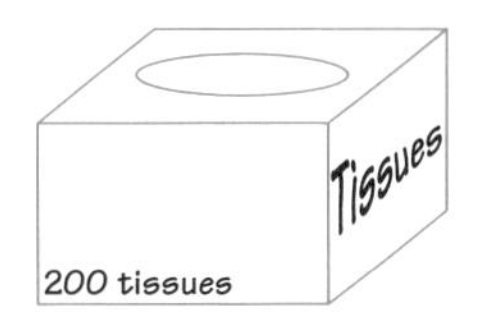

Toby takes $\frac{3}{5}$ of these tissues.

Work out how many tissues he takes. (2)

(Total 2 marks)

Edexcel, June 2008, Paper 13 Foundation, Question 5

8 A fruit punch was made using $\frac{1}{2}$ lemonade, $\frac{1}{5}$ orange juice with the rest lemon juice. What fraction of the drink is lemon juice?

9 Find $\frac{3}{5}$ of 45 kg.

10 Calculate the following, giving your answers as fractions in their simplest forms.

a $\frac{3}{8} + \frac{1}{8}$

b $\frac{9}{10} - \frac{1}{2}$

11 **a** Work out

i $\frac{3}{5}$ of 175 **ii** $\frac{3}{4} \times \frac{2}{3}$

b What fraction is 13 weeks out of a year of 52 weeks?

c What is $1 - \frac{2}{5}$?

12 Two-fifths of the price of a book goes to the bookshop. A book cost £12. How much goes to the bookshop?

13 When a cross is carved from a piece of wood, $\frac{4}{5}$ of the wood is cut away. The original block weighs 215 grams. What weight of wood is cut off?

14 **a** Write $\frac{5}{8}$ as a decimal.

b Write 0.6 as a fraction. Give your answer in its lowest terms.

15 **a** Work out $30 \times \frac{2}{3}$

b Work out the value of $\frac{14}{15} \times \frac{3}{4}$

Give your answer as a fraction in its simplest form.

16 The size of a packet of cereal is increased by one-fifth.

The new size is later reduced by one-fifth.

Is the new packet smaller, the same size or larger than the original packet?

Explain clearly how you worked out your answer.

17 Change the following fractions to decimals.

a $\frac{1}{5}$

b $\frac{1}{3}$

18 Packets of Wheetix used to contain 550 grams of cereal. New packets contain one-fifth more. How much does a new packet contain?

19 Change the following fractions to decimals.

a $\frac{1}{7}$

b $\frac{5}{13}$

20 **a** Write 24 as a fraction of 36

Give your answer in its simplest form. (2)

b Change $\frac{3}{5}$ into a decimal. (2)

(Total 4 marks)

Edexcel, May 2008, Paper 12 Foundation, Question 11

21 **a** Work out $\frac{1}{3} + \frac{1}{12}$ (2)

b Work out $\frac{3}{4} \times \frac{1}{5}$ (1)

(Total 3 marks)

Edexcel, May 2008, Paper 1 Foundation, Question 20

22 Work out $3\frac{2}{5} - 1\frac{2}{3}$

Worked Examination Questions

1 Fred gets 3 pints of milk delivered on Saturday.

He uses $\frac{5}{8}$ of a pint a day.

On which day will he run out of milk?

$3 \div \frac{5}{8}$ — You get 1 method mark for identifying this as a division problem.

$= 4.8$ or $4\frac{4}{5}$ — Use your calculator to work this out. This is worth 1 mark.

The milk runs out on Wednesday. — Count 5 days on from Saturday. The correct answer receives 1 mark.

Total: 3 marks

PS **2** There are 900 students in a college.

$\frac{1}{10}$ are under 16.

$\frac{1}{4}$ are over 18.

What fraction of the students are neither under 16 nor over 18?

First method:

$\frac{1}{10}$ of $900 = \frac{1}{10} \times 900$

$\frac{1}{4}$ of $900 = \frac{1}{4} \times 900$ — 1 method mark is available for identifying a method to work out the number under 16 or over 18.

$= 90$ and 225 — You get 1 mark for obtaining both of these answers.

$900 - 90 - 225 = 585$ — You get 1 mark for attempting to work out the number of people.

$\frac{585}{900}$ or $\frac{13}{20}$ — You get 1 accuracy mark for the correct answer.

Second method:

$\frac{1}{10} + \frac{1}{4}$ — 1 method mark is available for identifying the valid method of adding the given fractions.

$\frac{2}{20} + \frac{5}{20}$ — You get 1 method mark for getting a common denominator.

$\frac{7}{20}$ — You get 1 mark for correctly getting the fraction under 16 and over 18.

$\frac{13}{20}$ — You get 1 mark for the correct answer.

Total: 4 marks

2 Functional Maths
In the trenches

During World War I, soldiers were often faced with very difficult circumstances. For long periods of time, they would live in trenches that were wet, dirty, and full of rats and disease. During periods either just before or during a big battle, food was also scarce.

Getting started

Think about the following questions.

- What are ounces and gills?
- With a partner, work out the metric equivalents of a soldier's rations.
- How much would each soldier receive each week?

Is there any other information that you need to be able to complete your task? Where might you be able to find this information?

Below is a list of what each soldier might be expected to receive each day.

20 oz bread
3 oz cheese
$\frac{5}{8}$ oz tea
4 oz jam
8 oz fresh vegetables
$\frac{1}{3}$ oz chocolate
$\frac{1}{2}$ gill rum

Current supplies

$\frac{3}{4}$ ton bread
300 lb cheese
100 lb tea
355 lb jam
$\frac{1}{2}$ ton fresh vegetables
30 lb chocolate
20 gallons rum

Your task

The logistics corps faced one of the biggest challenges of the war. They were in charge of making sure that the soldiers in the trenches got the food they were entitled to. Imagine you were working in the logistics corps.

Using the information below, decide if you have everything you need to supply the 200 troops on your part of the front line for the next week. If you have shortages, decide how you are going to ration what you have.

Then, write a letter to the field commander, ordering your supplies, so that you have everything you need for the following week.

JAM
BISCUITS
3 OUNCES
HER ROYAL HIGHNESS
THE PRINCESS MARY'S
CHRISTMAS FUND
1914
SALT
H
FRAY
BENTOS
CORNED BEEF
WD14
This is your combat
issue of Tea–
Use it wisely!
BISCUITS
3 OUNCES

Why this chapter matters

Life is full of pairs: up and down, hot and cold, left and right, light and dark, rough and smooth, to name a few. One pairing that is particularly relevant to maths is positive and negative.

You are already familiar with positive numbers, including where they appear in real life and how to carry out calculations with them. However, sometimes we need to use a set of numbers in addition to the positive counting numbers. This set of numbers is known as the negative numbers. Here are some examples of where you might encounter negative numbers.

A negative number on a bank statement will show by how much you are overdrawn (or, how much money you have spent above what you have in your bank account).

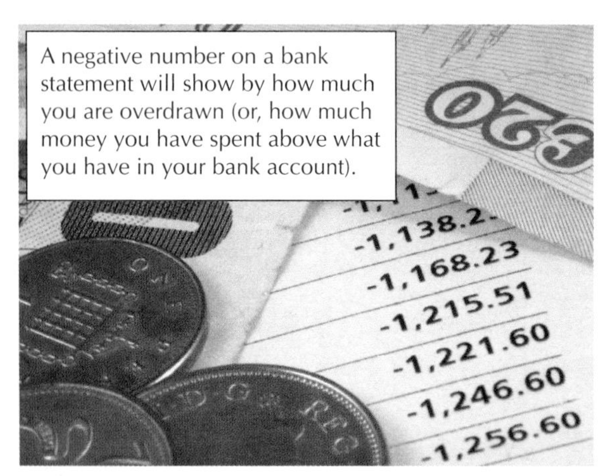

On the temperature scale of degrees Celsius zero is known as 'freezing point'. In many places – even in the UK! – temperatures fall below freezing point. At this stage we need negative numbers to represent the temperature.

Jet pilots experience g-forces when their aircraft accelerates or decelerates quickly. Negative g-forces can be felt when an object accelerates downwards very quickly and they are represented by negative numbers. These negative forces are responsible for the feeling of weightlessness that you have on rollercoasters!

$5 - 9 = -4$

When a bigger number is taken from a smaller one, the result is a negative number.

In lifts, negative numbers are used to represent floors below ground level. These are often called 'lower ground' floors.

Sea level can be given the value 'zero'. Mountains are described as being 'above sea level' and ocean floors as 'below sea level'. This means that when a submarine goes under the sea the depths that it reaches are given using negative numbers.

As you can see, negative numbers are just as important as positive numbers and you will encounter them in your everyday life.

Chapter 3

Number: Negative numbers

The grades given in this chapter are target grades.

1. Introduction to negative numbers
2. Everyday use of negative numbers
3. The number line
4. Arithmetic with negative numbers

This chapter will show you ...

- G how negative numbers are used in real life
- G what is meant by a negative number
- G how to use inequalities with negative numbers
- F how to do arithmetic with negative numbers

Visual overview

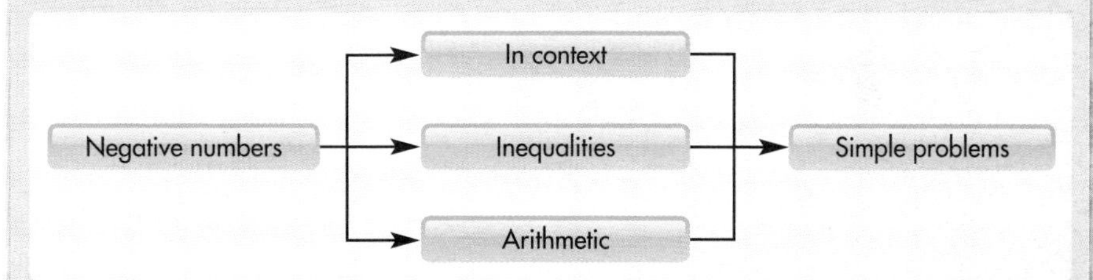

What you should already know

- What a negative number means **(KS3 level 3, GCSE grade G)**
- How to put numbers in order **(KS3 level 3, GCSE grade G)**

Quick check

Put the numbers in the following lists into order, smallest first.

1 8, 2, 5, 9, 1, 0, 4

2 14, 19, 11, 10, 17

3 51, 92, 24, 0, 32

4 87, 136, 12, 288, 56

5 5, 87, $\frac{1}{2}$, 100, 0, 50

3.1 Introduction to negative numbers

This section will show you how to:

- use negative numbers in real life

Key words
below
difference
negative number

Negative numbers are numbers **below** zero. You meet **negative numbers** often in winter when the temperature falls below freezing (0 °C).

The diagram below shows a thermometer with negative temperatures. The temperature is –3 °C. This means the temperature is three degrees below zero.

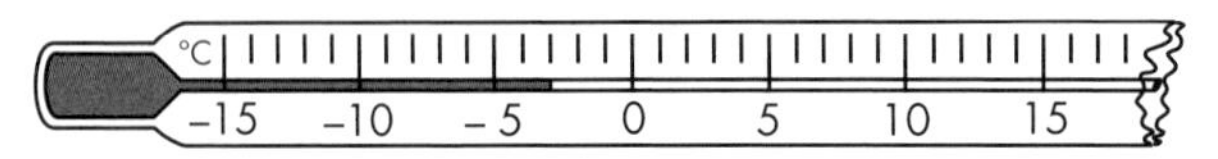

The number line below shows positive and negative numbers.

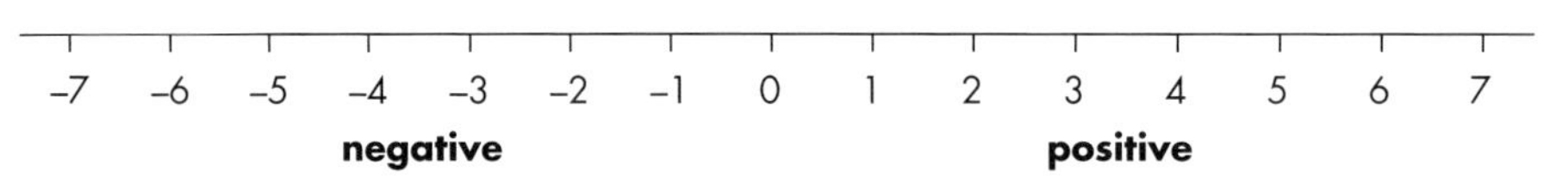

ACTIVITY

Seaport colliery

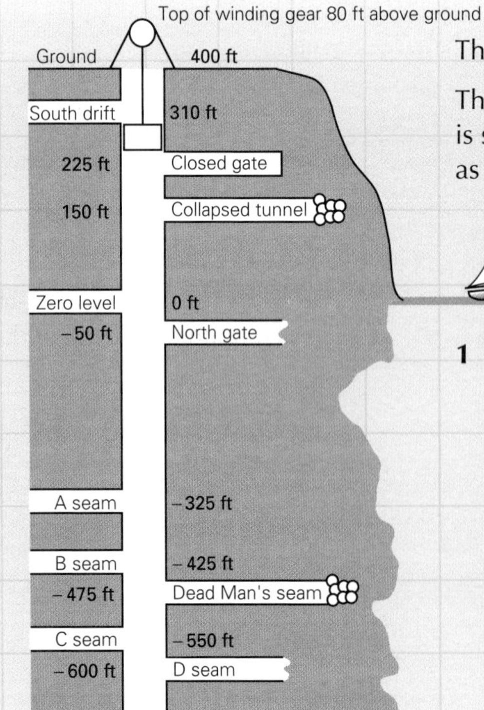

This is a section through Seaport Colliery.

The height above sea (zero) level for each tunnel is shown. Note that some of the heights are given as negative numbers.

The ground is 400 ft above sea level.

1 What is the **difference** in height between the following levels?

a South drift and C seam

b The ground and A seam

c The closed gate and the collapsed tunnel

d Zero level and Dead Man's seam

e Ground level and the bottom gate

f Collapsed tunnel and B seam

g North gate and Dead Man's seam

- **h** Zero level and the south drift
- **i** Zero level and the bottom gate
- **j** South drift and the bottom gate

2 How high above sea level is the top of the winding gear?

3 How high above the bottom gate is the top of the winding gear?

4 There are two pairs of tunnels that are 75 ft apart. Which two pairs are they?

5 How much cable does the engineman let out to get the cage from the south drift to D seam?

6 There are two pairs of tunnels that are 125 ft apart. Which two pairs are they?

7 Which two tunnels are 200 ft apart?

EXERCISE 3A

1 Write down the temperature for each thermometer.

a

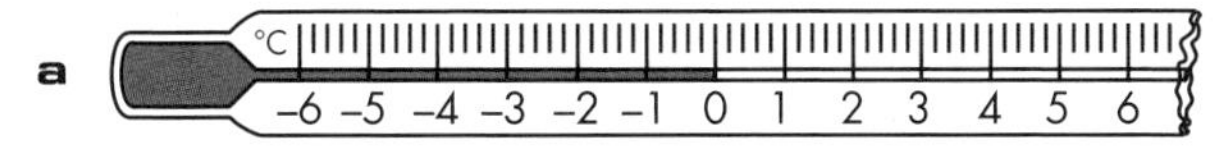

b

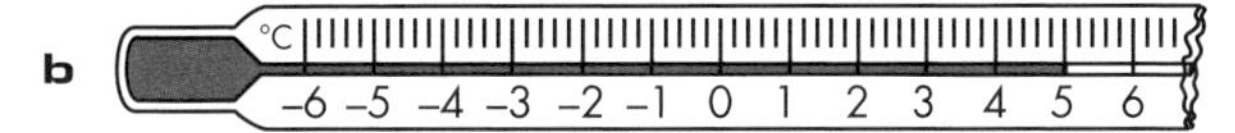

c

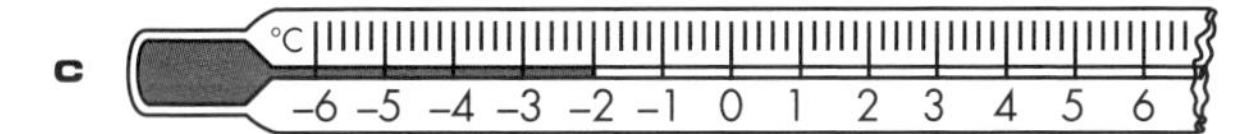

d

°C
–6 –5 –4 –3 –2 –1 0 1 2 3 4 5 6

e

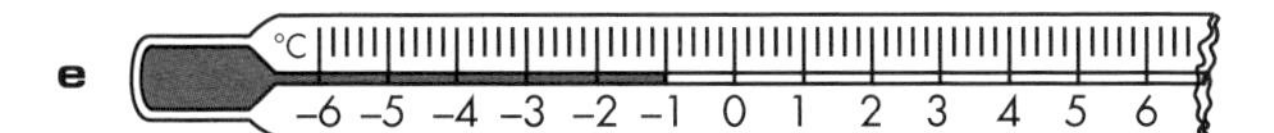

2

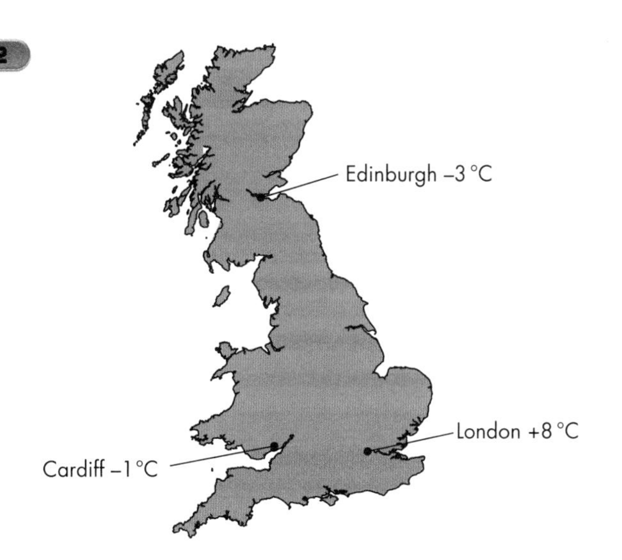

- **a** How much colder is it in Edinburgh than in London?
- **b** How much warmer is it in London than in Cardiff?

G

G

FM 3 The instructions on a bottle of de-icer say that it will stop water freezing down to –12 °C. The temperature is –4 °C.

How many more degrees does the temperature need to fall before the de-icer stops working?

F

FM 4 Chris Boddington, the famous explorer, is leading a joint expedition to the Andes mountains and cave system. The climbers establish camps at the various places shown and the cavers find the various caves shown.

To help him move supplies about, Chris needs a chart of the differences between the heights of the camps and caves. Help him by copying the chart below into your book and filling it in. (Some of the boxes have been done for you.)

	Sump	River Cave	Lost Cave	Echo Cave	Angel Cavern	Bat Cave	Base	Camp 1	Camp 2	Camp 3	Camp 4	Summit
Summit										670	220	0
Camp 4											0	
Camp 3										0		
Camp 2				2840					0			
Camp 1								0				
Base						245	0					
Bat Cave					130	0						
Angel Cavern					0							
Echo Cave				0								
Lost Cave	100	80	0									
River Cave	20	0										
Sump	0											

3.2 Everyday use of negative numbers

This section will show you how to:
- use positive and negative numbers in everyday life

Key words
above/below
after/before
credit/debit
negative number/positive number
profit/loss

There are many other situations where **negative numbers** are used. Here are three examples.

- When +15 m means 15 metres **above** sea level, then –15 m means 15 metres **below** sea level.
- When +2 h means 2 hours **after** midday, then –2 h means 2 hours **before** midday.
- When +£60 means a **profit** of £60, then –£60 means a **loss** of £60.

You also meet negative numbers on graphs, and you may already have plotted coordinates with negative numbers.

On bank statements and bills a negative number means you owe money. A **positive number** means they owe you money.

A **debit** is when you pay money out of your account.

A **credit** is when you pay money into your account.

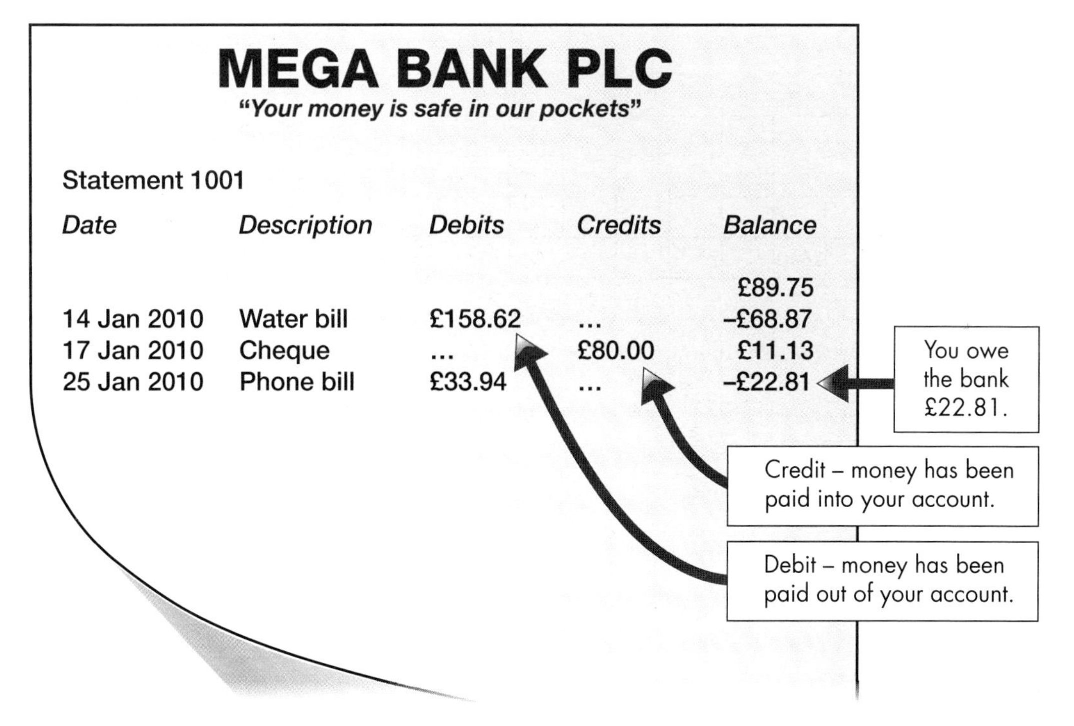

Date	Description	Debits	Credits	Balance
				£89.75
14 Jan 2010	Water bill	£158.62	...	–£68.87
17 Jan 2010	Cheque	...	£80.00	£11.13
25 Jan 2010	Phone bill	£33.94	...	–£22.81

EXERCISE 3B

Copy and complete each of the following.

G

1. If +£5 means a profit of five pounds, then …… means a loss of five pounds.
2. If +£9 means a profit of nine pounds, then a loss of £9 is …… .
3. If –£4 means a loss of four pounds, then +£4 means a …… of four pounds.
4. If +200 m means 200 metres above sea level, then …… means 200 metres below sea level.
5. If +50 m means fifty metres above sea level, then fifty metres below sea level is written …… .
6. If –100 m means one hundred metres below sea level, then +100 m means one hundred metres …… sea level.
7. If +3 h means three hours after midday, then …… means three hours before midday.
8. If +5 h means 5 hours after midday, then …… means 5 hours before midday.
9. If –6 h means six hours before midday, then +6 h means six hours …… midday.
10. If +2 °C means two degrees above freezing point, then …… means two degrees below freezing point.
11. If +8 °C means eight degrees above freezing point, then …… means eight degrees below freezing point.
12. If –5 °C means five degrees below freezing point, then +5 °C means five degrees …… freezing point.
13. If +70 km means 70 kilometres north of the equator, then …… means 70 kilometres south of the equator.
14. If +200 km means 200 kilometres north of the equator, then 200 kilometres south of the equator is written …… .
15. If –50 km means fifty kilometres south of the equator, then +50 km means fifty kilometres …… of the equator.
16. If 10 minutes before midnight is represented by –10 minutes, then five minutes after midnight is represented by …… .
17. If a car moving forwards at 10 miles per hour is represented by +10 mph, then a car moving backwards at 5 miles per hour is represented by …… .

G

18 In an office building, the third floor above ground level is represented by +3. So, the second floor below ground level is represented by …… .

FM **19**

MEGA BANK PLC

"Your money is safe in our pockets"

Statement 2010

Date	*Description*	*Debits*	*Credits*	*Balance*
				£320.45
20 Sept 2009	Gas bill	£410.17	...	–£89.72
23 Sept 2009	Cheque	...	£140.00	£50.28
25 Sept 2009	Mobile phone bill	£63.48	...	–£13.20

a What does –£89.72 mean?

b What is a debit?

c What is a credit?

20 The temperature on three days in Moscow was –7 °C, –5 °C and –11 °C.

a Which temperature is the lowest?

b What is the difference in temperature between the coldest and the warmest days?

21 **a** Which is the smallest number in the cloud?

b Which is the largest number in the cloud?

c What is the difference between the smallest and largest numbers in the cloud?

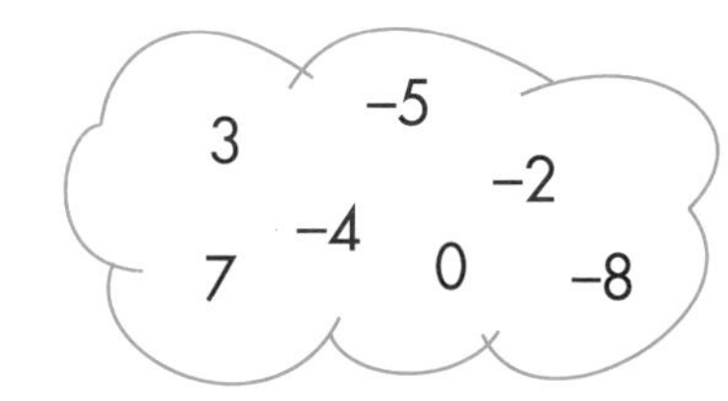

F

PS **22** Sydney the snail is at the bottom of a 10-foot well.

Each day he climbs 2 feet up the wall of the well.

Each night he slides 1 foot back down the wall of the well.

How many days does it take Sydney to reach the top of the well?

AU **23** A thermostat is set at 16 °C.

The temperature in a room at 1.00 am is –2 °C.

The temperature rises two degrees every 6 minutes.

At what time is the temperature on the thermostat reached?

3.3 The number line

This section will show you how to:
- use a number line to represent negative numbers
- use inequalities with negative numbers

Key words
greater than
inequality
less than
more than
negative
number line
positive

Look at the **number line**.

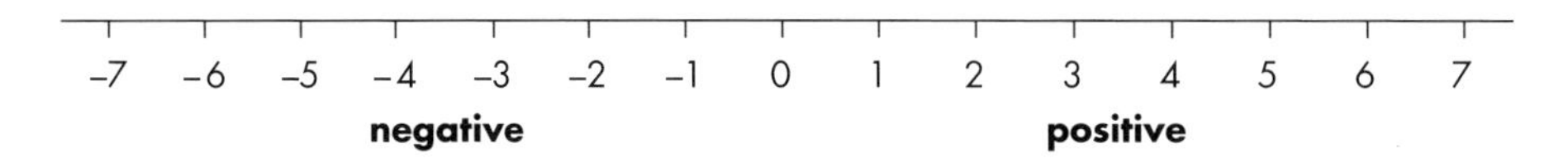

Notice that the **negative** numbers are to the left of 0 and the **positive** numbers are to the right of 0.

Numbers to the right of any number on the number line are always bigger than that number.

Numbers to the left of any number on the number line are always smaller than that number.

So, for example, you can see from a number line that:

2 is *smaller* than 5 because 2 is to the *left* of 5.

You can write this as $2 < 5$.

–3 is *smaller* than 2 because –3 is to the *left* of 2.

You can write this as $-3 < 2$.

7 is *bigger* than 3 because 7 is to the *right* of 3.

You can write this as $7 > 3$.

–1 is *bigger* than –4 because –1 is to the *right* of –4.

You can write this as $-1 > -4$.

Reminder The **inequality** signs:

$<$ means 'is **less than**'

$>$ means 'is **greater than**' or 'is **more than**'

EXERCISE 3C

G

1 Copy and complete each of the following by putting a suitable number in the box.

a □ is smaller than 3 **b** □ is smaller than 1 **c** □ is smaller than –3

d □ is smaller than –7 **e** –5 is smaller than □ **f** –1 is smaller than □

g 3 is smaller than □ **h** –2 is smaller than □ **i** □ is smaller than 0

j –4 is smaller than □ **k** □ is smaller than –8 **l** –7 is smaller than □

2 Copy and complete each of the following by putting a suitable number in the box.

a □ is bigger than –3 **b** □ is bigger than 1 **c** □ is bigger than –2

d □ is bigger than –1 **e** –1 is bigger than □ **f** –8 is bigger than □

g 1 is bigger than □ **h** –5 is bigger than □ **i** □ is bigger than –5

j 2 is bigger than □ **k** □ is bigger than –4 **l** –2 is bigger than □

3 Copy each of these and put the correct phrase in each space.

a –1 …… 3 **b** 3 …… 2 **c** –4 …… –1

d –5 …… –4 **e** 1 …… –6 **f** –3 …… 0

g –2 …… –1 **h** 2 …… –3 **i** 5 …… –6

j 3 …… 4 **k** –7 …… –5 **l** –2 …… –4

4

-1 $-\frac{3}{4}$ $-\frac{1}{2}$ $-\frac{1}{4}$ 0 $\frac{1}{4}$ $\frac{1}{2}$ $\frac{3}{4}$ 1

Copy each of these and put the correct phrase in each space.

a $\frac{1}{4}$ …… $\frac{3}{4}$ **b** $-\frac{1}{2}$ …… 0 **c** $-\frac{3}{4}$ …… $\frac{3}{4}$

d $\frac{1}{4}$ …… $-\frac{1}{2}$ **e** -1 …… $\frac{3}{4}$ **f** $\frac{1}{2}$ …… 1

5 In each case below, copy the statement and put the correct symbol, either $<$ or $>$, in the box.

a 3 □ 5 **b** –2 □ –5 **c** –4 □ 3 **d** 5 □ 9

e –3 □ 2 **f** 4 □ –3 **g** –1 □ 0 **h** 6 □ –4

i 2 □ –3 **j** 0 □ –2 **k** –5 □ –4 **l** 1 □ 3

m –6 □ –7 **n** 2 □ –3 **o** –1 □ 1 **p** 4 □ 0

G

PS 6 Copy these number lines and fill in the missing numbers.

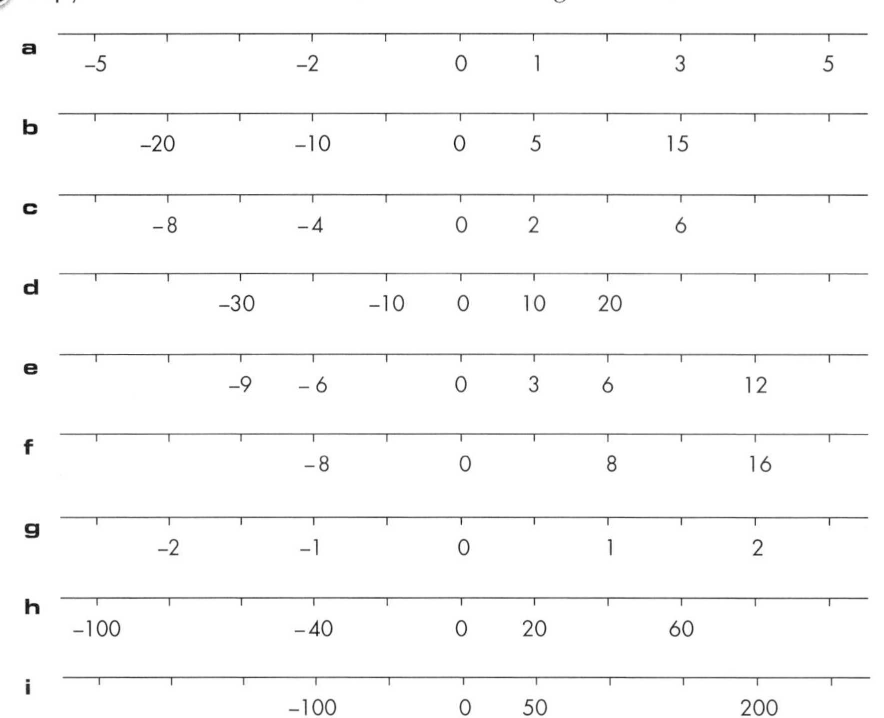

FM 7 Here are some temperatures.

2 °C   −2 °C   −4 °C   6 °C

Copy and complete the weather report, using these temperatures.

> The hottest place today is Barnsley with a temperature of ____, while in Eastbourne a ground frost has left the temperature just below zero at ____. In Bristol it is even colder at ____. Finally, in Tenby the temperature is just above freezing at ____.

AU 8 Here are some numbers.

$-4\frac{1}{2}$ $+3\frac{3}{4}$ $-\frac{1}{4}$

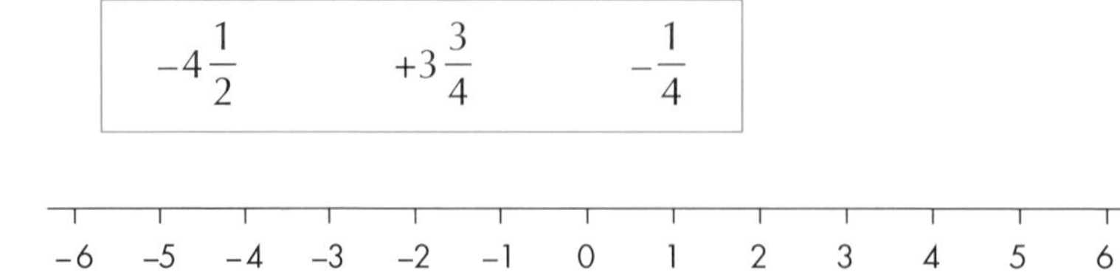

Copy the number line and mark the numbers on it.

3.4 Arithmetic with negative numbers

This section will show you how to:
- add and subtract positive and negative numbers to or from both positive and negative numbers

Key words
add
subtract

Adding and subtracting positive numbers

These two operations can be illustrated on a thermometer scale.

- **Adding** a positive number moves the marker *up* the thermometer scale. For example,

  $-2 + 6 = 4$

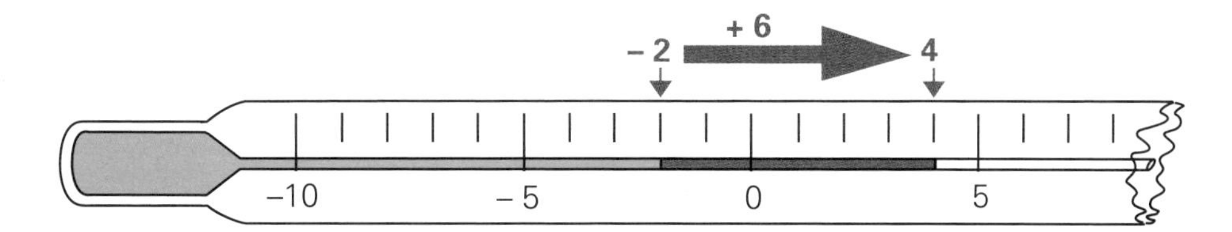

- **Subtracting** a positive number moves the marker *down* the thermometer scale. For example,

  $3 - 5 = -2$

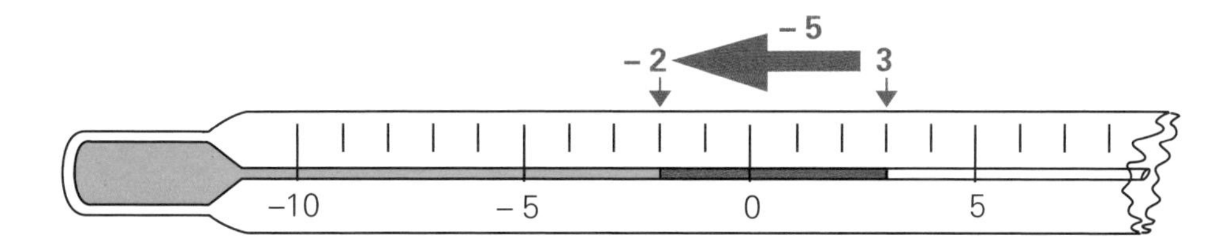

EXAMPLE 1

The temperature at midnight was 2 °C but then it fell by five degrees. What was the new temperature?

Falling five degrees means the calculation is 2 – 5, which is equal to –3. So, the new temperature is –3 °C.

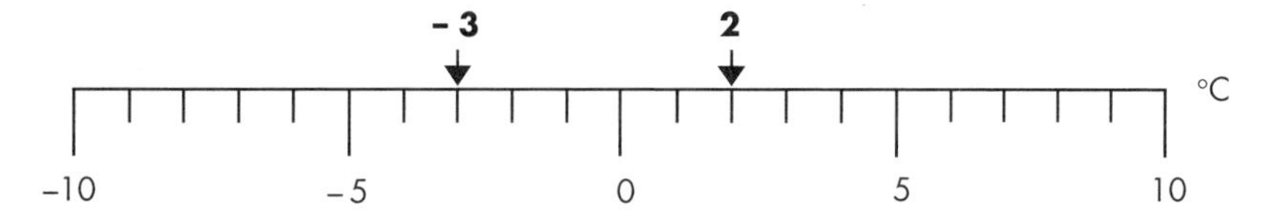

EXAMPLE 2

The temperature is -4 °C. It then falls by five degrees. What is the new temperature?

Falling five degrees means the calculation is $-4 - 5$, which is equal to -9. So, the new temperature is -9 °C.

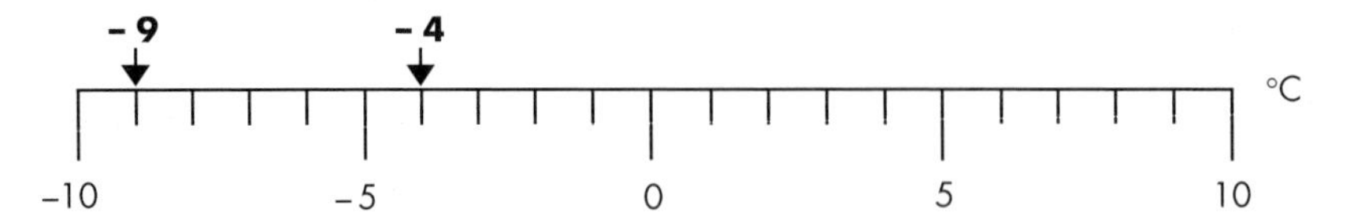

EXERCISE 3D

F

1 Use a thermometer scale to find the answer to each of the following.

a $2° - 4° =$ **b** $4° - 7° =$ **c** $3° - 5° =$ **d** $1° - 4° =$
e $6° - 8° =$ **f** $5° - 8° =$ **g** $-2 + 5 =$ **h** $-1 + 4 =$
i $-4 + 3 =$ **j** $-6 + 5 =$ **k** $-3 + 5 =$ **l** $-5 + 2 =$
m $-1 - 3 =$ **n** $-2 - 4 =$ **o** $-5 - 1 =$ **p** $3 - 4 =$
q $2 - 7 =$ **r** $1 - 5 =$ **s** $-3 + 7 =$ **t** $5 - 6 =$
u $-2 - 3 =$ **v** $2 - 6 =$ **w** $-8 + 3 =$ **x** $4 - 9 =$

2 Answer each of the following *without* the help of a thermometer scale.

a $5 - 9 =$ **b** $3 - 7 =$ **c** $-2 - 8 =$ **d** $-5 + 7 =$
e $-1 + 9 =$ **f** $4 - 9 =$ **g** $-10 + 12 =$ **h** $-15 + 20 =$
i $23 - 30 =$ **j** $30 - 42 =$ **k** $-12 + 25 =$ **l** $-30 + 55 =$
m $-10 - 22 =$ **n** $-13 - 17 =$ **o** $45 - 50 =$ **p** $17 - 25 =$
q $18 - 30 =$ **r** $-25 + 35 =$ **s** $-23 - 13 =$ **t** $31 - 45 =$
u $-24 + 65 =$ **v** $-19 + 31 =$ **w** $25 - 65 =$ **x** $199 - 300 =$

3 Work out each of the following.

a $8 + 3 - 5 =$ **b** $-2 + 3 - 6 =$ **c** $-1 + 3 + 4 =$
d $-2 - 3 + 4 =$ **e** $-1 + 1 - 2 =$ **f** $-4 + 5 - 7 =$
g $-3 + 4 - 7 =$ **h** $1 + 3 - 6 =$ **i** $8 - 7 + 2 =$
j $-5 - 7 + 12 =$ **k** $-4 + 5 - 8 =$ **l** $-4 + 6 - 8 =$
m $103 - 102 + 7 =$ **n** $-1 + 4 - 2 =$ **o** $-6 + 9 - 12 =$
p $-3 - 3 - 3 =$ **q** $-3 + 4 - 6 =$ **r** $-102 + 45 - 23 =$
s $8 - 10 - 5 =$ **t** $9 - 12 + 2 =$ **u** $99 - 100 - 46 =$

4 Use a calculator to check your answers to questions 1 to 3.

AU **5** At 5 am the temperature in London was –4 °C.

At 11 am the temperature was 3 °C.

a By how many degrees did the temperature rise?

b The temperature in Brighton was two degrees lower than in London at 5 am
What was the temperature is Brighton at 5 am?

PS **6** Here are five numbers.

4 7 8 2 5

a Use two of the numbers to make a calculation with an answer of –6.

b Use three of the numbers to make a calculation with an answer of –1.

c Use four of the numbers to make a calculation with an answer of –18.

d Use all five of the numbers to make a calculation with an answer of –12.

FM **7** A submarine is 1600 feet below sea level.

A radar system can detect submarines down to 900 feet below sea level.

To safely avoid detection, the submarine captain keeps the submarine 200 feet below the level of detection.

How many feet can the submarine climb to be safe from detection?

F

Adding and subtracting negative numbers

To *subtract a negative number* …

… treat the – – as a +

For example: 4 – (–2) = 4 + 2 = 6

To *add a negative number* …

… treat the + – as a –

For example: 3 + (–5) = 3 – 5 = –2

Using your calculator

Calculations involving negative numbers can be done by using the **(–)** key.

EXAMPLE 3

Work out –3 + 7.

Press **(–)** **3** **+** **7** **=**

The answer should be 4.

EXAMPLE 4

Work out $-6 - (-2)$.

Press (–) 6 – (–) 2 =

The answer should be -4.

EXERCISE 3E

F

1 Answer each of the following. Check your answers on a calculator.

a $2 - (-4) =$ **b** $4 - (-3) =$ **c** $3 - (-5) =$ **d** $5 - (-1) =$

e $6 - (-2) =$ **f** $8 - (-2) =$ **g** $-1 - (-3) =$ **h** $-4 - (-1) =$

i $-2 - (-3) =$ **j** $-5 - (-7) =$ **k** $-3 - (-2) =$ **l** $-8 - (-1) =$

m $4 + (-2) =$ **n** $2 + (-5) =$ **o** $3 + (-2) =$ **p** $1 + (-6) =$

q $5 + (-2) =$ **r** $4 + (-8) =$ **s** $-2 + (-1) =$ **t** $-6 + (-2) =$

u $-7 + (-3) =$ **v** $-2 + (-7) =$ **w** $-1 + (-3) =$ **x** $-7 + (-2) =$

2 Write down the answer to each of the following, then check your answers on a calculator.

a $-3 - 5 =$ **b** $-2 - 8 =$ **c** $-5 - 6 =$ **d** $6 - 9 =$

e $5 - 3 =$ **f** $3 - 8 =$ **g** $-4 + 5 =$ **h** $-3 + 7 =$

i $-2 + 9 =$ **j** $-6 + -2 =$ **k** $-1 + -4 =$ **l** $-8 + -3 =$

m $5 - -6 =$ **n** $3 - -3 =$ **o** $6 - -2 =$ **p** $3 - -5 =$

q $-5 - -3 =$ **r** $-2 - -1 =$ **s** $-4 - 5 =$ **t** $2 - 7 =$

u $-3 + 8 =$ **v** $-4 + - 5 =$ **w** $1 - -7 =$ **x** $-5 - -5 =$

3 The temperature at midnight was 4 °C. Find the temperature if it *fell* by:

a 1 degree **b** 4 degrees **c** 7 degrees

d 9 degrees **e** 15 degrees

4 What is the *difference* between the following temperatures?

a 4 °C and –6 °C

b –2 °C and –9 °C

c –3 °C and 6 °C

5 Rewrite the following list, putting the numbers in order of size, lowest first.

1 –5 3 –6 –9 8 –1 2

F

6 Write down the answers to each of the following, then check your answers on a calculator.

a $2 - 5 =$ **b** $7 - 11 =$ **c** $4 - 6 =$

d $8 - 15 =$ **e** $9 - 23 =$ **f** $-2 - 4 =$

g $-5 - 7 =$ **h** $-1 - 9 =$ **i** $-4 + 8 =$

j $-9 + 5 =$ **k** $9 - -5 =$ **l** $8 - -3 =$

m $-8 - -4 =$ **n** $-3 - -2 =$ **o** $-7 + -3 =$

p $-9 + 4 =$ **q** $-6 + 3 =$ **r** $-1 + 6 =$

s $-9 - -5 =$ **t** $9 - 17 =$

7 Find what you have to *add to* 5 to get:

a 7 **b** 2 **c** 0

d –2 **e** –5 **f** –15

8 Find what you have to *subtract from* 4 to get:

a 2 **b** 0 **c** 5

d 9 **e** 15 **f** –4

9 Find what you have to *add to* –5 to get:

a 8 **b** –3 **c** 0

d –1 **e** 6 **f** –7

10 Find what you have to *subtract from* –3 to get:

a 7 **b** 2 **c** –1

d –7 **e** –10 **f** 1

AU 11 Write down *ten* different addition sums that give the answer 1.

AU 12 Write down *ten* different subtraction calculations that give the answer 1. There must be *one negative number* in each calculation.

13 Use a calculator to work out each of these.

a $-7 + -3 - -5 =$ **b** $6 + 7 - 7 =$ **c** $-3 + -4 - -7 =$

d $-1 - 3 - -6 =$ **e** $8 - -7 + -2 =$ **f** $-5 - 7 - -12 =$

g $-4 + 5 - 7 =$ **h** $-4 + -6 - -8 =$ **i** $103 - -102 - -7 =$

j $-1 + 4 - -2 =$ **k** $6 - -9 - 12 =$ **l** $-3 - -3 - -3 =$

m $-45 + -56 - -34 =$ **n** $-3 + 4 - -6 =$ **o** $102 + -45 - 32 =$

F

14 Give the outputs of each of these function machines.

a −4, −3, −2, −1, 0 → [+ 3] → ?, ?, ?, ?, ?

b −4, −3, −2, −1, 0 → [− 2] → ?, ?, ?, ?, ?

c −4, −3, −2, −1, 0 → [+ 1] → ?, ?, ?, ?, ?

d −4, −3, −2, −1, 0 → [− 4] → ?, ?, ?, ?, ?

e −4, −3, −2, −1, 0 → [− 5] → ?, ?, ?, ?, ?

f −4, −3, −2, −1, 0 → [+ 7] → ?, ?, ?, ?, ?

g −10, −9, −8, −7, −6 → [− 2] → ?, ?, ?, ?, ?

h −10, −9, −8, −7, −6 → [− 6] → ?, ?, ?, ?, ?

i −5, −4, −3, −2, −1, 0 → [+ 3] → ?, ?, ?, ?, ?, ? → [− 2] → ?, ?, ?, ?, ?, ?

j −5, −4, −3, −2, −1, 0 → [− 7] → ?, ?, ?, ?, ?, ? → [− 2] → ?, ?, ?, ?, ?, ?

k −5, −4, −3, −2, −1, 0 → [+ 3] → ?, ?, ?, ?, ?, ? → [+ 2] → ?, ?, ?, ?, ?, ?

l −3, −2, −1, 0, 1, 2, 3 → [− 5] → ?, ?, ?, ?, ?, ? → [+ 3] → ?, ?, ?, ?, ?, ?

m −3, −2, −1, 0, 1, 2, 3 → [− 7] → ?, ?, ?, ?, ?, ? → [+ 9] → ?, ?, ?, ?, ?, ?

n −3, −2, −1, 0, 1, 2, 3 → [+ 6] → ?, ?, ?, ?, ?, ? → [− 8] → ?, ?, ?, ?, ?, ?

15 What numbers are missing from the boxes to make the number sentences true?

a $2 + -6 = \square$

b $4 + \square = 7$

c $-4 + \square = 0$

d $5 + \square = -1$

e $3 + 4 = \square$

f $\square - -5 = 7$

g $\square - 5 = 2$

h $6 + \square = 0$

i $\square - -5 = -2$

j $2 + -2 = \square$

k $\square - 2 = -2$

l $-2 + -4 = \square$

m $2 + 3 + \square = -2$

n $-2 + -3 + -4 = \square$

o $\square - 5 = -1$

p $\square - 8 = -8$

q $-4 + 2 + \square = 3$

r $-5 + 5 = \square$

s $7 - -3 = \square$

t $\square - -5 = 0$

u $3 - \square = 0$

v $-3 - \square = 0$

w $-6 + -3 = \square$

x $\square - 3 - -2 = -1$

y $\square - 1 = -4$

z $7 - \square = 10$

AU 16 You have the following cards.

a Which card should you choose to make the answer to the following sum as large as possible? What is the answer?

+6 + ☐ = ……

b Which card should you choose to make the answer to part **a** as small as possible? What is the answer?

c Which card should you choose to make the answer to the following subtraction as large as possible? What is the answer?

+6 – ☐ = ……

d Which card should you choose to make the answer to part **c** as small as possible? What is the answer?

AU 17 You have the following cards.

-9 -7 -5 -4 0 +1 +2 +4 +7

a Which cards should you choose to make the answer to the following calculation as large as possible? What is the answer?

+5 + ☐ – ☐ = ……

b Which cards should you choose to make the answer to part **a** as small as possible? What is the answer?

c Which cards should you choose to make the answer to the following number sentence zero? Give all possible answers.

☐ + ☐ = 0

FM AU 18 The thermometer in a car is inaccurate by up to two degrees.

An ice alert warning comes on at 3 °C, according to the thermometer temperature.

If the actual temperature is 2 °C, will the alert come on?

Explain how you decide.

PS 19 Two numbers have a sum of 5.

One of the numbers is negative.

The other number is even.

What are the two numbers if the even number is as large as possible?

Negative magic squares

Make your own magic square with negative numbers. You need nine small square cards and two pens or pencils of different colours.

This is perhaps the best known magic square.

8	3	4
1	5	9
6	7	2

But magic squares can be made from many different sets of numbers, as shown by this second square.

This square is now used to show you how to make a magic square with negative numbers. But the method works with any magic square. So, if you can find one of your own, use it!

8	13	6
7	9	11
12	5	10

- Arrange the nine cards in a square and write on them the numbers of the magic square. Picture **a** below.
- Rearrange the cards in order, lowest number first, to form another square. Picture **b** below.
- Keeping the cards in order, turn them over so that the blank side of each card is face up. Picture **c** below.

a

8	13	6
7	9	11
12	5	10

b

5	6	7
8	9	10
11	12	13

c

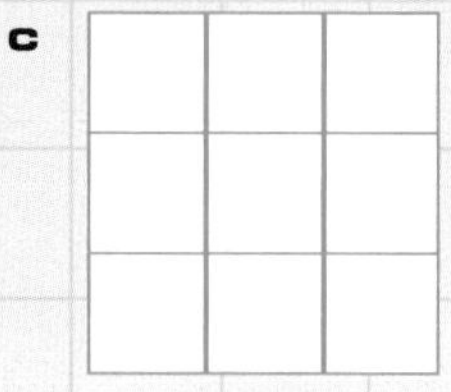

- Now use a different coloured pen to avoid confusion.
- Choose any number (say four) for the top left-hand corner of the square. Picture **d** below.
- Choose another number (say three) and subtract it from each number in the first row to get the next number. Picture **e** below.
- Now choose a third number (say two) and subtract it from each number in the top row to make the second row, and then again from each number in the second row. Picture **f** below.

d

4		

e

4	1	–2

f

4	1	–2
2	–1	–4
0	–3	–6

- Turn the cards over. Picture **g**.
- Rearrange the cards into the original magic square. Picture **h**.
- Turn them over again. Picture **i**.

g

5	6	7
8	9	10
11	12	13

h

8	13	6
7	9	11
12	5	10

i

2	–6	1
–2	–1	0
–3	4	–4

You should have a **magic square of negative numbers**.

Try it on any square. It works even with squares bigger than 3 × 3. Try it on this 4 × 4 square.

2	13	9	14
16	7	11	4
15	8	12	3
5	10	6	17

EXERCISE 3F

PS Copy and complete each of these magic squares. In each case, write down the 'magic number'.

1

–1		
–5	–4	–3
		–7

2

	–4	3
		–2
	4	–1

3

–6	–5	–4
		–10

4

2		
–4		
–7	6	–8

5

		–9
–3	–6	–9

6

		–1
	–7	
–13		–12

7

–4		
–8		–6
–9		

8

2	1	–3
	0	

9

–2		
	–5	
–7		–8

10

–8			–14
–8	–9		
		–4	–5
1	–10	–12	–5

11

–7		2	–16
	–8		
–11	–3	0	–2
		–13	–1

SUMMARY

GRADE BOOSTER

- **G** You can use negative numbers in context
- **G** You can use < for less than and > for greater than
- **F** You can add positive and negative numbers to positive and negative numbers
- **F** You can subtract positive and negative numbers from positive and negative numbers
- **E** You can solve problems involving simple negative numbers

What you should know now

- How to order positive and negative numbers
- How to add and subtract positive and negative numbers
- How to use negative numbers in practical situations
- How to use a calculator when working with negative numbers

1 The temperature in a school yard was measured at 9am each morning for one week.

Day	Monday	Tuesday	Wednesday	Thursday	Friday
9am temperature	−1	−3	−2	1	2

a Which day was the coldest at 9am?

b Which day was the warmest at 9am?

2 The table shows the temperatures at midnight in 6 cities during one night in 2006.

City	Temperature
Berlin	5°C
London	10°C
Moscow	−3°C
New York	2°C
Oslo	−8°C
Paris	7°C

a Write down the city which had the lowest temperature. (1)

b Work out the difference in temperature between London and Moscow. (2)

(Total 3 marks)

Edexcel, June 2008, Paper 10 Foundation, Unit 3 Test, Question 2

3

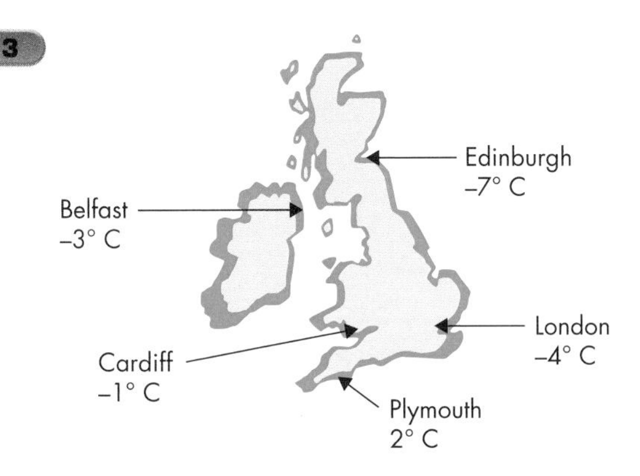

The map shows the temperature in five cities, at midnight, one night last year.

a Write down the name of the city with the highest temperature. (1)

b Write down the name of the city with the lowest temperature. (1)

(Total 2 marks)

Edexcel, June 2007, Paper 10 Foundation, Unit 3 Test, Question 3

4 Write down the temperature shown. (1)

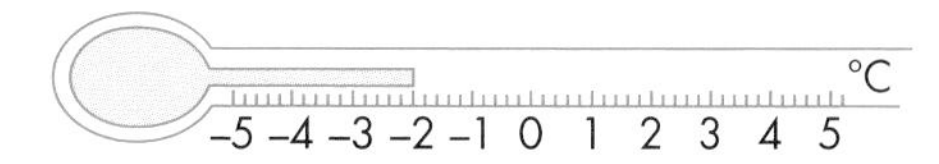

(Total 1 mark)

Edexcel, November 2007, Paper 10 Foundation, Unit 3 Test, Question 1

5 **a** Work out $-2 + 5$ (1)

b Work out $-3 - 5$ (1)

(Total 2 marks)

Edexcel, May 2008, Paper 12 Foundation, Question 3

6 The table shows the temperatures in four cities at noon one day.

Oslo	−13°C
New York	−5°C
Cape Town	9°C
London	2°C

a Write down the **highest** temperature. (1)

b Work out the difference in temperature between Oslo and New York. (1)

At 8 pm the temperature in London was 3°C lower than the temperature at noon.

c Work out the temperature in London at 8 pm. (1)

(Total 3 marks)

Edexcel, June 2008, Paper 13 Foundation, Question 2

7 Write out and complete the following to make each statement correct.

a $3 - \square = -5$ **b** $\square + 4 = -5$

c $1 - \square = 8$

8 The table shows the temperature on the surface of each of five planets.

Planet	Temperature
Venus	480 °C
Mars	–60 °C
Jupiter	–150 °C
Saturn	–180 °C
Uranus	–210 °C

a Work out the difference in temperature between Mars and Jupiter. (1)

b Work out the difference in temperature between Venus and Mars. (1)

c Which planet has a temperature 30 °C higher than the temperature on Saturn? (1)

The temperature on Pluto is 20 °C lower than the temperature on Uranus.

d Work out the temperature on Pluto. (1)

(Total 4 marks)

Edexcel, June 2005, Paper 2 Foundation, Question 8

9 The temperatures of the first few days of January were recorded as

1 °C, –1 °C, 0 °C and –2 °C

a Write down the four temperatures, in order, with the lowest first.

b What is the difference between the coldest and the warmest of these four days?

10 You have the following cards.

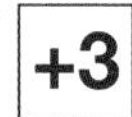 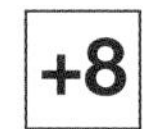  

a i What card should you choose to make the answer to the following sum as large as possible?

+1 + ☐ = ...

ii What is the answer to the sum in **i**?

iii What card would you have chosen to make the sum as small as possible?

b i What card should you choose to make the answer to the following subtraction as large as possible?

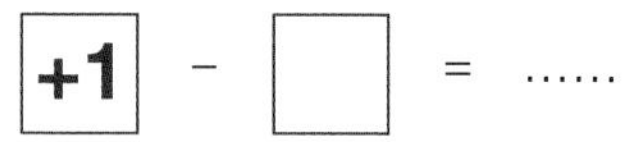

ii What is the answer to the subtraction sum in **i**?

iii What card would you have chosen to make the subtraction as small as possible?

11 Nitrogen gas makes up most of the air we breathe.

Nitrogen freezes under –210 °C and is a gas above –196 °C. In between it is liquid.

Write down a possible temperature where nitrogen is

i a liquid **ii** a gas **iii** frozen.

12 The most common rocket fuel is liquid hydrogen and liquid oxygen. These two gases are kept in storage, as a liquid, at the following temperatures:

Liquid hydrogen –253 °C

Liquid oxygen –183 °C

a i Which of the two gases is kept at the coldest temperature?

ii What is the difference between the two storage temperatures?

b Scientists are experimenting with liquid methane as its liquid storage temperature is only –162 °C. How much warmer is the stored liquid methane than the

i stored liquid oxygen?

ii stored liquid hydrogen?

Worked Examination Questions

AU **1** The number $2\frac{1}{2}$ is halfway between 1 and 4.

What number is halfway between:

a –8 and –1 **b** $\frac{1}{4}$ and $1\frac{1}{4}$?

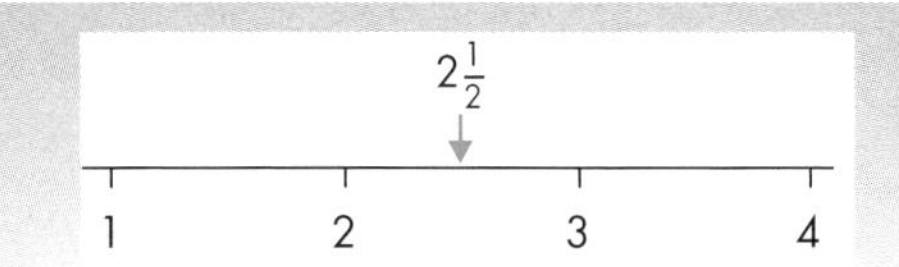

a The number halfway between –8 and –1 is $-4\frac{1}{2}$.

1 mark

The hint in the question is to sketch the numbers on a number line.

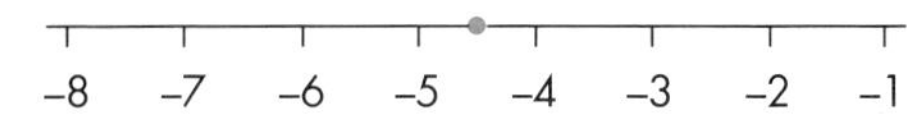

Just by counting from each end you can find the middle value. You get 1 mark for the correct answer.

b The number halfway between $\frac{1}{4}$ and $1\frac{1}{4}$ is $\frac{3}{4}$.

1 mark

Total: 2 marks

Sketch the number line and mark on the quarters.

$\frac{1}{4}$ $\frac{1}{2}$ $\frac{3}{4}$ 1 $1\frac{1}{4}$

Count from each end to identify the middle value to get 1 mark.

2 The table shows the temperatures in four cities.

a Write down the lowest temperature.

b Work out the difference in temperature between Birmingham and London.

c Three hours later the temperature in Manchester had risen by two degrees. Work out the new temperature in Manchester.

Birmingham	5 °C
London	–1 °C
Manchester	–3 °C
Exeter	1 °C

a Lowest temperature is –3 °C in Manchester

1 mark

From lowest to highest the temperatures are –3 °C, –1 °C, 1 °C, 5 °C.
You get 1 mark for the correct answer.

b 6 degrees

1 mark

Find the difference between temperatures in Birmingham (5°C) and London (–1 °C).
You get 1 mark for independent workings.

c –1 °C

1 mark

You get 1 mark for independent workings for the correct calculation –3 °C + 2 °C.

Total: 3 marks

FM **3** Maisie has £103.48 in her bank account.

On 11th December 2009 she takes £20 out of a cash machine and uses her debit card to pay for petrol that costs £25.60.

	Transaction	Paid In	Paid Out	Balance
11/12/09	Carried forward			£103.48
11/12/09	Cash ATM		£20	
11/12/09	Sovereign garage		£25.60	
11/12/09	Hair n' Nails	£152.70		
11/12/09	Closing Balance			

On the same day her weekly wages of £152.70 are paid in.

Fill in the statement to work out how much she has at the end of the day.

£83.48, £77.48, £230.58

Total: 3 marks

Subtract any values in the 'Paid out' column and add any values in the 'Paid in' column. You get 1 mark for each correct answer.

3 Problem Solving
The negative numbers game

In some sports you are required to keep a score card as you play and each player is responsible for filling in their card. In some sports these cards are handed in at the end of the game for checking.

Start
–1
+5
+5
0
–4
+6
–2
+8
–8
+3
+2
–2
–5
–1
+2
+7
–5
+4
–3
–2
+10
–6
–2
–1
+6
–3
–2
+5
–10
+9
–8
+6
–7
–2
–3
0
–2
–4
–6
+20
–20
Finish

Your task

You are going to fill in a score card of your own but instead of positive numbers, your card will involve negative numbers.

How to play

- In groups of three or four, discuss the rules given below and check that you all understand how to play the game.
- Together, decide how many errors you can make before you are disqualified.
- Once you are sure of the rules, each player takes a score card and throws the dice. The person who rolls the highest number goes first.
- After all the players have reached the finish line, compare final scores.
- You should consider:
  - who has won
  - what the most popular scores in the game were
  - if there were some numbers that were better to roll than others.

Resources required

Coloured counters
A dice
Score cards

The rules

- Each player starts with 10 points.
- Each player must keep a record of their total but they must not reveal it to their opponents until the end of the game.
- In turn, players throw the dice and move their counter forward that number of spaces.
- Each time they land on a number they have to add it to their score.
- At the finish the player with the highest score wins.
- Other players have to check the winner's score card. Any errors should be corrected but too many errors can lead to a player being disqualified!

Score card

Number landed on	Running total
Start	+10

Why this chapter matters

You will have seen percentages in many places and situations.

Why use percentages?

Because:

- basic percentages are quite easy to understand
- they are a good way of comparing fractions
- percentages are used a lot in everyday life.

Who uses them?

Shops often have sales where they offer a certain percentage off their standard prices. By understanding percentages you will be able to quickly work out where the real bargains are.

Each month (or year) banks pay their customers interest on their savings. The interest rate the banks offer is expressed as a percentage. Also, banks charge interest on loans. Again, the rate of interest is also expressed as a percentage. With loans, banks add a percentage of the total money owed onto the debt each month so unless you pay off the loan, the money you owe will continue to get bigger.

Some salespeople, such as this car salesman, earn commission on every sale they make. Their commission is paid as a percentage of the retail price of every item they sell. Commission acts as an incentive to sell lots of items – all those percentages can add up to a lot of money!

The government will often use percentages to demonstrate changes in social and economic circumstances. For example, they may tell the public that 7% of the population of working age is unemployed or that value added tax (VAT) will be set at 17.5%.

You will have often received marks on a test in the form of a percentage.

Can you think of other examples? You will find several everyday uses in this chapter.

Chapter 4

Number: Percentages and ratio

The grades given in this chapter are target grades.

1 Equivalent percentages, fractions and decimals

2 Calculating a percentage of a quantity

3 Ratio

4 Best buys

5 Speed, time and distance

This chapter will show you ...

- G what is meant by percentage
- G how to convert between decimals and fractions
- E how to do calculations involving percentages
- E what a ratio is
- E how to use your calculator to work out percentages by using a multiplier
- E to D how to compare prices to find the 'best buy'
- D how to calculate speeds, distances and times
- D to C how to divide an amount according to a given ratio

Visual overview

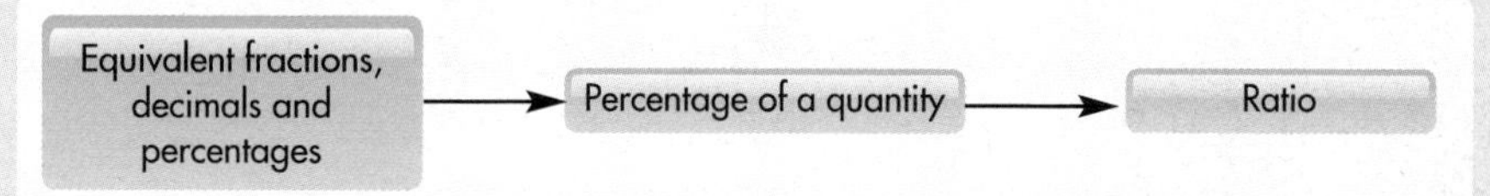

What you should already know

- Multiplication tables up to 10 × 10 **(KS3 level 4, GCSE grade G)**
- How to simplify fractions **(KS3 level 5, GCSE grade G)**
- How to calculate with fractions **(KS3 level 5, GCSE grade F)**
- How to multiply decimals by 100 (move the digits two places to the left) **(KS3 level 5, GCSE grade G)**
- How to divide decimals by 100 (move the digits two places to the right) **(KS3 level 5, GCSE grade G)**

Quick check

1 Simplify these fractions.

a $\frac{12}{32}$ **b** $\frac{20}{45}$ **c** $\frac{28}{48}$ **d** $\frac{36}{60}$

2 Work out these amounts.

a $\frac{2}{3}$ of 27 **b** $\frac{5}{8}$ of 32 **c** $\frac{1}{4} \times 76$ **d** $\frac{3}{5} \times 45$

3 Work out these amounts.

a 12×100 **b** $34 \div 100$ **c** 0.23×100 **d** $4.7 \div 100$

4.1 Equivalent percentages, fractions and decimals

This section will show you how to:
- convert percentages to fractions and decimals and vice versa

Key words
decimal
decimal equivalents
fraction
percentage

100% means the *whole* of something. So if you want to, you can express *part* of the whole as a **percentage**.

Per cent means 'out of 100'.

So, any percentage can be converted to a **fraction** with denominator 100.

For example:

$32\% = \frac{32}{100}$ which can be simplified by cancelling to $\frac{8}{25}$

Also, any percentage can be converted to a **decimal** by dividing the percentage number by 100. This means moving the digits two places to the right.

For example:

$65\% = 65 \div 100 = 0.65$

Any decimal can be converted to a percentage by multiplying by 100%.

For example:

$0.43 = 0.43 \times 100\% = 43\%$

Any fraction can be converted to a percentage by making the denominator into 100 and taking the numerator as the percentage.

For example:

$\frac{2}{5} = \frac{40}{100} = 40\%$

Fractions can also be converted to percentages by dividing the numerator by the denominator and multiplying by 100%.

For example:

$\frac{2}{5} = 2 \div 5 \times 100\% = 40\%$

Knowing the percentage and **decimal equivalents** of the common fractions is extremely useful. So, do try to learn them.

$\frac{1}{2} = 0.5 = 50\%$ $\frac{1}{4} = 0.25 = 25\%$ $\frac{3}{4} = 0.75 = 75\%$ $\frac{1}{8} = 0.125 = 12.5\%$

$\frac{1}{10} = 0.1 = 10\%$ $\frac{1}{5} = 0.2 = 20\%$ $\frac{1}{3} = 0.33 = 33\frac{1}{3}\%$ $\frac{2}{3} = 0.67 = 67\%$

The following table shows how to convert from one to the other.

Convert from percentage to:	
Decimal	**Fraction**
Divide the percentage by 100, for example $52\% = 52 \div 100$ $= 0.52$	Make the percentage into a fraction with a denominator of 100 and simplify by cancelling down if possible, for example $52\% = \frac{52}{100} = \frac{13}{25}$

Convert from decimal to:	
Percentage	**Fraction**
Multiply the decimal by 100%, for example $0.65 = 0.65 \times 100\%$ $= 65\%$	If the decimal has 1 decimal place put it over the denominator 10, if it has 2 decimal places put it over the denominator 100, etc. Then simplify by cancelling down if possible, for example $0.65 = \frac{65}{100} = \frac{13}{20}$

Convert from fraction to:	
Percentage	**Decimal**
If the denominator is a factor of 100 multiply numerator and denominator to make the denominator 100, then the numerator is the percentage, for example $\frac{3}{20} = \frac{15}{100} = 15\%$ or convert to a decimal and change the decimal to a percentage, for example $\frac{7}{8} = 7 \div 8 = 0.875 = 87.5\%$	Divide the numerator by the denominator, for example $\frac{9}{40} = 9 \div 40 = 0.225$

EXAMPLE 1

Convert the following to decimals: **a** 78% **b** 35% **c** $\frac{3}{25}$ **d** $\frac{7}{40}$.

a $78\% = 78 \div 100 = 0.78$

b $35\% = 35 \div 100 = 0.35$

c $\frac{3}{25} = 3 \div 25 = 0.12$

d $\frac{7}{40} = 7 \div 40 = 0.175$

EXAMPLE 2

Convert the following to percentages: **a** 0.85 **b** 0.125 **c** $\frac{7}{20}$ **d** $\frac{3}{8}$.

a $0.85 = 0.85 \times 100\% = 85\%$

b $0.125 = 0.125 \times 100\% = 12.5\%$

c $\frac{7}{20} = \frac{35}{100} = 35\%$

d $\frac{3}{8} = 3 \div 8 \times 100\% = 0.375 \times 100\% = 37.5\%$

EXAMPLE 3

Convert the following to fractions: **a** 0.45 **b** 0.4 **c** 32% **d** 15%.

a $0.45 = \frac{45}{100} = \frac{9}{20}$

b $0.4 = \frac{4}{10} = \frac{2}{5}$

c $32\% = \frac{32}{100} = \frac{8}{25}$

d $15\% = \frac{15}{100} = \frac{3}{20}$

EXERCISE 4A

G

1 Write each percentage as a fraction in its simplest form.

a 8% **b** 50% **c** 25%

d 35% **e** 90% **f** 75%

2 Write each percentage as a decimal.

a 27% **b** 85% **c** 13%

d 6% **e** 80% **f** 32%

3 Write each decimal as a fraction in its simplest form.

a 0.12 **b** 0.4 **c** 0.45

d 0.68 **e** 0.25 **f** 0.625

4 Write each decimal as a percentage.

a 0.29 **b** 0.55 **c** 0.03

d 0.16 **e** 0.6 **f** 1.25

5 Write each fraction as a percentage.

a $\frac{7}{25}$ **b** $\frac{3}{10}$ **c** $\frac{19}{20}$

d $\frac{17}{50}$ **e** $\frac{11}{40}$ **f** $\frac{7}{8}$

6 Write each fraction as a decimal.

a $\frac{9}{15}$ **b** $\frac{3}{40}$ **c** $\frac{19}{25}$

d $\frac{5}{16}$ **e** $\frac{1}{20}$ **f** $\frac{1}{8}$

G

7 Of the 300 members of a social club 50% are men. How many members are women?

8 Gillian came home and told her dad that she got 100% of her spellings correct. She told her mum that there were 25 spellings to learn. How many spellings did Gillian get wrong?

9 Every year a school library likes to replace 1% of its books. One year the library had 2000 books. How many did it replace?

10 **a** If 23% of pupils go home for lunch, what percentage do not go home for lunch?

b If 61% of the population takes part in the National Lottery, what percentage do not take part?

c If 37% of members of a gym are males, what percentage of the members are females?

11 I calculated that 28% of my time is spent sleeping and 45% is spent working. How much time is left to spend doing something else?

12 In one country, 24.7% of the population is below the age of 16 and 13.8% of the population is aged over 65. How much of the population is aged from 16 to 65 inclusive?

13 Approximately what percentage of each bottle is filled with water?

a

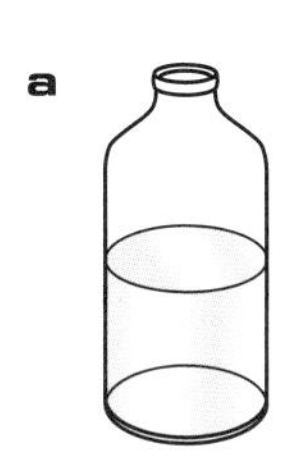

b

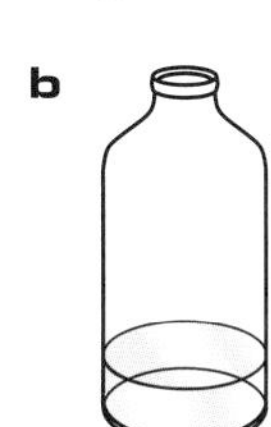

c

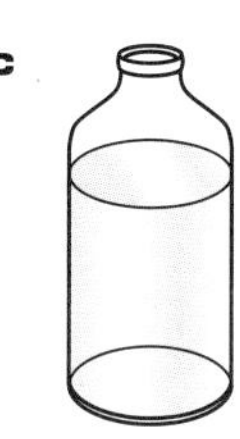

PS **14** Helen made a cake for James. The amount of cake left each day is shown in the diagram.

What percentage has been eaten each day?

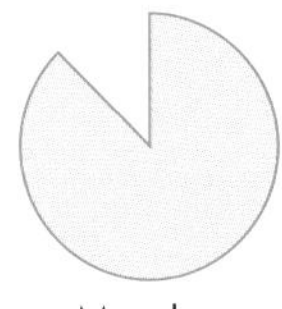

Monday

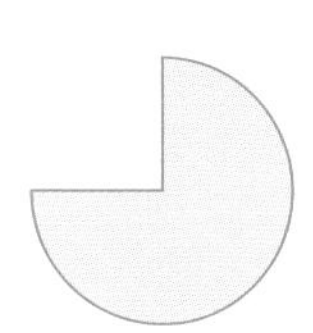

Tuesday

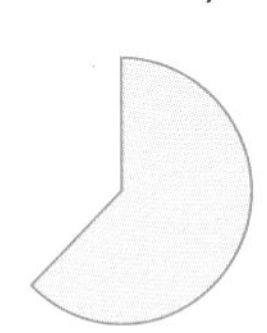

Wednesday

Thursday

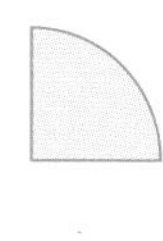

Friday

15 Convert each fraction into a percentage.

a $\frac{1}{5}$ **b** $\frac{1}{4}$ **c** $\frac{3}{4}$ **d** $\frac{9}{20}$ **e** $\frac{7}{50}$

f $\frac{1}{2}$ **g** $\frac{3}{5}$ **h** $\frac{7}{40}$ **i** $\frac{11}{20}$ **j** $\frac{13}{10}$

16 Convert each fraction into a percentage. Give your answers to one decimal place.

a $\frac{1}{3}$ **b** $\frac{1}{6}$ **c** $\frac{2}{3}$ **d** $\frac{5}{6}$ **e** $\frac{2}{7}$

f $\frac{47}{60}$ **g** $\frac{31}{45}$ **h** $\frac{8}{9}$ **i** $\frac{73}{90}$ **j** $\frac{23}{110}$

17 Change each of these decimals into a percentage.

a 0.07 **b** 0.8 **c** 0.66 **d** 0.25 **e** 0.545

f 0.82 **g** 0.3 **h** 0.891 **i** 1.2 **j** 2.78

18 Chris scored 24 marks out of a possible 40 in a maths test.

a Write this score as a fraction.

b Write this score as a decimal.

c Write this score as a percentage.

19 **a** Convert each of the following test scores into a percentage. Give each answer to the nearest whole number.

Subject	Result	Percentage
Mathematics	38 out of 60	
English	29 out of 35	
Science	27 out of 70	
History	56 out of 90	
Technology	58 out of 75	

b If all the tests are of the same standard, which was the best result?

20 There were two students missing from my class of 30. What percentage of my class were away?

21 In one season, Paulo Di Canio had 110 shots at goal. He scored with 28 of these shots. What percentage of his shots resulted in goals?

22 Copy and complete the table.

Percentage	Decimal	Fraction
34%		
	0.85	
		$\frac{3}{40}$
45%		
	0.3	
		$\frac{2}{3}$
84%		
	0.45	
		$\frac{3}{8}$

FM 23 The manager of a garage wants to order 27 000 litres of fuel. A fuel tanker holds 30 000 litres when full. How much fuel should the manager order as a fraction or percentage of a full tanker load?

Percentage dominoes

You need a piece of card about 30 cm by 30 cm.

On the card, draw a grid of 49 rectangles, as shown below. Each rectangle measures 4 cm by 2 cm, and is divided into two parts by a dashed line.

In each rectangle, write the appropriate percentage and fraction as given in the diagram.

The fractions in the top row of rectangles are the correct equivalents of the percentages. Use a light blue pencil or crayon to colour in the top row, so that the rectangles can be easily spotted during the game.

50%	$\frac{1}{2}$	25%	$\frac{1}{4}$	10%	$\frac{1}{10}$	75%	$\frac{3}{4}$	20%	$\frac{1}{5}$	40%	$\frac{2}{5}$	80%	$\frac{4}{5}$
50%	$\frac{1}{4}$	25%	$\frac{1}{10}$	10%	$\frac{3}{4}$	75%	$\frac{1}{5}$	20%	$\frac{2}{5}$	40%	$\frac{4}{5}$	80%	$\frac{1}{2}$
50%	$\frac{1}{10}$	25%	$\frac{3}{4}$	10%	$\frac{1}{5}$	75%	$\frac{2}{5}$	20%	$\frac{4}{5}$	40%	$\frac{1}{2}$	80%	$\frac{1}{4}$
50%	$\frac{3}{4}$	25%	$\frac{1}{5}$	10%	$\frac{2}{5}$	75%	$\frac{4}{5}$	20%	$\frac{1}{2}$	40%	$\frac{1}{4}$	80%	$\frac{1}{10}$
50%	$\frac{1}{5}$	25%	$\frac{2}{5}$	10%	$\frac{4}{5}$	75%	$\frac{1}{2}$	20%	$\frac{1}{4}$	40%	$\frac{1}{10}$	80%	$\frac{3}{4}$
50%	$\frac{2}{5}$	25%	$\frac{4}{5}$	10%	$\frac{1}{2}$	75%	$\frac{1}{4}$	20%	$\frac{1}{10}$	40%	$\frac{3}{4}$	80%	$\frac{1}{5}$
50%	$\frac{4}{5}$	25%	$\frac{1}{2}$	10%	$\frac{1}{4}$	75%	$\frac{1}{10}$	20%	$\frac{3}{4}$	40%	$\frac{1}{5}$	80%	$\frac{2}{5}$

Carefully cut out all the rectangles to form a set of 'dominoes' that you can play with.

Now play a game of dominoes, putting *equivalents* next to each other.
For example:

50%	$\frac{1}{10}$	10%	$\frac{1}{10}$	10%	$\frac{4}{5}$

Do *not* put identical percentages or identical fractions next to each other.

4.2 Calculating a percentage of a quantity

This section will show you how to:
- calculate a percentage of a quantity

Key words
multiplier
quantity

To calculate a percentage of a **quantity**, you multiply the quantity by the percentage. The percentage may be expressed as either a fraction or a decimal. When finding percentages without a calculator, base the calculation on 10% (or 1%) as these are easy to calculate.

EXAMPLE 4

Calculate: **a** 10% of 54 kg **b** 15% of 54 kg.

a 10% is $\frac{1}{10}$ so $\frac{1}{10}$ of 54 kg = 54 kg ÷ 10 = 5.4 kg

b 15% is 10% + 5% = 5.4 kg + 2.7 kg = 8.1 kg

EXAMPLE 5

Calculate 12% of £80.

10% of £80 is £8 and 1% of £80 is £0.80

12% = 10% + 1% + 1% = £8 + £0.80 + £0.80 = £9.60

Using a percentage multiplier

You have already seen that percentages and decimals are equivalent so it is easier, particularly when using a calculator, to express a percentage as a decimal and use this to do the calculation.

For example, 13% is a **multiplier** of 0.13, 20% a multiplier of 0.2 (or 0.20) and so on.

EXAMPLE 6

Calculate 45% of 160 cm.

45% = 0.45, so 45% of 160 = 0.45 × 160 = 72 cm

Find 52% of £460.

52% = 0.52

So, 0.52 × 460 = 239.2

This gives £239.20.

Remember to always write a money answer with 2 decimal places.

EXERCISE 4B

1 What multiplier is equivalent to a percentage of:

a 88% **b** 30% **c** 25% **d** 8% **e** 115%?

2 What percentage is equivalent to a multiplier of:

a 0.78 **b** 0.4 **c** 0.75 **d** 0.05 **e** 1.1?

3 Calculate the following.

a 15% of £300 **b** 6% of £105 **c** 23% of 560 kg

d 45% of 2.5 kg **e** 12% of 9 hours **f** 21% of 180 cm

g 4% of £3 **h** 35% of 8.4 m **i** 95% of £8

j 11% of 308 minutes **k** 20% of 680 kg **l** 45% of £360

FM **4** The manager of a school canteen estimates that 40% of students will buy a school lunch. There are 1200 students in the school. She knows that her estimates are usually accurate to within 2%. Including this 2% figure, what is the greatest number of lunches she will have left over?

5 An estate agent charges 2% commission on every house he sells. How much commission will he earn on a house that he sells for £120 500?

6 A department store had 250 employees. During one week of a flu epidemic, 14% of the store's employees were absent.

a What percentage of the employees went into work?

b How many of the employees went into work?

7 It is thought that about 20% of fans at a rugby match are women. For a match at Twickenham there were 42 600 fans. How many of these do you think would be women?

8 At St Pancras railway station, in one week 350 trains arrived. Of these trains, 5% arrived early and 13% arrived late. How many arrived on time?

AU **9** For the FA Cup Final that was held at Wembley, each year the 90 000 tickets were split up as follows.

Each of the teams playing received 30% of the tickets.

The referees' association received 1% of the tickets.

The other 90 teams received 10% of the tickets among them.

The FA associates received 20% of the tickets among them.

The rest were for the special celebrities.

How many tickets went to each set of people?

E

FM **10** A school estimates that for a school play 60% of the students will attend. There are 1500 students in the school. The caretaker is told to put out one seat for each person expected to attend plus an extra 10% of that amount in case more attend. How many seats does he need to put out?

HINTS AND TIPS

It is not 70% of the number of the students in the school.

11 A school had 850 pupils and the attendance record in the week before Christmas was:

Monday 96% Tuesday 98% Wednesday 100% Thursday 94% Friday 88%

How many pupils were present each day?

12 Calculate the following.

a 12.5% of £26 **b** 6.5% of 34 kg **c** 26.8% of £2100

d 7.75% of £84 **e** 16.2% of 265 m **f** 0.8% of £3000

13 Air consists of 80% nitrogen and 20% oxygen (by volume). A man's lungs have a capacity of 600 cm^3. How much of each gas will he have in his lungs when he has just taken a deep breath?

14 A factory estimates that 1.5% of all the garments it produces will have a fault in them. One week the factory produces 850 garments. How many are likely to have a fault?

15 An insurance firm sells house insurance and the annual premiums are usually set at 0.3% of the value of the house. What will be the annual premium for a house valued at £90 000?

D

PS **16** Average prices in a shop went up by 3% last year and 3% this year. Did the actual average price of items this year rise by more, the same amount, or less than last year?

Explain how you decided.

4.3 Ratio

This section will show you how to:
- simplify a ratio
- express a ratio as a fraction
- divide amounts according to ratios
- complete calculations from a given ratio and partial information

Key words
cancel
common unit
ratio
simplest form

A **ratio** is a way of comparing the sizes of two or more quantities.

A ratio can be expressed in a number of ways. For example, if Joy is five years old and James is 20 years old, the ratio of their ages is:

	Joy's age : James's age
which is:	5 : 20
which simplifies to:	1 : 4 (dividing both sides by 5)

A ratio is usually given in one of these three ways.

Joy's age : James's age	or	5 : 20	or	1 : 4
Joy's age to James's age	or	5 to 20	or	1 to 4
$\frac{\text{Joy's age}}{\text{James's age}}$	or	$\frac{5}{20}$	or	$\frac{1}{4}$

Common units

When working with a ratio involving different units, *always convert them to a* **common unit**. A ratio can be simplified only when the units of each quantity are the *same*, because the ratio itself has no units. Once the units are the same, the ratio can be simplified or **cancelled**.

For example, the ratio of 125 g to 2 kg must be converted to the ratio of 125 g to 2000 g, so that you can simplify it.

	125 : 2000	
Divide both sides by 25:	5 : 80	
Divide both sides by 5:	1 : 16	The ratio 125 : 2000 can be simplified to 1 : 16.

EXAMPLE 7

Express 25 minutes : 1 hour as a ratio in its simplest form.

The units must be the same, so change 1 hour into 60 minutes.

25 minutes : 1 hour	=	25 minutes : 60 minutes	
	=	25 : 60	Cancel the units (minutes)
	=	5 : 12	Divide both sides by 5

So, 25 minutes : 1 hour simplifies to 5 : 12

Ratios as fractions

A ratio in its **simplest form** can be expressed as portions of a quantity by expressing the whole numbers in the ratio as fractions with the same denominator (bottom number).

EXAMPLE 8

A garden is divided into lawn and shrubs in the ratio 3 : 2.

What fraction of the garden is covered by **a** lawn, **b** shrubs?

The denominator (bottom number) of the fraction comes from adding the number in the ratio (that is, 2 + 3 = 5).

a the lawn covers $\frac{3}{5}$ of the garden

b and the shrubs cover $\frac{2}{5}$ of the garden.

EXERCISE 4C

E

1 Express each of the following ratios in its simplest form.

a 6 : 18 **b** 15 : 20 **c** 16 : 24 **d** 24 : 36

e 20 to 50 **f** 12 to 30 **g** 25 to 40 **h** 125 to 30

i 15 : 10 **j** 32 : 12 **k** 28 to 12 **l** 100 to 40

m 0.5 to 3 **n** 1.5 to 4 **o** 2.5 to 1.5 **p** 3.2 to 4

2 Write each of the following ratios of quantities in its simplest form. (Remember to always express both parts in a common unit before you simplify.)

a £5 to £15 **b** £24 to £16 **c** 125 g to 300 g

d 40 minutes : 5 minutes **e** 34 kg to 30 kg **f** £2.50 to 70p

g 3 kg to 750 g **h** 50 minutes to 1 hour **i** 1 hour to 1 day

j 12 cm to 2.5 mm **k** 1.25 kg : 500 g **l** 75p : £3.50

m 4 weeks : 14 days **n** 600 m: 2 km **o** 465 mm : 3 m

p 15 hours : 1 day

D

3 A length of wood is cut into two pieces in the ratio 3 : 7. What fraction of the original length is the longer piece?

4 Jack and Thomas find a bag of marbles that they share between them in the ratio of their ages. Jack is 10 years old and Thomas is 15 years old. What fraction of the marbles did Jack get?

5 Dave and Sue share a pizza in the ratio 2 : 3. They eat it all.

a What fraction of the pizza did Dave eat?

b What fraction of the pizza did Sue eat?

D

6 A camp site allocates space to caravans and tents in the ratio 7 : 3. What fraction of the total space is given to:

a the caravans

b the tents?

7 Two sisters, Amy and Katie, share a packet of sweets in the ratio of their ages. Amy is 15 and Katie is 10. What fraction of the sweets does each sister get?

8 The recipe for a fruit punch is 1.25 litres of fruit crush to 6.75 litres of lemonade. What fraction of the punch is each ingredient?

PS **9** Three cows, Gertrude, Gladys and Henrietta, produced milk in the ratio 2 : 3 : 4. Henrietta produced $1\frac{1}{2}$ litres more than Gladys. How much milk did the three cows produce altogether?

10 In a safari park at feeding time, the elephants, the lions and the chimpanzees are given food in the ratio 10 to 7 to 3. What fraction of the total food is given to:

a the elephants **b** the lions **c** the chimpanzees?

11 Three brothers, James, John and Joseph, share a huge block of chocolate in the ratio of their ages. James is 20, John is 12 and Joseph is 8. What fraction of the bar of chocolate does each brother get?

12 The recipe for a pudding is 125 g of sugar, 150 g of flour, 100 g of margarine and 175 g of fruit. What fraction of the pudding is each ingredient?

C

PS **13** June wins three-quarters of her bowls matches. She loses the rest.

What is the ratio of wins to losses?

AU **14** Three brothers share some cash.
The ratio of Mark's and David's share is 1 : 2
The ratio of David's and Paul's share is 1 : 2

What is the ratio of Mark's share to Paul's share?

Dividing amounts in a given ratio

To divide an amount in a given ratio, you first look at the ratio to see how many parts there are altogether.

For example, 4 : 3 has 4 parts and 3 parts giving 7 parts altogether.

7 parts is the whole amount.

1 part can then be found by dividing the whole amount by 7.

3 parts and 4 parts can then be worked out from 1 part.

EXAMPLE 9

Divide £28 in the ratio 4 : 3

$4 + 3 = 7$ parts altogether

So 7 parts = £28

Dividing by 7:

1 part = £4

4 parts = 4 × £4 = £16 and 3 parts = 3 × £4 = £12

So £28 divided in the ratio 4 : 3 = £16 : £12

To divide an amount in a given ratio you can also use fractions. You first express the whole numbers in the ratio as fractions with the same common denominator. Then you multiply the amount by each fraction.

EXAMPLE 10

Divide £40 between Peter and Hitan in the ratio 2 : 3

Changing the ratio to fractions gives:

$$\text{Peter's share} = \frac{2}{(2+3)} = \frac{2}{5}$$

$$\text{Hitan's share} = \frac{3}{(2+3)} = \frac{3}{5}$$

So Peter receives £40 × $\frac{2}{5}$ = £16 and Hitan receives £40 × $\frac{3}{5}$ = £24.

EXERCISE 4D

C

1 Divide the following amounts according to the given ratios.

a 400 g in the ratio 2 : 3

b 280 kg in the ratio 2 : 5

c 500 in the ratio 3 : 7

d 1 km in the ratio 19 : 1

e 5 hours in the ratio 7 : 5

f £100 in the ratio 2 : 3 : 5

g £240 in the ratio 3 : 5 : 12

h 600 g in the ratio 1 : 5 : 6

i £5 in the ratio 7 : 10 : 8

j 200 kg in the ratio 15 : 9 : 1

2 The ratio of female to male members of Lakeside Gardening Club is 7 : 3. The total number of members of the group is 250.

a How many members are female?

PS **b** What percentage of members are male?

C

3 A supermarket aims to stock branded goods and their own goods in the ratio 2 : 3. They stock 500 kg of breakfast cereal.

a What percentage of the cereal stock is branded?

b How much of the cereal stock is their own?

4 The Illinois Department of Health reported that, for the years 1981 to 1992 when they tested a total of 357 horses for rabies, the ratio of horses with rabies to those without was 1 : 16.

How many of these horses had rabies?

5 Being overweight increases the chances of an adult suffering from heart disease. A way to test whether an adult has an increased risk is shown below:

For women, there is increased risk when $W/H > 0.8$

For men, there is increased risk when $W/H > 1.0$

W = waist measurement
H = hip measurement

a Find whether the following people have an increased risk of heart disease.

Miss Mott: waist 26 inches, hips 35 inches
Mrs Wright: waist 32 inches, hips 37 inches
Mr Brennan: waist 32 inches, hips 34 inches
Ms Smith: waist 31 inches, hips 40 inches
Mr Kaye: waist 34 inches, hips 33 inches

b Give three examples of waist and hip measurements that would suggest no risk of heart disease for a man, but would suggest a risk for a woman.

6 Rewrite the following scales as ratios as simply as possible.

a 1 cm to 4 km **b** 4 cm to 5 km **c** 2 cm to 5 km

d 4 cm to 1 km **e** 5 cm to 1 km **f** 2.5 cm to 1 km

g 8 cm to 5 km **h** 10 cm to 1 km **i** 5 cm to 3 km

7 A map has a scale of 1 cm to 10 km.

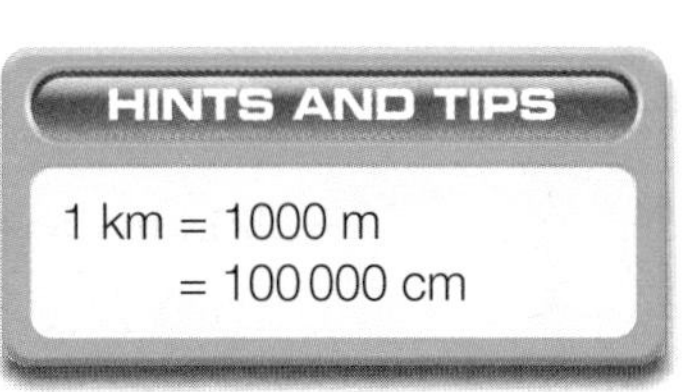

a Rewrite the scale as a ratio in its simplest form.

b What is the actual length of a lake that is 4.7 cm long on the map?

c How long will a road be on the map if its actual length is 8 km?

8 A map has a scale of 2 cm to 5 km.

a Rewrite the scale as a ratio in its simplest form.

b How long is a path that measures 0.8 cm on the map?

c How long should a 12 km road be on the map?

9 The scale of a map is 5 cm to 1 km.

a Rewrite the scale as a ratio in its simplest form.

b How long is a wall that is shown as 2.7 cm on the map?

c The distance between two points is 8 km; how far will this be on the map?

10 You can simplify a ratio by changing it into the form 1 : n. For example, 5 : 7 can be rewritten as

$$\frac{5}{5} : \frac{7}{5} = 1 : 1.4$$

Rewrite each of the following ratios in the form 1 : n.

a 5 : 8

b 4 : 13

c 8 : 9

d 25 : 36

e 5 : 27

f 12 : 18

g 5 hours : 1 day

h 4 hours : 1 week

i £4 : £5

Calculating with ratios when only part of the information is known

EXAMPLE 11

A fruit drink is made by mixing orange squash with water in the ratio 2 : 3
How much water needs to be added to 5 litres of orange squash to make the drink?

2 parts is 5 litres

Dividing by 2:

1 part is 2.5 litres

3 parts = 2.5 litres × 3 = 7.5 litres

So 7.5 litres of water is needed to make the drink.

EXAMPLE 12

Two business partners, Lubna and Adama, divided their total profit in the ratio 3 : 5.
Lubna received £2100. How much did Adama get?

Lubna's £2100 was $\frac{3}{8}$ of the total profit. (Check that you know why.)
$\frac{1}{8}$ of the total profit = £2100 ÷ 3 = £700

So Adama's share, which was $\frac{5}{8}$, amounted to £700 × 5 = £3500.

EXERCISE 4E

1 Derek, aged 15, and Ricki, aged 10, shared all the conkers they found in the woods in the same ratio as their ages. Derek had 48 conkers.

- **a** Simplify the ratio of their ages.
- **b** How many conkers did Ricki have?
- **c** How many conkers did they find altogether?

2 Two types of crisps, plain and salt 'n' vinegar, were bought for a school party in the ratio 5 : 3. The school bought 60 packets of salt 'n' vinegar crisps.

- **a** How many packets of plain crisps did they buy?
- **b** How many packets of crisps altogether did they buy?

3 Robin is making a drink from orange juice and lemon juice in the ratio 9 : 1. If Robin has only 3.6 litres of orange juice, how much lemon juice does he need to make the drink?

4 When I picked my strawberries, I found some had been spoilt by snails. The rest were good. These were in the ratio 3 : 17. 18 of my strawberries had been spoilt by snails. How many good strawberries did I find?

5 A blend of tea is made by mixing Lapsang with Assam in the ratio 3 : 5. I have a lot of Assam tea but only 600 g of Lapsang. How much Assam do I need to make the blend using all the Lapsang?

6 The ratio of male to female spectators at ice hockey games is 4 : 5. At the Steelers' last match, 4500 men watched the match. What was the total attendance at the game?

7 'Proper tea' is made by putting milk and tea together in the ratio 2 : 9. How much 'proper tea' can be made if you have 1 litre of milk?

8 A teacher always arranged the content of each of his lessons to Year 10 as 'teaching' and 'practising learnt skills' in the ratio 2 : 3.

- **a** If a lesson lasted 35 minutes, how much teaching would he do?
- **b** If he decided to teach for 30 minutes, how long would the lesson be?

9 A 'good' children's book is supposed to have pictures and text in the ratio 17 : 8. In a book I have just looked at, the pictures occupy 23 pages.

- **a** Approximately how many pages of text should this book have to be deemed a 'good' children's book?
- **b** What percentage of a 'good' children's book will be text?

10 Three business partners, Kevin, John and Margaret, put money into a business in the ratio 3 : 4 : 5. They shared any profits in the same ratio. Last year, Margaret made £3400 out of the profits. How much did Kevin and John make last year?

11 The soft drinks Cola, Orange fizz and Vimto were bought for the school disco in the ratio 10 : 5 : 3. The school bought 80 cans of Orange fizz.

a How much Cola did they buy? **b** How much Vimto did they buy?

AU 12 **a** Iqra is making a drink from lemonade, orange and ginger ale in the ratio 40 : 9 : 1. If Iqra has only 4.5 litres of orange, how much of the other two ingredients does she need to make the drink?

b Another drink made from lemonade, orange and ginger ale uses the ratio 10 : 2 : 1.

Which drink has a larger proportion of ginger ale, Iqra's or this one? Show how you work out your answer.

PS 13 In a factory the ratio of female employees to male employees is 3 : 8. In the factory there are 85 **more** males than females. How many females work in the factory?

PS 14 There is a group of boys and girls waiting for school buses. 25 girls get on the first bus. The ratio of boys to girls at the stop is now 3 : 2. 15 boys get on the second bus. There are now the same number of boys and girls at the bus stop. How many students altogether were originally at the bus stop?

PS 15 A jar contains 100 cc of a mixture of oil and water in the ratio 1 : 4. Enough oil is added to make the ratio of oil to water 1 : 2. How much water must be added to make the ratio of oil to water 1 : 3?

4.4 Best buys

This section will show you how to:
- find the cost per unit weight
- find the weight per unit cost
- use the above to find which product is the cheaper

Key words
best buy
better value
value for money

When you wander around a supermarket and see all the different prices for the many different-sized packets, it is rarely obvious which are the '**best buys**'. However, with a calculator you can easily compare **value for money** by finding either:

the cost per unit weight **or** the weight per unit cost

To find:

- *cost per unit weight,* divide *cost by weight*
- *weight per unit cost,* divide *weight by cost.*

The next two examples show you how to do this.

EXAMPLE 13

A 300 g tin of cocoa costs £1.20. Find the cost per unit weight and the weight per unit cost.

First change £1.20 to 120p. Then divide, using a calculator, to get:

Cost per unit weight $120 \div 300 = 0.4$p per gram

Weight per unit cost $300 \div 120 = 2.5$ g per penny

EXAMPLE 14

A supermarket sells two different-sized packets of Whito soap powder. The medium size contains 800 g and costs £1.60 and the large size contains 2.5 kg and costs £4.75. Which is the better buy?

Find the weight per unit cost for both packets.

Medium: $800 \div 160 = 5$ g per penny

Large: $2500 \div 475 = 5.26$ g per penny

From these it is clear that there is more weight per penny with the large size, which means that the large size is the better buy.

Sometimes it is easier to use a scaling method to compare prices and find **better value**.

EXAMPLE 15

Which of these boxes of fish fingers is better value?

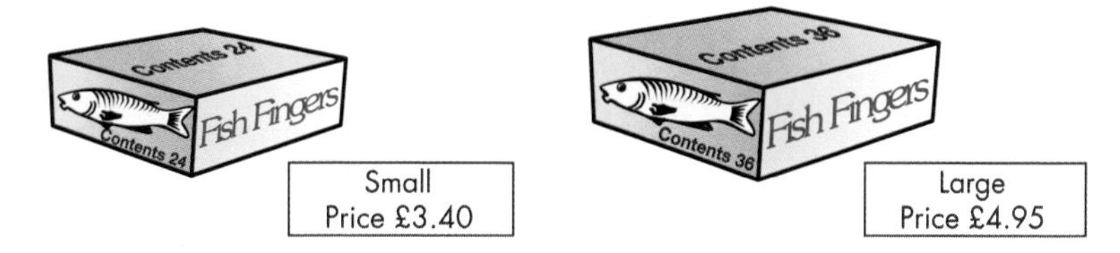

12 is a common factor of 24 and 36 so work out the cost of 12 fish fingers.

For the small box, 12 fish fingers cost £3.40 ÷ 2 = £1.70
For the large box, 12 fish fingers cost £4.95 ÷ 3 = £1.65

So the large box is better value.

EXAMPLE 16

Which of these packs of yoghurt is better value?

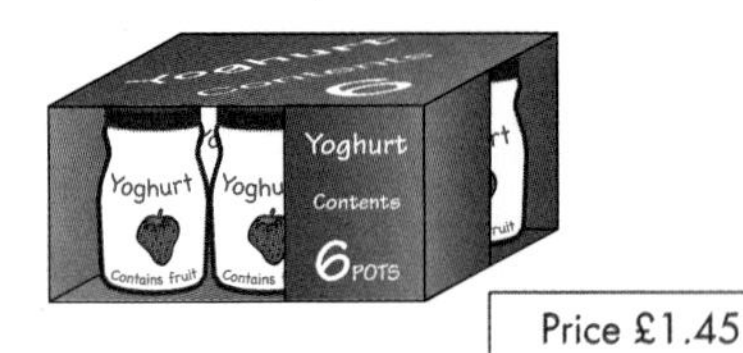

30 is the least common multiple of 5 and 6 so work out the cost of 30 yoghurts.

For the six-pack the cost of 30 yoghurts is £1.45 × 5 = £7.25
For the five-pack the cost of 30 yoghurts is £1.20 × 6 = £7.20

So the five-pack is better value.

EXERCISE 4F

E

1 Compare the prices of the following pairs of products and state which, if any, is the better buy.

a Chocolate bars: £2.50 for a 5-pack, £4.50 for a 10-pack

b Eggs: £1.08 for 6, £2.25 for 12

c Car shampoo: £4.99 for 2 litres, £2.45 for 1 litre

d Dishwasher tablets: £7.80 for 24, £3.90 for 12

e Carrots: 29p for 250 grams, 95p for 750 grams

f Bread rolls: £1.39 for a pack of 6, £5.60 for a pack of 24

g Juice: £1.49 for 1 carton, £4 for 3 cartons

D

FM AU **2** Compare the following pairs of products and state which is the better buy. Explain why.

a Coffee: a medium jar which contains 140 g for £1.10 or a large jar which contains 300 g for £2.18

b Beans: a 125 g tin at 16p or a 600 g tin at 59p

c Flour: a 3 kg bag at 75p or a 5 kg bag at £1.20

d Toothpaste: a large tube containing 110 ml for £1.79 or a medium tube containing 75 ml for £1.15

e Frosted Flakes: a large box which contains 750 g for £1.64 or a medium box which contains 500 g for £1.10

f Rice Crisp: a medium box which contains 440 g for £1.64 or a large box which contains 600 g for £2.13

g Hair shampoo: a bottle containing 400 ml for £1.15 or a bottle containing 550 ml for £1.60

D

FM **3** Julie wants to respray her car with yellow paint. In the local automart, she sees the following tins:

Small tin	350 ml at a cost of £1.79
Medium tin	500 ml at a cost of £2.40
Large tin	1.5 litres at a cost of £6.70

a What is the cost per litre of paint in the small tin?

b Which tin is offered at the lowest price per litre?

FM **4** Tisco's sells bottled water in three sizes.

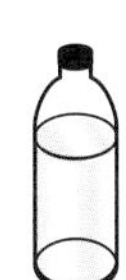
Handy size 40 cl
£0.38

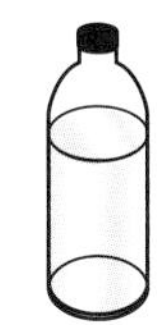
Family size 2 litres
£0.98

Giant size 5 litres
£2.50

a Work out the cost per litre of the 'handy' size.

b Which bottle is the best value for money?

PS **5** Two drivers are comparing the petrol consumption of their cars.

Ahmed says, 'I get 320 miles on a tank of 45 litres.'
Bashir says, 'I get 230 miles on a tank of 32 litres.'

Whose car is the more economical?

PS **6** Mary and Jane are arguing about which of them is better at mathematics.

Mary scored 49 out of 80 on a test.
Jane scored 60 out of 100 on a test of the same standard.

Who is better at mathematics?

PS AU **7** Paula and Kelly are comparing their running times.

Paula completed a 10-mile run in 65 minutes.
Kelly completed a 10-kilometre run in 40 minutes.

Given that 8 kilometres are equal to 5 miles, which girl has the greater average speed?

4.5 Speed, time and distance

This section will show you how to:
- recognise the relationship between speed, distance and time
- calculate average speed from distance and time
- calculate distance travelled from the speed and the time taken
- calculate the time taken on a journey from the speed and the distance

Key words
average
distance
speed
time

The relationship between **speed**, **time** and **distance** can be expressed in three ways:

$$\text{speed} = \frac{\text{distance}}{\text{time}} \qquad \text{distance} = \text{speed} \times \text{time} \qquad \text{time} = \frac{\text{distance}}{\text{speed}}$$

In problems relating to speed, you usually mean **average** speed, as it would be unusual to maintain one exact speed for the whole of a journey.

This diagram will help you remember the relationships between distance (D), time (T) and speed (S).

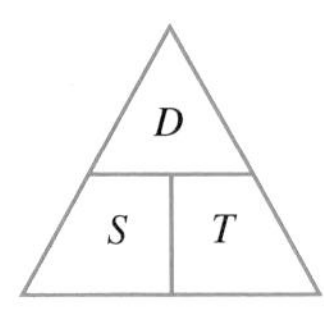

$$D = S \times T \qquad S = \frac{D}{T} \qquad T = \frac{D}{S}$$

EXAMPLE 17

Paula drove a distance of 270 miles in 5 hours. What was her average speed?

$$\text{Paula's average speed} = \frac{\text{distance she drove}}{\text{time she took}} = \frac{270}{5} = 54 \text{ miles per hour (mph)}$$

EXAMPLE 18

Sarah drove from Sheffield to Peebles in $3\frac{1}{2}$ hours at an average speed of 60 mph. How far is it from Sheffield to Peebles?

Since:

distance = speed × time

the distance from Sheffield to Peebles is given by:

$60 \times 3.5 = 210$ miles

Note: You need to change the time to a decimal number and use 3.5 (*not* 3.30).

EXAMPLE 19

Sean is going to drive from Newcastle upon Tyne to Nottingham, a distance of 190 miles. He estimates that he will drive at an average speed of 50 mph. How long will it take him?

$$\text{Sean's time} = \frac{\text{distance he covers}}{\text{his average speed}} = \frac{190}{50} = 3.8 \text{ hours}$$

Change the 0.8 hour to minutes by multiplying by 60, to give 48 minutes.

So, the time for Sean's journey will be 3 hours 48 minutes.

Remember: When you calculate a time and get a decimal answer, as in Example 19, *do not mistake* the decimal part for minutes. You must either:

- leave the time as a decimal number and give the unit as hours, or
- change the decimal part to minutes by multiplying it by 60 (1 hour = 60 minutes) and give the answer in hours and minutes.

EXERCISE 4G

1 A cyclist travels a distance of 90 miles in 5 hours. What was her average speed?

2 How far along a motorway would you travel if you drove at 70 mph for 4 hours?

3 I drive to Bude in Cornwall from Sheffield in about 6 hours. The distance from Sheffield to Bude is 315 miles. What is my average speed?

4 The distance from Leeds to London is 210 miles. The train travels at an average speed of 90 mph. If I catch the 9.30 am train in London, at what time should I expect to arrive in Leeds?

5 How long will an athlete take to run 2000 m at an average speed of 4 metres per second?

HINTS AND TIPS

Remember to convert time to a decimal if you are using a calculator, for example, 8 hours 30 minutes is 8.5 hours.

HINTS AND TIPS

km/h means kilometres per hour.
m/s means metres per second.

D

D

6 Copy and complete the following table.

	Distance travelled	Time taken	Average speed
a	150 miles	2 hours	
b	260 miles		40 mph
c		5 hours	35 mph
d		3 hours	80 km/h
e	544 km	8 hours 30 minutes	
f		3 hours 15 minutes	100 km/h
g	215 km		50 km/h

7 Eliot drove from Sheffield to Inverness, a distance of 410 miles, in 7 hours 45 minutes.

a Change the time 7 hours 45 minutes to a decimal.

b What was the average speed of the journey? Round your answer to 1 decimal place.

8 Colin drives home from his son's house in 2 hours 15 minutes. He says that he drives at an average speed of 44 mph.

a Change the 2 hours 15 minutes to a decimal.

b How far is it from Colin's home to his son's house?

9 The distance between Paris and Le Mans is 200 km. The express train between Paris and Le Mans travels at an average speed of 160 km/h.

a Calculate the time taken for the journey from Paris to Le Mans, giving your answer as a decimal number of hours.

b Change your answer to part **a** to hours and minutes.

C

FM **10** The distance between Sheffield and Land's End is 420 miles.

a What is the average speed of a journey from Sheffield to Land's End that takes 8 hours 45 minutes?

b If Sam covered the distance at an average speed of 63 mph, how long would it take him?

FM **11** A train travels at 50 km/h for 2 hours, then slows down to do the last 30 minutes of its journey at 40 km/h.

a What is the total distance of this journey?

b What is the average speed of the train over the whole journey?

FM **12** Jade runs and walks the 3 miles from home to work each day. She runs the first 2 miles at a speed of 8 mph, then walks the next mile at a steady 4 mph.

a How long does it take Jade to get to work?

b What is her average speed?

13 Change the following speeds to metres per second.

a 36 km/h **b** 12 km/h **c** 60 km/h

d 150 km/h **e** 75 km/h

14 Change the following speeds to kilometres per hour.

a 25 m/s **b** 12 m/s **c** 4 m/s

d 30 m/s **e** 0.5 m/s

PS **15** A train travels at an average speed of 18 m/s.

a Express its average speed in km/h.

b Find the approximate time the train would take to travel 500 m.

c The train set off at 7.30 on a 40 km journey. At approximately what time will it reach its destination?

16 A cyclist is travelling at an average speed of 24 km/h.

a What is this speed in metres per second?

b What distance does he travel in 2 hours 45 minutes?

c How long does it take him to travel 2 km?

d How far does he travel in 20 seconds?

AU **17** How much longer does it take to travel 100 miles at 65 mph than at 70 mph?

HINTS AND TIPS

Remember that there are 3600 seconds in an hour and 1000 metres in a kilometre. So to change from km/h to m/s multiply by 1000 and divide by 3600.

HINTS AND TIPS

To change from m/s to km/h multiply by 3600 and divide by 1000.

HINTS AND TIPS

To convert a decimal fraction of an hour to minutes, just multiply by 60.

C

GRADE BOOSTER

- **G** You can find equivalent fractions, decimals and percentages
- **E** You can find simple percentages of a quantity
- **E** You can find any percentages of a quantity
- **E** You can simplify a ratio
- **D** You can calculate speeds, distances and times
- **D** You can compare prices to find the 'best buy'
- **C** You can solve problems, using ratio

What you should know now

- How to find equivalent percentages, decimals and fractions
- How to calculate percentages
- How to divide any amount according to a given ratio

1 **a** Write $\frac{7}{16}$ as a decimal.

b Write 27% as a decimal.

2 Andrew got 42 out of 50 marks in a history test.

He got 48 out of 60 marks in a geography test.

The marks for each test were changed to a percentage.

In which test did Andrew get the higher percentage mark?

You must show all your calculations. (4)

(Total 4 marks)

Edexcel, June 2008, Paper 13 Foundation, Question 12

3 There are 75 penguins at a zoo. There are 15 baby penguins.

What percentage of the penguins are babies?

4 Breakfast cereal is sold in two sizes of packet.

The small packet holds 500 grams and costs £2.10.

The large packet holds 875 grams and costs £3.85.

Which packet is better value for money?
You *must* show all your working.

5 Mr and Mrs Jones are buying a tumble dryer that normally costs £250. They save 12% in a sale.

a What is 12% of £250?

b How much do they pay for the tumble dryer?

6 **a** Write 92% as a decimal. (1)

b Write 3% as a fraction. (1)

c Work out 5% of 400 grams. (2)

(Total 4 marks)

Edexcel, May 2008, Paper 1 Foundation, Question 16

7 Work out 28% of £85 000. (2)

(Total 2 marks)

Edexcel, November 2008, Paper 2 Foundation, Question 16

8 Work out 35% of £400 (2)

(Total 2 marks)

Edexcel, November 2007, Paper 10 Foundation, Unit 3 Test, Question 4

9 A television reporter did a survey.

She asked people to name their favourite sport.

The table gives some information about the answers she got.

Favourite Sport	Percentage
Football	30%
Cricket	14%
Hockey	9%
Snooker	8%
Tennis	4%
Other	....

a Complete the table. (1)

b Write 30% as a fraction.
Give your answer in its simplest form. (2)

2000 people took part in the survey.

c Work out the number of people who said cricket. (2)

40 people said golf.

d Work out 40 out of 2000 as a percentage. (2)

(Total 7 marks)

Edexcel, November 2008, Paper 13 Foundation, Question 12

10 **a** Write 37% as a fraction. (1)

b Work out 37% of £415 (2)

(Total 3 marks)

Edexcel, March 2007, Paper 10 Foundation, Unit 3 Test, Question 6

11 Supermarkets often make 'Buy one, get one free' offers. What percentage saving is this?

10%, 50%, 100% or 200%

12 A tin of cat food costs 40p.

A shop has a special offer on the cat food.

Julie wants 12 tins of cat food.

a Work out how much she pays. (3)

9 of the 12 tins are tuna.

b Write 9 out of 12 as a percentage. (2)

The normal price of a cat basket is £20

In a sale, the price of the cat basket is reduced by 15%.

c Work out the sale price of the cat basket. (3)

(Total 8 marks)

Edexcel, November 2008, Paper 1 Foundation, Question 17

13 **a** Brian travels 234 miles by train. His journey takes $2\frac{1}{2}$ hours.

What is the average speed of the train?

b Val drives 234 miles at an average speed of 45 mph.

How long does her journey take?

14 Joe travelled 60 miles in 1 hour 30 minutes.

Work out Joe's average speed.

Give your answer in miles per hour. (2)

(Total 2 marks)

Edexcel, June 2007, Paper 10 Foundation, Unit 3 Test, Question 8

15 A car travels for 3 hours.

Its average speed is 75 km/h.

Work out the total distance the car travels. (2)

(Total 2 marks)

Edexcel, November 2007, Paper 10 Foundation, Unit 3 Test, Question 7

16 Stuart drives 180 km in 2 hours 15 minutes.

Work out Stuart's average speed. (3)

(Total 3 marks)

Edexcel, November 2008, Paper 10 Foundation, Unit 3 Test, Question 8

17 A country walk is 15 miles long. A leaflet states that this walk can be done in 4 hours.

a Calculate the average speed required to complete the walk in the time stated.

b A walker completes the route in 4 hours. She averages 5 miles an hour for the first hour.

Calculate her average speed for the remainder of the journey.

18 The only pets a pet shop sells are hamsters and fish. The ratio of the number of hamsters to the number of fish is 12 : 28

a What fraction of these pets are hamsters? Give your fraction in its simplest form. (2)

The only fish the pet shop sells are goldfish and tropical fish.

The ratio of goldfish to tropical fish is 1 : 4

The shop has 280 fish.

b Work out the number of goldfish the shop has. (2)

(Total 4 marks)

Edexcel, Question 2, Paper 12A Intermediate, March 2005

19 The length of a coach is 15 metres. Jonathan makes a model of the coach. He uses a scale of 1 : 24 (2)

Work out the length, in centimetres, of the model coach. (2)

(Total 4 marks)

Edexcel, June 2005, Paper 4 Intermediate, Question 2

20 **a** The most popular picture frames are those for which the ratio of width to length is 5 : 8.

Which of these frames are in the ratio 5 : 8?

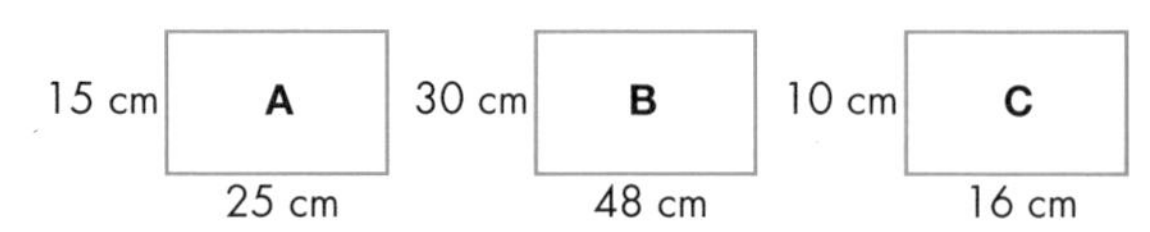

b There are 52 cards in a normal pack of cards. For a game, Dad shares the pack between Jack and Kenny in the ratio of 6 : 7

How many cards does each player receive?

21 There are 40 chocolates in a box.
12 chocolates are plain chocolates.
The remaining chocolates are milk chocolates.

a Work out the ratio of the number of plain chocolates to the number of milk chocolates in the box. Give your ratio in its simplest form. (2)

Some plain chocolates are added to the box so that the ratio of the number of plain chocolates to the number of milk chocolates is 1 : 2

b Work out how many plain chocolates are added to the box. (2)

(Total 4 marks)

Edexcel, January 2005, Paper 12B Intermediate, Question 3

Worked Examination Questions

FM **1** The land area of a farm is 385 acres.

Two-fifths of the land is used to grow barley.

96 acres is pasture.

22% is used to keep livestock.

The rest is unused.

How many acres are unused?

$\frac{2}{5} \times 385 = 154$ — You will get 1 mark for setting up the calculation and 1 mark for the answer. You could also do this as 0.4×385.

$0.22 \times 385 = 84.7$ acres — You will get 1 mark for setting up the calculation and 1 mark for the answer.

$385 - 96 - 154 - 84.7 = 50.3$ or 50 acres — Work out the area by subtracting all the areas from 385. You will get 1 mark for this.

Total: 5 marks

PS AU **2** David is mixing compost with soil.
He mixes 2 kg of compost with 1 kg of soil.

Kayren is also mixing compost with soil.
She mixes 3 kg of compost with 2 kg of soil.

Which mixture has the greater percentage of compost?
You **must** show your working.

David uses 2 kg of compost out of 3 kg altogether.

Percentage $= \frac{2}{3} \times 100\%$ — You will get 1 mark for method, used at least once.

$= 66.6\%$ or 67% — You will get 1 mark for accuracy of 66.6% (67%) or 60%.

Kayren uses 3 kg of compost out of 5 kg altogether.

Percentage $= \frac{3}{5} \times 100\%$

$= 60\%$

So David's mixture has the greater percentage of compost. — You will get 1 mark for accuracy of the other percentage and stating David's mixture. Stating David's mixture without showing working scores no marks.

Total: 3 marks

Functional Maths
Conserving water

We use water every day: it is vital to our survival and important in making our lives more comfortable.

Your task

Ushma has noticed that her water bill has increased over the last few months. The rates charged by the water company have not changed, so Ushma knows that she is paying more money because she is using more water. She is not only concerned that she is spending too much money on water: she is also worried about the impact that her increased water use is having on the environment.

1 Using the information opposite, write a report analysing Ushma's use of water.

In your report you should think about:

- the percentage of water that Ushma uses for different domestic activities
- the cost of each domestic activity that involves water
- where she could realistically reduce her water usage and by how much
- the impact that reducing her usage would have on her water bill
- what her total percentage decrease in water usage would be
- how best to represent your findings.

2 Seeing her water bill inspires Ushma to research how water is used on a global scale. The pie chart and table show the information she discovered.

What can Ushma tell from this information about water supplies throughout the world? Can she work out how reducing her water usage will impact on global water supplies?

Express your ideas as they might be seen in a newspaper.

Handy hint

Use the internet to look up facts to do with water usage. How do water companies and global organisations represent their 'water facts'?

Ushma's use of water

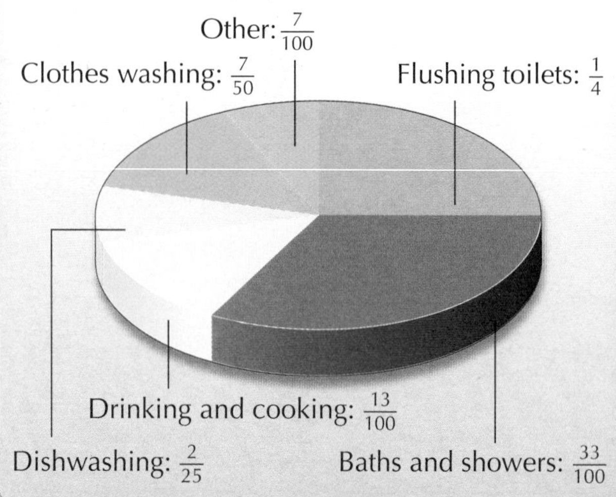

Global usage of freshwater

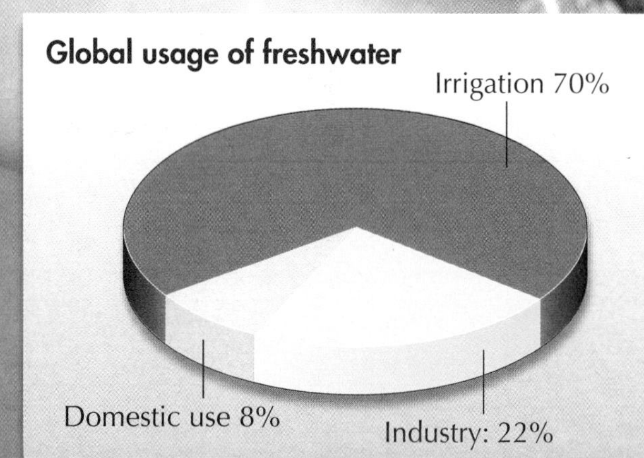

Global distribution of water	
Location	**Volume (km^3)**
Oceans	1 338 000 000
Ice	24 364 000
Groundwater	23 400 000
Lakes and reservoirs	176 400
Soil moisture	16 500
Water vapour	12 900
Rivers	2120
Swamp water	11 470

Handy hint

Water bills include a 'standing charge', which is a fixed amount paid by all customers regardless of how much water they have used. They also include a variable charge, which will change according to how much water the household has used.

Account number 12349876001	Date: 15-03-2010
Bill period January – March 2010	Meter No. 367X93007

Bill summary	
Water supply standing charge	£35.80
Water supply	£301.10
Surface drainage	£39.78
Sewerage standing charge	£1.68
Sewerage	£321.45
Total	**£699.81**
Payment due by: 14 April 2010	

contact us at: www.south-ukwater.co.uk

Water utility bill

Account number 12349876001	Date: 15-12-2009
Bill period October – December 2009	Meter No. 367X93007

Bill summary	
Water supply standing charge	£35.80
Water supply	£258.00
Surface drainage	£39.78
Sewerage standing charge	£1.68
Sewerage	£294.48
Total	**£629.74**
Payment due by: 14 Jan 2010	

contact us at: www.south-ukwater.co.uk

Why this chapter matters

Most jobs will require you to use some mathematics on a day-to-day basis. Being competent in all your number skills will therefore help you to be more successful in your job.

The mathematics used in the world of work will range from simple mental arithmetic calculations such as addition, subtraction, multiplication and division, to more complex calculations involving calculators, negative numbers and approximation. It will be up to you to select the mathematics that you need to carry out your job. Understanding essential mathematical techniques and how to apply these to a real-life context will make this much easier.

Jobs using mathematics

How many jobs can you think of that require some mathematics?

Here are a few ideas.

Accountant – How much profit have they made?

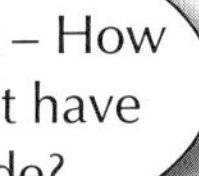

Pilot – How much fuel do I need?

Engineer – What measurements do I need to take? How much of each type of material will be needed?

Delivery driver – What is the best route?

Cashier – What coins do I need to give as change? What is the best price to sell my goods at?

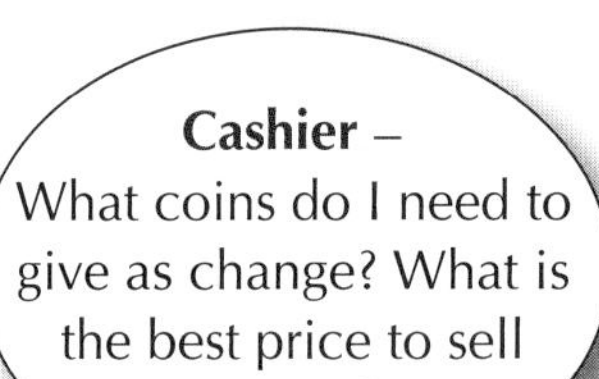

Doctor – How much medicine should I prescribe?

Sports commentator – How many minutes are left in the game? What is his batting average?

Baker – What quantity of flour should I order?

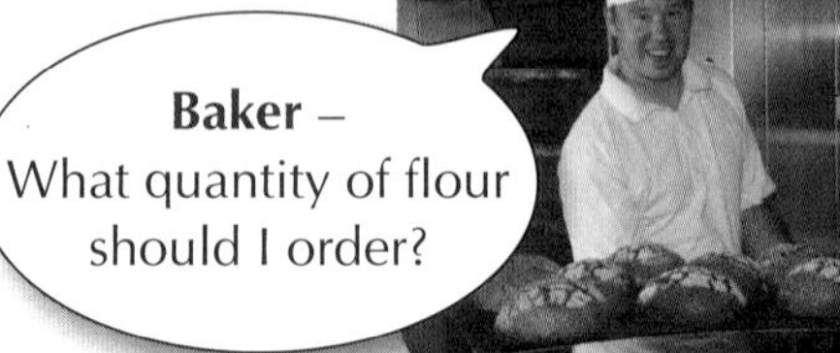

If you already have a career in mind that you would like to do, think of the questions you will need to ask to carry out your job successfully and the mathematics you will require to find the answers to those questions.

Chapter 5

Number: Further number skills

The grades given in this chapter are target grades.

1. Long multiplication
2. Long division
3. Arithmetic with decimal numbers
4. Multiplying and dividing with negative numbers
5. Approximation of calculations

This chapter will show you …

- **F** a reminder of the ways you can multiply a three-digit number by a two-digit number
- **F** a reminder of long division
- **E** how to calculate with decimal numbers
- **E** how to multiply and divide negative numbers
- **C** how to use decimal places and significant figures to make approximations
- **C** sensible rounding methods

Visual overview

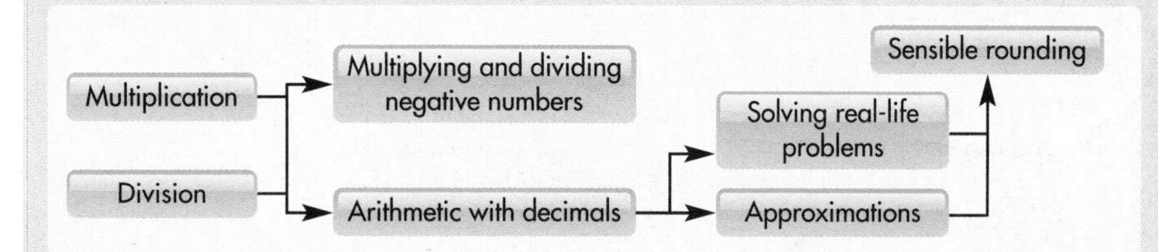

What you should already know

- Multiplication tables up to 10 × 10 **(KS3 level 4, GCSE grade G)**
- How to simplify fractions **(KS3 level 5, GCSE grade G)**

Quick check

1 Write down the first five multiples of 6.

2 Write down the first five multiples of 8.

3 Write down a number that is both a multiple of 3 and a multiple of 5.

4 Write down the smallest number that is a multiple of 4 and a multiple of 5.

5 Write down the smallest number that is a multiple of 4 and a multiple of 6.

6 Simplify the following fractions.

a $\frac{8}{10}$ **b** $\frac{5}{20}$ **c** $\frac{4}{16}$ **d** $\frac{32}{100}$ **e** $\frac{36}{100}$ **f** $\frac{16}{24}$ **g** $\frac{16}{50}$

5.1 Long multiplication

This section will show you how to:

- multiply a three-digit number (e.g. 358) by a two-digit number (e.g. 74) using
  - the grid method (or box method)
  - the column method (or traditional method)
  - the partition method

Key words

column method (or traditional method)
grid method (or box method)
partition method

When you are asked to do long multiplication on the GCSE non-calculator paper, you will be expected to use an appropriate method. The three most common are:

- The **grid method** (or box method), see Example 1 below.
- The **column method** (or traditional method), see Example 2 below.
- The **partition method**, see Example 3 below.

EXAMPLE 1

Work out 243×68 without using a calculator.

Using the grid method split the two numbers into hundreds, tens and units and write them in a grid like the one below. Multiply all the pairs of numbers.

×	200	40	3
60	12 000	2400	180
8	1600	320	24

Add the separate answers to find the total.

```
 12000
  2400
   180
  1600
   320
    24
 -----
 16524
  1 1
```

So, $243 \times 68 = 16524$

Note the use of carry marks to help with the calculation. Always try to write carried marks much smaller than the other numbers, so that you don't confuse them with the main calculation.

EXAMPLE 2

Work out 357 × 24 without using a calculator.

There are several different ways to do long multiplication, but the following is perhaps the method that is most commonly used. This is the column method.

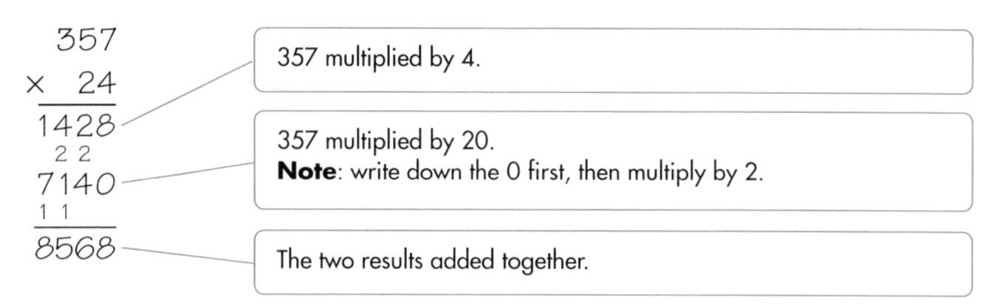

So, 357 × 24 = 8568

EXAMPLE 3

Work out 358 × 74 by the partition method.

Set out a grid as shown.

3	5	8	
2 / 1	3 / 5	5 / 6	7
1 / 2	2 / 0	3 / 2	4

5
+1 ← 14
2 6 4 9 2

- Put the larger number along the top and the smaller number down the right-hand side.
- Multiply each possible pair in the grid, putting the numbers into each half as shown.
- Add up the numbers in each diagonal. If a total is larger than nine (in this example there is a total of 14), split the number and put the 1 in the next column on the left ready to be added in that diagonal.
- When you have completed the totalling, the number you are left with is the answer to the multiplication.

So, 358 × 74 = 26 492

EXERCISE 5A

1 Use your preferred method to calculate the following without using a calculator.

a 357 × 34	**b** 724 × 63	**c** 714 × 42	**d** 898 × 23
e 958 × 54	**f** 676 × 37	**g** 239 × 81	**h** 437 × 29
i 539 × 37	**j** 477 × 55	**k** 371 × 85	**l** 843 × 93
m 507 × 34	**n** 810 × 54	**o** 905 × 73	**p** 1435 × 72
q 2504 × 56	**r** 4037 × 23	**s** 8009 × 65	**t** 2070 × 38

F

F

2 There are 48 cans of soup in a crate. A supermarket had a delivery of 125 crates of soup. The supermarket shelves will hold a total of 2500 cans. How many cans will be in the store room?

FM **3** Greystones Primary School has 12 classes, each of which has 26 students. The school hall will hold 250 students. Can all the students fit in the hall for an assembly?

FM **4** Suhail walks to school each day, there and back. The distance to school is 450 m. Suhail claims he walks over a marathon in distance, to and from school, in a school term of 64 days. A marathon is 42.1 km.

Is he correct?

5 On one page of a newspaper there are seven columns. In each column there are 172 lines, and in each line there are 50 letters. The newspaper has the equivalent of 20 pages of print, excluding photographs and adverts. Are there more than a million letters in the paper?

6 A tank of water was emptied into casks. Each cask held 81 litres. 71 casks were filled and there were 68 litres left over. How much water was there in the tank to start with?

FM **7** Joy was going to do a sponsored walk to raise money for the Macmillan Nurses. She managed to get 18 people to sponsor her, each for 35p per kilometre. She walked a total of 48 km. Did Joy reach her target of £400?

5.2 Long division

This section will show you how to:

- divide, without a calculator, a three-digit or four-digit number by a two-digit number, e.g. $840 \div 24$

Key words
long division
remainder

There are several different ways of doing **long division**. It is acceptable to use any of them, provided it gives the correct answer and you can show all your working clearly. Two methods are shown in this book. Example 4 shows the *Italian method*, sometimes called the DMSB method (Divide, Multiply, Subtract and Bring down). It is the most commonly used way of doing long division.

Example 5 shows a method of repeated subtraction, which is sometimes called the *chunking method*.

EXAMPLE 4

Work out 840 ÷ 24.

It is a good idea to jot down the appropriate times table before you start the long division. In this case, it will be the 24 times table.

1	2	3	4	5	6	7	8	9
24	48	72	96	120	144	168	192	216

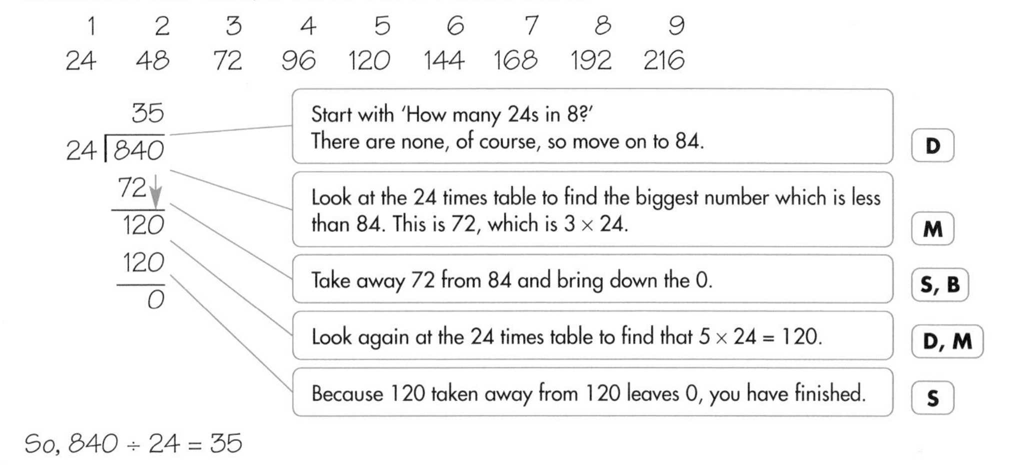

So, 840 ÷ 24 = 35

You may do a division as a short division, without writing down all the numbers. It will look like this:

$$\begin{array}{r} 3\ 5 \\ 24 \overline{)\ 8\ 4\ {}^{12}0} \end{array}$$

Notice how the **remainder** from 84 is placed in front of the 0 to make it 120.

EXAMPLE 5

Work out 1655 ÷ 35.

Jot down some of the multiples of 35 that may be useful.

$1 \times 35 = 35 \quad 2 \times 35 = 70 \quad 5 \times 35 = 175 \quad 10 \times 35 = 350 \quad 20 \times 35 = 700$

Working	Multiple	Explanation
1655		
− 700	20 × 35	From 1655, subtract a large multiple of 35, such as 20 × 35 = 700.
955		
− 700	20 × 35	From 955, subtract a large multiple of 35, such as 20 × 35 = 700.
255		
− 175	5 × 35	From 255, subtract a multiple of 35, such as 5 × 35 = 175.
80		
− 70	2 × 35	From 80, subtract a multiple of 35, such as 2 × 35 = 70.
10	47	

Once the remainder of 10 has been found, you cannot subtract any more multiples of 35. Add up the multiples to see how many times 35 has been subtracted.

So, 1655 ÷ 35 = 47, remainder 10

Sometimes, as here, you will not need a whole multiplication table, and so you could jot down only those parts of the table that you will need. But, don't forget, you are going to have to work *without* a calculator, so you do need all the help you can get.

EXAMPLE 6

Naseema is organising a coach trip for 640 people. Each coach will carry 46 people. How many coaches are needed?

You need to divide the number of people (640) by the number of people in a coach (46).

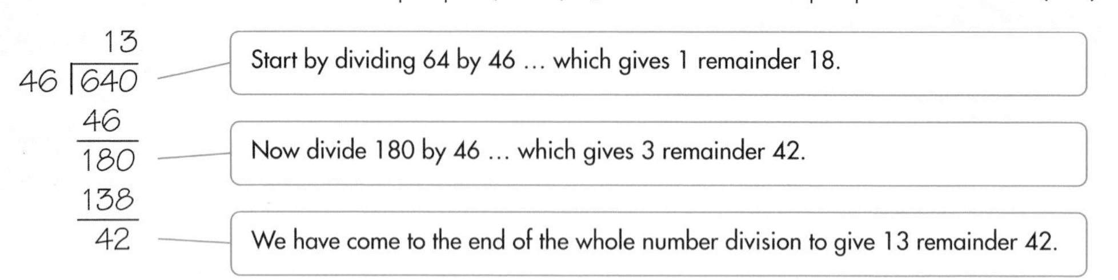

This tells Naseema that 14 coaches are needed to take all 640 passengers. (There will be $46 - 42 = 4$ spare seats)

EXERCISE 5B

1 Solve the following by long division.

a $525 \div 21$ **b** $480 \div 32$ **c** $925 \div 25$ **d** $645 \div 15$

e $621 \div 23$ **f** $576 \div 12$ **g** $1643 \div 31$ **h** $728 \div 14$

i $832 \div 26$ **j** $2394 \div 42$ **k** $829 \div 22$ **l** $780 \div 31$

m $895 \div 26$ **n** $873 \div 16$ **o** $875 \div 24$ **p** $225 \div 13$

q $759 \div 33$ **r** $1478 \div 24$ **s** $756 \div 18$ **t** $1163 \div 43$

2 3600 supporters of Barnsley Football Club want to go to an away game by coach. Each coach can hold 53 passengers. How many coaches will they need altogether?

3 How many stamps costing 26p each can I buy for £10?

FM 4 Kirsty is collecting a set of 40 porcelain animals. Each animal costs 45p. Each month she spends up to £5 of her pocket money on these animals. How many months will it take her to buy the full set?

FM 5 Amina wanted to save up to see a concert. The cost of a ticket was £25. She was paid 75p per hour to mind her little sister. For how many hours would Amina have to mind her sister to be able to afford the ticket?

AU 6 The magazine *Teen Dance* costs £2.20 in a newsagent. It costs £21 to get the magazine on a yearly subscription. How much cheaper is the magazine each month if it is bought by subscription?

F

5.3 Arithmetic with decimal numbers

This section will show you how to:
- identify the information that a decimal number shows
- round a decimal number
- identify decimal places
- add and subtract two decimal numbers
- multiply and divide a decimal number by a whole number less than 10
- multiply a decimal number by a two-digit number
- multiply a decimal number by another decimal number

Key words
decimal fraction
decimal place
decimal point
digit
evaluate
round

The number system is extended by using decimal numbers to represent fractions.

The **decimal point** separates the **decimal fraction** from the whole-number part.

For example, the number 25.374 means:

Tens	Units		Tenths	Hundredths	Thousandths
10	1		$\frac{1}{10}$	$\frac{1}{100}$	$\frac{1}{1000}$
2	**5**	**.**	**3**	**7**	**4**

You already use decimal notation to express amounts of money. For example:

£32.67 means $3 \times$ £10
$2 \times$ £1
$6 \times$ £0.10 (10 pence)
$7 \times$ £0.01 (1 penny)

Decimal places

When a number is written in decimal form, the **digits** to the right of the decimal point are called **decimal places**. For example:

79.4 is written 'with one decimal place'

6.83 is written 'with two decimal places'

0.526 is written 'with three decimal places'.

To **round** a decimal number to a particular number of decimal places, take these steps:

- Count along the decimal places from the decimal point and look at the first digit to be removed.
- When the value of this digit is less than five, just remove the unwanted places.

- When the value of this digit is five or more, add 1 onto the digit in the last decimal place then remove the unwanted places.

Here are some examples.

5.852 rounds to 5.85 to two decimal places

7.156 rounds to 7.16 to two decimal places

0.274 rounds to 0.3 to one decimal place

15.3518 rounds to 15.4 to one decimal place

EXERCISE 5C

F

1 Round each of the following numbers to one decimal place.

a 4.83 **b** 3.79 **c** 2.16 **d** 8.25

e 3.673 **f** 46.935 **g** 23.883 **h** 9.549

i 11.08 **j** 33.509 **k** 7.054 **l** 46.807

m 0.057 **n** 0.109 **o** 0.599 **p** 64.99

q 213.86 **r** 76.07 **s** 455.177 **t** 50.999

HINTS AND TIPS

Just look at the value of the digit in the second decimal place.

2 Round each of the following numbers to two decimal places.

a 5.783 **b** 2.358 **c** 0.977 **d** 33.085 **e** 6.007

f 23.5652 **g** 91.7895 **h** 7.995 **i** 2.3076 **j** 23.9158

k 5.9999 **l** 1.0075 **m** 3.5137 **n** 96.508 **o** 0.009

p 0.065 **q** 7.8091 **r** 569.897 **s** 300.004 **t** 0.0099

3 Round each of the following to the number of decimal places (dp) indicated.

a 4.568 (1 dp) **b** 0.0832 (2 dp) **c** 45.715 93 (3 dp)

d 94.8531 (2 dp) **e** 602.099 (1 dp) **f** 671.7629 (2 dp)

g 7.1124 (1 dp) **h** 6.903 54 (3 dp) **i** 13.7809 (2 dp)

j 0.075 11 (1 dp) **k** 4.001 84 (3 dp) **l** 59.983 (1 dp)

m 11.9854 (2 dp) **n** 899.995 85 (3 dp) **o** 0.0699 (1 dp)

p 0.009 87 (2 dp) **q** 6.0708 (1 dp) **r** 78.392 5 (3 dp)

s 199.9999 (2 dp) **t** 5.0907 (1 dp)

4 Round each of the following to the nearest whole number.

a 8.7 **b** 9.2 **c** 2.7 **d** 6.5 **e** 3.28

f 7.82 **g** 3.19 **h** 7.55 **i** 6.172 **j** 3.961

k 7.388 **l** 1.514 **m** 46.78 **n** 23.19 **o** 96.45

p 32.77 **q** 153.9 **r** 342.5 **s** 704.19 **t** 909.5

FM **5** Belinda puts the following items in her shopping basket: bread (£1.09), meat (£6.99), cheese (£3.91) and butter (£1.13).

By rounding each price to the nearest pound, work out an estimate of the total cost of the items.

AU **6** Which of the following are correct roundings of the number 3.456?

3   3.0   3.4   3.40   3.45   3.46   3.47   3.5   3.50

PS **7** When a number is rounded to three decimal places the answer is 4.728

Which of these could be the number?

4.71   4.7275   4.7282   4.73

F

E

Adding and subtracting with decimals

When you are working with decimals, you must *always* set out your work carefully.

Make sure that the decimal points are in line underneath the first point and each digit is in its correct place or column.

Then you can add or subtract just as you have done before. The decimal point of the answer will be placed directly underneath the other decimal points.

EXAMPLE 7

Work out 4.72 + 13.53.

```
   4.72
+ 13.53
  18.25
   1
```

So, 4.72 + 13.53 = 18.25

Notice how to deal with 7 + 5 = 12, the 1 carrying forward into the next column.

EXAMPLE 8

Work out 7.3 – 1.5.

```
  ⁶7.¹3
–  1.5
   5.8
```

So, 7.3 – 1.5 = 5.8

Notice how to deal with the fact that you cannot take 5 from 3. You have to take one of the units from 7, replace the 7 with a 6 and make the 3 into 13.

Hidden decimal point

Whole numbers are usually written without decimal points. Sometimes you *do* need to show the decimal point in a whole number (see Example 9), in which case it is placed at the right-hand side of the number, followed by a zero.

EXAMPLE 9

Work out 4.2 + 8 + 12.9.

$$\begin{array}{r} 4.2 \\ 8.0 \\ +\,12.9 \\ \hline 25.1 \\ \hline {\scriptstyle 1\,1} \end{array}$$

So, 4.2 + 8 + 12.9 = 25.1

EXERCISE 5D

1 Work out each of these.

a 47.3 + 2.5 **b** 16.7 + 4.6 **c** 43.5 + 4.8

d 28.5 + 4.8 **e** 1.26 + 4.73 **f** 2.25 + 5.83

g 83.5 + 6.7 **h** 8.3 + 12.9 **i** 3.65 + 8.5

j 7.38 + 5.7 **k** 7.3 + 5.96 **l** 6.5 + 17.86

HINTS AND TIPS

When the numbers to be added or subtracted do not have the same number of decimal places, put in extra zeros, for example:

$$\begin{array}{r} 3.65 \\ +\ 8.50 \\ \hline \end{array} \qquad \begin{array}{r} 8.25 \\ -\ 4.50 \\ \hline \end{array}$$

2 Work out each of these.

a 3.8 – 2.4 **b** 4.3 – 2.5 **c** 7.6 – 2.8

d 8.7 – 4.9 **e** 8.25 – 4.5 **f** 19.7 – 13.8

g 9.4 – 5.7 **h** 8.62 – 4.85 **i** 8 – 4.3

j 9 – 7.6 **k** 15 – 3.2 **l** 24 – 8.7

3 **Evaluate** each of the following. (Take care – they are a mixture.)

a 23.8 + 6.9 **b** 8.3 – 1.7 **c** 9 – 5.2 **d** 12.9 + 3.8

e 17.4 – 5.6 **f** 23.4 + 6.8 **g** 35 + 8.3 **h** 9.54 – 2.81

i 34.8 + 3.15 **j** 8.1 – 3.4 **k** 12.5 – 8.7 **l** 198.5 + 12

AU 4 Viki has £4.75 in her purse after buying a watch for £11.99.

a How much did she have in her purse before she bought the watch?

b She then goes home by bus. After paying her bus fare she has £3.35 left in her purse. How much was the bus fare?

F

F E D

FM 5 Pipes are sold in 5.3 m lengths.

Will three pipes be enough to make this arrangement (excluding the corner pieces)? Justify your answer.

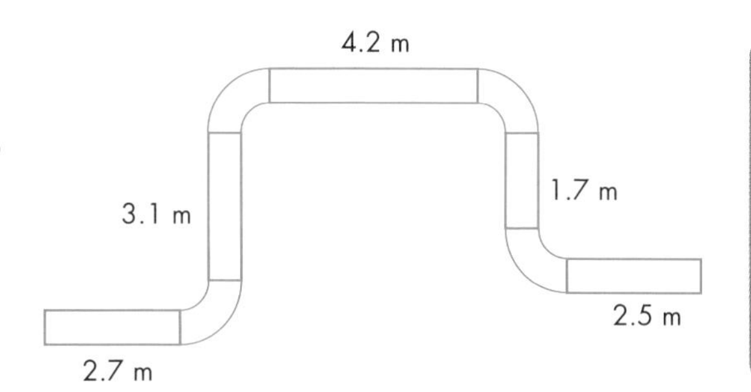

AU 6 Copy and complete the following.

a 12.5 + …… = 17.8 **b** 31.6 + …… = 38.3 **c** 7.16 + …… = 9.21

d …… + 4.2 = 6.1 **e** …… + 10.21 = 15.16 **f** …… + 27.54 = 31.25

AU 7 Copy and complete the following.

a 13.2 – …… = 11.6 **b** 81.4 – …… = 38.7 **c** 7.51 – …… = 3.22

d …… – 3.1 = 9.7 **e** …… – 14.6 = 7.8 **f** …… – 17.32 = 34.65

PS 8 Mark went shopping. He went into three shops and bought one item from each shop.

In total he spent £43.97.

Music Store	
CDs:	£5.98
DVDs:	£7.99

Clothes Store	
Shirt:	£12.50
Jeans	£32.00

Book store	
Magazine:	£2.25
Pen:	£3.98

What did he buy?

Multiplying and dividing decimals by single-digit numbers

You can carry out these operations in exactly the same way as with whole numbers, as long as you remember to put each digit in its correct column.

Again, the decimal point is kept in line underneath or above the first point.

EXAMPLE 10

Work out 4.5 × 3.

$$\begin{array}{r} 4.5 \\ \times \quad 3 \\ \hline 13.5 \\ \hline {}_{1}\ \ \end{array}$$

So, 4.5 × 3 = 13.5

EXAMPLE 11

Work out $8.25 \div 5$.

$$\begin{array}{r} 1.\,6\,5 \\ 5\,\overline{)\,8.^{3}2^{2}5} \end{array}$$

So, $8.25 \div 5 = 1.65$

EXAMPLE 12

Work out $5.7 \div 2$.

$$\begin{array}{r} 2.\,8\,5 \\ 2\,\overline{)\,5.^{1}7^{1}0} \end{array}$$

So, $5.7 \div 2 = 2.85$

HINTS AND TIPS

We add a 0 after the 5.7 in order to continue dividing.
We do not use remainders with decimal places.

EXERCISE 5E

F

1 Evaluate each of these.

a 2.4×3	**b** 3.8×2	**c** 4.7×4	**d** 5.3×7
e 6.5×5	**f** 3.6×8	**g** 2.5×4	**h** 9.2×6
i 12.3×5	**j** 24.4×7	**k** 13.6×6	**l** 19.3×5

2 Evaluate each of these.

a 2.34×4	**b** 3.45×3	**c** 5.17×5	**d** 4.26×3
e 0.26×7	**f** 0.82×4	**g** 0.56×5	**h** 0.92×6
i 6.03×7	**j** 7.02×8	**k** 2.55×3	**l** 8.16×6

3 Evaluate each of these.

a $3.6 \div 2$	**b** $5.6 \div 4$	**c** $4.2 \div 3$	**d** $8.4 \div 7$
e $4.26 \div 2$	**f** $3.45 \div 5$	**g** $8.37 \div 3$	**h** $9.68 \div 8$
i $7.56 \div 4$	**j** $5.43 \div 3$	**k** $1.32 \div 4$	**l** $7.6 \div 4$

4 Evaluate each of these.

a $3.5 \div 2$ **b** $6.4 \div 5$ **c** $7.4 \div 4$ **d** $7.3 \div 2$

e $8.3 \div 5$ **f** $5.8 \div 4$ **g** $7.1 \div 5$ **h** $9.2 \div 8$

i $6.7 \div 2$ **j** $4.9 \div 5$ **k** $9.2 \div 4$ **l** $7.3 \div 5$

HINTS AND TIPS

Remember to keep the decimal points in line.

5 Evaluate each of these.

a $7.56 \div 4$ **b** $4.53 \div 3$ **c** $1.32 \div 5$ **d** $8.53 \div 2$

e $2.448 \div 2$ **f** $1.274 \div 7$ **g** $0.837 \div 9$ **h** $16.336 \div 8$

i $9.54 \div 5$ **j** $14 \div 5$ **k** $17 \div 4$ **l** $37 \div 2$

6 Soup is sold in packs of five for £3.25 and packs of eight for £5. Which is the cheaper way of buying soup?

7 **a** Mike took his wife and four children to a theme park. The tickets were £13.25 for each adult and £5.85 for each child. How much did all the tickets cost Mike?

b While in the park, the children wanted ice creams.

Ice creams cost £1.60 each for large ones or £1.20 each for small ones.
Their mother only has £5.

Which size should she buy? Give a reason for your answer.

PS **8** Mary was laying a path through her garden. She bought nine paving stones, each 1.35 m long. She wanted the path to run straight down the garden, which is 10 m long. Has Mary bought too many paving stones? Show all your working.

Long multiplication with decimals

As before, you must put each digit in its correct column and keep the decimal point in line.

EXAMPLE 13

Evaluate 4.27×34.

$$\begin{array}{rr} & 4.27 \\ \times & 34 \\ \hline & 17.08 \\ & {\scriptstyle 1\ 2} \\ & 128.10 \\ & {\scriptstyle 2} \\ \hline & 145.18 \\ \hline & {\scriptstyle 1} \end{array}$$

So, $4.27 \times 34 = 145.18$

E

EXERCISE 5F

1 Evaluate each of these.

a 3.72×24 **b** 5.63×53 **c** 1.27×52 **d** 4.54×37

e 67.2×35 **f** 12.4×26 **g** 62.1×18 **h** 81.3×55

i 5.67×82 **j** 0.73×35 **k** 23.8×44 **l** 99.5×19

2 Find the total cost of each of the following purchases.

a 18 ties at £12.45 each

b 25 shirts at £8.95 each

c 13 pairs of tights at £2.30 a pair

HINTS AND TIPS

When the answer is an amount of money in pounds, you must write it with two decimal places. Writing £224.1 will lose you a mark. It should be £224.10.

PS 3 Theo says that he can change multiplications into easier multiplications with the same answer by doubling one number and halving the other number.

For example to work out 8.4×12 he does 16.8×6

Use his method to work out 2.5×14.

D

FM 4 A party of 24 scouts and their leader went into a zoo. The cost of a ticket for each scout was £2.15 and the cost of a ticket for the leader was £2.60. What was the total cost of entering the zoo?

AU FM 5 **a** A market gardener bought 35 trays of seedlings. Each tray cost £3.45. What was the total cost of the trays of seedlings?

b There were 20 seedlings on each tray which are sold at 85p each. How much profit did the market gardener make per tray?

Multiplying two decimal numbers together

Follow these steps to multiply one decimal number by another decimal number.

- First, complete the whole calculation as if the decimal points were not there.
- Then, count the total number of decimal places in the two decimal numbers. This gives the number of decimal places in the answer.

EXAMPLE 14

Evaluate 3.42×2.7.

Ignoring the decimal points gives the following calculation:

```
    342
×    27
   2394
    2 1
   6840
   9234
   1 1
```

Now, 3.42 has two decimal places (.42) and 2.7 has one decimal place (.7). So, the total number of decimal places in the answer is three.

So, $3.42 \times 2.7 = 9.234$

EXERCISE 5G

1 Evaluate each of these.

a 2.4×0.2 **b** 7.3×0.4 **c** 5.6×0.2 **d** 0.3×0.4

e 0.14×0.2 **f** 0.3×0.3 **g** 0.24×0.8 **h** 5.82×0.52

i 5.8×1.23 **j** 5.6×9.1 **k** 0.875×3.5 **l** 9.12×5.1

FM 2 Jerome is making a jacket. He has £30 to spend and needs 3.4 m of cloth. The cloth is £8.75 per metre.

Can Jerome afford the cloth?

3 For each of the following:

i estimate the answer by first rounding each number to the nearest whole number

ii calculate the exact answer, and then calculate the difference between this and your answers to part **i**.

a 4.8×7.3 **b** 2.4×7.6 **c** 15.3×3.9 **d** 20.1×8.6

e 4.35×2.8 **f** 8.13×3.2 **g** 7.82×5.2 **h** 19.8×7.1

AU PS 4 **a** Use any method to work out 26×22

b Use your answer to part **a** to work out:

i 2.6×2.2

ii 1.3×1.1

iii 2.6×8.8.

E

D

5.4 Multiplying and dividing with negative numbers

This section will show you how to:
- multiply and divide with negative numbers

Key word
negative

The rules for multiplying and dividing with **negative** numbers are very easy.

- When the signs of the numbers are the *same*, the answer is *positive*.
- When the signs of the numbers are *different*, the answer is *negative*.

Here are some examples.

$2 \times 4 = 8$ $\quad 12 \div -3 = -4$ $\quad -2 \times -3 = 6$ $\quad -12 \div -3 = 4$

A common error is to confuse, for example, -3^2 and $(-3)^2$.

$-3^2 = -3 \times 3 = -9$

but,

$(-3)^2 = -3 \times -3 = +9$

So, this means that if a variable is introduced, for example, $a = -5$, the calculation would be as follows:

$a^2 = -5 \times -5 = +25$

EXAMPLE 15

$a = -2$ and $b = -6$

Work out

a a^2 **b** $a^2 + b^2$ **c** $b^2 - a^2$ **d** $(a - b)^2$

a $a^2 = -2 \times -2 = +4$

b $a^2 + b^2 = +4 + -6 \times -6 = 4 + 36 = 40$

c $b^2 - a^2 = 36 - 4 = 32$

d $(a - b) = (-2 - -6)^2 = (-2 + 6)^2 = (4)^2 = 16$

EXERCISE 5H

E

1 Write down the answers to the following.

a -3×5	**b** -2×7	**c** -4×6	**d** -2×-3	**e** -7×-2
f $-12 \div -6$	**g** $-16 \div 8$	**h** $24 \div -3$	**i** $16 \div -4$	**j** $-6 \div -2$
k 4×-6	**l** 5×-2	**m** 6×-3	**n** -2×-8	**o** -9×-4
p $24 \div -6$	**q** $12 \div -1$	**r** $-36 \div 9$	**s** $-14 \div -2$	**t** $100 \div 4$
u -2×-9	**v** $32 \div -4$	**w** 5×-9	**x** $-21 \div -7$	**y** -5×8

E

2 Write down the answers to the following.

a $-3 + -6$ **b** -2×-8 **c** $2 + -5$ **d** 8×-4 **e** $-36 \div -2$

f -3×-6 **g** $-3 - -9$ **h** $48 \div -12$ **i** -5×-4 **j** $7 - -9$

k $-40 \div -5$ **l** $-40 + -8$ **m** $4 - -9$ **n** $5 - 18$ **o** $72 \div -9$

p $-7 - -7$ **q** $8 - -8$ **r** 6×-7 **s** $-6 \div -1$ **t** $-5 \div -5$

u $-9 - 5$ **v** $4 - -2$ **w** $4 \div -1$ **x** $-7 \div -1$ **y** -4×0

3 What number do you multiply by –3 to get the following?

a 6 **b** –90 **c** –45 **d** 81 **e** 21

4 What number do you divide –36 by to get the following?

a –9 **b** 4 **c** 12 **d** –6 **e** 9

5 Evaluate the following.

a $-6 + (4 - 7)$ **b** $-3 - (-9 - -3)$ **c** $8 + (2 - 9)$

6 Evaluate the following.

a $4 \times (-8 \div -2)$ **b** $-8 -(3 \times -2)$ **c** $-1 \times (8 - -4)$

7 What do you get if you divide –48 by the following?

a –2 **b** –8 **c** 12 **d** 24

AU 8 Write down six different multiplications that give the answer –12.

AU 9 Write down six different divisions that give the answer –4.

10 Find the answers to the following.

a -3×-7 **b** $3 + -7$ **c** $-4 \div -2$ **d** $-7 - 9$ **e** $-12 \div -6$

f $-12 - -7$ **g** 5×-7 **h** $-8 + -9$ **i** $-4 + -8$ **j** $-3 + 9$

k -5×-9 **l** $-16 \div 8$ **m** $-8 - -8$ **n** $6 \div -6$ **o** $-4 + -3$

p -9×4 **q** $-36 \div -4$ **r** -4×-8 **s** $-1 - -1$ **t** $2 - 67$

AU 11 **a** Work out 6×-2

b The average temperature drops by 2 °C every day for six days. How much has the temperature dropped altogether?

c The temperature drops by 6 °C for each of the next three days. Write down the calculation to work out the total drop in temperature over these three days.

PS 12 Put these calculations in order from lowest to highest.

-5×4 $-20 \div 2$ $-16 \div -4$ 3×-6

13 $x = -2$, $y = -3$ and $z = -4$. Work out

a x^2 **b** $y^2 + z^2$ **c** $z^2 - x^2$ **d** $(x - y)^2$

D

5.5 Approximation of calculations

This section will show you how to:
- identify significant figures
- round to one significant figure
- approximate the result before multiplying two numbers together
- approximate the result before dividing two numbers
- round a calculation, at the end of a problem, to give what is considered to be a sensible answer

Key words
approximate
round
significant figure

Rounding to significant figures

You will often use **significant figures** when you want to **approximate** a number with quite a few digits in it.

The following table illustrates some numbers written to one, two and three significant figures (sf).

One sf	8	50	200	90 000	0.000 07	0.003	0.4
Two sf	67	4.8	0.76	45 000	730	0.006 7	0.40
Three sf	312	65.9	40.3	0.0761	7.05	0.003 01	0.400

In the GCSE examination you usually only have to **round** numbers to one significant figure.

The steps taken to round a number to one significant figure are very similar to those used for decimal places.

- From the left, find the second digit. If the original number is less than one, start counting from the first non-zero digit.
- When the value of the second digit is less than five, leave the first digit as it is.
- When the value of the second digit is equal to or greater than five, add 1 to the first digit.
- Put in enough zeros at the end to keep the number the right size.

For example, the following tables show some numbers rounded to one significant figure.

Number	Rounded to 1 sf
78	80
32	30
0.69	0.7
1.89	2
998	1000
0.432	0.4

Number	Rounded to 1 sf
45 281	50 000
568	600
8054	8000
7.837	8
99.8	100
0.078	0.08

EXERCISE 5I

1 Round each of the following numbers to 1 significant figure.

a 46 313	**b** 57 123	**c** 30 569	**d** 94 558	**e** 85 299
f 54.26	**g** 85.18	**h** 27.09	**i** 96.432	**j** 167.77
k 0.5388	**l** 0.2823	**m** 0.005 84	**n** 0.047 85	**o** 0.000 876
p 9.9	**q** 89.5	**r** 90.78	**s** 199	**t** 999.99

2 Round each of the following numbers to 1 significant figure.

a 56 147	**b** 26 813	**c** 79 611	**d** 30 578	**e** 14 009
f 5876	**g** 1065	**h** 847	**i** 109	**j** 638.7
k 1.689	**l** 4.0854	**m** 2.658	**n** 8.0089	**o** 41.564
p 0.8006	**q** 0.458	**r** 0.0658	**s** 0.9996	**t** 0.009 82

3 Write down the smallest and the greatest numbers of sweets that can be found in each of these jars.

a

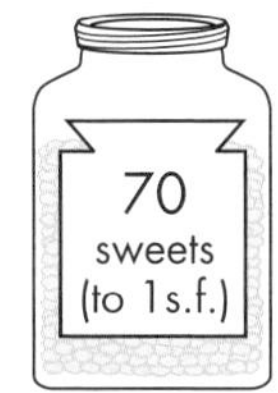

b

c

4 Write down the smallest and the greatest numbers of people that might live in these towns.

Elsecar	population 800 (to one significant figure)
Hoyland	population 1000 (to one significant figure)
Barnsley	population 200 000 (to one significant figure)

PS 5 A joiner estimates that he has 20 pieces of skirting board in stock. He is correct to one significant figure. He uses three pieces and now has 10 left to one significant figure.

How many could he have had to start with?
Work out all possible answers.

AU 6 There are 500 fish in a pond to one significant figure.

What is the least possible number of fish that could be taken from the pond so that there are 400 fish in the pond to one significant figure?

E D C

Approximation of calculations

How would you approximate the value of a calculation? What would you actually do when you try to approximate an answer to a problem?

For example, what is the approximate answer to 35.1×6.58?

To find the approximate answer, you simply round each number to 1 significant figure, then complete the calculation. So in this case, the approximation is:

$35.1 \times 6.58 \approx 40 \times 7 = 280$

Note the symbol $\approx$ which means 'approximately equal to'.

For the division $89.1 \div 2.98$, the approximate answer is $90 \div 3 = 30$.

Sometimes when dividing it can be sensible to round to 2 sf instead of 1 sf. For example,

$24.3 \div 3.87$ using $24 \div 4$ gives an approximate answer of 6

whereas

$24.3 \div 3.87$ using $20 \div 4$ gives an approximate answer of 5.

Both of these answers would be acceptable in the GCSE examination as they are both sensible answers, but generally rounding to one significant figure is easier.

A quick approximation is always a great help in any calculation since it often stops you giving a silly answer.

EXERCISE 5J

D

1 Find approximate answers to the following.

a 5435×7.31 **b** 5280×3.211 **c** $63.24 \times 3.514 \times 4.2$

d 3508×2.79 **e** $72.1 \times 3.225 \times 5.23$ **f** $470 \times 7.85 \times 0.99$

g $354 \div 79.8$ **h** $36.8 \div 1.876$ **i** $5974 \div 5.29$

Check your answers on a calculator to see how close you were.

2 Find the approximate monthly pay of the following people whose annual salaries are given.

a Paul £35 200 **b** Michael £25 600

c Jennifer £18 125 **d** Ross £8420

3 Find the approximate annual pay of the following people who earn:

a Kevin £270 a week **b** Malcolm £1528 a month **c** David £347 a week

AU 4 A farmer bought 2713 kg of seed at a cost of £7.34 per kg. Find the approximate total cost of this seed.

D

5 By rounding, find an approximate answer to each of the following.

a $\frac{573 + 783}{107}$ **b** $\frac{783 - 572}{24}$ **c** $\frac{352 + 657}{999}$ **d** $\frac{1123 - 689}{354}$

e $\frac{589 + 773}{658 - 351}$ **f** $\frac{793 - 569}{998 - 667}$ **g** $\frac{354 + 656}{997 - 656}$ **h** $\frac{1124 - 661}{355 + 570}$

i $\frac{28.3 \times 19.5}{97.4}$ **j** $\frac{78.3 \times 22.6}{3.69}$ **k** $\frac{3.52 \times 7.95}{15.9}$ **l** $\frac{11.78 \times 77.8}{39.4}$

C

6 Find the approximate answer to each of the following.

a $208 \div 0.378$ **b** $96 \div 0.48$ **c** $53.9 \div 0.58$

d $14.74 \div 0.285$ **e** $28.7 \div 0.621$ **f** $406.9 \div 0.783$

Check your answers on a calculator to see how close you were.

7 A litre of paint will cover an area of about 8.7 m^2. Approximately how many litre cans will I need to buy to paint a room with a total surface area of 73 m^2?

8 By rounding, find the approximate answer to each of the following.

a $\frac{84.7 + 12.6}{0.483}$ **b** $\frac{32.8 \times 71.4}{0.812}$ **c** $\frac{34.9 - 27.9}{0.691}$ **d** $\frac{12.7 \times 38.9}{0.42}$

FM 9 It took me 6 hours and 40 minutes to drive from Sheffield to Bude, a distance of 295 miles. My car uses petrol at the rate of about 32 miles per gallon. The petrol cost £3.51 per gallon.

a Approximately how many miles did I travel each hour?

b Approximately how many gallons of petrol did I use in going from Sheffield to Bude?

c What was the approximate cost of all the petrol for my journey to Bude and back again?

PS 10 Kirsty arranges for magazines to be put into envelopes. She sorts out 178 magazines between 10.00 am and 1.00 pm. Approximately how many magazines will she be able to sort in a week in which she works for 17 hours?

11 An athlete runs 3.75 km every day. Approximately how far does he run in:

a a week **b** a month **c** a year?

AU 12 1 kg = 1000 g

A box full of magazines weighs 8 kg. One magazine weighs about 15 g. Approximately how many magazines are there in the box?

13 An apple weighs about 280 grams.

a What is the approximate weight of a bag containing a dozen apples?

b Approximately how many apples will there be in a sack weighing 50 kg?

Sensible rounding

Sensible rounding is simply writing or saying answers to questions which have a real-life context so that the answer makes sense and is the sort of thing someone might say in a normal conversation.

For example:

The distance from Rotherham to Sheffield is 9 miles is a sensible statement.

The distance from Rotherham to Sheffield is 8.7864 miles is not sensible.

6 tins of paint is sensible.

5.91 tins of paint is not sensible.

As a general rule if it sounds sensible it will be acceptable.

EXERCISE 5K

D

1 Round each of the following to give sensible answers.

- **a** I am 1.7359 m tall.
- **b** It took me 5 minutes 44.83 seconds to mend the television.
- **c** My kitten weighs 237.97 g.
- **d** The correct temperature at which to drink Earl Grey tea is 82.739 °C.
- **e** The distance from Wath to Sheffield is 15.528 miles.
- **f** The area of the floor is 13.673 m^2.

FM **2** Rewrite the following article using sensible amounts.

It was a hot day, the temperature was 81.699 °F and still rising. I had now walked 5.3289 km in just over 113.98 minutes. But I didn't care since I knew that the 43 275 people watching the race were cheering me on. I won by clipping 6.2 seconds off the record time. This was the 67th time the race had taken place since records first began in 1788. Well, next year I will only have 15 practice walks beforehand as I strive to beat the record by at least another 4.9 seconds.

C

AU **3** A lorry load of scrap metal weighs 39.715 tonnes.
It is worth £20.35 per tonne.

Approximately how much is the load worth?

PS **4** The accurate temperature is 18.2 °C.
David rounds the temperature to the nearest 5 °C.
David says the temperature is about 20 °C.

How much would the temperature rise for David to say that the temperature is about 25 °C?

GRADE BOOSTER

- **F** You can multiply a three-digit number by a two-digit number without using a calculator
- **F** You can divide a three- or four-digit number by a two-digit number
- **F** You can solve real problems involving multiplication and division
- **E** You can round decimal numbers to a specific number of places
- **E** You can evaluate calculations involving decimal numbers
- **E** You can multiply and divide negative numbers
- **E** You can round numbers to one significant figure
- **D** You can estimate the approximate value of a calculation before calculating
- **C** You can give sensible answers to real-life questions

What you should know now

- How to do long multiplication
- How to do long division
- How to perform calculations with decimal numbers
- How to round to a specific number of decimal places
- How to round to a specific number of significant figures
- How to multiply and divide with negative numbers
- How to make estimates by suitable rounding

1

Gift shop

Price list

Key ring	£3.20
Hat	£3.99
Pencil case	£2.70
Ruler	45p
Pen	60p
Pencil	

Keith buys 3 pens.

a Work out the total cost. (2)

Simon buys a pencil case, a ruler and a pen.

He pays with a £5 note.

b Work out how much change he should get. (3)

The gift shop also sells pencils.

The price of a pencil is $\frac{2}{3}$ of the price of a pen.

c Work out the price of a pencil. (2)

(Total 7 marks)

Edexcel, November 2008, Paper 1 Foundation, Question 9

2 Work out the following. Be careful as they are a mixture of addition, subtraction, multiplication and division problems. Decide what the calculation is and use a column method to work it out.

a How much change do I get from a £20 note if I spend £13.45?

b I buy three pairs of socks at £2.46 each. How much do I pay altogether?

c A burger costs £1.65, fries cost 98p and a drink is 68p. How much will a burger, fries and a drink cost altogether?

3 A coach costs £345 to hire.

There are 55 seats on the coach.

Work out a sensible seat price assuming at least 45 seats are sold.

4 450 people go on a football trip. Each coach will seat 54 people.

a How many coaches are needed?

b How many seats will be empty?

5 Josh buys 40 litres of milk.

The total cost is £33.20

Work out the cost of 1 litre of the milk. (3)

(Total 3 marks)

Edexcel, June 2008, Paper 13 Foundation, Question 4

6 Estimate the value of these expressions.

a 15.7×29.2

b 143.1×17.8

7 **a** Work out $13 \times 17 - 11 \times 17$

b Find an approximate value of $\frac{51 \times 250}{82}$

You **must** show **all** your working.

8 Work out 25.6×1.6

You **must** show **all** your working. (3)

(Total 3 marks)

Edexcel, June 2007, Paper 10 Foundation, Unit 3 Test, Question 8

9 Three pupils use calculators to work out

$$\frac{32.7 + 14.3}{1.28 - 0.49}$$

Arnie gets 43.4, Bert gets 36.2 and Chuck gets 59.5. Use approximations to show which one of them is correct.

10 Use approximations to estimate the value of

$$\sqrt{\frac{323\,407}{0.48}}$$

11 Work out an estimate for

$$\frac{29.8 \times 4.1}{0.21}$$ (3)

(Total 3 marks)

Edexcel, November 2007, Paper 10 Foundation, Unit 3 Test, Question 10

Worked Examination Questions

FM **1** In a **survey** the number of visitors to a website was recorded daily. Altogether it was visited 30 million times. Each day it was visited 600 000 times.

Based on this information, for how many days did the survey last?

$$30\,000\,000 \div 600\,000 = \frac{30\,000\,000}{600\,000} = \frac{300}{6} = 50 \text{ days}$$

You need to find the number of days using the following calculation:
total times visited ÷ number of visits a day
You get 1 method mark for showing the calculation.

You can cancel the five zeros on the top and the bottom.

Check that your final answer is sensible. You get 1 accuracy mark for the correct answer of 50 days.

Total: 2 marks

PS **2** A theatre has 400 tickets available each night for a show on Friday and Saturday nights. The manager of the theatre wants to raise £1500 from ticket sales each night. She expects to sell all the tickets for Saturday but only three-quarters of the tickets for Friday. If the ticket price is the same for both days, how much should she charge?

$\frac{3}{4} \times 400 = 300$

You get 1 method mark for working out that she sells $\frac{3}{4}$ of 400 tickets on Friday.

300 + 400 = 700 tickets altogether sold

£3000 needed

You get 1 mark for realising that £3000 (£1500 × 2) needs to raised.

3000 ÷ 700 = 4.28

For attempting to work out the exact amount per seat (3000 ÷ 700 or £4.28) you get 1 method mark.

Round up to ensure a profit,
so for example £4.50 or £5 per ticket

Give a sensible price (e.g. to nearest 10p or with a built-in profit margin) to get 1 mark for accuracy.

Total: 4 marks

5 Functional Maths
In the gym

Healthy eating and regular exercise form part of a balanced lifestyle. Health clubs and gyms are a popular way to keep fit. Exercise machines in gyms often show how many calories are burnt off during a workout.

This activity investigates the relationship between calorie intake and burn-off rates.

Remember

We must not burn off all the calories that we consume: the body needs calories as energy.

Getting started

Before you begin your main task, use the information given on these pages to work out:

- the weight category for each person
- the calories they take in at breakfast and the amount they burn off through exercise.

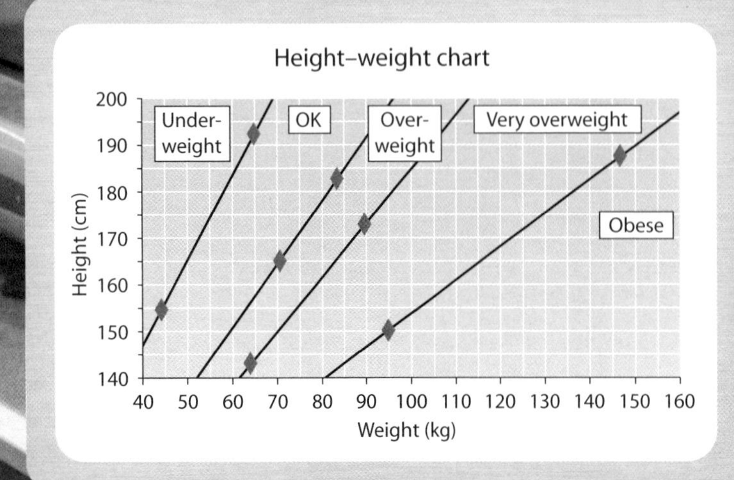

Facts

Calories per 100 g

Apple	46
Bacon	440
Banana	76
Bread	246
Butter	740
Cornflakes	370
Eggs	148
Porridge	368
Sausages	186
Yoghurt	62
Apple juice (100 ml)	41
Orange juice (100 ml)	36
Semi-skimmed milk (100 ml)	48
Skimmed milk (100 ml)	34
Sugar (1 teaspoonful)	20
Tea or coffee (black)	0

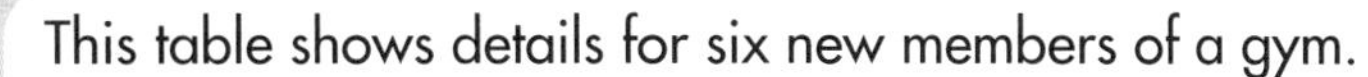

This table shows details for six new members of a gym.

Person	Height & weight	Breakfast	Exercise machine
Lynn	height 162 cm weight 60 kg	150 g toast 20 g butter 250 ml orange juice	cross-trainer (9 cal/min) for 15 minutes
Jess	height 168 cm weight 82 kg	50 g bacon 200 g sausages 50 g eggs tea with 40 ml semi-skimmed milk	walking (6 cal/min) for 30 minutes
Andy	height 180 cm weight 75 kg	100 g apple 100 g banana 150 g yoghurt 200 ml apple juice	running (8 cal/min) for 20 minutes
Dave	height 192 cm weight 95 kg	50 g bacon 150 g bread 20 g butter 200 ml semi-skimmed milk	exercise bike (7 cal/min) for 20 minutes
Pete	height 175 cm weight 60 kg	30 g cornflakes 300 ml semi-skimmed milk tea with 40 ml skimmed milk 2 teaspoons sugar	step machine (9 cal/min) for 15 minutes
Ola	height 165 cm weight 45 kg	75 g porridge 300 ml semi-skimmed milk black coffee 1 teaspoon sugar	rowing (8 cal/min for 20 minutes

Your task

Design a healthy breakfast and then look at how long it would take to burn off the calories contained in this breakfast on each exercise machine.

Use these points to help you in your task:

- Think about the different breakfast options that you find in a canteen, at home or in a cafe.
- Try to calculate the calories in some of these breakfast options.
- Now think about how much exercise you should do to burn off these calories.

Why this chapter matters

Simon Stevin was a Dutch mathematician. He was born in Bruges (now in Belgium) in 1584. He added greatly to the study of several areas of mathematics, including trigonometry and mechanics, the study of motion. He was also influential in the fields of geography, navigation, architecture and musical theory.

Simon Stevin has several other claims to fame.

- He invented a carriage with sails, a little model of which was preserved at Scheveningen until 1802. The carriage itself had been lost long before, but we know that in about 1600 Stevin, with Prince Maurice of Orange and 26 others, used it on the seashore between Scheveningen and Petten. It moved only by the force of the wind and, at its top speed, it was just a bit quicker than horses.
- He had the idea of explaining the tides by the attraction of the moon.

His greatest success, however, was a small pamphlet, first published in Dutch in 1585, in which he put forward the use of a decimal system for measurement throughout. He declared that the universal introduction of decimal coinage, measures and weights was only a question of time.

In the UK, we used imperial units such as, feet, inches, ounces and pounds for many years. Your grandparents will remember them well.

In the later part of the 20th century, the British Government started trying to ensure that we all use metric units here, as much as possible. This is now common practice, apart for the use of the **mile**.

The metric system is now predominantly used in the UK and most of the world, apart from the USA.

Road signs in the UK give distances in miles, but in Europe distances are given in kilometres.

A brief history of the metric system

1585	Simon Stevin suggested a decimal system for weights and measures.
1790	Thomas Jefferson proposed a decimal-based measurement system for the USA. A vote in the USA congress to replace the UK imperial system by a metric system was lost by just one vote.
1795	The metric system became the official system of measurement in France.
1959	The UK and USA redefined the inch to be 2.54 cm.
1963	The UK redefined the pound to be 0.453 592 37 kilograms.
1985	The UK redefined the gallon to be 4.546 09 litres.
Now	The metric system has been adopted by virtually every country, with the only notable exception being the USA.

It is vital that you understand and can use the metric system, but you should also have some understanding of the older imperial units. This will help you to understand what your grandparents sometimes say and the signs that are still in imperial units in the UK today.

Chapter

Measures: Systems of measurement

The grades given in this chapter are target grades.

This chapter will show you ...

- **G** how to read scales
- **G** which units to use when measuring length, weight and capacity
- **F** how to make estimations
- **F** how to convert from one metric unit to another
- **F** to **C** how to convert from imperial units to metric units
- **E** how to convert from one imperial unit to another

Visual overview

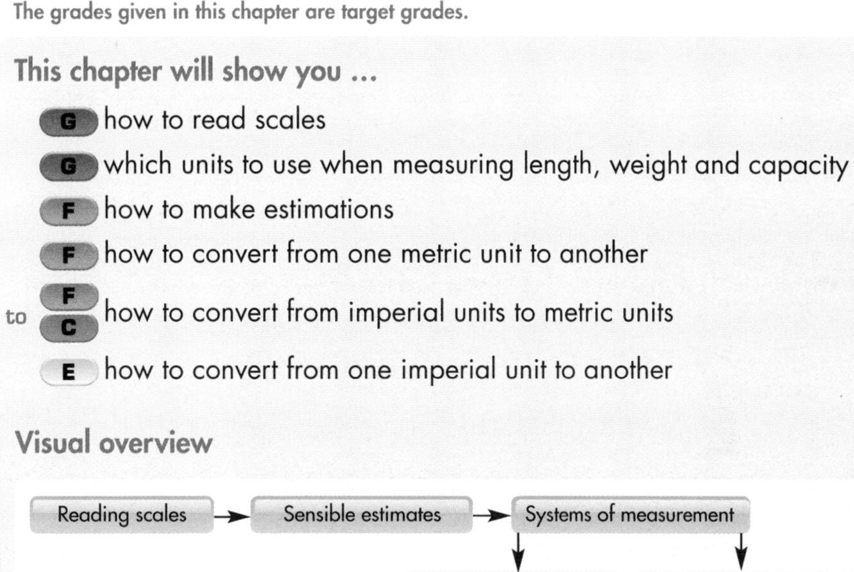

What you should already know

- The names of common 3D shapes **(KS3 level 3, GCSE grade G)**
- How to measure lengths of lines **(KS3 level 4, GCSE grade G)**
- The metric units for length **(KS3 level 4, GCSE grade G)**

Quick check

Name these 3D shapes.

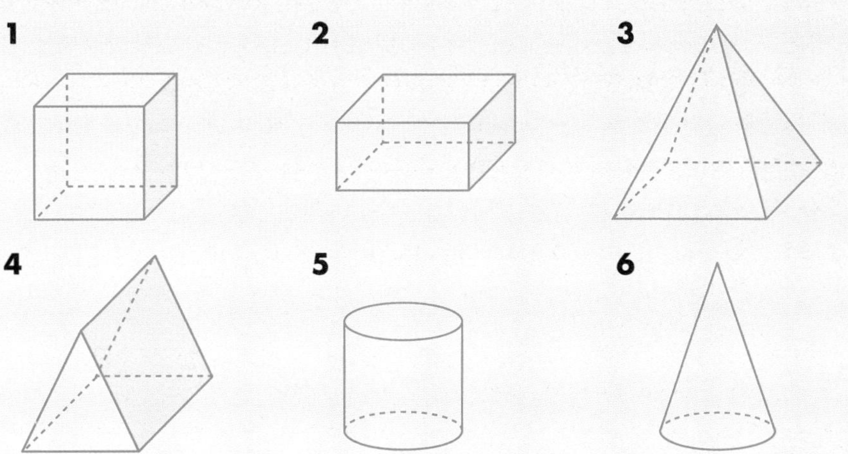

6.1 Reading scales

This section will show you how to:
- read and interpret scales

Key words
divisions
scales
units

You will come across **scales** in a lot of different places.

For example, there are scales on thermometers, car speedometers and weighing scales. It is important that you can read scales accurately.

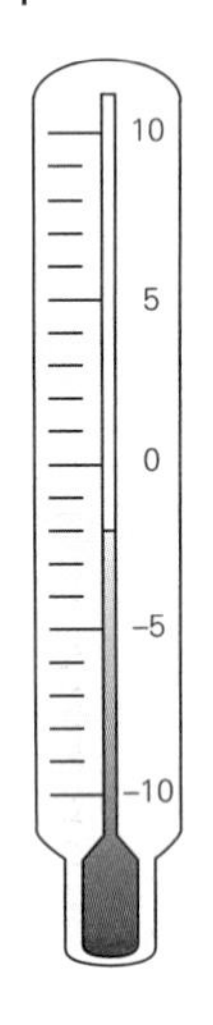

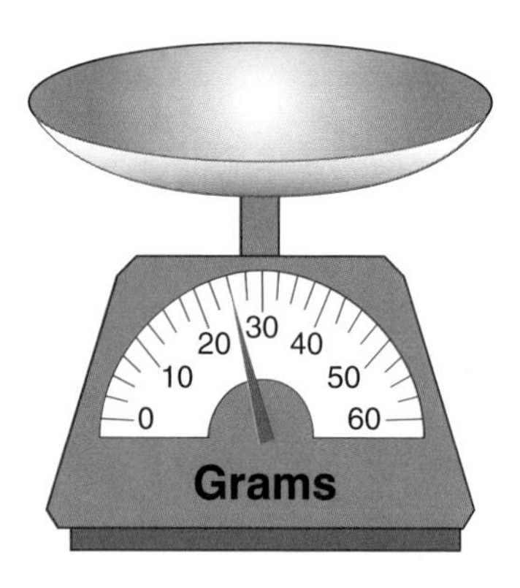

There are two things to do when reading a scale. First, make sure that you know what each **division** on the scale represents. Second, make sure you read the scale in the right direction, for example some scales read from right to left.

Also, make sure you note the **units**, if given, and include them in your answer.

EXAMPLE 1

Read the values from the following scales.

a 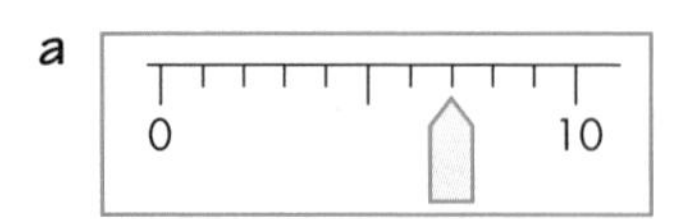

b 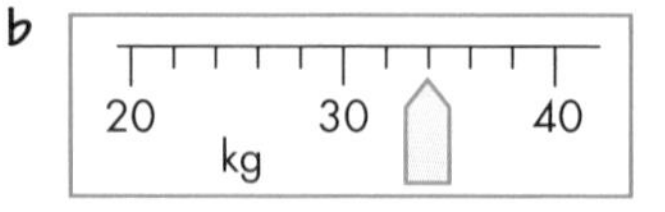

c 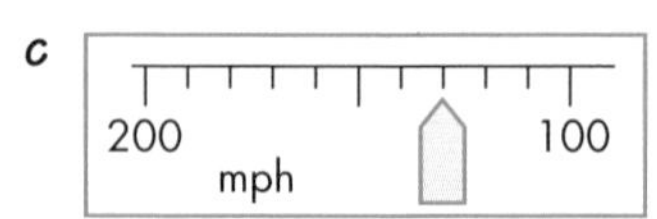

a The scale shows 7. This is a very straightforward scale. It reads from left to right and each division is worth 1 unit.

b The scale shows 34 kg. The scale reads from left to right and each division is worth 2 units.

c The scale shows 130 mph. The scale reads from right to left and each division is worth 10 units. You should know that mph stands for miles per hour. This is a unit of speed found on most British car speedometers.

EXERCISE 6A

1 Read the values from the following scales.
Remember to state the units if they are shown.

> **HINTS AND TIPS**
>
> Remember to check what each division is worth.

a **i**

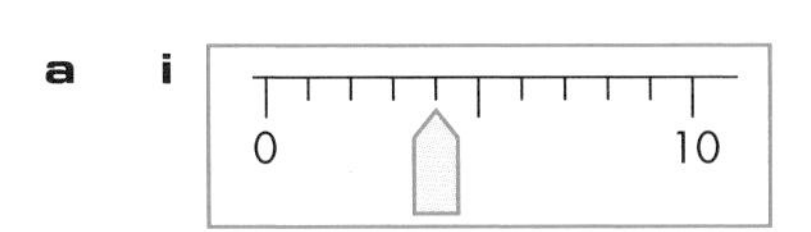

ii

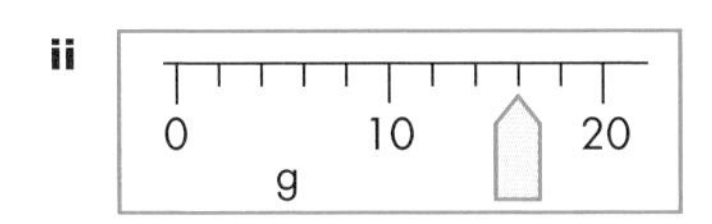

iii

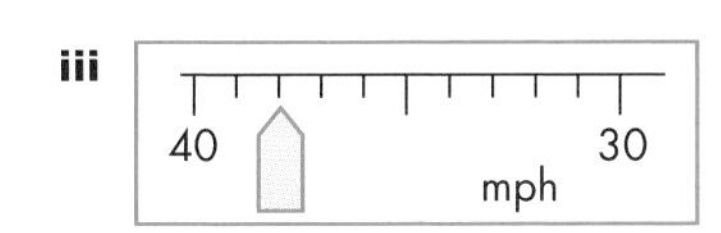

b **i**

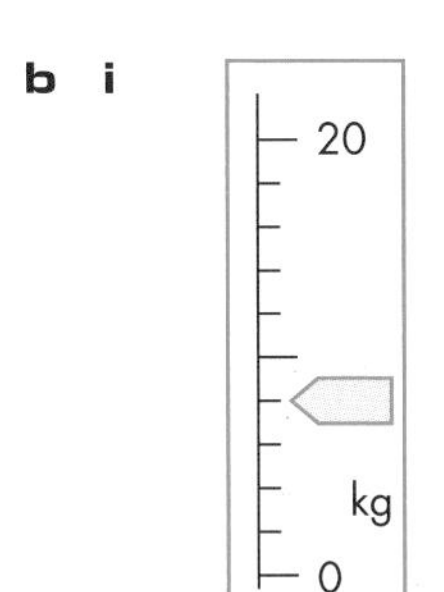

ii

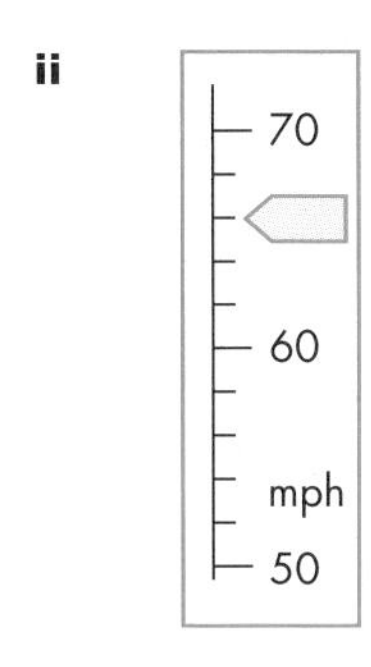

iii

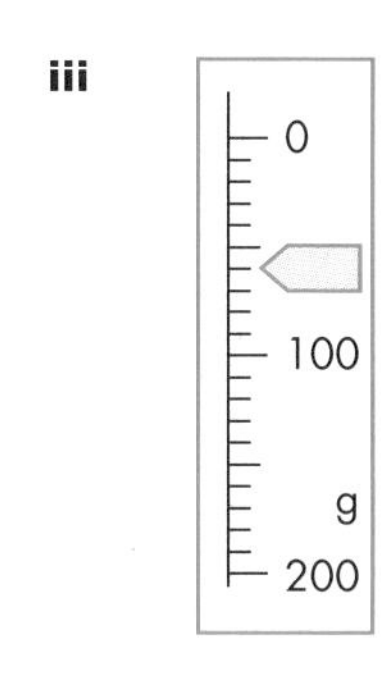

c **i**

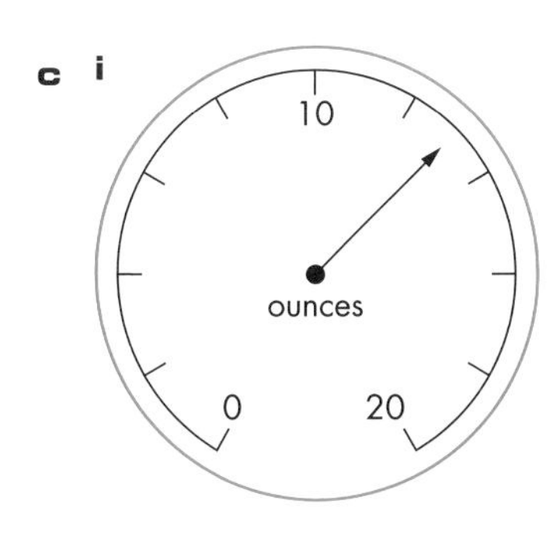

ii

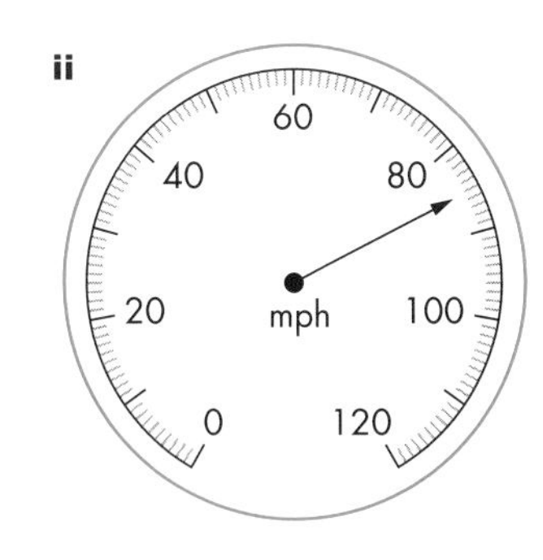

iii

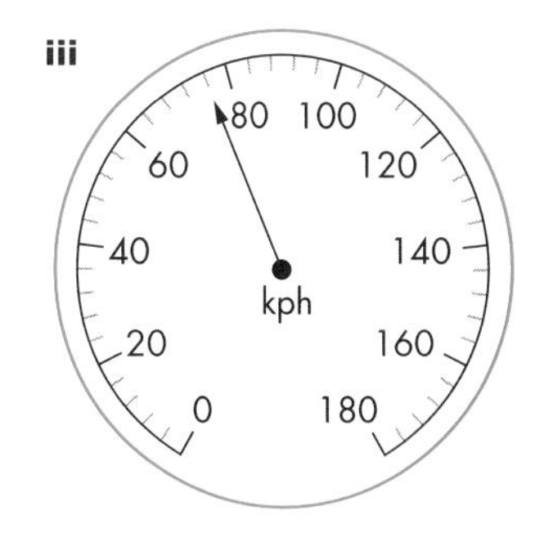

d **i**

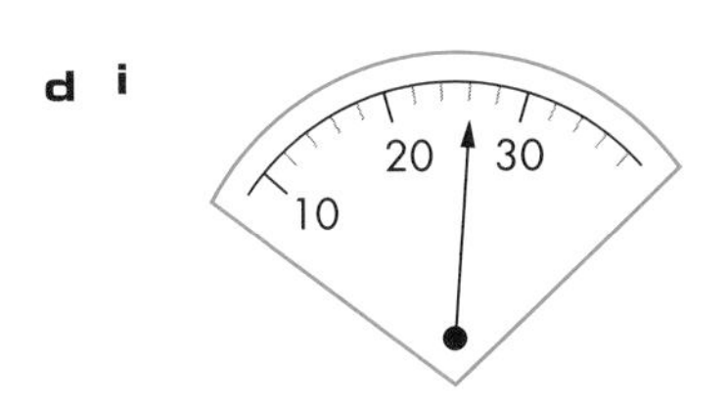

ii

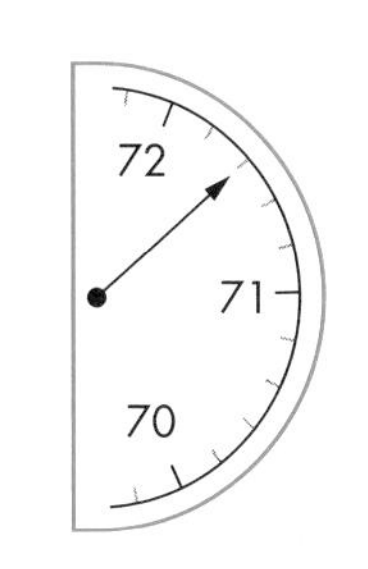

iii

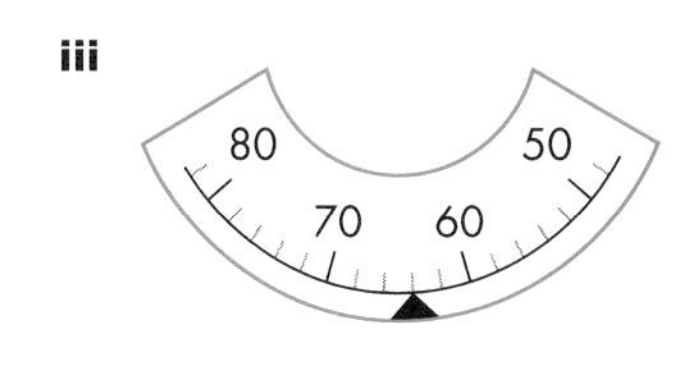

2 Copy (or trace) the following dials and mark on the values shown.

a

7 kg

b

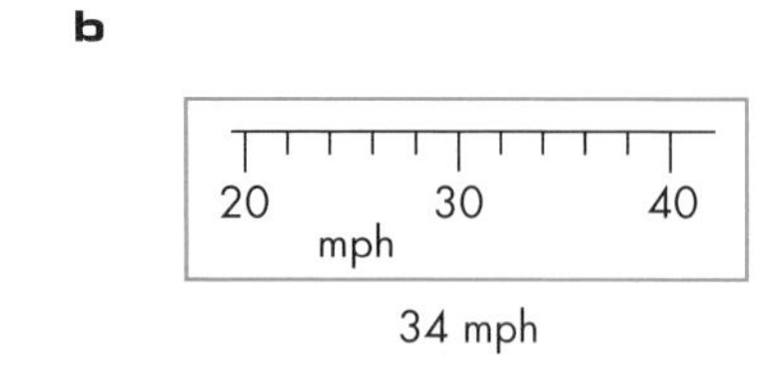

34 mph

c

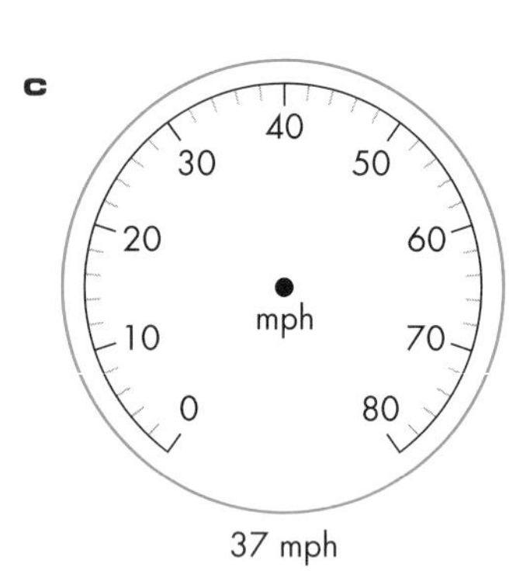

37 mph

d

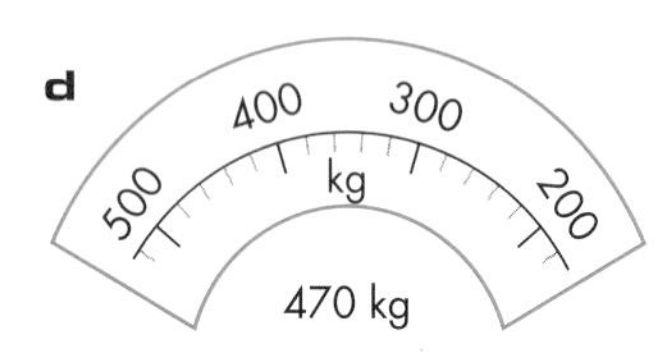

470 kg

e

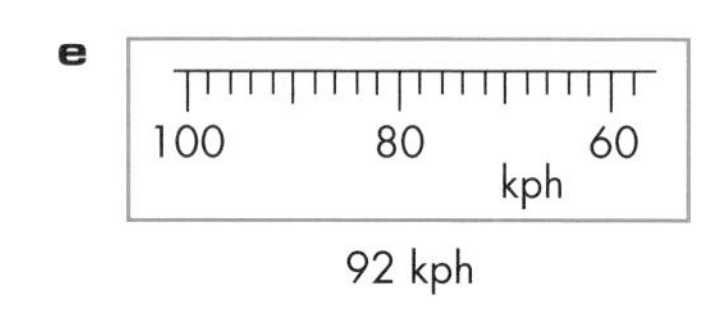

92 kph

f

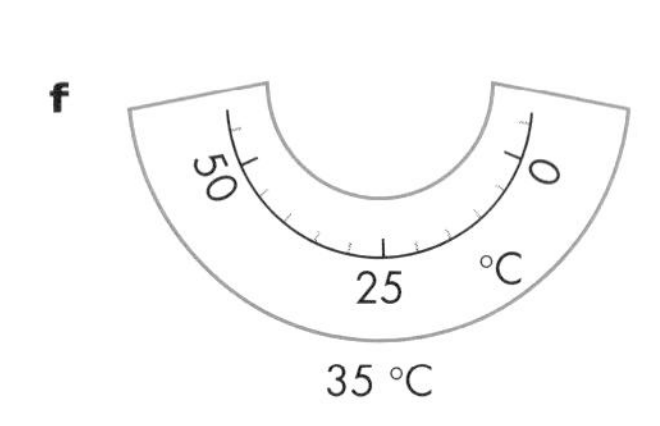

35 °C

3 Read the temperatures shown by each of these thermometers.

a

b

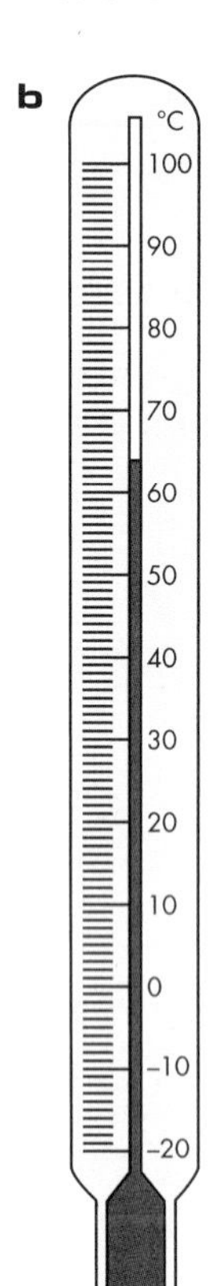

c

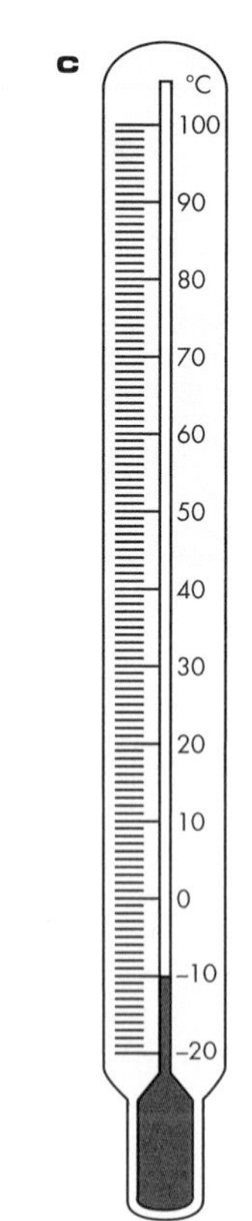

d

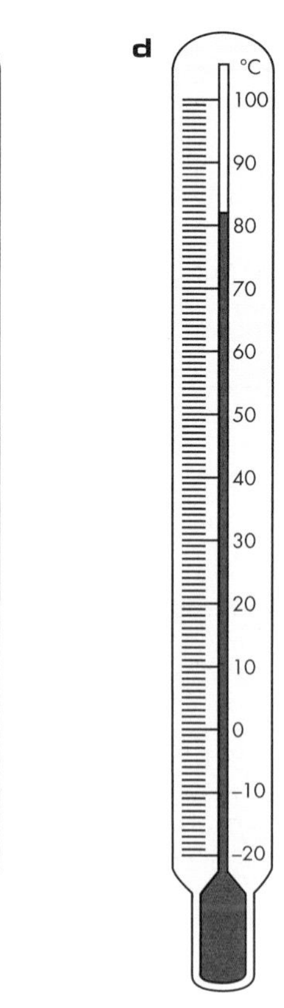

e

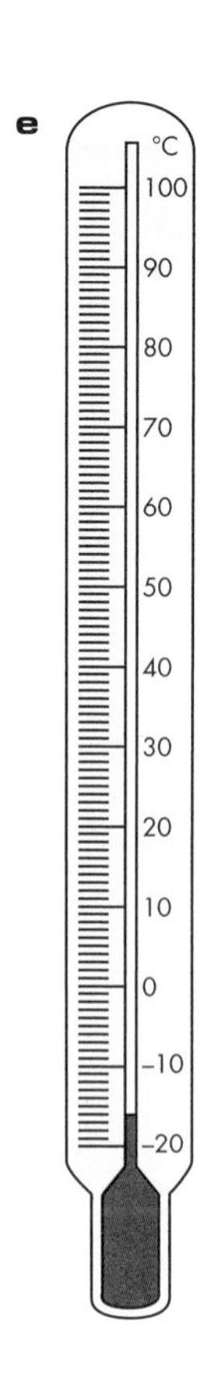

G

4 Read the values shown on these scales.

a

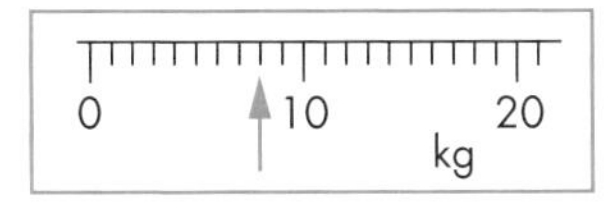

b

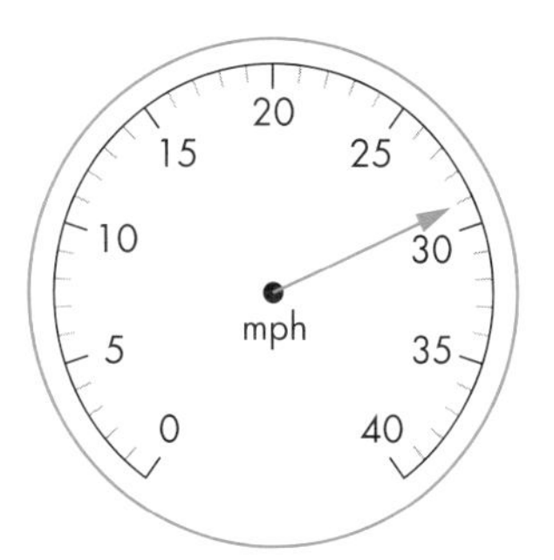

c

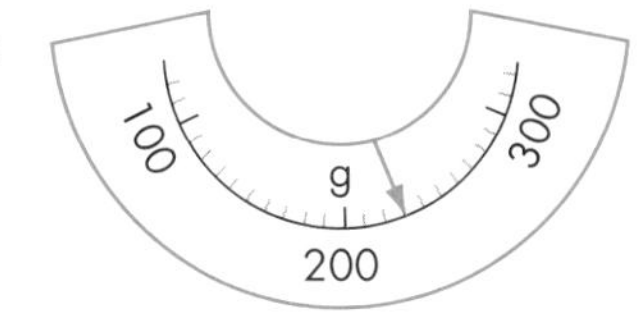

d

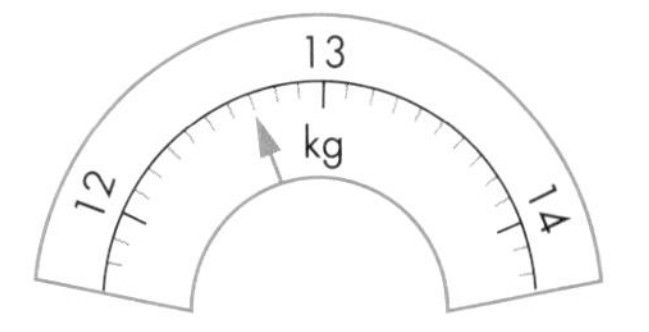

5 Susie is using kitchen scales to weigh out flour.

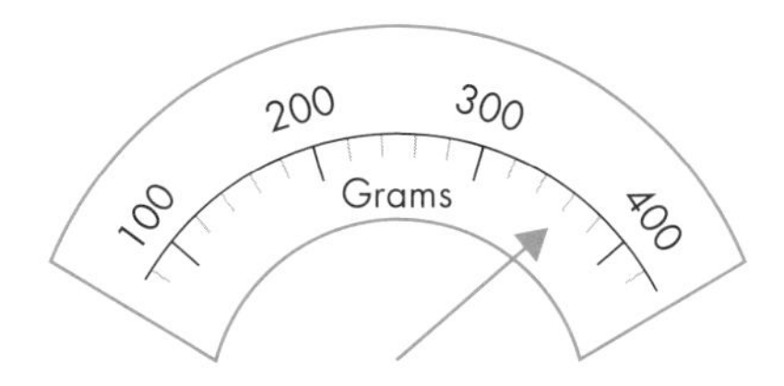

a What is the weight of the flour shown on the scales?

b These scales can weigh items up to 400 g. Susie needs to weigh 700 g of currants using these scales. Explain how she could do this.

F

6 A pineapple was weighed.

A pineapple and an orange were weighed together.

a How much does the pineapple weigh?
Give your answer in kilograms.

b How much does the orange weigh?
Give your answer in grams.

F

PS 7 **a** What speed is shown on the scale?

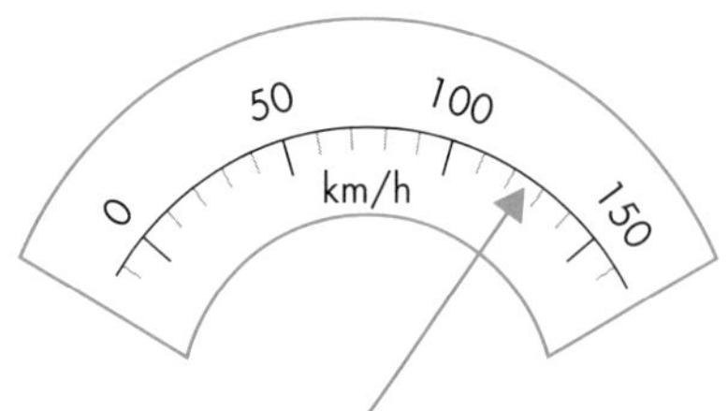

b Copy this scale and show the same speed as in part **a**.

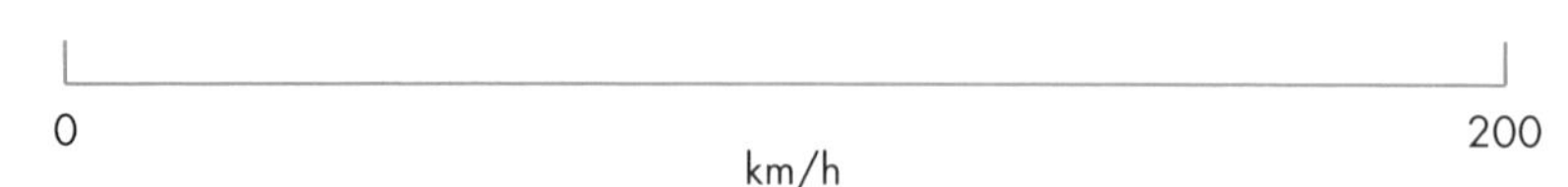

AU 8 Dean says that the arrow on this scale is pointing to 7.49 m.

Dean is not correct. Explain the mistake that he has made.

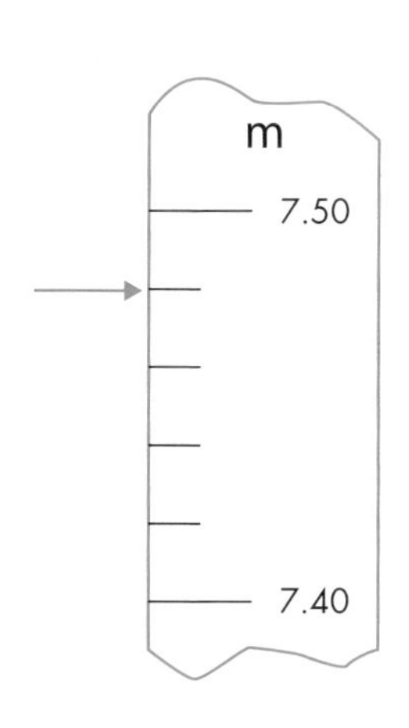

6.2 Sensible estimates

This section will show you how to:

- make sensible estimates using standard measures

Key word
estimate

The average height of a man is 1.78 m. For a sensible **estimate**, we would usually say that the height of a man is about 1.8 m. From this information, you can estimate the lengths or heights of other objects.

EXAMPLE 2

Look at the picture.

Estimate, in metres, the height of the lamppost and the length of the bus.

Assume the man is about 1.8 m tall. The lamppost is about three times as high as he is. **Note:** One way to check this is to use tracing paper to mark off the height of the man and then measure the other lengths against this. This makes the lamppost about 5.4 m high, or close to 5 m high. As it is an estimate, there is no need for an exact value.

The bus is about four times as long as the man so the bus is about 7.2 m long, or close to 7 m long.

EXAMPLE 3

Look at the picture.

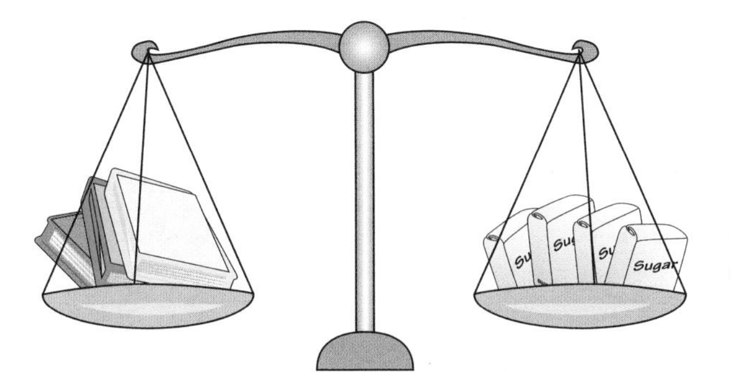

It shows three maths textbooks balanced by four bags of sugar. Estimate the mass of one textbook.

You should know that a bag of sugar weighs 1 kilogram, so the three maths books weigh 4000 grams. This means that each one weighs about 1333 grams or about 1.3 kg.

EXERCISE 6B

F

1 The car in the picture is 4 metres long. Use this to estimate the length of the bicycle, bus and the train.

2 Estimate the greatest height and length of the whale.

HINTS AND TIPS

Remember, a man is about 1.8 m tall.

3 Estimate the weight of one apple.

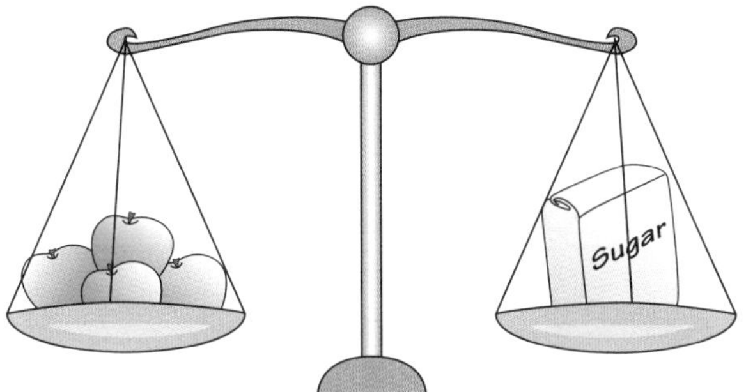

HINTS AND TIPS

Remember, a bag of sugar weighs 1 kg.

4

Estimate the following.

a the height of the traffic lights

b the width of the road

c the height of the flagpole

5

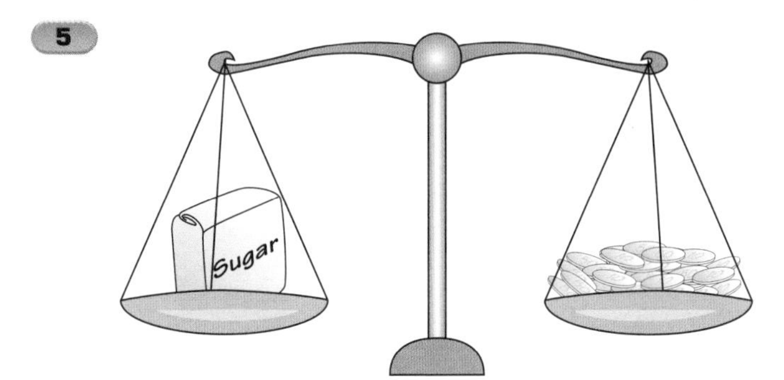

A charity collection balances pound coins against a bag of sugar. It take £105 to balance the bag of sugar. Estimate the weight of one pound coin.

F

6 This is an illustration of Joel standing next to a statue.

Joel's height is 1.5 m. Explain how he could use this information to estimate the height of the statue.

AU 7 Estimate the height of Tyrannosaurus Rex.

HINTS AND TIPS

Remember, a man is about 1.8 m tall.

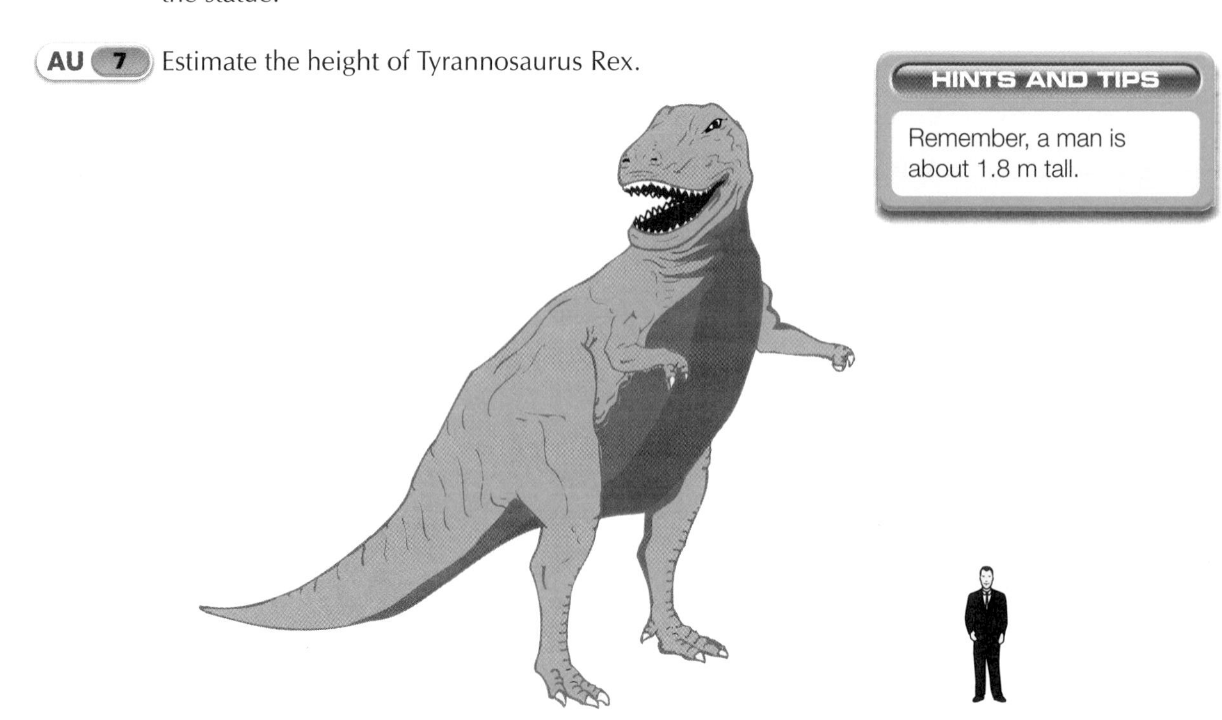

6.3 Systems of measurement

This section will show you how to:

- decide which units to use when measuring length, weight and capacity

Key words
capacity
imperial
length
metric
volume
weight

There are two systems of measurement currently in use in Britain: the **imperial** system and the **metric** system.

The imperial system is based on traditional units of measurement, many of which were first introduced several hundred years ago. It is gradually being replaced by the metric system, which is used throughout Europe and in many other parts of the world.

The main disadvantage of the imperial system is that it has a lot of awkward conversions, such as 12 inches = 1 foot. The metric system has the advantage that it is based on powers of 10, namely 10, 100, 1000 and so on, so it is much easier to use when you do calculations.

It will be many years before all the units of the imperial system disappear, so you have to know units in both systems.

System	Unit	How to estimate it
	Length	
Metric system	1 metre	A long stride for an average person
	1 kilometre	Two and a half times round a school running track
	1 centimetre	The distance across a fingernail
Imperial system	1 foot	The length of an A4 sheet of paper
	1 yard	From your nose to your fingertips when you stretch out your arm
	1 inch	The length of the top joint of an adult's thumb
	Weight	
Metric system	1 gram	A 1p coin weighs about 4 grams
	1 kilogram	A bag of sugar
	1 tonne	A saloon car
Imperial system	1 pound	A jar full of jam
	1 stone	A bucket three-quarters full of water
	1 ton	A saloon car
	Volume/Capacity	
Metric system	1 litre	A full carton of orange juice
	1 centilitre	A small wine glass is about 10 centilitres
	1 millilitre	A full teaspoon is about 5 millilitres
Imperial system	1 pint	A full bottle of milk
	1 gallon	A half-full bucket of water

Volume and capacity

The term 'capacity' is normally used to refer to the volume of a liquid or a gas.

For example, when referring to the volume of petrol that a car's fuel tank will hold, people may say its capacity is 60 litres or 13 gallons.

In the metric system, there is an equivalence between the units of capacity and volume, as you can see on page 161.

EXAMPLE 4

Choose an appropriate metric unit for each of the following.

a your own height
b the thickness of this book
c the distance from home to school
d your own weight
e the weight of a coin
f the weight of a bus
g a large bottle of lemonade
h a dose of medicine
i a bottle of wine

a metres or centimetres
b millimetres
c kilometres
d kilograms
e grams
f tonnes
g litres
h millilitres
i centilitres

EXERCISE 6C

G

1 Decide the metric unit you would be most likely to use to measure each of the following.

a The height of your classroom
b The distance from London to Barnsley
c The thickness of your little finger
d The weight of this book
e The amount of water in a fish tank
f The weight of water in a fish tank
g The weight of an aircraft
h A spoonful of medicine
i The amount of wine in a standard bottle
j The length of a football pitch
k The weight of your head teacher
l The amount of water in a bath
m The weight of a mouse
n The amount of tea in a teapot
o The thickness of a piece of wire

F

2 Estimate the approximate metric length, weight or capacity of each of the following.

- **a** This book (both length and weight)
- **b** The length of your school hall
- **c** The capacity of a milk bottle
- **d** A brick (length, width and weight)
- **e** The diameter of a 10p coin, and its weight
- **f** The distance from your school to Manchester
- **g** The weight of a cat
- **h** The amount of water in one raindrop
- **i** The dimensions of the room you are in
- **j** Your own height and weight

FM 3 Bob was asked to put up some decorative bunting from the top of each lamp post in his street. He had three sets of ladders he could use, a two metre, a 3.5 metre and a five metre ladder.

He looked at the lamp posts and estimated that they were about three times his height. He is slightly below average height for an adult male.

Which of the ladders should he use? Give a reason for your choice.

AU 4 The distance from Bournemouth to Basingstoke is shown on a website as 55 miles.

Explain why this unit is used instead of inches, feet or yards.

6.4 Metric units

This section will show you how to:

- convert from one metric unit to another

Key words

centilitre (cl)
centimetre (cm)
gram (g)
kilogram (kg)
kilometre (km)
litre (l)
metre (m)
millilitre (ml)
millimetre (mm)
tonne (t)

You should already know the relationships between these metric units.

Length

10 **millimetres** = 1 **centimetre**

1000 millimetres = 100 centimetres
= 1 **metre**

1000 metres = 1 **kilometre**

Weight

1000 **grams** = 1 **kilogram**

1000 kilograms = 1 **tonne**

Capacity

10 **millilitres** = 1 **centilitre**

1000 millilitres = 100 centilitres
= 1 **litre**

Volume

1000 litres = 1 metre^3

1 millilitre = 1 centimetre^3

Note the equivalence between the units of capacity and volume:

1 litre = 1000 cm^3 which means 1 ml = 1 cm^3

You need to be able to convert from one metric unit to another.

Since the metric system is based on powers of 10, you should be able easily to multiply or divide to change units. Work through the following examples.

EXAMPLE 5

To change *small* units to *larger* units, always *divide*.

Change:

a 732 cm to metres

$732 \div 100 = 7.32$ m

b 410 mm to centimetres

$410 \div 10 = 41$ cm

c 840 mm to metres

$840 \div 1000 = 0.84$ m

d 450 cl to litres

$450 \div 100 = 4.5$ l

EXAMPLE 6

To change *large* units to *smaller* units, always *multiply*.

Change:

a 1.2 m to centimetres

$1.2 \times 100 = 120$ cm

b 0.62 cm to millimetres

$0.62 \times 10 = 6.2$ mm

c 3 m to millimetres

$3 \times 1000 = 3000$ mm

d 75 cl to millilitres

$75 \times 10 = 750$ ml

EXERCISE 6D

1 Fill in the gaps, using the information in this section.

a 125 cm = … m
b 82 mm = … cm
c 550 mm = … m
d 2100 m = … km
e 208 cm = … m
f 1240 mm = … m
g 4200 g = … kg
h 5750 kg = … t
i 85 ml = … cl
j 2580 ml = … l
k 340 cl = … l
l 600 kg = … t
m 755 g = … kg
n 800 ml = … l
o 200 cl = … l
p 630 ml = … cl
q 8400 l = … m^3
r 35 ml = … cm^3
s 1035 l = … m^3
t 530 l = … m^3
u 34 km = … m

2 Fill in the gaps, using the information in this section.

a 3.4 m = … mm
b 13.5 cm = … mm
c 0.67 m = … cm
d 7.03 km = … m
e 0.72 cm = … mm
f 0.25 m = … cm
g 0.64 km = … m
h 2.4 l = … ml
i 5.9 l = … cl
j 8.4 cl = … ml
k 5.2 m^3 = … l
l 0.58 kg = … g
m 3.75 t = … kg
n 0.94 cm^3 = … l
o 21.6 l = … cl
p 15.2 kg = … g
q 14 m^3 = … l
r 0.19 cm^3 = … ml

FM 3 Sarif was planning to do some DIY. He wanted to buy two lengths of wood, each 2 m long, and 1.5 cm by 2 cm. He went to the local store where the types of wood were described as:

2000 mm × 15 mm × 20 mm

200 mm × 15 mm × 20 mm

200 mm × 150 mm × 2000 mm

1500 mm × 2000 mm × 20 000 mm

Should he choose any of these? If so, which one?

AU 4 1 litre is equivalent to 1000 millilitres.

Referring to centimetres, explain how you know this.

PS 5 How many square millimetres are there in a square kilometre?

HINTS AND TIPS

The answer is not 1 000 000.

F

E

6.5 Imperial units

This section will show you how to:

- convert from one imperial unit to another

Key words
foot (ft)
gallon (gal)
inch (in)
mile (m)
ounce (oz)
pint (pt)
pound (lb)
stone (st)
ton (T)
yard (yd)

You need to be familiar with imperial units that are still in daily use. The main ones are:

Length	12 **inches**	=	1 **foot**
	3 feet	=	1 **yard**
	1760 yards	=	1 **mile**
Weight	16 **ounces**	=	1 **pound**
	14 pounds	=	1 **stone**
	2240 pounds	=	1 **ton**
Capacity	8 **pints**	=	1 **gallon**

Examples of the everyday use of imperial measures are:

miles for distances by road

gallons for petrol (in conversation)

feet and inches for people's heights

pints for milk

pounds for the weight of babies (in conversation)

ounces for the weight of food ingredients in a food recipe

EXAMPLE 7

- *To change large units to smaller units, always multiply.*
- *To change small units to larger units, always divide.*

Change:

a 4 feet to inches

$4 \times 12 = 48$ inches

b 5 gallons to pints

$5 \times 8 = 40$ pints

c 36 feet to yards

$36 \div 3 = 12$ yards

d 48 ounces to pounds

$48 \div 16 = 3$ pounds

EXERCISE 6E

1 Fill in the gaps, using the information in this section.

a 2 feet = … inches
b 4 yards = … feet
c 2 miles = … yards
d 5 pounds = … ounces
e 4 stone = … pounds
f 3 tons = … pounds
g 5 gallons = … pints
h 4 feet = … inches
i 1 yard = … inches
j 10 yards = … feet
k 4 pounds = … ounces
l 60 inches = … feet
m 5 stone = … pounds
n 36 feet = … yards
o 1 stone = … ounces

2 Fill in the gaps, using the information in this section.

a 8800 yards = … miles
b 15 gallons = … pints
c 1 mile = … feet
d 96 inches = … feet
e 98 pounds = … stones
f 56 pints = … gallons
g 32 ounces = … pounds
h 15 feet = … yards
i 11 200 pounds = … tons
j 1 mile = … inches
k 128 ounces = … pounds
l 72 pints = … gallons
m 140 pounds = … stones
n 15 840 feet = … miles
o 1 ton = … ounces

FM 3 Andrew was asked to do some shopping for his grandmother. She sent him out to get a two-pound bag of sugar from the market. When Andrew got to the market, the only bags of sugar that he saw were:

8-ounce bags, 16-ounce bags, 32-ounce bags and 40-ounce bags

Which bag should he take back for his grandmother?

PS 4 How many square inches are there in a square mile?

AU 5 1 kilogram is approximately 2.2 pounds.

Explain how you know that 1 ton is heavier than 1 tonne.

E

6.6 Conversion factors

This section will show you how to:
- use the approximate conversion factors to change between imperial units and metric units

Key words
conversion factor
imperial
metric

You need to know the approximate conversions between certain **imperial** units and **metric** units.

The **conversion factors** you should be familiar with are given below.

The symbol '≈' means 'is approximately equal to'.

Those you do need to know for your examination are in **bold** type.

Length	1 inch	≈	2.5 centimetres	**Weight**	1 pound	≈	450 grams
	1 foot	≈	**30 centimetres**		**2.2 pounds**	≈	**1 kilogram**
	1 mile	≈	1.6 kilometres				
	5 miles	≈	**8 kilometres**				
Capacity	1 pint	≈	570 millilitres				
	1 gallon	≈	**4.5 litres**				
	$1\frac{3}{4}$ pints	≈	**1 litre**				

EXAMPLE 8

Use the conversion factors above to find the following approximations.

a Change 5 gallons into litres.

$5 \times 4.5 \approx 22.5$ litres

b Change 45 miles into kilometres.

45×1.6 kilometres ≈ 72 kilometres

c Change 5 pounds into kilograms.

$5 \div 2.2 \approx 2.3$ kilograms (rounded to 1 decimal place)

Note: An answer should be rounded when it has several decimal places, since it is only an approximation.

EXERCISE 6F

1 Fill in the gaps to find the approximate conversions for the following. Use the conversion factors on page 168.

a 8 inches = … cm
b 6 kg = … pounds
c 30 miles = … km
d 15 gallons = … litres
e 5 pints = … ml
f 45 litres = … gallons
g 30 cm = … inches
h 80 km = … miles
i 11 pounds = … kg
j 1710 ml = … pints
k 100 miles = … km
l 56 kg = … pounds
m 40 gallons = … litres
n 200 pounds = … kg
o 1 km = … yards
p 1 foot = … cm
q 1 stone = … kg
r 1 yard = … cm

FM **2** Which is heavier, a tonne or a ton? Show your working clearly.

FM **3** Which is longer, a metre or a yard? Show your working clearly.

FM **4** The weight of 1 cm^3 of water is about 1 gram.

a What is the weight of 1 litre of water:

i in grams **ii** in kilograms?

b What is the approximate weight of 1 gallon of water:

i in grams **ii** in kilograms?

FM **5** While on holiday in France, I saw a sign that said: 'Paris 216 km'. I was travelling on a road that had a speed limit of 80 km/h.

a Approximately how many miles was I from Paris?

b What was the approximate speed limit in miles per hour?

c If I travelled at the top speed all the way, how long would it take me to get to Paris? Give your answer in hours and minutes.

PS **6** While cycling on holiday in France, Tom had to cover a 200-km stretch in one day. He knew that, at home, he averaged 30 mph on the roads.

How long would he expect the journey to take, with no stops?

AU **7** A cowboy's 'ten-gallon' hat could actually hold only 1 gallon of water.

How many cubic inches could a 'ten-gallon' hat hold?

GRADE BOOSTER

- **G** You can read and interpret different scales
- **G** You can make sensible estimates
- **F** You can convert from one metric unit to another
- **F** You can use the approximate conversion factors to change from imperial units to metric units
- **E** You can convert from one imperial unit to another
- **C** You can solve problems, using conversion factors

What you should know now

- How to read a variety of scales
- How to make estimates
- How to convert from one metric unit to another
- How to convert from one imperial unit to another
- How to use conversion factors to change imperial units into metric units
- How to solve problems, using metric units and imperial units

1 Copy and complete this table. Write a sensible unit for each measurement.

	Metric	Imperial
The length of a football pitch	....	Yards
The weight of a newborn baby	....	Pounds
The length of this book	Centimetres	....

2 The diagram shows a 1 litre measuring flask.

700 ml of milk are needed for a recipe.

Copy the scale. Draw an arrow to show where 700 ml is on the scale.

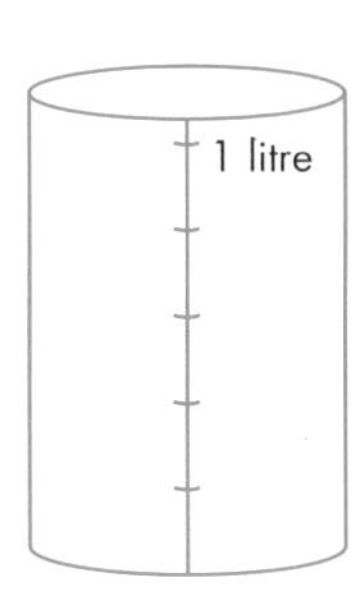

3 This is part of a ruler.

a Write down the length marked with an arrow. (1)

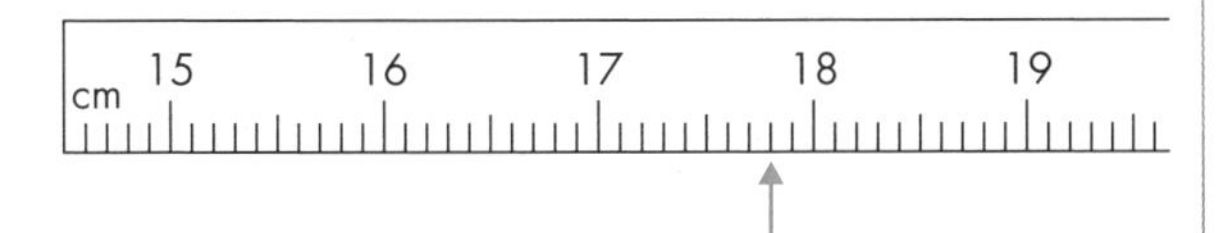

This is a thermometer.

b Write down the temperature shown. (1)

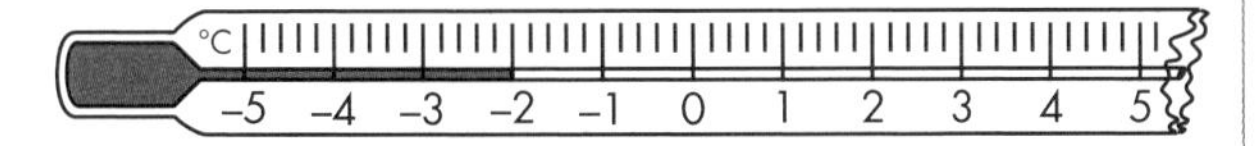

This is a parcel on some scales.

c Write down the weight of the parcel. (1)

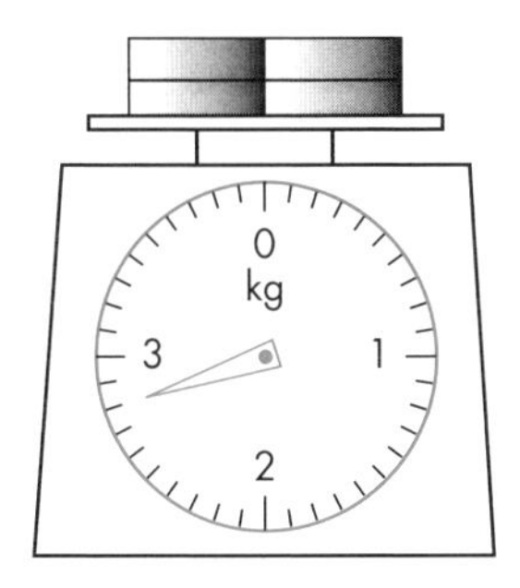

(Total 3 marks)

Edexcel, November 2007, Paper 10 Foundation, Unit 3 Test, Question 1

4 Two villages are 40 km apart.

a Change 40 km into metres.

b How many miles are the same as 40 km?

5 A school canteen orders 30 litres of milk, but 30 pints of milk are delivered instead. Does the canteen have enough milk?

6

The diagram shows a building and a man.

The man is of normal height.

The man and the building are drawn to the same scale.

a Write down an estimate for the height of the man. (1)

b Write down an estimate for the height of the building. (2)

(Total 3 marks)

Edexcel, November 2008, Paper 2 Foundation, Question 10

7 A model of the Eiffel Tower is made to a scale of 2 millimetres to 1 metre.

The width of the base of the real Eiffel Tower is 125 metres.

a Work out the width of the base of the model.

Give your answer in millimetres. (2)

Eiffel Tower

The height of the model is 648 millimetres.

b Work out the height of the real Eiffel Tower.

Give your answer in metres. (2)

(Total 4 marks)

Edexcel, June 2008, Paper 2 Foundation, Question 17(b), (c)

8 **a** Complete the table by writing a **sensible** metric unit for each measurement.

The first one has been done for you. (3)

The length of the river Nile	6700 kilometres
The height of the world's tallest tree	110 ....
The weight of a chicken's egg	70 ....
The amount of petrol in a full petrol tank of a car	40 ....

b Change 4 metres to centimetres. (1)

c Change 1500 grams to kilograms. (1)

(Total 5 marks)

Edexcel, May 2008, Paper 1 Foundation, Question 13

9 **a** Write down the weight in kg shown on this scale. (1)

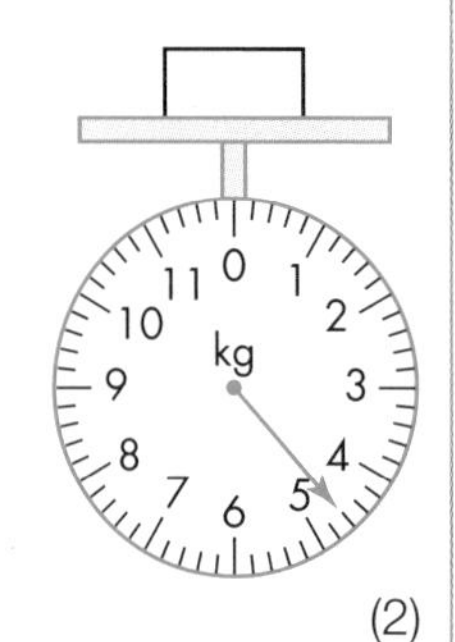

b **i** How many pounds are there in 1 kg?

The weight of a baby is 5 kg.

ii Change 5 kg to pounds. (2)

(Total 3 marks)

Edexcel, November 2008, Paper 2 Foundation, Question 9

10 Brian is driving through France. The speed limit on the motorway is 130 kilometres per hour.

Change 130 km/h to miles per hour.

11 The diagram below shows the dimensions of a bookcase. The thickness of all the wood used is 30 mm.

a Calculate the height of the bookcase, giving your answer in metres.

b Calculate the length of the bookcase, giving your answer in metres.

AU 12 The distance from Sheffield to Birmingham is shown on a website as 140 kilometres. Explain why this unit is used instead of centimetres.

FM 13 Freyja and Agnes took their grandmother on a self-catering holiday in France. They settled into a cottage and travelled 12 km to the nearest supermarket.

Their grandmother asked them to get 5 lbs of potatoes, 8 oz of butter and 4 pts of milk.

a Their grandmother asked how many miles away the supermarket was. What answer should Freyja and Agnes give her?

b The supermarket only sold goods in metric units. Convert their grandmother's shopping list into metric units.

Worked Examination Questions

1 Draw a net of an open box that has measurements 2 cm by 3 cm by 6 cm.

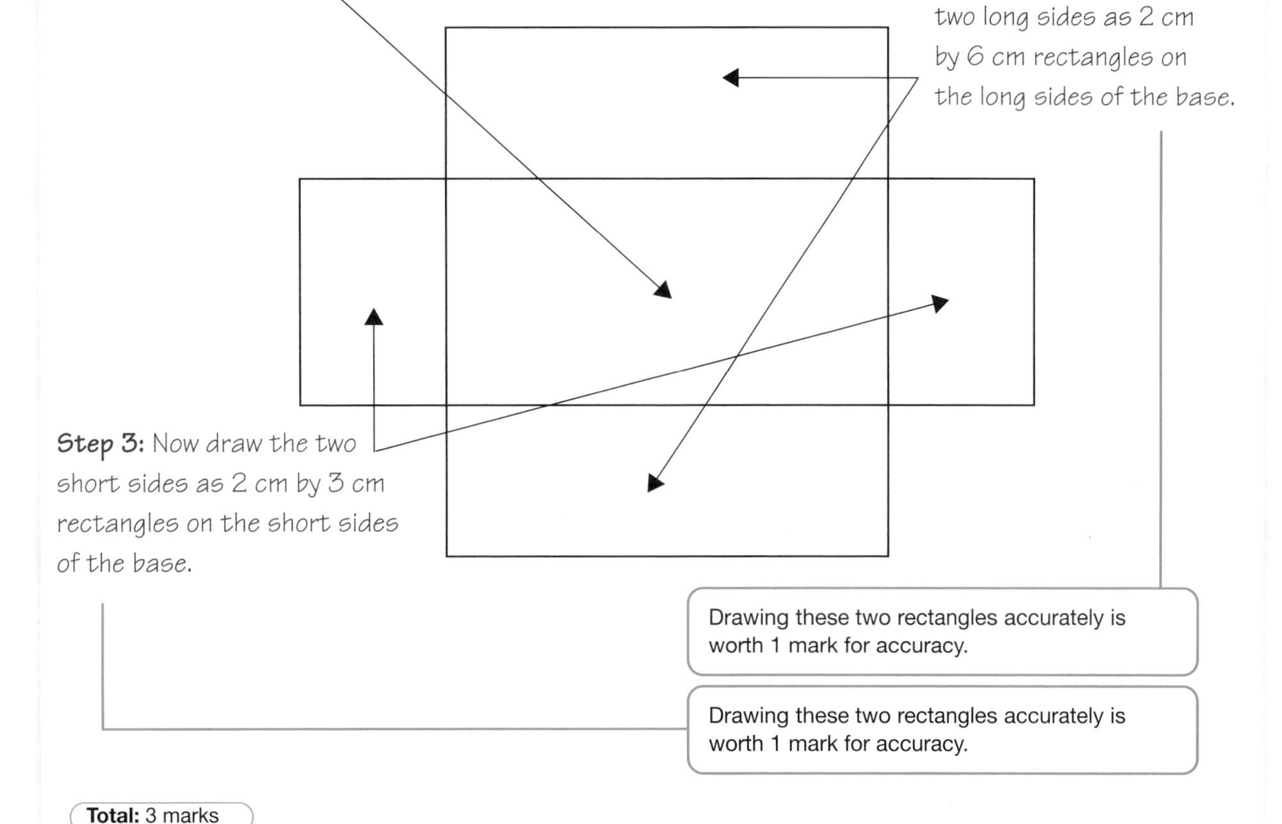

Total: 3 marks

Worked Examination Questions

EQ 2 The diagram shows seven **cubes** arranged to make a **3D shape**

Study the diagrams below.

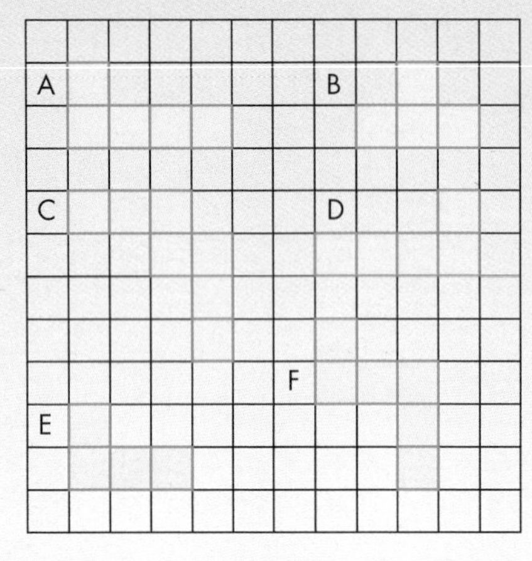

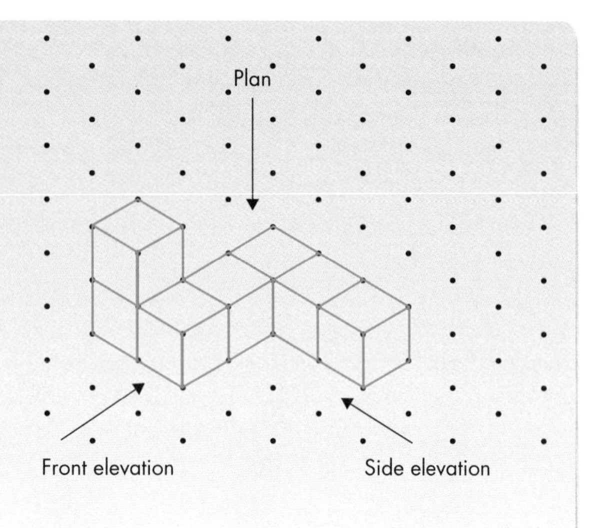

a Which is the **plan** view?

b Which is the **front elevation** view?

c Which is the **side elevation view**?

a F

1 mark

You get 1 mark for accuracy for the correct answer.

b A

1 mark

You get 1 mark for accuracy for the correct answer.

c E

1 mark

You get 1 mark for accuracy for the correct answer.

Total: 3 marks

Worked Examination Questions

FM **3** Aleks is planning a holiday in Madrid.

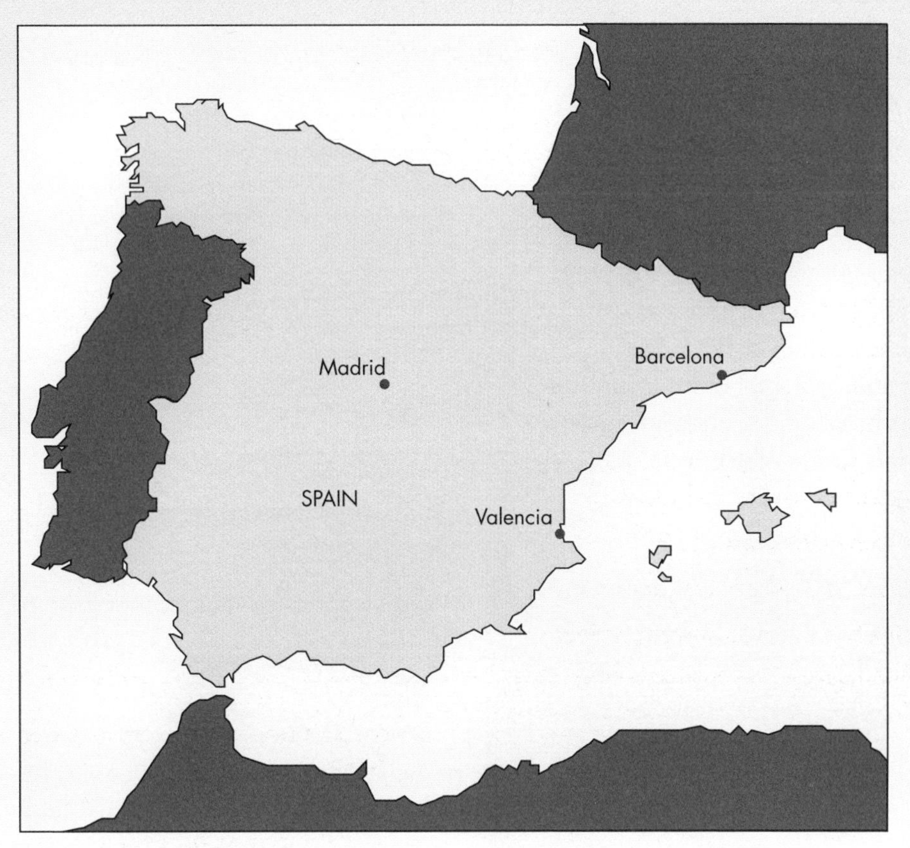

He looks at a map of Spain. The scale of the map is 1 : 10 000 000

Work out the direct distance from Madrid to

a Valencia

b Barcelona

1 cm = 10 000 000 cm = 100 000 m = 100 km — Convert the scale into a suitable format.

According to this, 1 cm represents 100 km — You earn 1 method mark for identifying the key fact that 1 cm = 100 km.

1 mark

a The direct distance from Madrid to Valencia is 310 km — On the map, Madrid to Valencia measures 3.1 cm. So, as 1 cm = 100 km, 3.1 cm must be multiplied by 100 to give the direct distance in kilometres. You earn 1 mark for accuracy for the correct answer.

1 mark

b The direct distance from Madrid to Barcelona is 460 km — On the map, Madrid to Barcelona measures 4.6 cm. As above, multiply 4.6 cm by 100 to give the direct distance in kilometres. You earn 1 mark for accuracy for the correct answer.

1 mark

Total: 3 marks

6 Functional Maths
Olympic Games

The first Olympic Games were held in Greece in 776 BC, but they were very different to the Games that we know today. In 1896 the first Olympic Games of modern times were held in Athens. Since then, the Games have gradually evolved, giving us our current system of international Olympic Games.

The Olympic Games show us that the way we measure things, from distance to weight, have changed over time. For example, over the centuries many countries have moved from using the imperial system of measurements to the metric system. This has meant that in order to maintain world records and make comparisons, measurements must be converted.

Your task

On the right are results from Olympic Games in two different years.

Write a report for team Great Britain, comparing the two sets of data. In your report you should:

- Show who was the best athlete overall in each category
- State any assumptions that you have made when completing your calculations, comparing the data and making statements.

Getting started

Use these questions to remind yourself of the key metric and imperial conversions.

1 Name three metric units of length.
2 Name three imperial units of length.
3 State the metric-to-metric conversions for length and use these in three conversions.
4 State the imperial-to-imperial conversion for length and use these in three conversions.
5 State the conversion facts that you know for metric-to-imperial measurements of length.

Extension

Design a stadium for the next Olympic Games.

- Plan the dimensions of your stadium using imperial and metric units, and draw it to scale.
- You will need to research past Olympic stadiums in order to find out the approximate sizes and requirements of an Olympic Stadium.

Year 1

Event	Winner (Women)	Result
100-yard sprint	Kathryn Ball	10.6 s
220-yard race	Kathryn Ball	25.2 s
880-yard race	Dianne Edwards	2 min 10 s
80-yards hurdle	Cho Ming	15.8 s
440-yards hurdle	Alejandra Lopez	58.3 s
High jump	Janet Guggiani	4 ft 5 in
Long jump	Barbara Charlton	5 yd 1 ft 8 in
Discus	Ife Adebayor	26 yd 2 ft 3 in
Javelin	Paula Ivan	25 yd 2 ft 11 in

Year 2

Event	Winner (Women)	Result
100-metre sprint	Aneta Jarzebska	13.1 s
200-metre race	Karim Djebli	28.3 s
800-metre race	Marie Auvergne	2 min 25 s
80-metre hurdle	Li Du	14.3 s
300-metre hurdle	Gabriela Lopez	56.7 s
High jump	Rachel Evans	1.45 m
Long jump	Brittany Banks	5.03 m
Discus	Ola Kubot	24.31 m
Javelin	Naveen Challa	23.72 m

Why this chapter matters

It is essential that we understand angles. They help us to construct everything, from a building to a table. So, angles literally shape our world.

Ancient measurement of angles

Ancient civilisations used **right angles** in surveying and in constructing buildings, however, not everything can be measured in right angles. There is a need for a smaller, more useful unit. The ancient Babylonians chose a unit angle that led to the development of the **degree**, which is what we still use now.

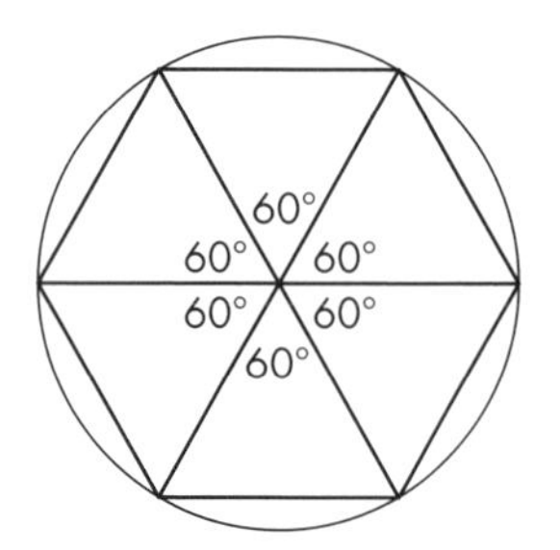

Most historians think that the ancient Babylonians believed that the 'circle' of the year consisted of 360 days. Mathematics historians also generally believe that the ancient Babylonians knew that the side of a **regular hexagon** inscribed in a circle is equal to the **radius** of the circle. This may have led to the division of the full circle (360 'days') into six equal parts, each part consisting of 60 'days', as shown opposite. They divided one angle of an **equilateral triangle** into 60 equal parts, now called degrees, then further subdivided a degree into 60 equal parts, called **minutes**, and a minute into 60 equal parts, called **seconds**.

The divisions 'minutes' and 'second' are also used in time-keeping.

Modern theodolite.

Modern measurement of angles

Modern surveyors use a **theodolite** for measuring angles. A modern theodolite comprises a movable telescope mounted within horizontal and a vertical axis. When the telescope is pointed at a desired object, the angle of each of these axes can be measured with great precision, typically on the scale of **arcseconds**. (There are 3600 arcseconds in 1°.)

This can be used for measuring both horizontal and vertical angles. It is a key tool in surveying and engineering work, particularly on inaccessible ground, but theodolites have been adapted for other specialised purposes in fields such as meteorology and rocket-launch technology.

Chapter 7 Geometry and measures: Angles

The grades given in this chapter are target grades.

This chapter will show you …

- F how to measure and draw angles
- F how to find angles on a line and at a point
- E to D how to find angles in a triangle
- D how to calculate angles in parallel lines
- D how to calculate angles in quadrilaterals

Visual overview

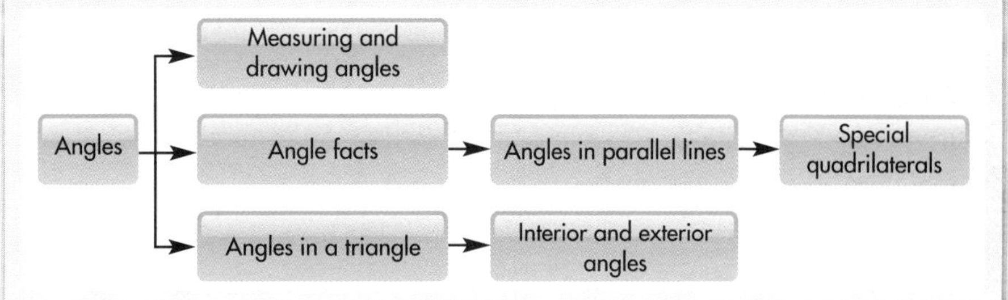

What you should already know

- How to use a protractor to measure an acute angle **(KS3 level 5, GCSE grade F)**
- The meaning of the terms 'acute', 'obtuse', 'reflex', 'right' and how to use these terms to describe angles **(KS3 level 5, GCSE grade G)**
- The meaning of the terms 'parallel lines' and 'perpendicular lines' **(KS3 level 5, GCSE grade G)**

Quick check

State whether these angles are acute, obtuse or reflex.

1 135° **2** 68° **3** 202° **4** 98° **5** 315°

7.1 Measuring and drawing angles

This section will show you how to:
- measure and draw an angle of any size

Key words
acute angle
obtuse angle
protractor
reflex angle

When you are using a **protractor**, it is important that you:

- place the centre of the protractor *exactly* on the corner (vertex) of the angle
- lay the base-line of the protractor *exactly* along one side of the angle.

You must follow these two steps to obtain an accurate value for the angle you are measuring.

You should already have discovered how easy it is to measure **acute angles** and **obtuse angles**, using the common semicircular protractor.

EXAMPLE 1

Measure the angles ABC, DEF and GHI in the diagrams below.

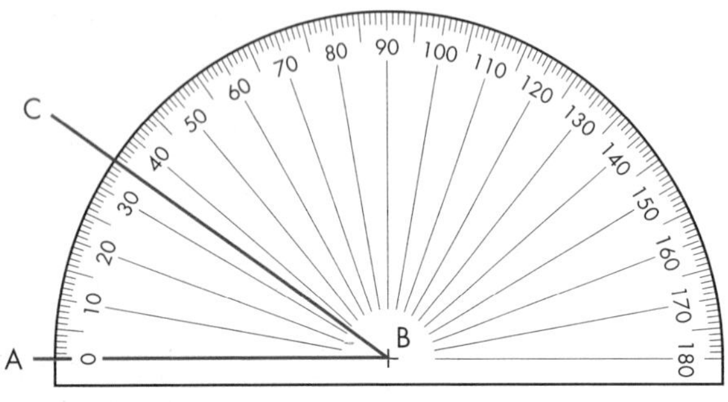

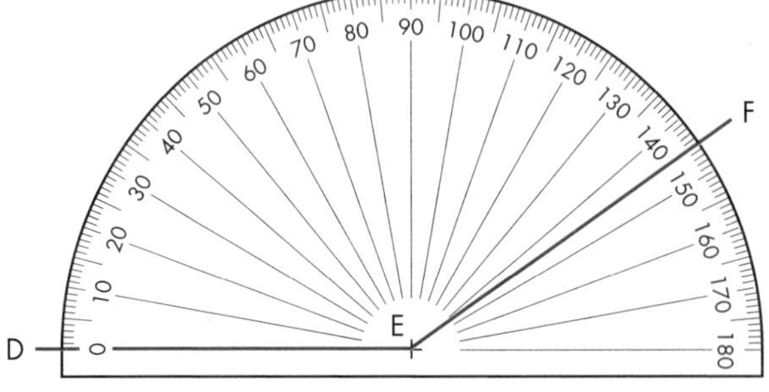

Acute angle ABC is 35° and obtuse angle DEF is 145°.

To measure **reflex angles**, such as angle GHI, it is easier to use a circular protractor if you have one.

Note the notation for angles.

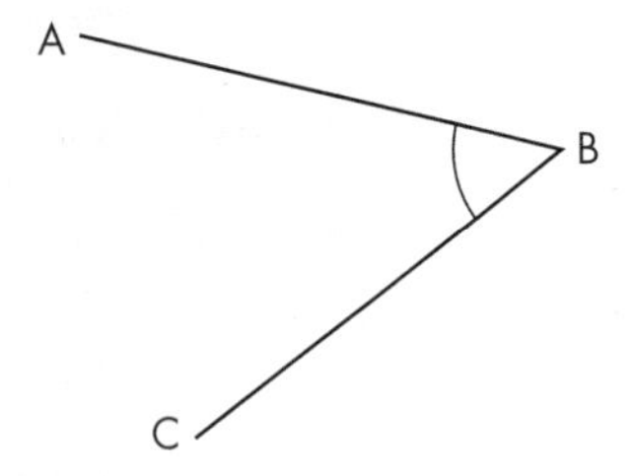

Angle ABC, or ∠ABC, means the angle at B between the lines AB and BC.

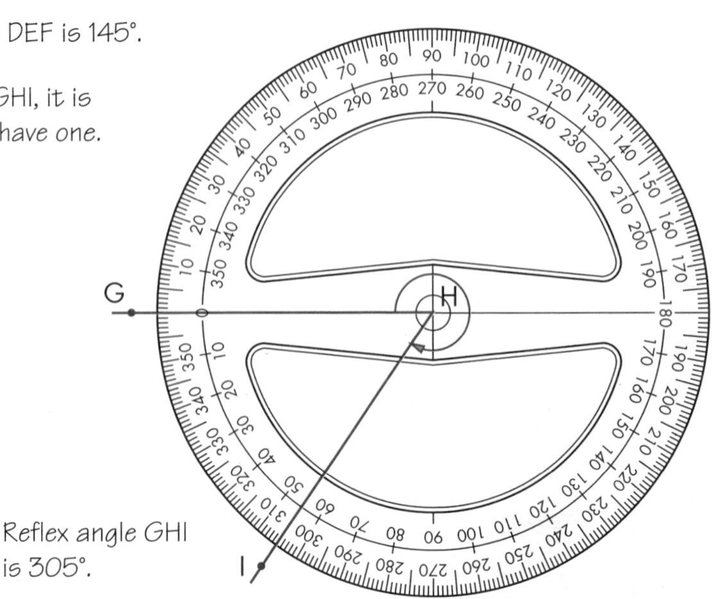

Reflex angle GHI is 305°.

EXERCISE 7A

1 Use a protractor to measure the size of each marked angle.

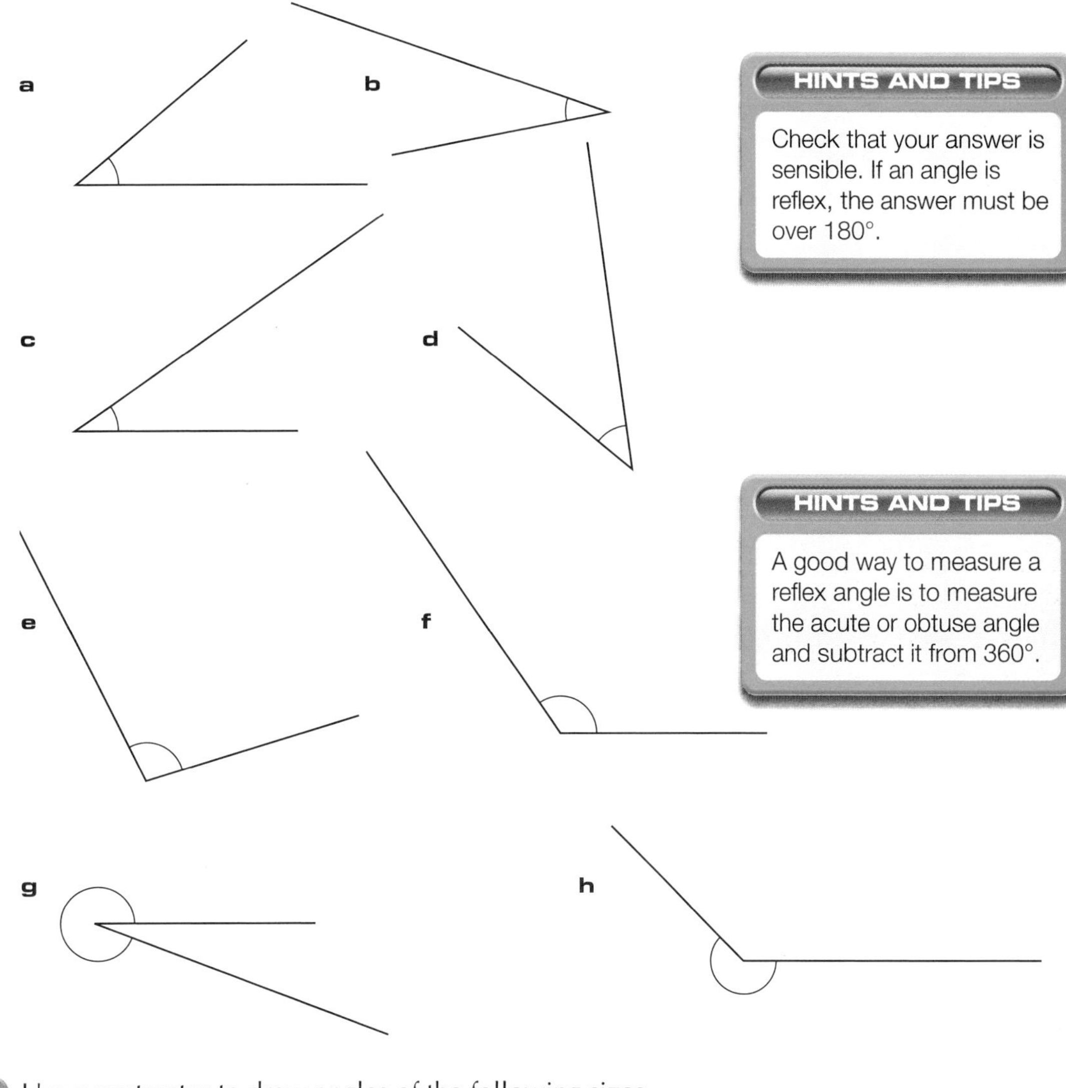

HINTS AND TIPS

Check that your answer is sensible. If an angle is reflex, the answer must be over 180°.

HINTS AND TIPS

A good way to measure a reflex angle is to measure the acute or obtuse angle and subtract it from 360°.

2 Use a protractor to draw angles of the following sizes.

a 30° **b** 60° **c** 90° **d** 10° **e** 20° **f** 45° **g** 75°

3 **a** **i** Draw any three acute angles.

ii Estimate their sizes. Record your results.

iii Measure the angles. Record your results.

iv Work out the difference between your estimate and your measurement for each angle. Add all the differences together. This is your total error.

b Repeat parts **i** to **iv** of part **a** for three obtuse angles.

c Repeat parts **i** to **iv** of part **a** for three reflex angles.

d Which type of angle are you most accurate with, and which type are you least accurate with?

F

FM 4 It is only safe to climb this ladder if the angle between the ground and the ladder is between 72° and 78°.

Is it safe for Oliver to climb the ladder?

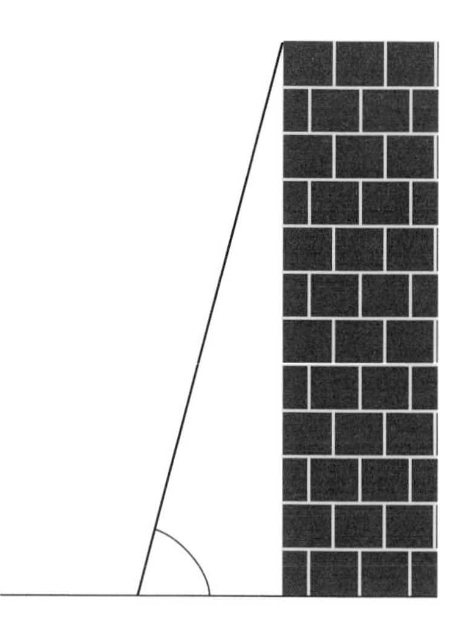

PS 5 An obtuse angle is 10° more than an acute angle.
Write down a possible value for the size of the obtuse angle.

AU 6 Which angle is the odd one out?

Give a reason for your answer.

a

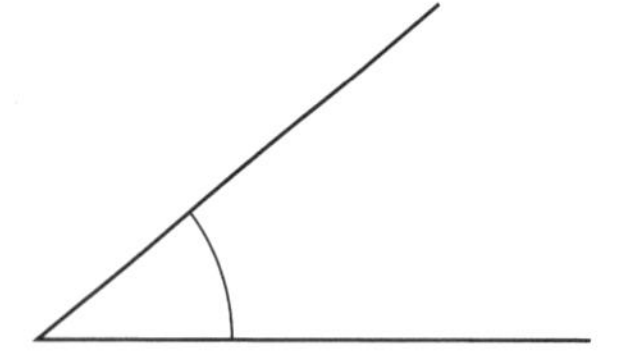

b

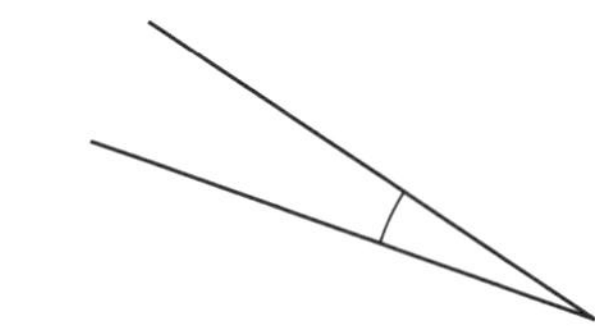

c

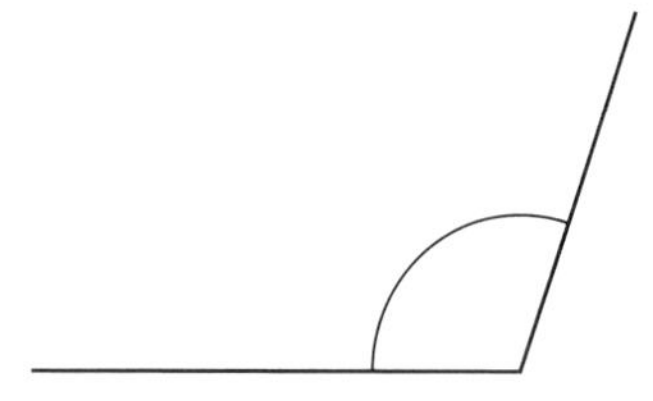

d

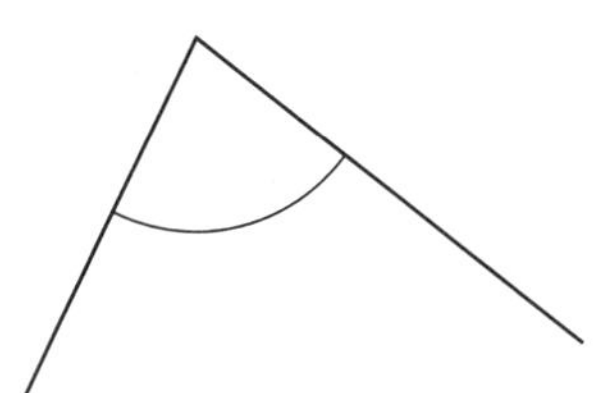

E

7 Use a ruler and a protractor to draw these triangles accurately. Then measure the unmarked angle in each one.

a

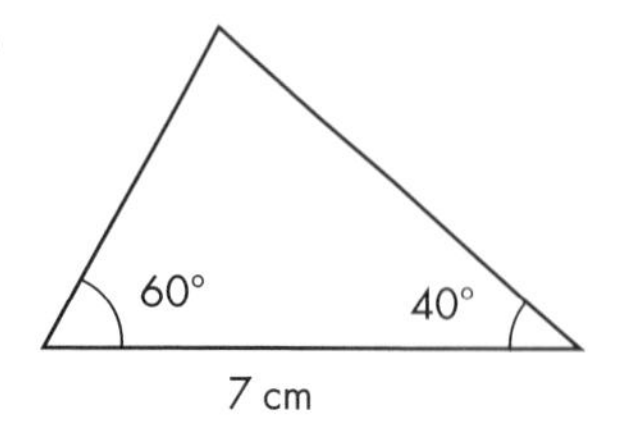

b

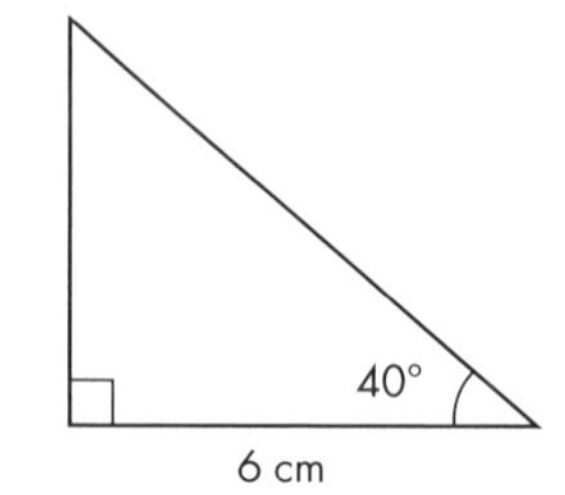

c

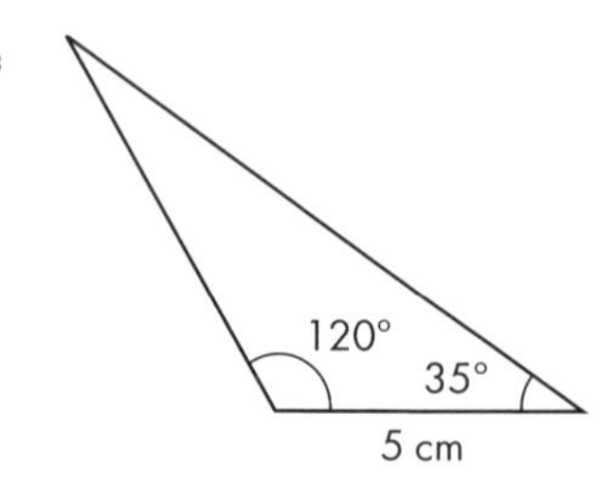

7.2 Angle facts

This section will show you how to:
- calculate angles on a straight line and angles around a point and use opposite angles

Key words
angles around a point
angles on a straight line
opposite angles

Angles on a line

The **angles on a straight line** add up to 180°.

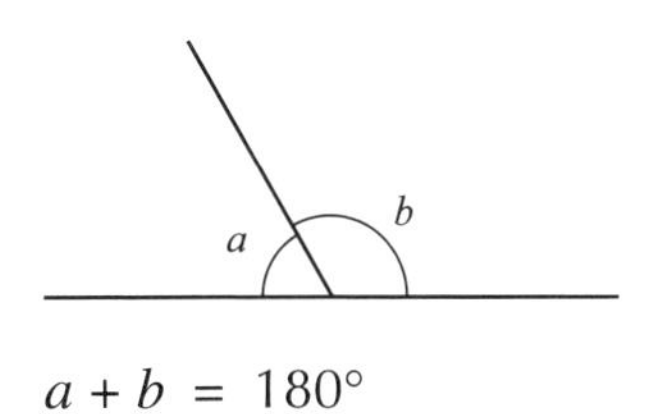

$a + b = 180°$

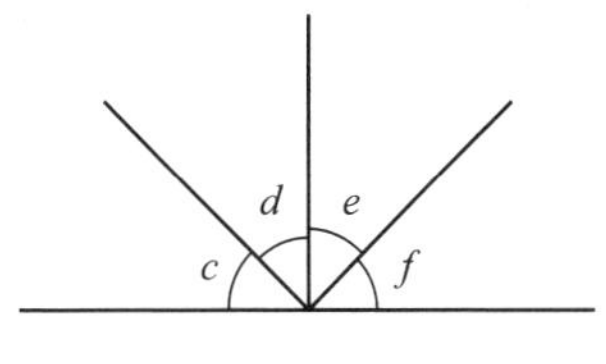

$c + d + e + f = 180°$

Draw an example for yourself (and measure *a* and *b*) to show that the statement is true.

Angles around a point

The sum of the **angles around a point** is 360°. For example:

$a + b + c + d + e = 360°$

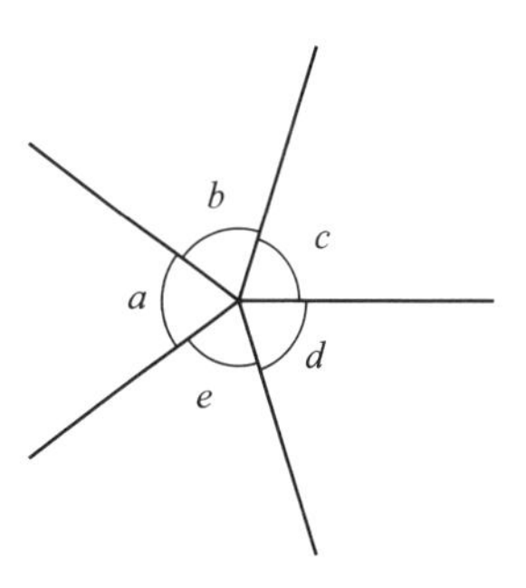

Again, check this for yourself by drawing an example and measuring the angles.

EXAMPLE 2

Find the size of angle x in the diagram.

Angles on a straight line add up to 180°.

$x + 72° = 180°$

So, $x = 180° - 72°$

$x = 108°$

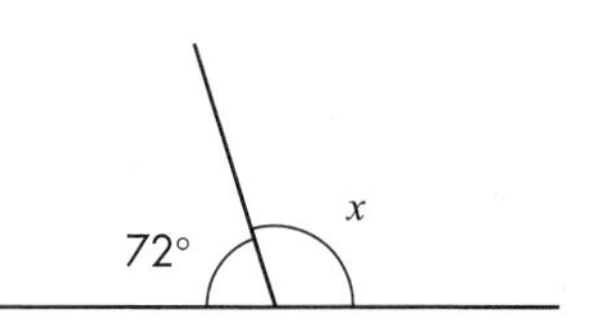

Sometimes equations can be used to solve angle problems.

EXAMPLE 3

Find the value of x in the diagram.

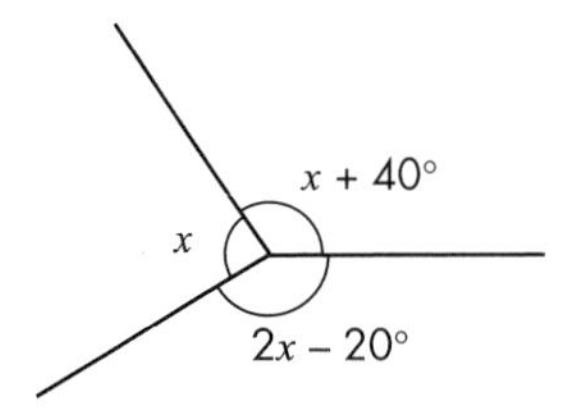

These angles are around a point, so they must add up to 360°.

$$
\begin{aligned}
\text{Therefore, } x + x + 40° + 2x - 20° &= 360° \\
4x + 20° &= 360° \\
4x &= 340° \\
x &= 85°
\end{aligned}
$$

Opposite angles

Opposite angles are equal.

So $a = c$ and $b = d$.

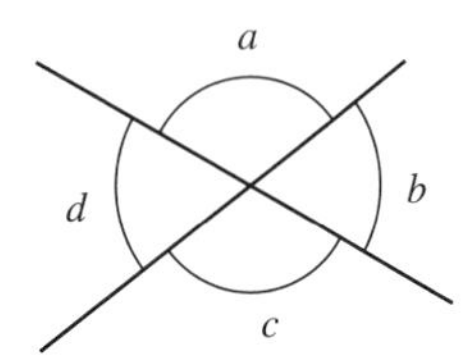

Sometimes opposite angles are called **vertically opposite angles**.

EXAMPLE 4

Find the value of x in the diagram.

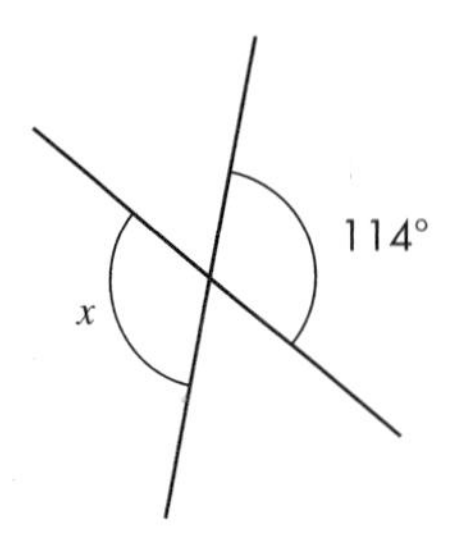

The two angles are opposite, so $x = 114°$.

EXERCISE 7B

Calculate the size of the angle marked x in each of these examples.

1 **a**

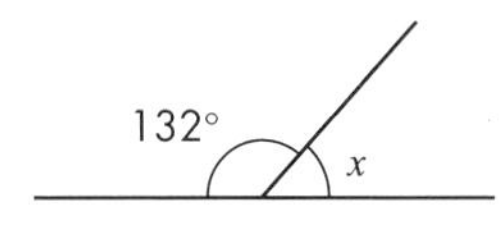

b

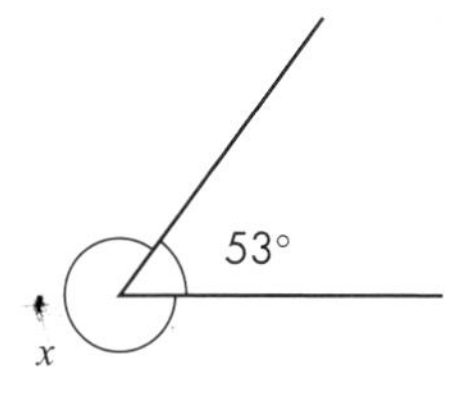

c

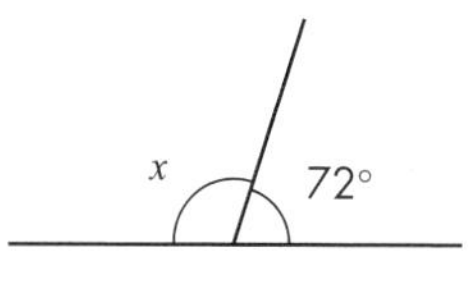

d

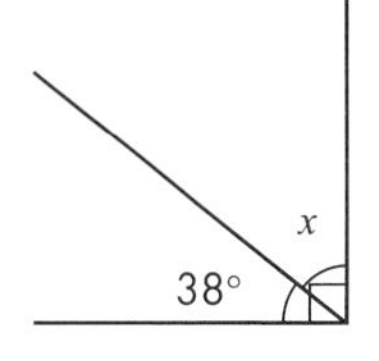

e

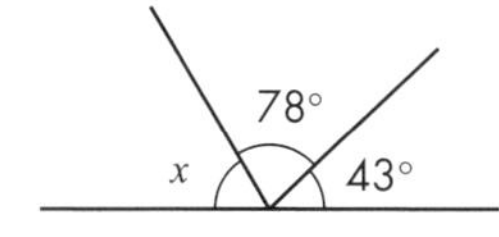

f

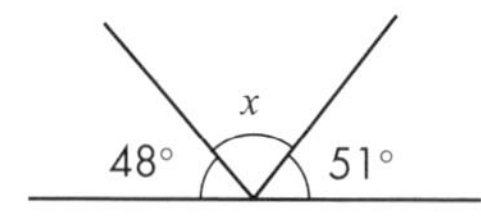

g

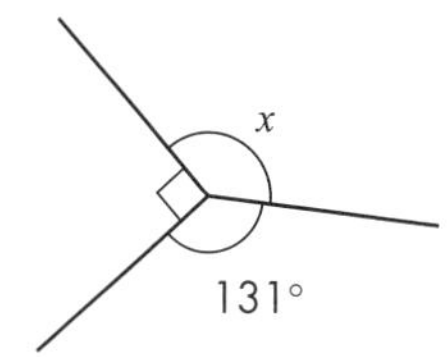

h

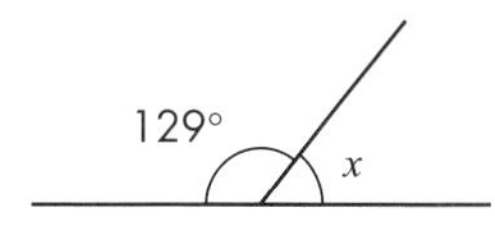

i

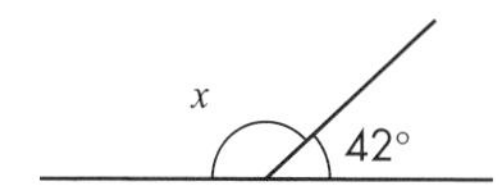

j

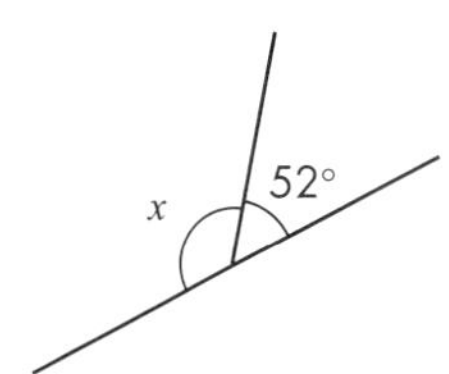

k

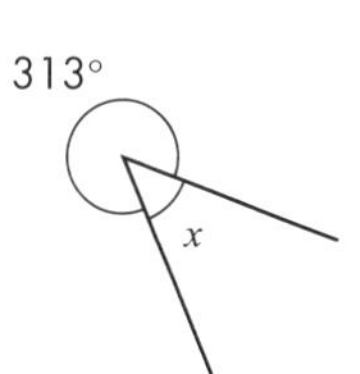

l

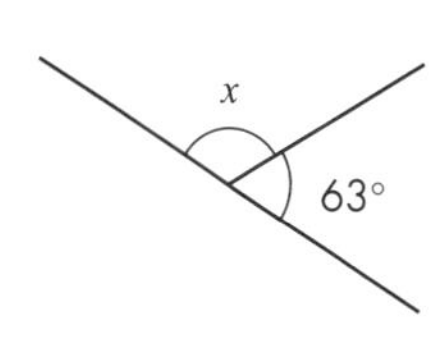

m

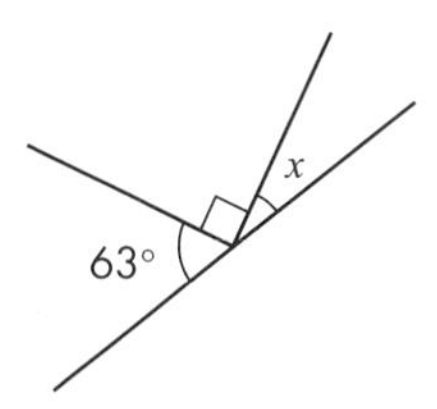

n

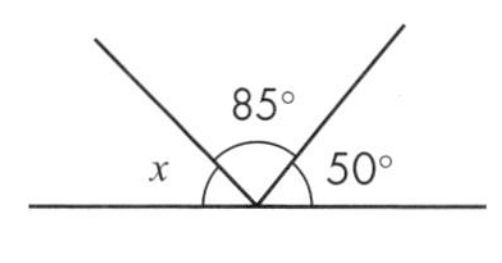

o

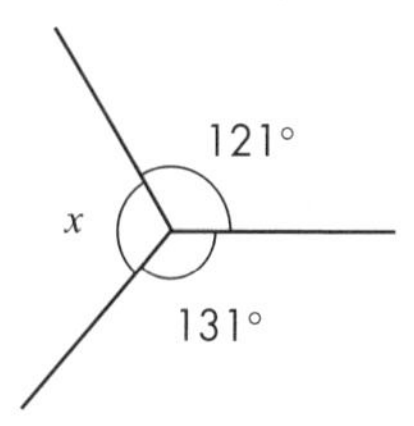

p

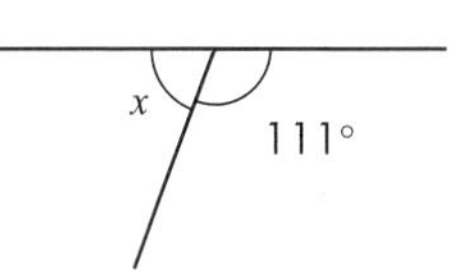

q

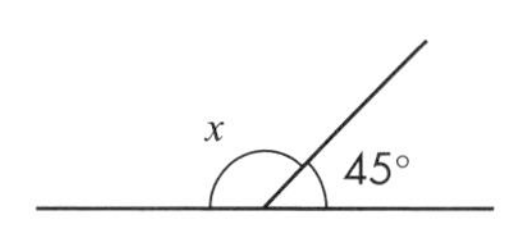

r

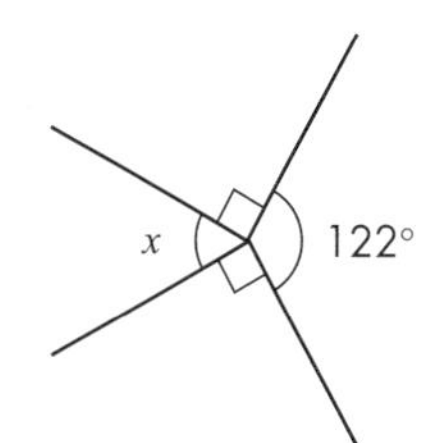

F

s

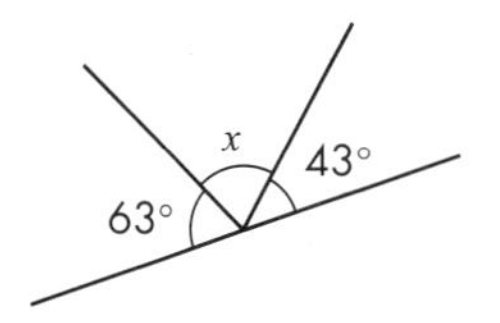

t

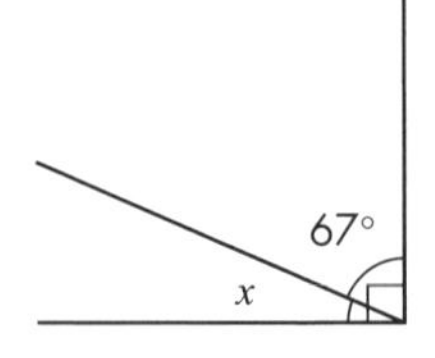

u

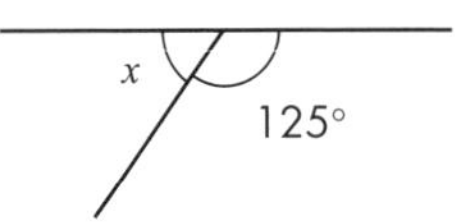

v

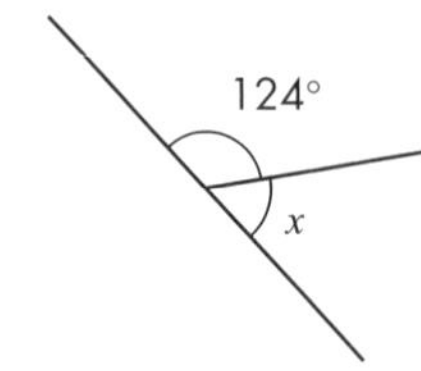

HINTS AND TIPS

Never measure angles in questions like these. Diagrams in examinations are not drawn accurately. Always calculate angles unless you are told to measure them.

2 Write down the value of x in each of these diagrams.

a

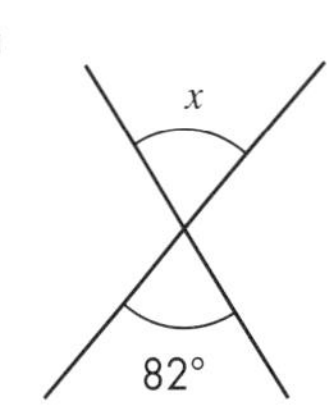

b

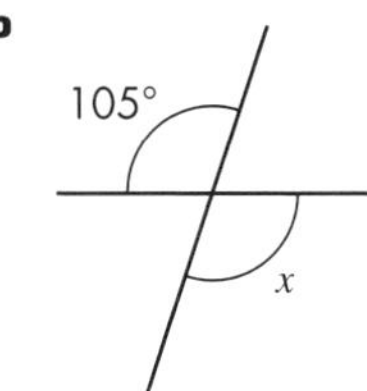

c

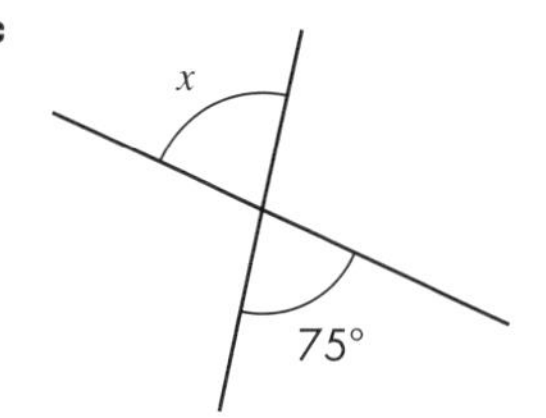

AU 3 In the diagram, angle ABD is 45° and angle CBD is 125°.

Decide whether ABC is a straight line. Write down how you decided.

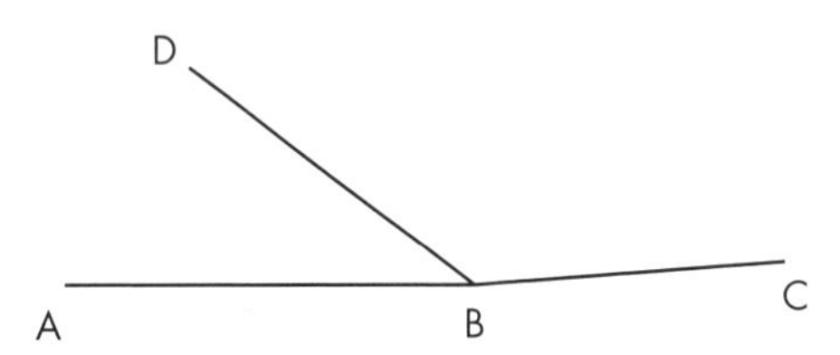

E

4 Calculate the value of x in each of these examples.

a

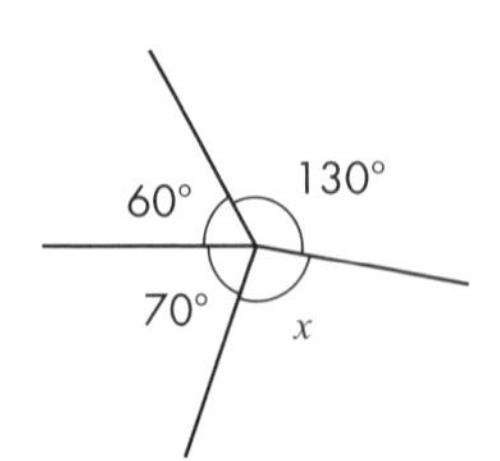

b

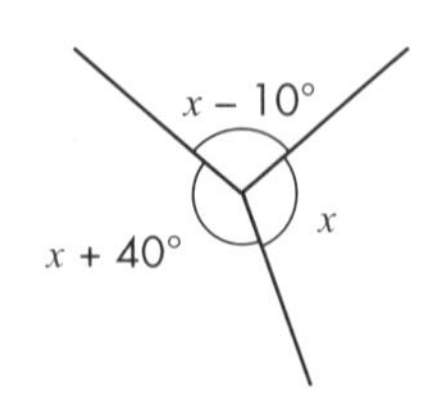

c

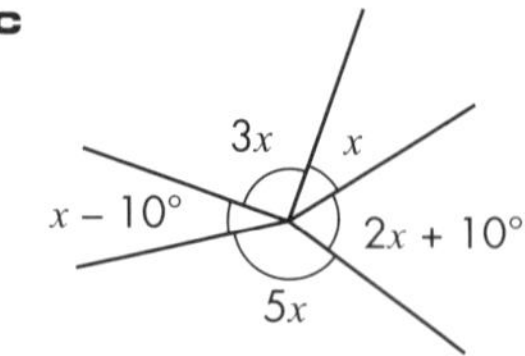

5 Calculate the value of x in each of these examples.

a

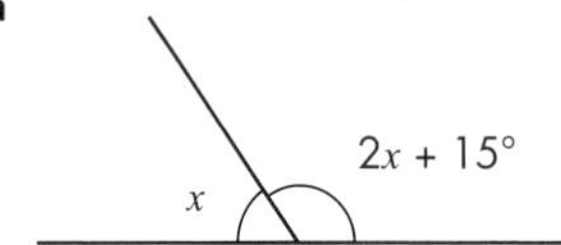

b

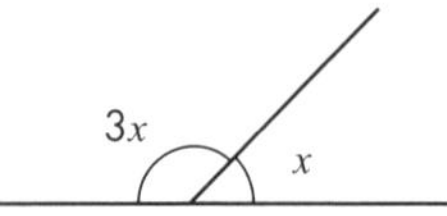

c

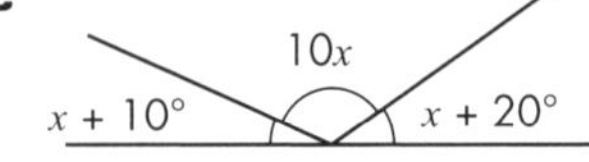

E

6 Calculate the value of x first and then calculate the value of y in each of these examples.

a

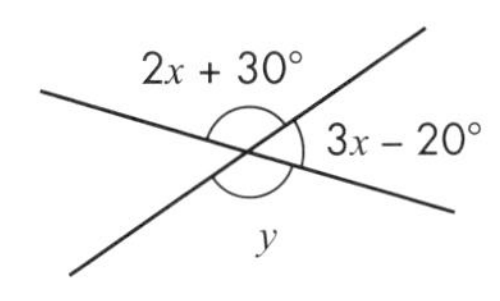

b

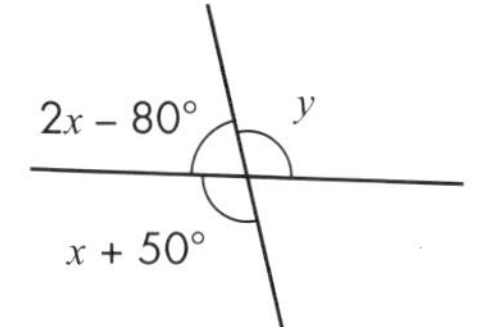

c

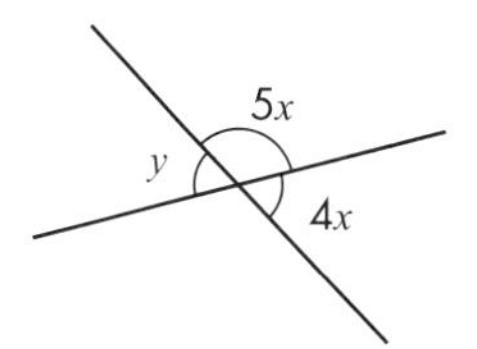

AU 7 Ella has a collection of tiles. They are all equilateral triangles and are all the same size.

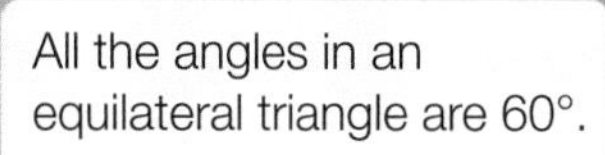

All the angles in an equilateral triangle are 60°.

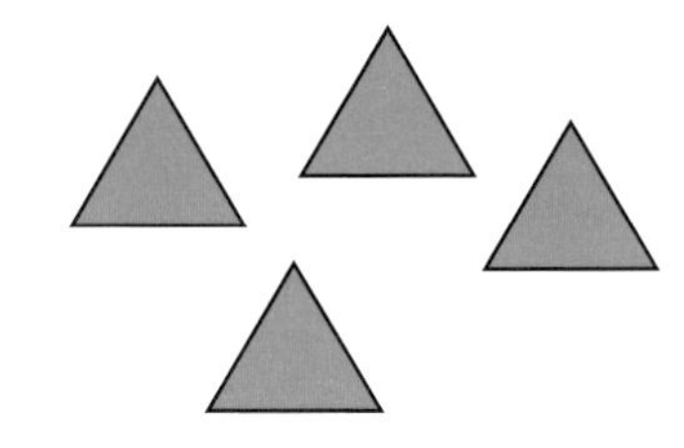

She says that six of the tiles will fit together and leave no gaps.

Explain why Ella is correct.

PS 8 Work out the value of y in the diagram.

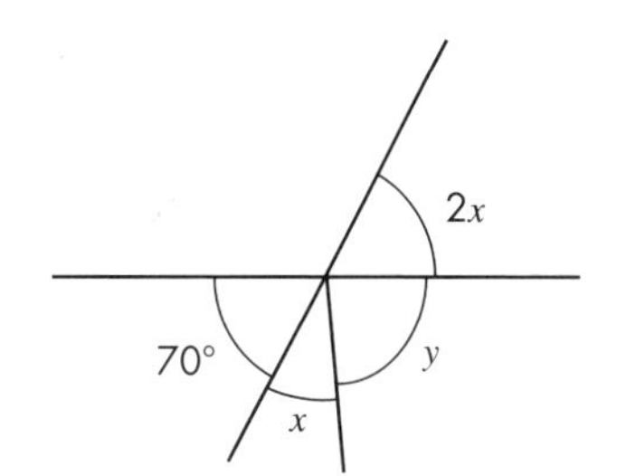

7.3 Angles in a triangle

This section will show you how to:

- calculate the size of angles in a triangle

Key words

angles in a triangle
equilateral triangle
exterior angle
interior angle
isosceles triangle
right-angled triangle

ACTIVITY

Angles in a triangle

You need a protractor.

Draw a triangle. Label the corners (vertices) A, B and C.

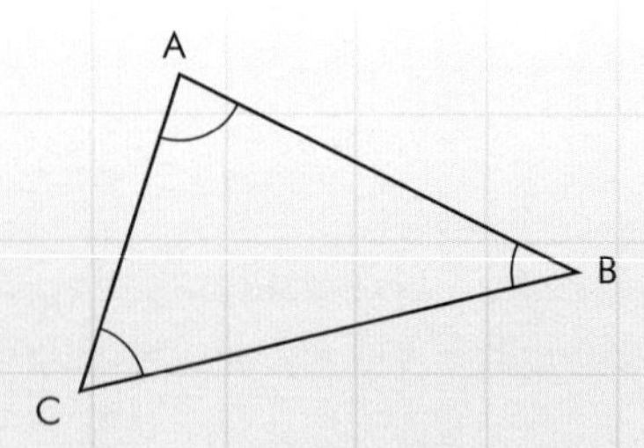

Use a ruler and make sure that the corners of your triangle form proper angles.

Like this. Not like this … … or this.

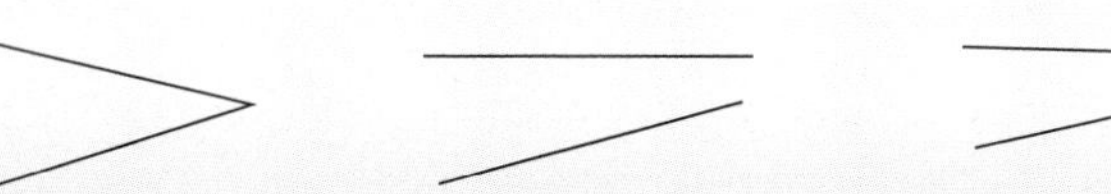

Measure each angle, A, B and C.

Write them down and add them up:

Angle A = ……°

Angle B = ……°

Angle C = ……°

Total =

Repeat this for five more triangles, including at least one with an obtuse angle.

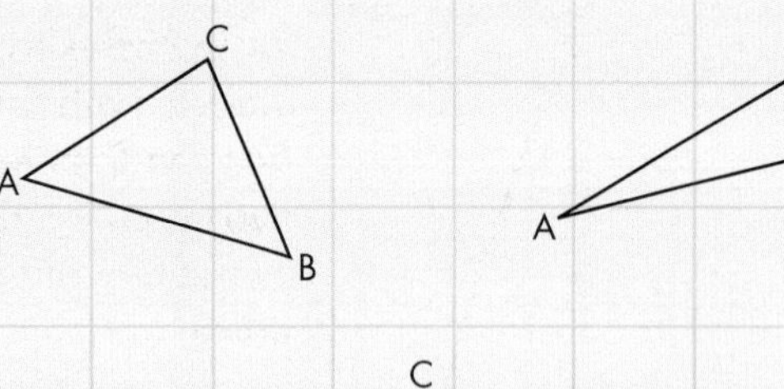

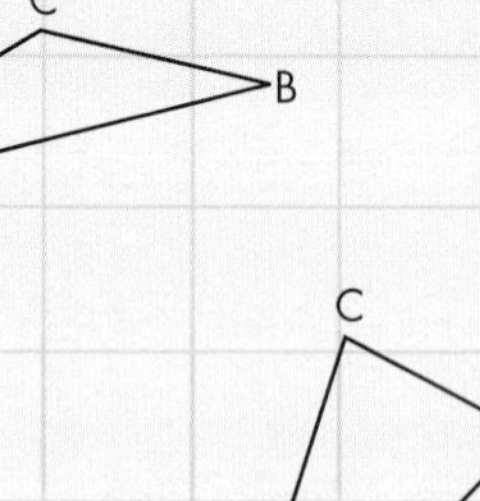

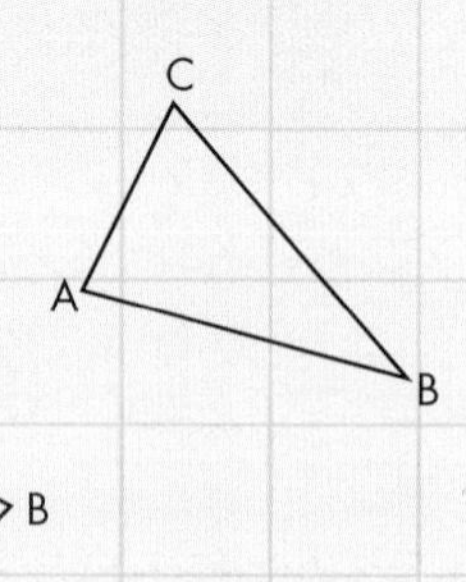

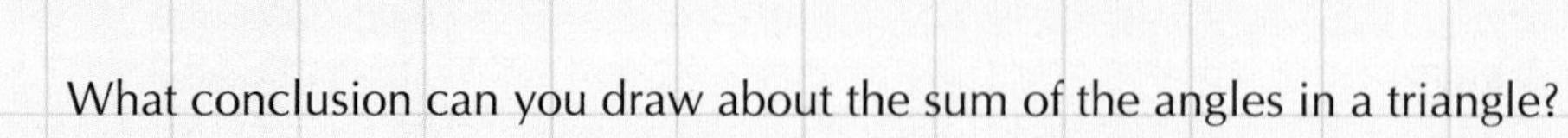

What conclusion can you draw about the sum of the angles in a triangle?

Remember:

You will not be able to measure with total accuracy.

You should have discovered that the three **angles in a triangle** add up to 180°.

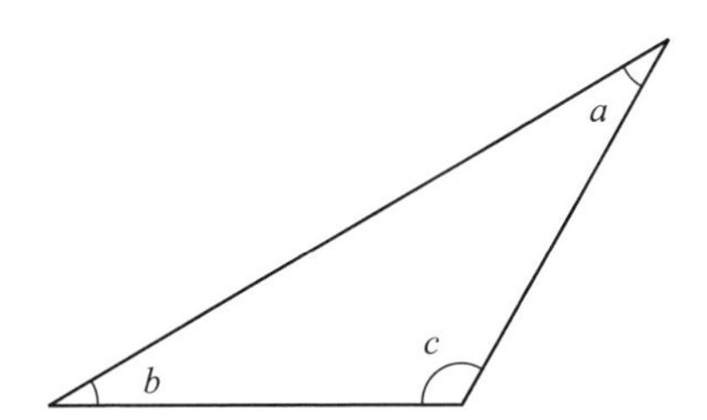

$a + b + c = 180°$

EXAMPLE 5

Calculate the size of angle a in the triangle below.

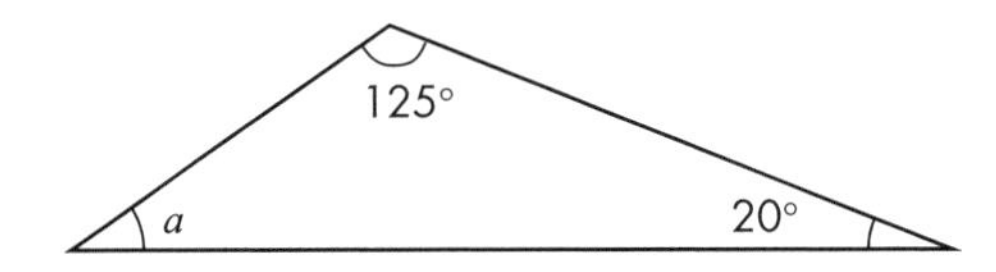

Angles in a triangle add up to 180°

Therefore, $a + 20° + 125° = 180°$

$a + 145° = 180°$

So $a = 35°$

Special triangles

Equilateral triangle

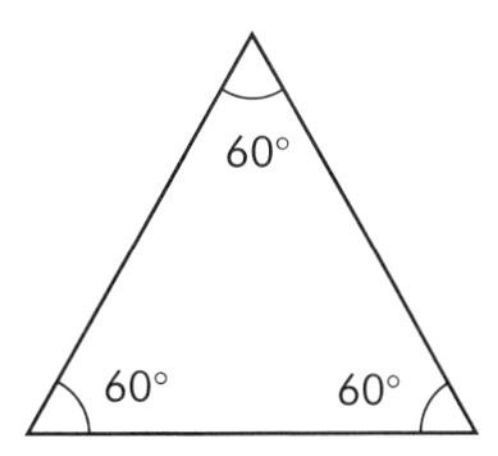

An **equilateral triangle** is a triangle with all its sides equal. Therefore, all three **interior angles** are 60°.

Isosceles triangle

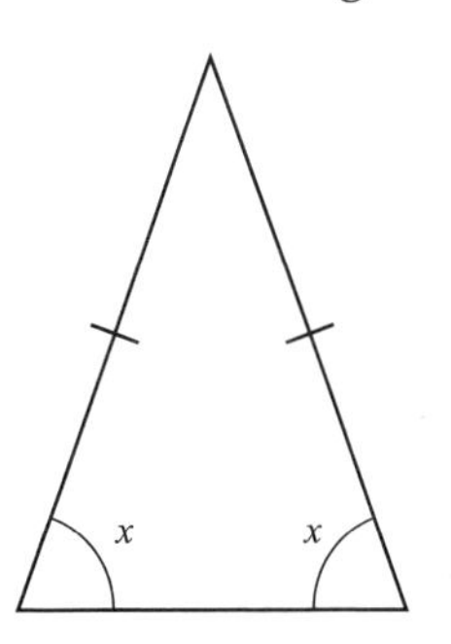

An **isosceles triangle** is a triangle with two equal sides and, therefore, with two equal interior angles (at the foot of the equal sides).

Notice how to mark the equal sides and equal angles.

Right-angled triangle

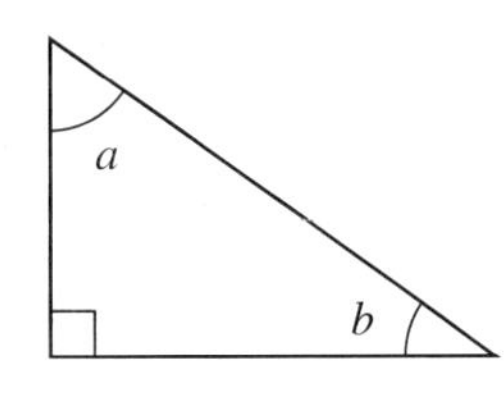

A **right-angled triangle** has an interior angle of 90°.
$a + b = 90°$

EXERCISE 7C

1 Find the size of the angle marked with a letter in each of these triangles.

a

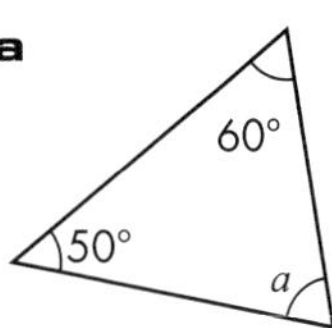

b

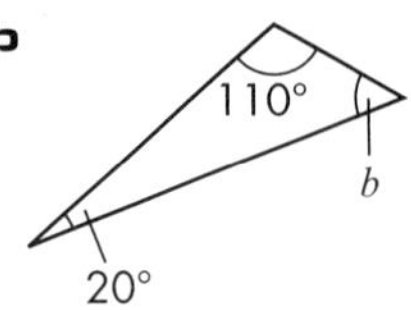

c

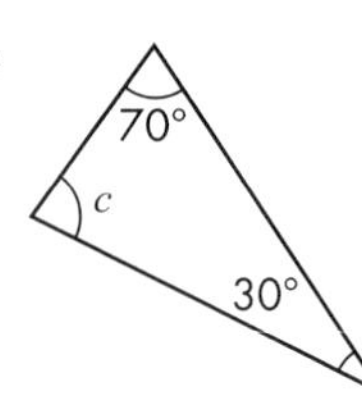

d

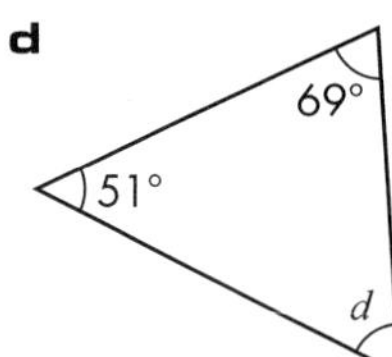

e

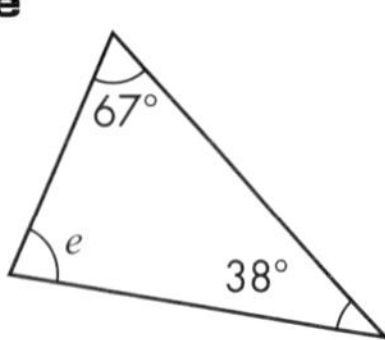

f

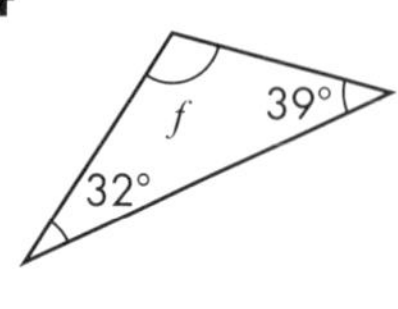

g

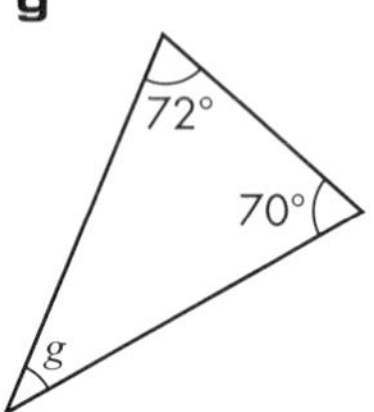

h

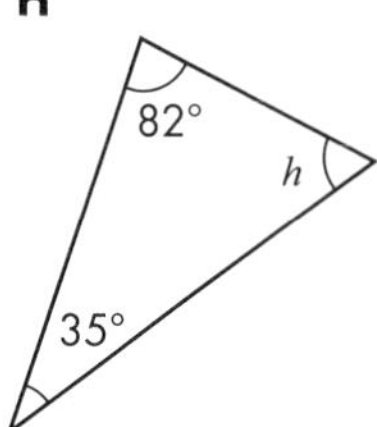

2 Do any of these sets of angles form the three angles of a triangle? Explain your answer.

a 35°, 75°, 80° **b** 50°, 60°, 70°

c 55°, 55°, 60° **d** 60°, 60°, 60°

e 35°, 35°, 110° **f** 102°, 38°, 30°

3 Two interior angles of a triangle are given in each case. Find the third one indicated by a letter.

a 20°, 80°, a **b** 52°, 61°, b

c 80°, 80°, c **d** 25°, 112°, d

e 120°, 50°, e **f** 122°, 57°, f

4 In the triangle on the right, all the interior angles are the same.

a What is the size of each angle?

b What is the name of a special triangle like this?

c What is special about the sides of this triangle?

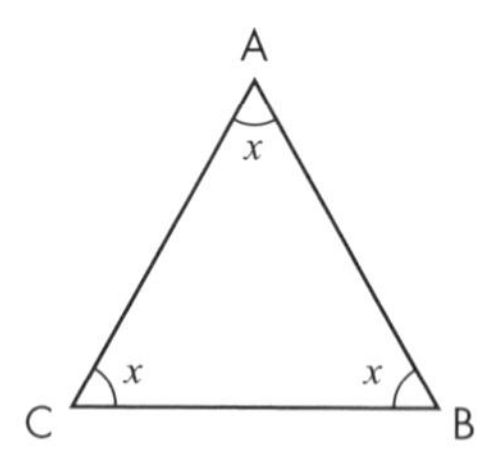

5 In the triangle on the right, two of the angles are the same.

a Work out the size of the lettered angles.

b What is the name of a special triangle like this?

c What is special about the sides AC and AB of this triangle?

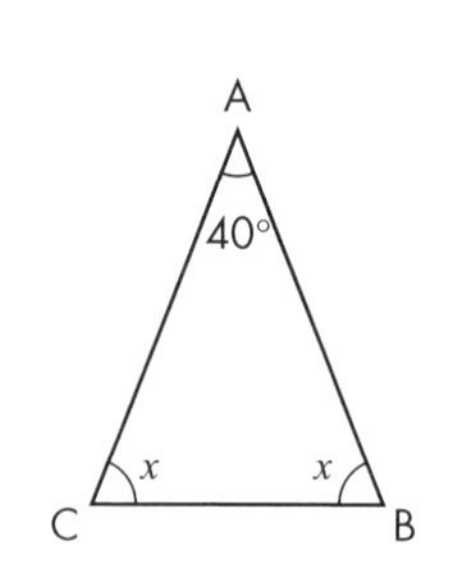

AU 6 In the triangle on the right, the angles at B and C are the same. Write down the size of the lettered angles.

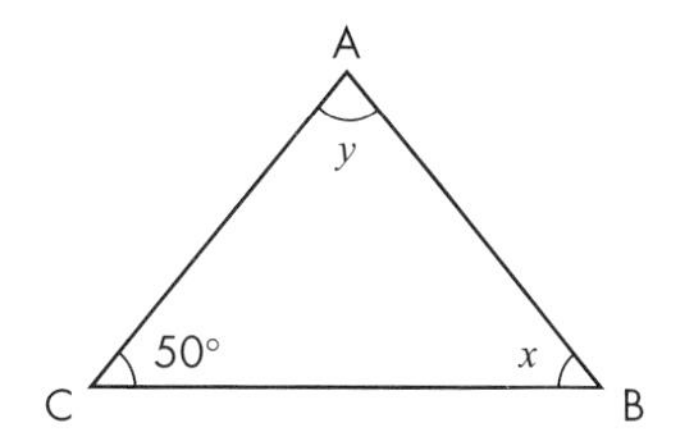

7 Find the size of the **exterior angle** marked with a letter in each of these diagrams.

a

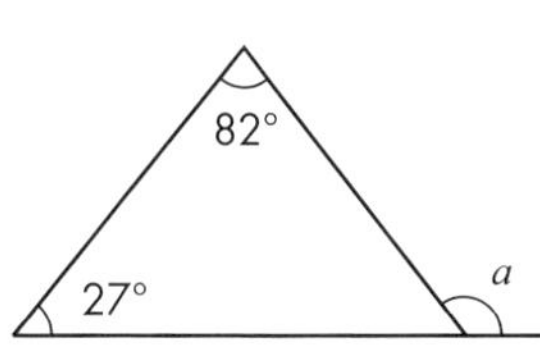

b

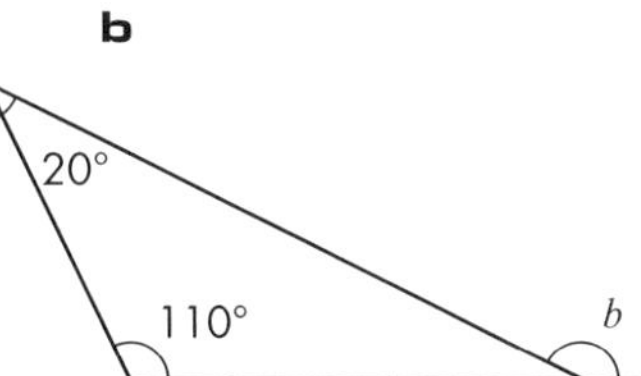

c

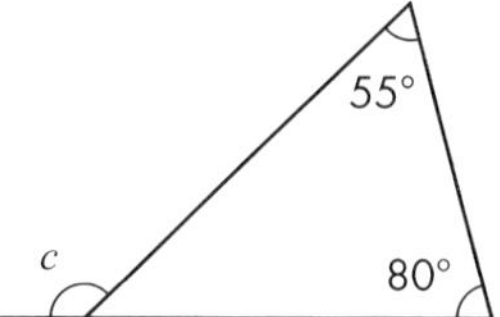

FM 8 A town planner has drawn this diagram to show three paths in a park but they have missed out the angle marked x.

Work out the value of x.

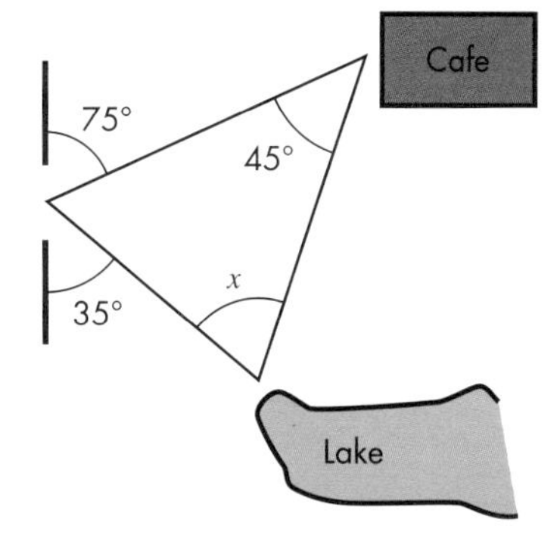

PS 9 What is the special name for triangle DEF?

Show all your working to explain your answer.

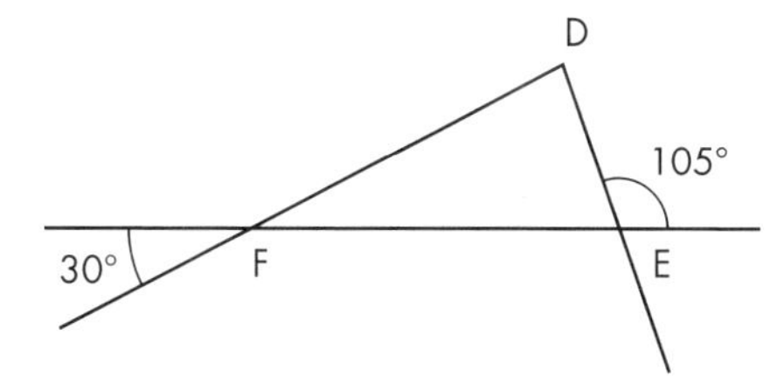

AU 10 The diagram shows three intersecting straight lines.

Work out the values of a, b and c.

Give reasons for your answers.

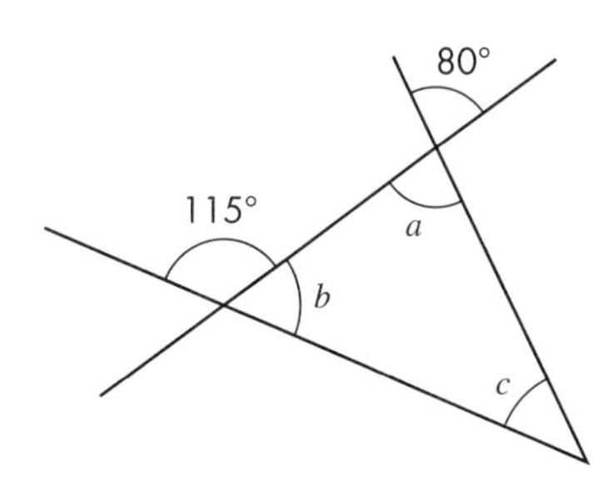

11 By using algebra, show that $x = a + b$.

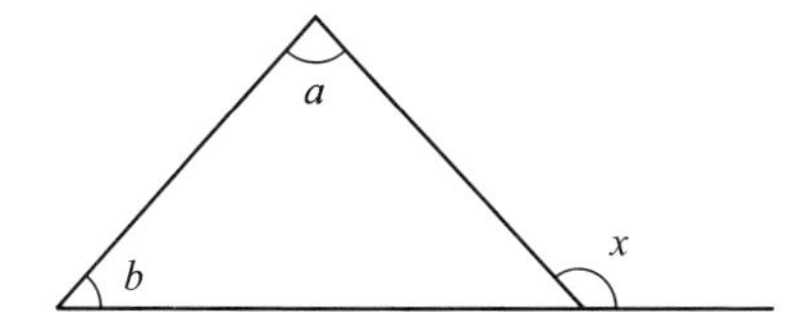

7.4 Parallel lines

This section will show you how to:
- find angles in parallel lines

Key words
allied angles
alternate angles
corresponding angles
interior angles

ACTIVITY

Angles in parallel lines

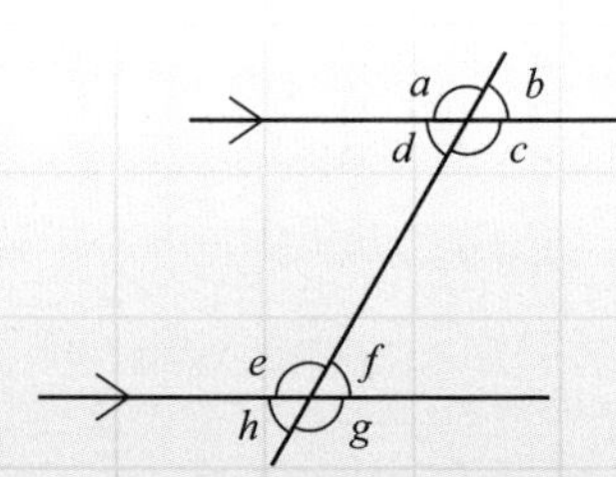

You need tracing paper or a protractor.

Draw two parallel lines about 5 cm apart and a third line that crosses both of them.

The arrowheads indicate that the lines are parallel and the line that crosses the parallel lines is called a *transversal*.

Notice that eight angles are formed. Label these a, b, c, d, e, f, g and h.

Measure or trace angle a. Find all the angles on the diagram that are the same size as angle a.

Measure or trace angle b. Find all the angles on the diagram that are the same size as angle b.

What is the sum of $a + b$?

Find all the pairs of angles on the diagram that add up to 180°.

Angles like these

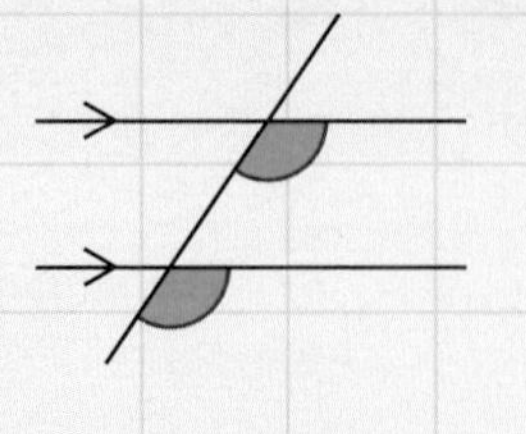

are called **corresponding angles** (Look for the letter F).

Corresponding angles are equal.

Angles like these

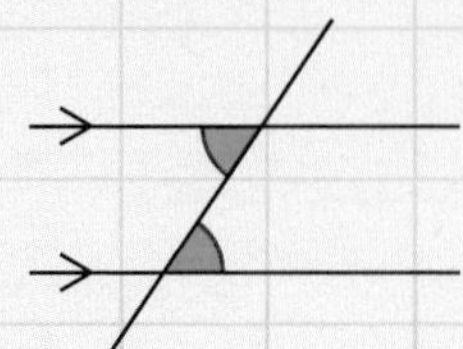

are called **alternate angles** (Look for the letter Z).

Alternate angles are equal.

Angles like these are called **allied angles** or **interior angles** (Look for the letter C).

Allied angles add to 180°.

Copy and complete these statements to make them true.

1 Angles h and ...... are corresponding angles.

2 Angles d and ...... are alternate angles.

3 Angles e and ...... are allied angles.

4 Angles b and ...... are corresponding angles.

5 Angles c and ...... are allied angles.

6 Angles c and ...... are alternate angles.

Note that in examinations you should use the correct terms for types of angles. Do *not* call them F, Z or C angles as you will lose marks for quality of written communication.

EXAMPLE 6

State the size of each of the lettered angles in the diagram.

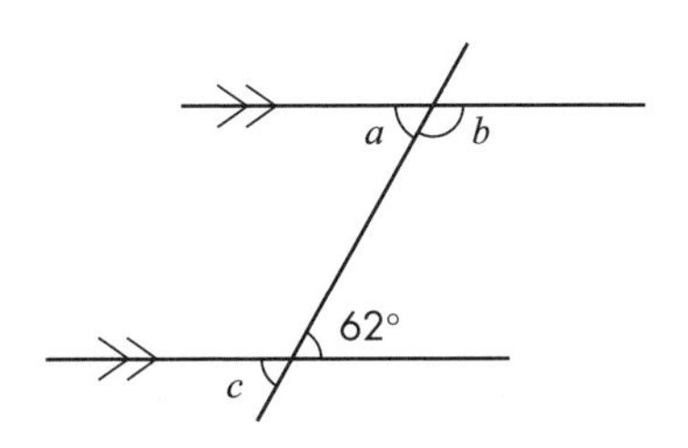

a = 62° (alternate angle)

b = 118° (allied angle or angles on a line)

c = 62° (vertically opposite angle)

EXERCISE 7D 

D

1 State the sizes of the lettered angles in each diagram.

a

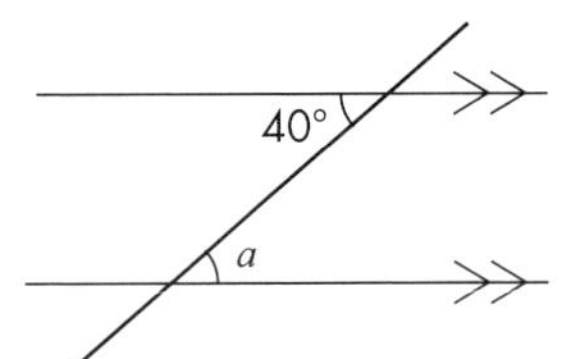

b

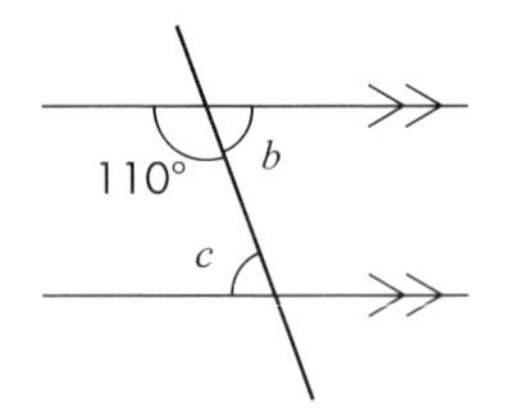

c

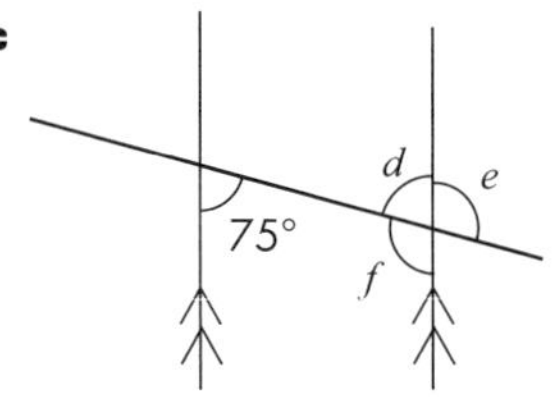

d

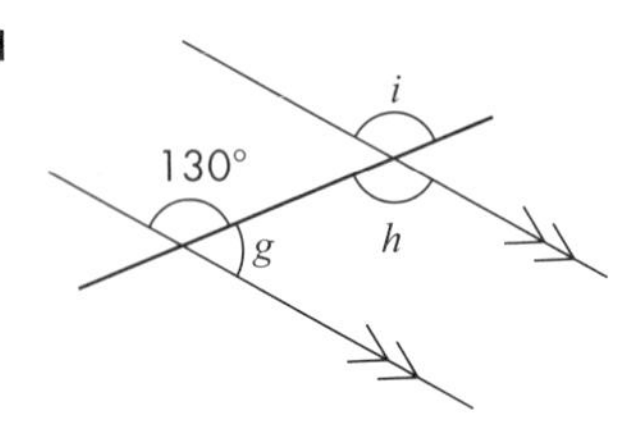

e

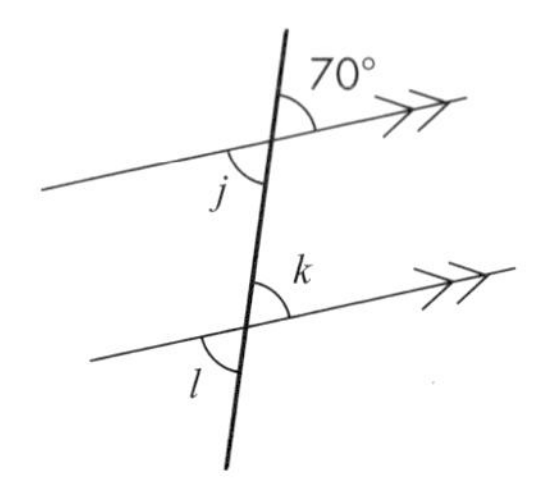

f

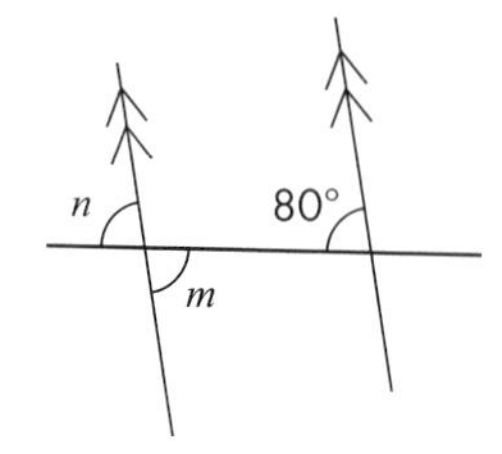

2 State the sizes of the lettered angles in each diagram.

a

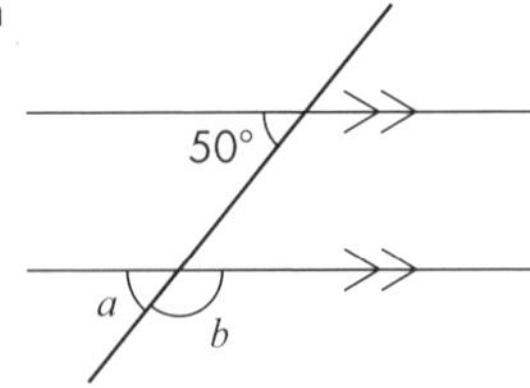

b

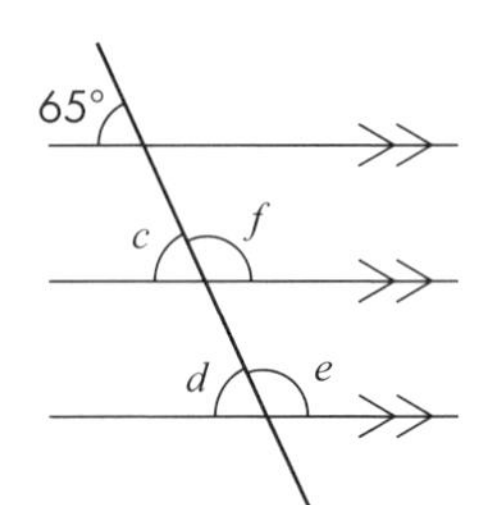

c

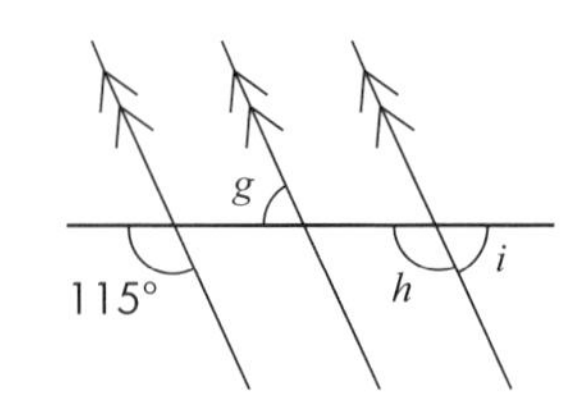

d

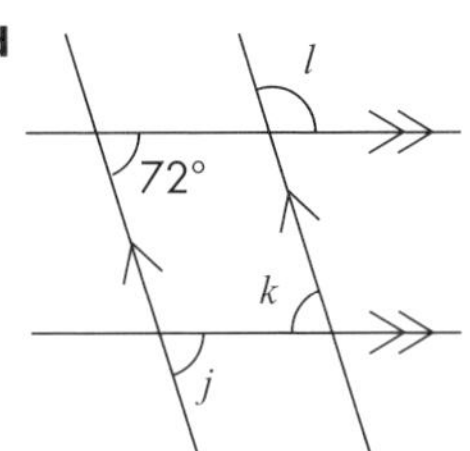

e

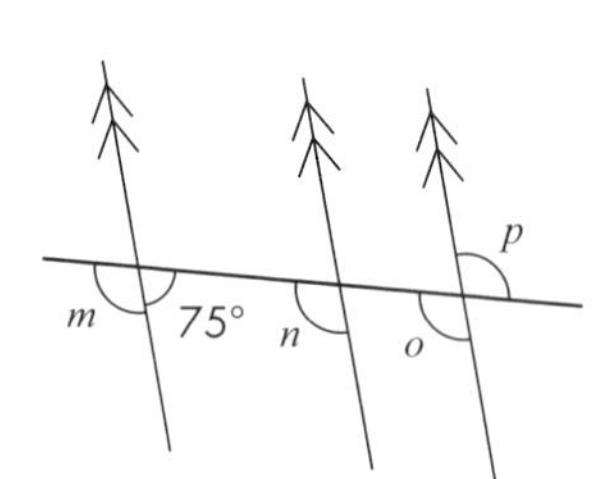

f

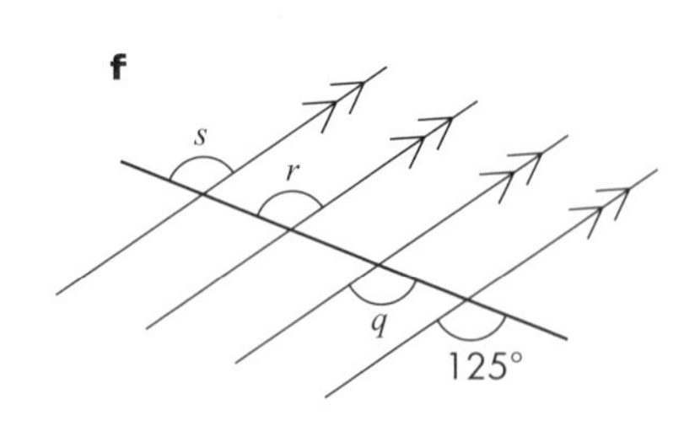

3 State the sizes of the lettered angles in these diagrams.

a

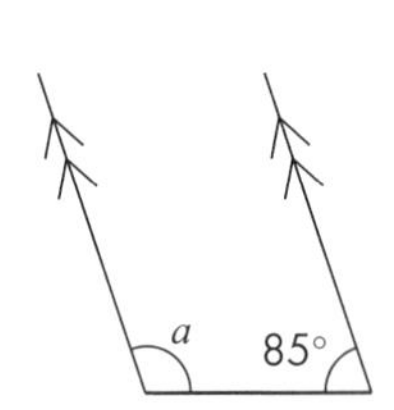

b

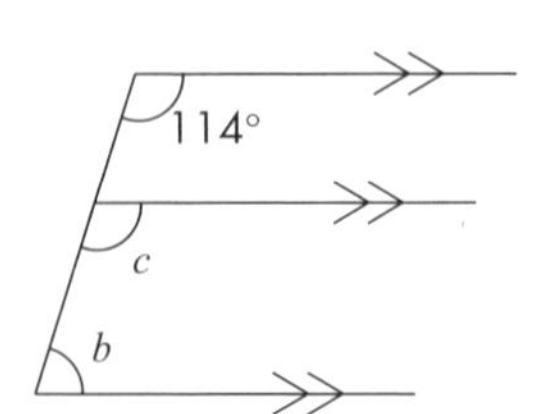

D

4 Calculate the values of x and y in these diagrams.

a

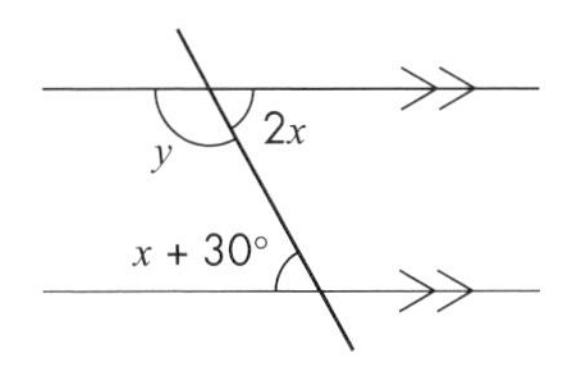

b

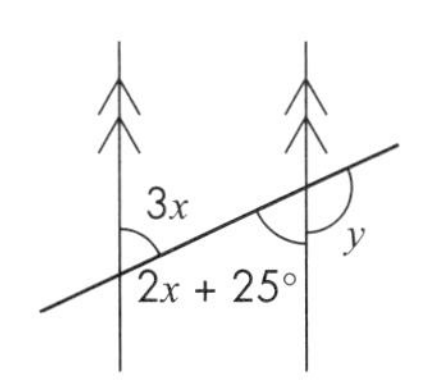

c

3x + 10°
4x – 20°
y

5 Calculate the values of x and y in these diagrams.

a

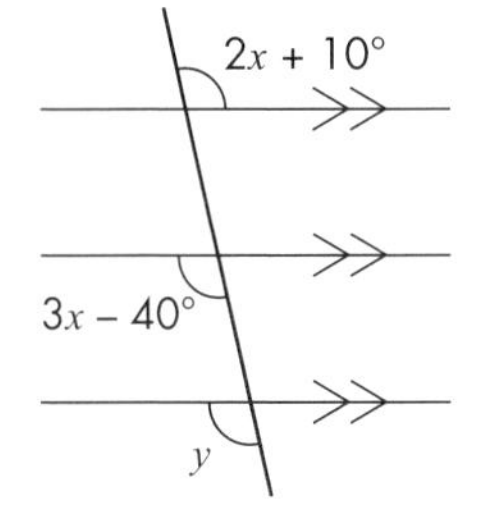

b

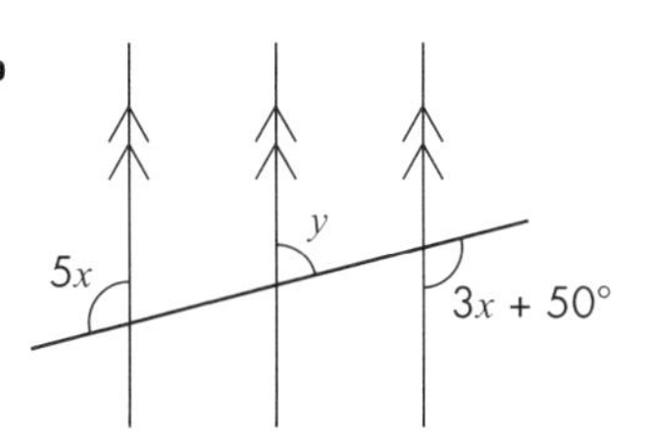

c

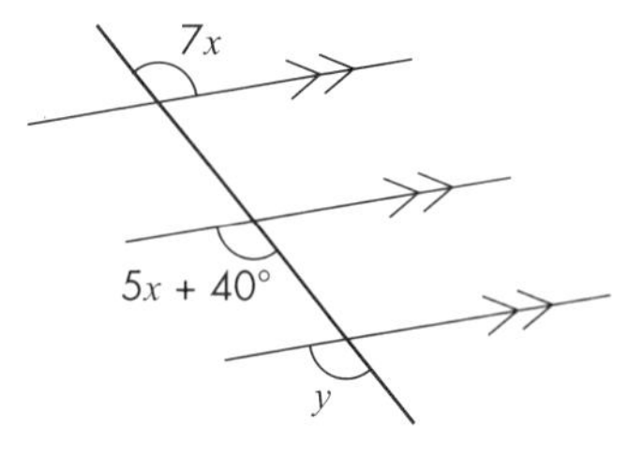

FM 6 A company makes signs in the shape of a chevron.

This is one of their signs. It has one line of symmetry.

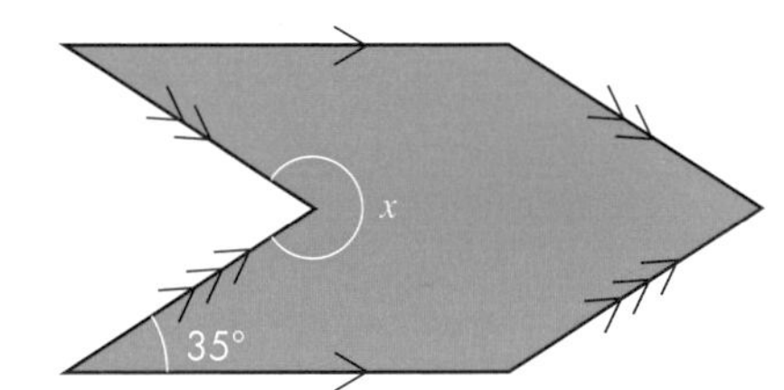

HINTS AND TIPS

Draw the line of symmetry on the shape first.

The designer for the company needs to know the size of the angle marked x on the diagram.

Work out the size of angle x.

PS 7 In the diagram, AE is parallel to BD.

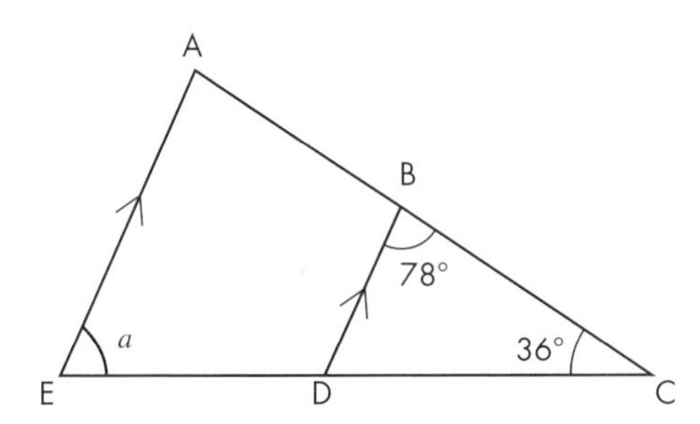

Work out the size of angle a.

Give clear reasons as to how you obtained your answers.

D

AU **8** Lizzie is writing out a solution to this question.

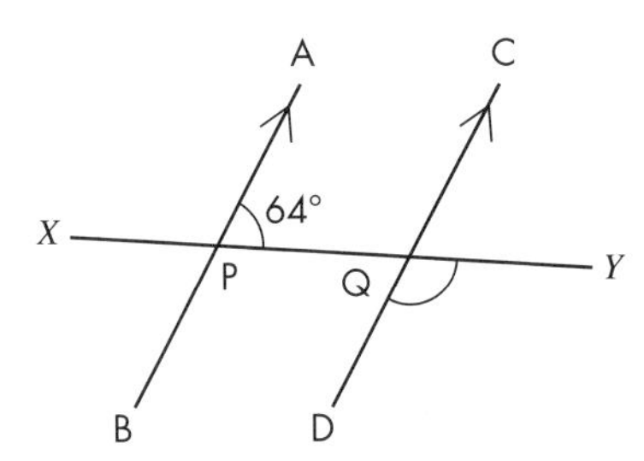

The line XY crosses the parallel lines AB and CD at P and Q.

Work out the size of angle DQY.

Give reasons for your answer.

This is her solution.

Angle PQD = 64° (corresponding angles)

So angle DQY = 124° (angles on a line = 190°)

Lizzie has made a number of errors in her solution.

Write out a correct solution for the question.

C

9 Use the diagram to prove that the three angles in a triangle add up to 180°.

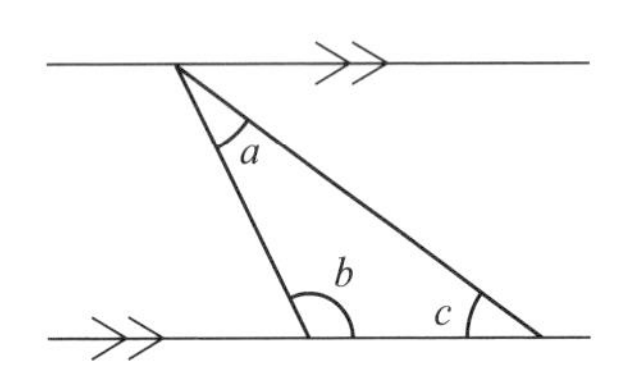

10 Prove that $p + q + r = 180°$.

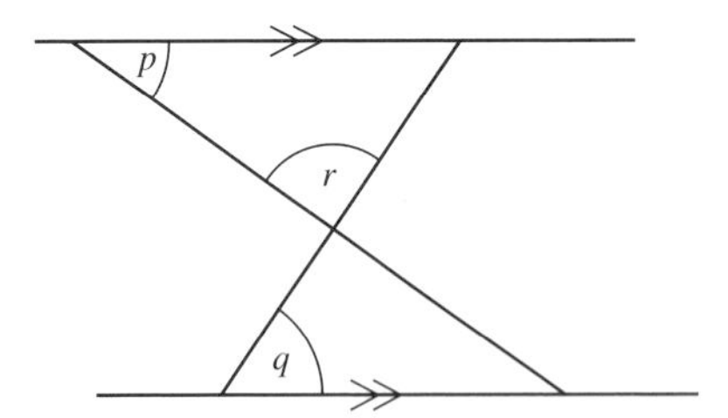

7.5 Special quadrilaterals

This section will show you how to:

- use angle properties in quadrilaterals

Key words
kite
parallelogram
rhombus
trapezium

You should know the names of the following quadrilaterals, be familiar with their angle properties and know how to use the three-letter notation to describe any angle.

Parallelogram

- A **parallelogram** has opposite sides parallel.
- Its opposite sides are equal.
- Its diagonals bisect each other.
- Its opposite angles are equal. That is:

  angle BAD = angle BCD
  angle ABC = angle ADC

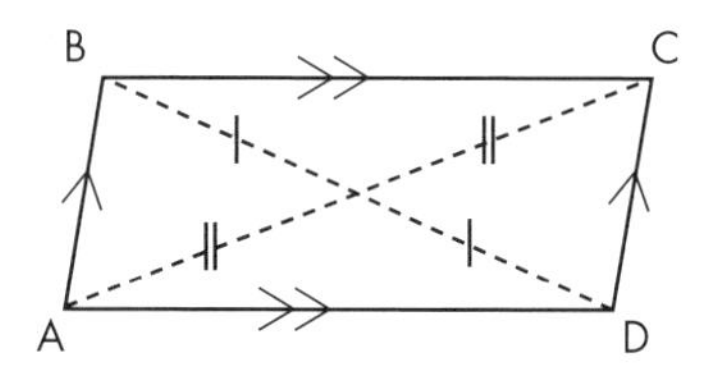

Rhombus

- A **rhombus** is a parallelogram with all its sides equal.
- Its diagonals bisect each other at right angles.
- Its diagonals also bisect the angles.

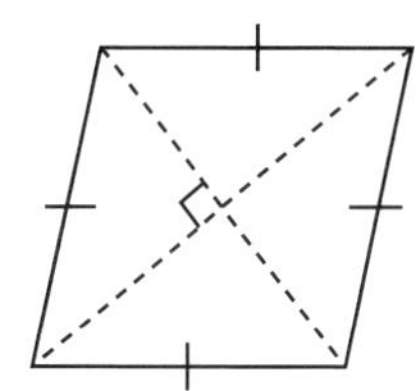

Kite

- A **kite** is a quadrilateral with two pairs of equal adjacent sides.
- Its longer diagonal bisects its shorter diagonal at right angles.
- The opposite angles between the sides of different lengths are equal.

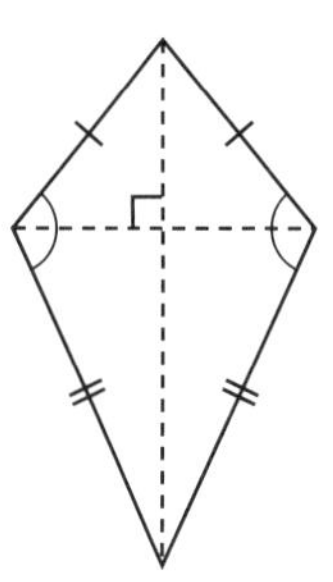

Trapezium

- A **trapezium** has two parallel sides.
- The sum of the interior angles at the ends of each non-parallel side is 180°. That is:

  angle BAD + angle ADC = 180°
  angle ABC + angle BCD = 180°

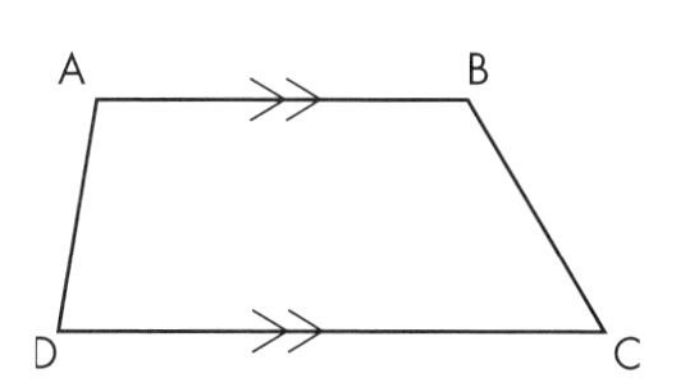

EXERCISE 7E

1 For each of the trapeziums, calculate the sizes of the lettered angles.

a

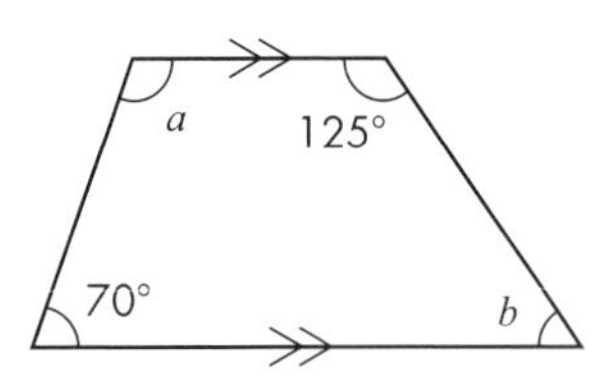

b

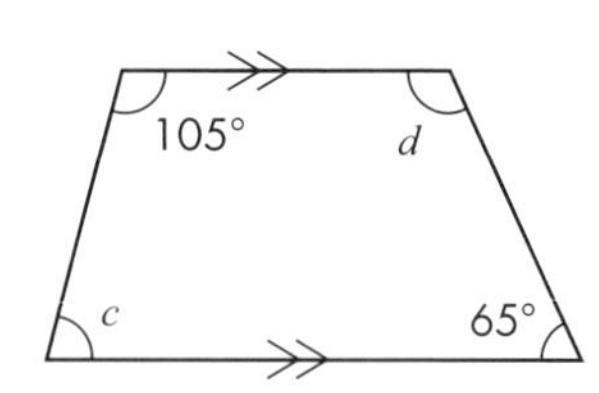

c

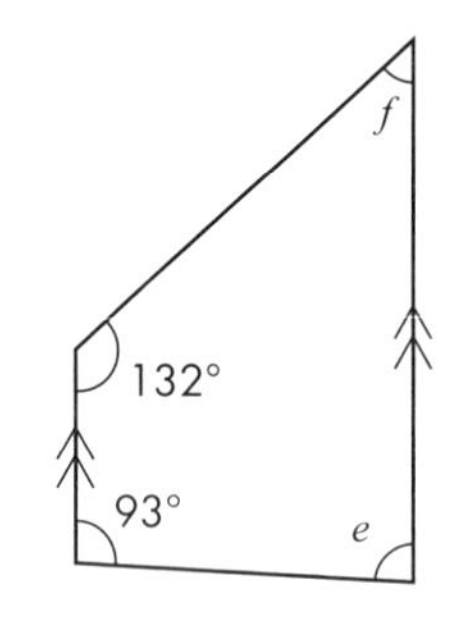

2 For each of these parallelograms, calculate the sizes of the lettered angles.

a

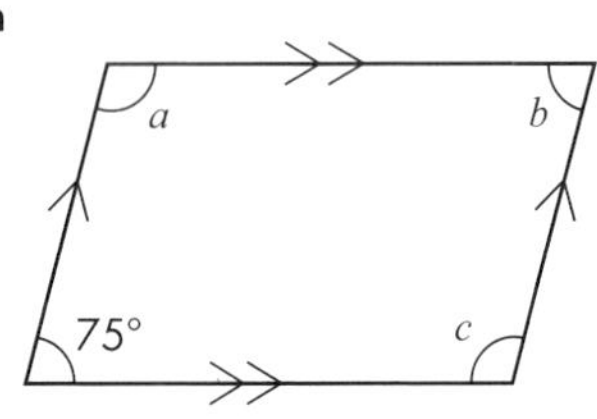

b

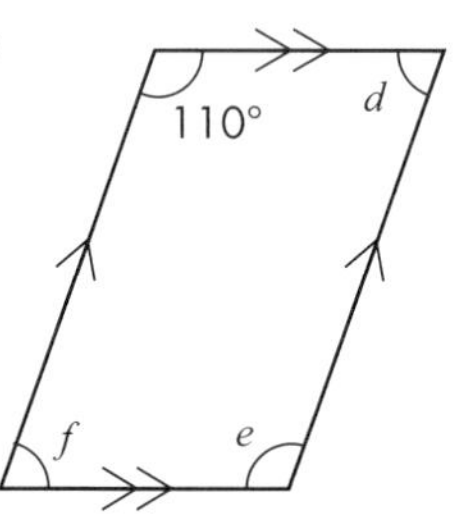

c

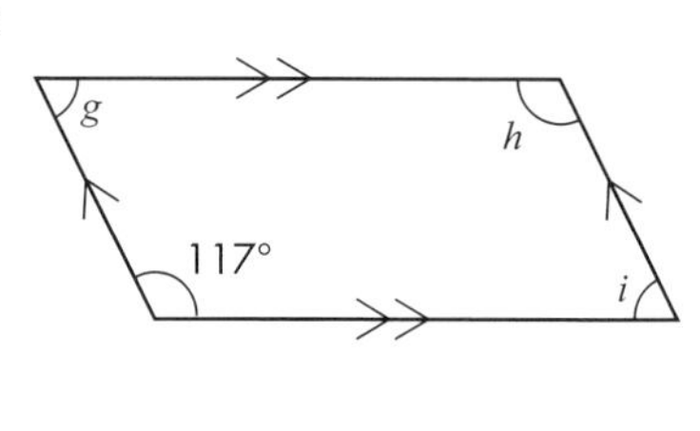

3 For each of these kites, calculate the sizes of the lettered angles.

a

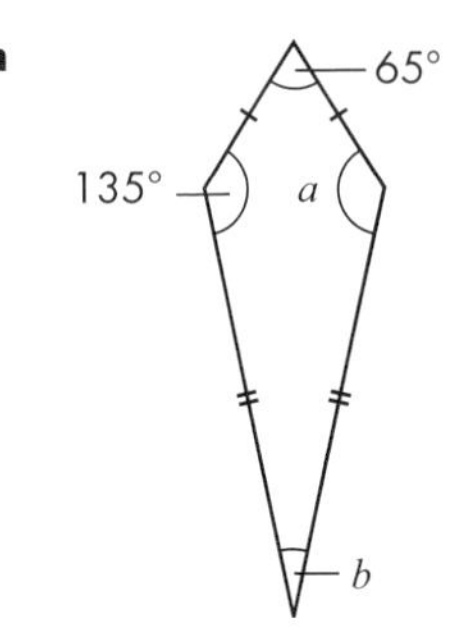

b

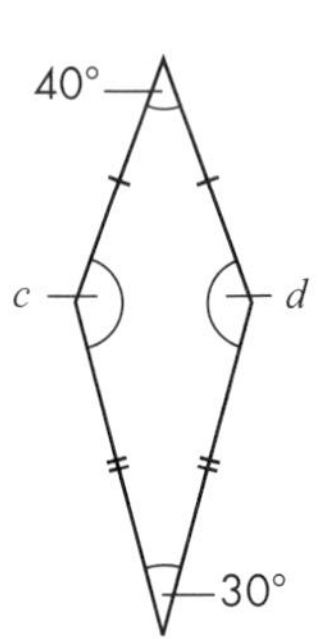

c

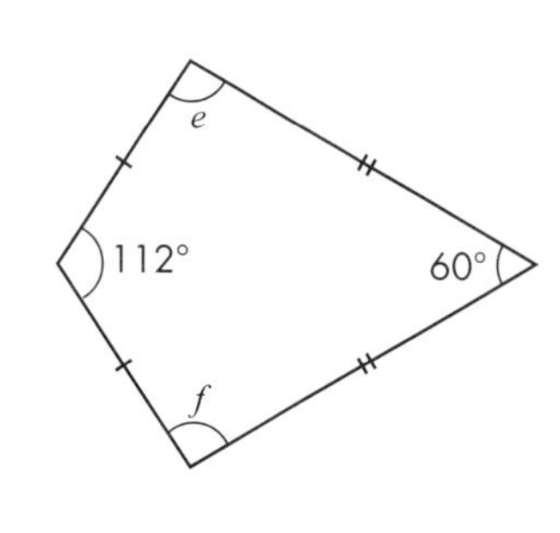

4 For each of these rhombuses, calculate the sizes of the lettered angles.

a

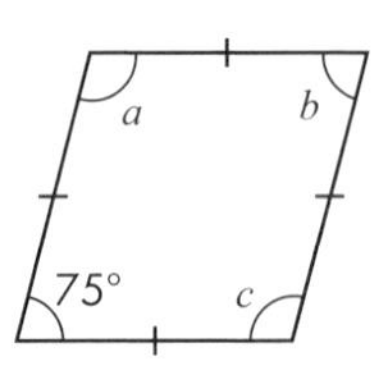

b

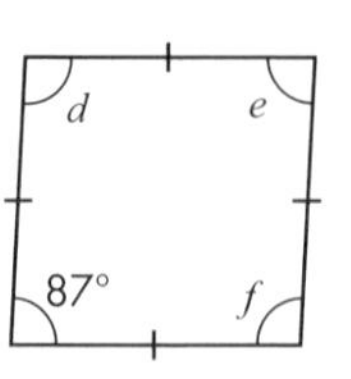

c

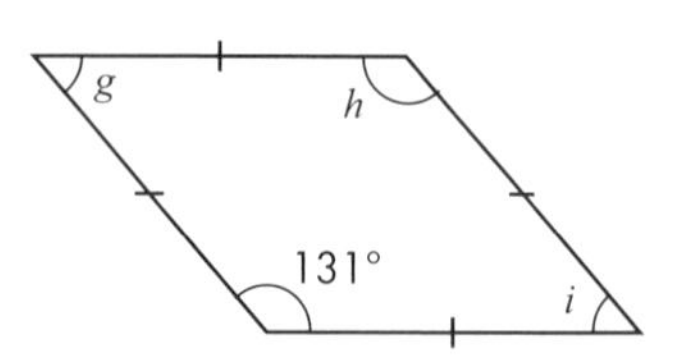

5 For each of these shapes, calculate the sizes of the lettered angles.

a

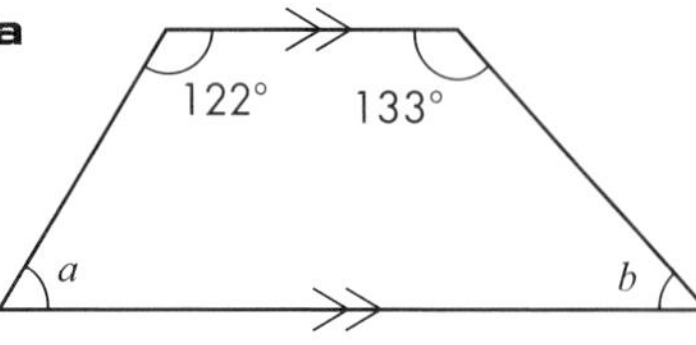

b

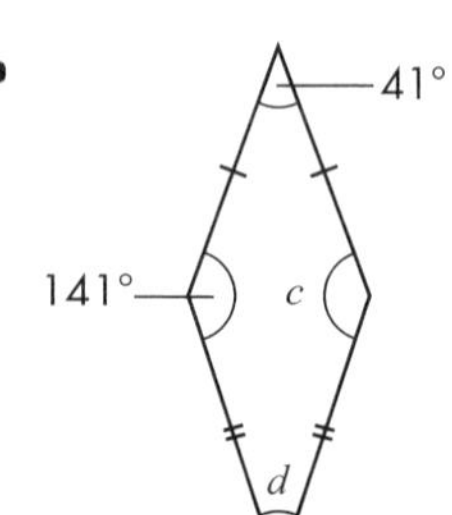

c

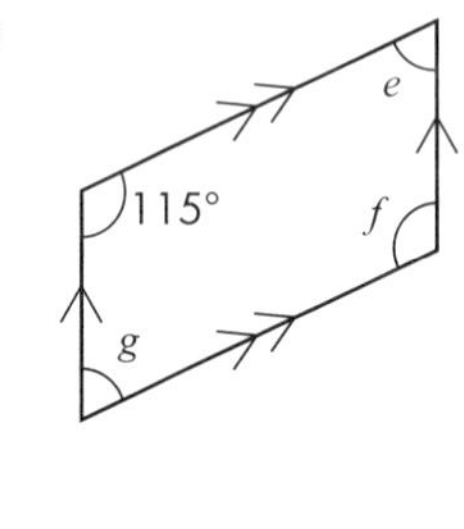

D

6 Dani is making a kite.

She needs angle C to be half the size of angle A.

Work out the size of angles B and D.

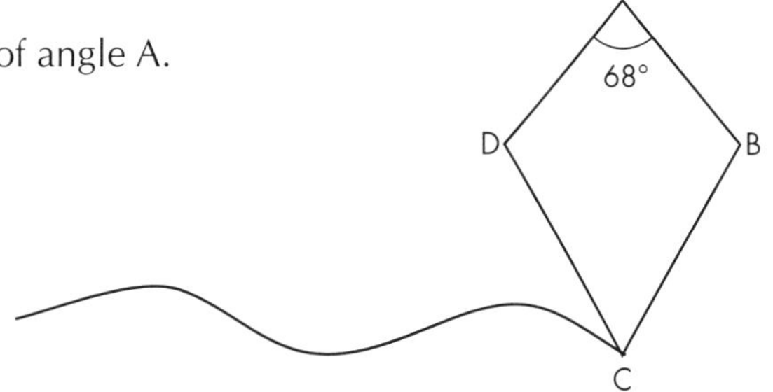

AU PS 7 David says that a parallelogram is a special type of rectangle.

Marie says that he is wrong and that a rectangle is a special type of parallelogram.

Who is correct?

Give a reason for your answer.

AU 8 The diagram shows a quadrilateral ABCD.

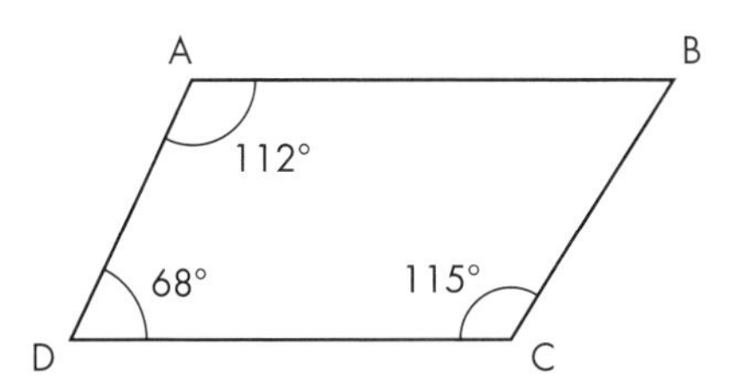

a Calculate the size of angle B.

b What special name is given to the quadrilateral ABCD? Explain your answer.

GRADE BOOSTER

- **F** You can measure and draw angles
- **F** You know that the sum of the angles on a line is 180°
- **F** You know that the sum of the angles at a point is 360°
- **E** You know that the sum of the angles in a triangle is 180°
- **E** You can find the exterior angle of a triangle
- **D** You can find angles in parallel lines
- **D** You know all the properties of special quadrilaterals

What you should know now

- How to measure and draw angles
- How to find angles on a line or at a point
- How to find angles in triangles

1 Which of the marked angles is an acute angle? (1)

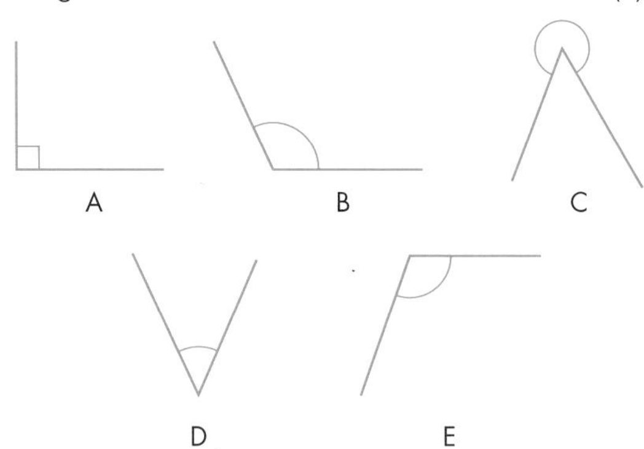

(Total 1 mark)

Edexcel, March 2009, Paper 7 Foundation, Unit 2 Stage 1, Question 5

2 The diagram shows three angles on a straight line AB.

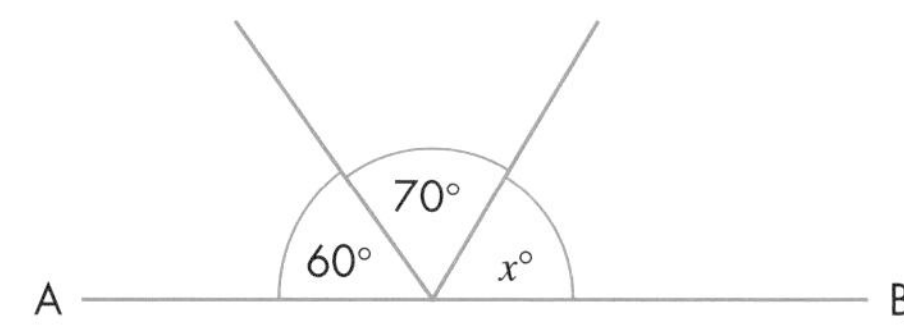

Work out the value of x.

3

x° 30°

Diagram **not** accurately drawn

a **i** Write down the value of x. (1)

ii Give a reason for your answer. (1)

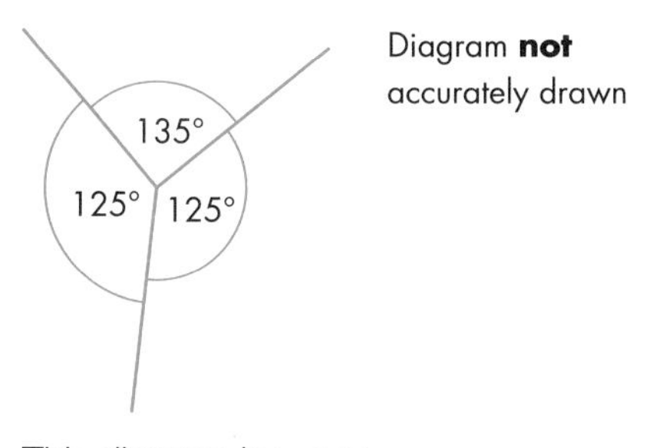

Diagram **not** accurately drawn

This diagram is wrong.

b Explain why. (1)

(Total 3 marks)

Edexcel, June 2008, Paper 2 Foundation, Question 14

4

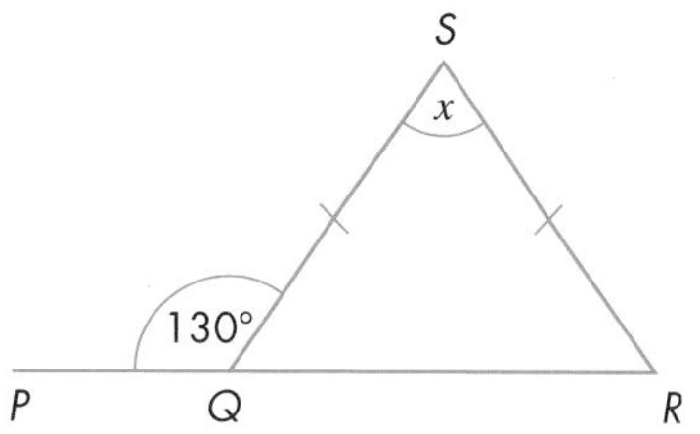

Diagram **not** accurately drawn

PQR is a straight line.

SQ = *SR*.

What is the size of the angle marked x? (1)

50°	130°	70°	80°	60°
A	**B**	**C**	**D**	**E**

(Total 1 mark)

Edexcel, March 2009, Paper 7 Foundation, Unit 2 Stage 1, Question 20

5

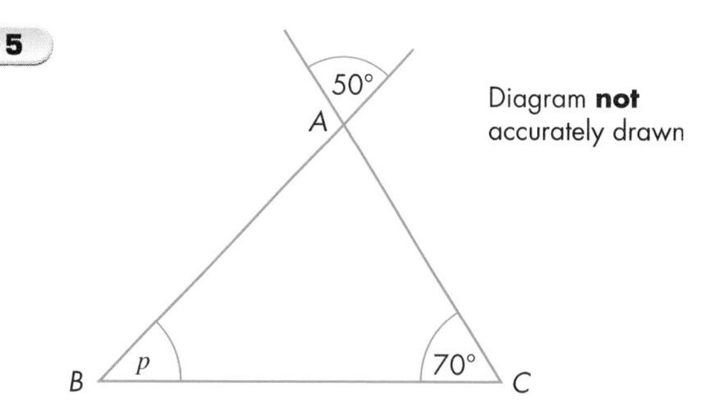

Diagram **not** accurately drawn

ABC is a triangle.

Work out the size of the angle marked p. (2)

(Total 2 marks)

Edexcel, June 2008, Paper 10 Foundation, Unit 3 Test, Question 7

6 The diagram shows a kite.

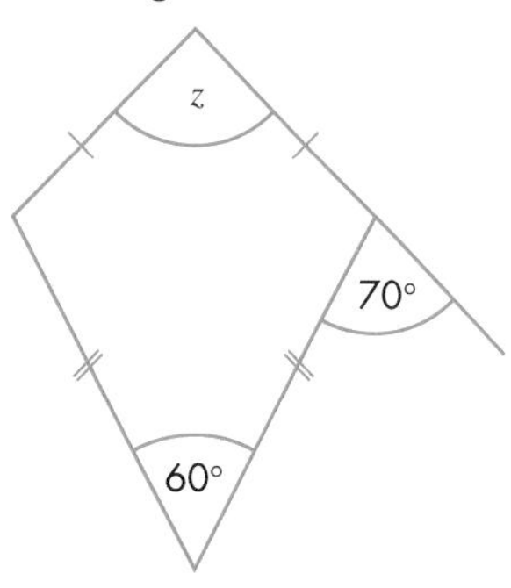

Calculate the size of the angle marked z.

7 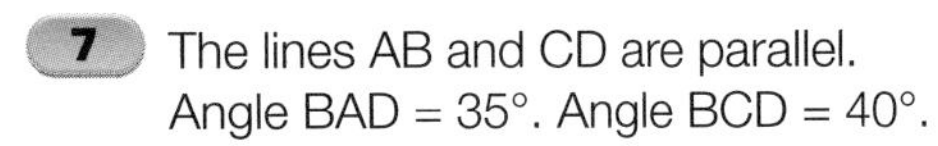
The lines AB and CD are parallel.
Angle BAD = 35°. Angle BCD = 40°.

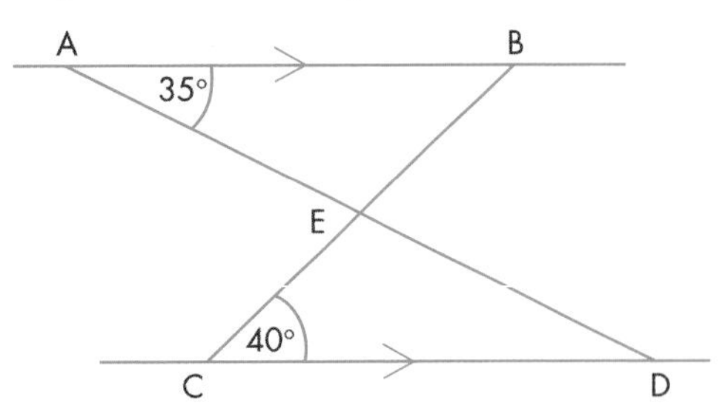

Show clearly why angle AEC is 75°.

8

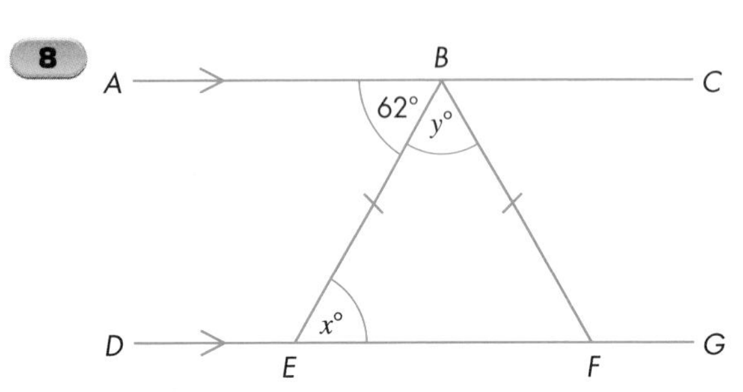

Diagram **not** accurately drawn

ABC and *DEFG* are straight lines.

AC is parallel to *DG*.

BE = *BF*.

Angle *ABE* = 62°.

a **i** Find the value of x. (1)

ii Give a reason for your answer. (1)

b Work out the value of y. (2)

(Total 4 marks)

Edexcel, March 2007, Paper 10 Foundation, Unit 3 Test, Question 7

9

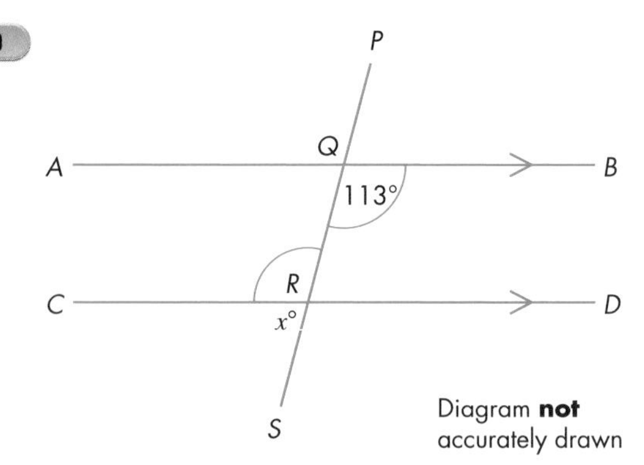

Diagram **not** accurately drawn

AQB, *CRD* and *PQRS* are straight lines.

AB is parallel to *CD*.

Angle *BQR* = 113°.

Work out the value of x. (2)

(Total 2 marks)

Edexcel, November 2007, Paper 10 Foundation, Unit 3 Test, Question 6

Worked Examination Questions

1 The lines AB and CD are **parallel**.

a Write down the value of a.
Give a reason for your answer.

b Write down the value of b.
Give a reason for your answer.

c Work out the value of c.
Give a reason for your answer.

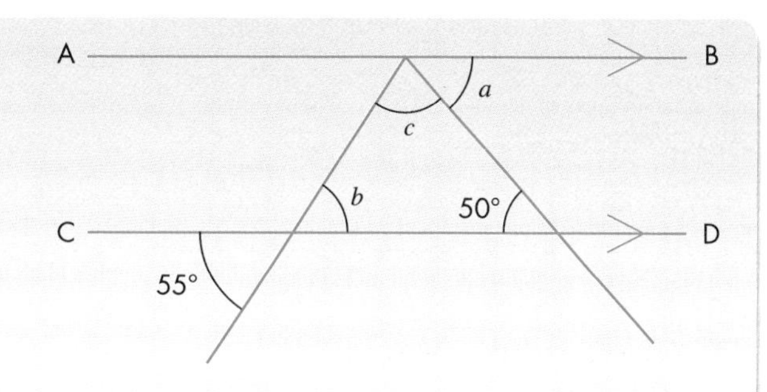

a $a = 50°$

It is an *alternate angle* between the parallel lines.

2 marks

b $b = 55°$

It is an *opposite angle* in two intersecting lines.

2 marks

c $c = 75°$

The three angles in the triangle add up to 180°.

2 marks

Total: 6 marks

Never measure the angles with a protractor as the diagrams in examinations are not drawn accurately.

You get 1 accuracy mark for the correct answer.

You get 1 mark for a correct reason.
Stating "It is a Z angle" is not acceptable in examinations. You must identify it as an alternate angle to get a mark for quality of written communication.

You get 1 accuracy mark for the correct answer.

You get 1 mark for a correct reason.

You get 1 accuracy mark for the correct answer.

You get 1 mark for a correct reason.

Functional Maths
Origami

Product designers will make a model of their product before the final product is manufactured. They use these models to trial products and to ensure that their design is suitable and fully-functioning.

In this task you are going to make a model of a chip holder, following instructions and using your knowledge of angles and properties of shapes to calculate unknown angles.

Getting started

- What is an angle?
- If one of the angles in a triangle is 36°, what could the other two angles be?
- Name some quadrilaterals that have at least one pair of parallel sides.
  - How many can you name?
  - What other properties do they have?

Your task

Alan owns a fish and chip shop. He uses plastic containers for his chips, but he wants to switch to paper holders. He has seen a design for a chip holder on the internet and thinks that it will work for him.

1 Follow the instructions to make the chip holder. Write a summary, giving the name of the shape, the effect of changing the size of the holder on the angles, and a description of how effective the holder would be.

2 Alan wants the chip holders to sit in a rack. To do this, he needs to work out the angles of his chip holder. However, he does not have a protractor.

Use your knowledge of angles and properties of shape to calculate the angles for Alan. Then, design a rack to fit the chip holders. You must include the rack's dimensions, angles and at least two different elevations (views) of the final product design.

Instructions for Alan's design

1 Fold a square piece of paper in half along its diagonal.

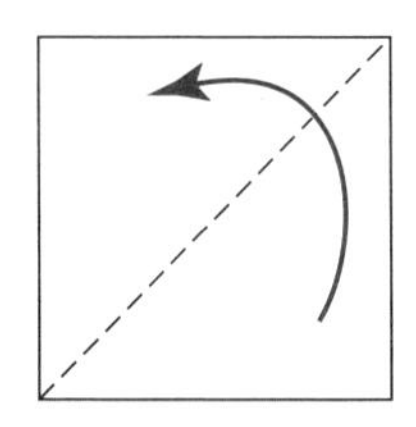

to

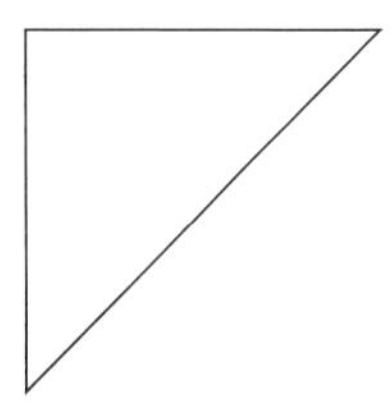

2 Turn the paper so that the longest side is along the bottom. Fold corner A to meet the opposite side, so that the top and bottom edges are parallel.

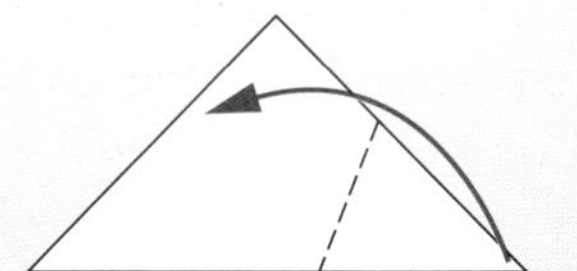

to

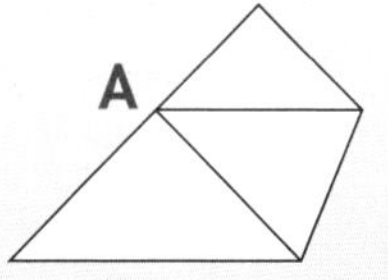

3 Fold corner B to meet point C.

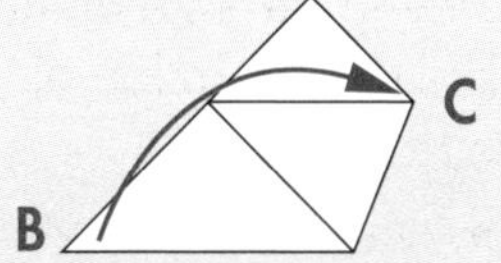

to

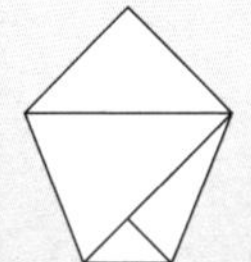

4 Fold the top layer down so that it lies over the folded sides.

to

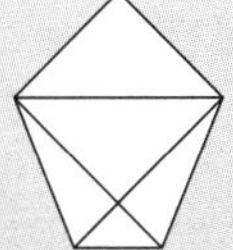

5 Turn over and repeat step 4 for this side.

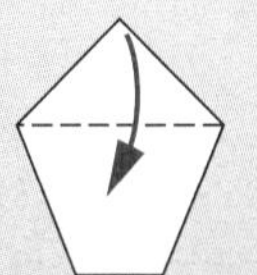

to

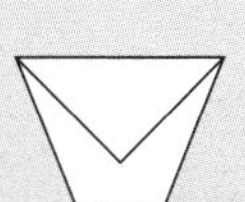

Extension

- Find a way to check the angles that you have worked out.
- In step 2 of the instructions, you need to fold point B. There is a way to deduce the position to which you need to fold point B, without relying on estimating by eye. What is this and why does it work?

Why this chapter matters

If you were asked to circle one of these to describe mathematics, which would it be?

Art   Science   Sport   Language

In fact, you could circle them all.

Art

Mathematics is important in art. The *Mona Lisa*, probably the most famous painting in the world, uses the proportions of the 'golden ratio' (approximately 1.618) as shown by the red rectangles marked on this copy of the painting. This 'golden ratio' is supposed to be particularly attractive to the human eye.

Science

Obviously, you cannot do much science without using mathematics. In 1962, a *Mariner* space probe went off course and had to be destroyed because someone had used a wrong symbol in a mathematical formula that was part of its programming.

Sport

Is mathematics a sport? There are national and international competitions each year that use mathematics. For example, there is a world Suduko championship each year and university students compete in the annual 'Mathematics Olympiad'.

Language

But perhaps the most important description in the list above is mathematics as language. As we saw in Chapter 1, mathematics is the only universal language. If you write the equation $3x = 9$, it will be understood by people in all countries.

Algebra is the way that the language of mathematics is expressed.

Algebra comes from the Arabic *al-jabr* which means something similar to 'completion'. It was used in a book written in 820AD by a Persian mathematician called al-Khwarizmi.

The use of symbols then developed until the middle of the 17th century, when René Descartes developed what is regarded as the basis of the algebra we use today.

Algebra: Basic algebra

1 The language of algebra

2 Simplifying expressions

3 Expanding brackets

4 Factorisation

5 Substitution

The grades given in this chapter are target grades.

This chapter will show you …

- **F** how to substitute numbers into expressions and formulae
- **F** how to use letters to represent numbers
- **F** how to form simple algebraic expressions
- **E** how to simplify expressions by collecting like terms
- **D** how to factorise expressions
- **D** how to express simple rules in algebraic form

Visual overview

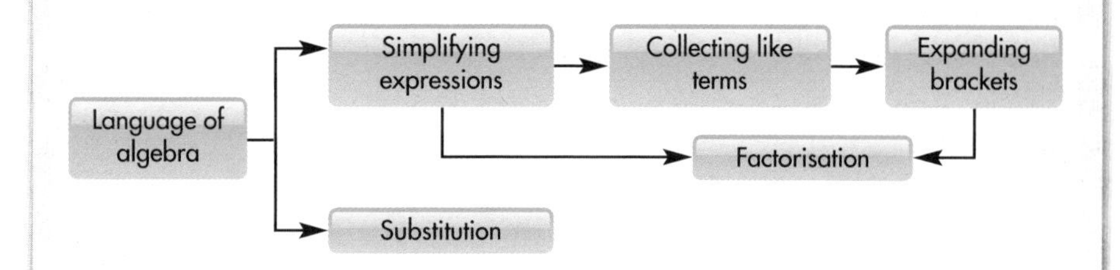

What you should already know

- The **BIDMAS/BODMAS** rule, which gives the order in which you must do the operations of arithmetic when they occur together **(KS3 level 5, GCSE grade F)**

Quick check

1 Write the answer to each expression.

a $(5 - 1) \times 2$

b $5 - (1 \times 2)$

2 Work out $(7 - 5) \times (5 + 4 - 2)$.

3 **a** Put brackets in the calculation to make the answer 40.

$2 + 3 + 5 \times 4$

b Put brackets in the calculation to make the answer 34.

$2 + 3 + 5 \times 4$

8.1 The language of algebra

This section will show you how to:

- use letters, numbers and mathematical symbols to write algebraic expressions and formulae

Key words
equation
expression
formula
solve
symbol
term
variable

Algebra is based on the idea that if something works with numbers, it will work with letters. The main difference is that when you work only with numbers, the answer is also a number. When you work with letters, you get an **expression** as the answer.

Algebra follows the same rules as arithmetic, and uses the same **symbols** (+, –, × and ÷). Below are seven important algebraic rules.

- Write '4 more than x' as $4 + x$ or $x + 4$.
- Write '6 less than p' or 'p minus 6' as $p - 6$.
- Write '4 times y' as $4 \times y$ or $y \times 4$ or $4y$. The last one of these is the neatest way to write it.
- Write 'b divided by 2' as $b \div 2$ or $\frac{b}{2}$.
- When a number and a letter or a letter and a letter appear together, there is a hidden multiplication sign between them. So, $7x$ means $7 \times x$ and ab means $a \times b$.
- Always write '$1 \times x$' as x.
- Write 't times t' as $t \times t$ or t^2.

Here are some algebraic words that you need to know.

Variable: This is what the letters used to represent numbers are called. These letters can take on any value, so they are said to 'vary'.

Expression: This is any combination of letters and numbers. For example, $2x + 4y$ and $\frac{p-6}{5}$ are expressions.

Equation: An equation contains an equals sign and at least one variable. The important fact is that a value can be found for the variable. This is called **solving** the equation. You will learn more about equations in chapter 13.

Formula: These are like equations in that they contain an equals sign, but there is more than one variable and they are rules for working out amounts such as area or the cost of taxi fares.

For example, $V = x^3$, $A = \frac{1}{2}bh$ and $C = 3 + 4m$ are all formulae.

Term: These are the separate parts of expressions, equations or formula.

For example, in $3x + 2y - 7$, there are three terms: $3x$, $+2y$ and -7.

Identify: An identity is similar to an equation, but is true for all values of the variable(s). Instead of the usual equals (=) sign, ≡ is used. For example, $2x \equiv 7x - 5x$.

EXAMPLE 1

What is the area of each of these rectangles?

a 4 cm by 6 cm **b** 4 cm by w cm **c** l cm by w cm

The rule for working out the area of a rectangle is:

area = length × width

So, the area of rectangle **a** is $4 \times 6 = 24$ cm^2

The area of rectangle **b** is $4 \times w = 4w$ cm^2

The area of rectangle **c** is $l \times w = lw$ cm^2

Now, if A represents the area of rectangle **c**:

$A = lw$

This is an example of a rule expressed algebraically.

EXAMPLE 2

What is the perimeter of each of these rectangles?

a 6 cm by 4 cm **b** 4 cm by w cm **c** l cm by w cm

The rule for working out the perimeter of a rectangle is:

perimeter = twice the longer side + twice the shorter side

So, the perimeter of rectangle **a** is $2 \times 6 + 2 \times 4 = 20$ cm

The perimeter of rectangle **b** is $2 \times 4 + 2 \times w = 8 + 2w$ cm

The perimeter of rectangle **c** is $2 \times l + 2 \times w = 2l + 2w$ cm

Now, let P represent the perimeter of rectangle **c**, so:

$P = 2l + 2w$

which is another example of a rule expressed algebraically.

Expressions such as $A = lw$ and $P = 2l + 2w$ are **formulae**.

As the two examples above show, a formula states the connection between two or more quantities, each of which is represented by a different letter.

In a formula, the letters are replaced by numbers when a calculation has to be made. This is called *substitution* and is explained on page 226.

EXAMPLE 3

Say if the following are expressions (E), equations (Q) or formula (F).

a $x - 5 = 7$ **b** $P = 4x$ **c** $2x - 3y$

a is an **equation** (Q) as it can be solved to give $x = 12$.

b is a **formula** (F). This is the formula for the perimeter of a square with a side of x.

c is an **expression** (E) with two terms.

EXERCISE 8A

1 Write down the algebraic expression for:

a 2 more than x

b 6 less than x

c k more than x

d x minus t

e x added to 3

f d added to m

g y taken away from b

h p added to t added to w

i 8 multiplied by x

j h multiplied by j

k x divided by 4

l 2 divided by x

m y divided by t

n w multiplied by t

o a multiplied by a

p g multiplied by itself.

2 Here are four squares.

i

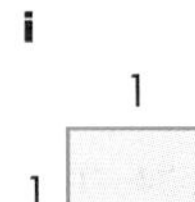

ii

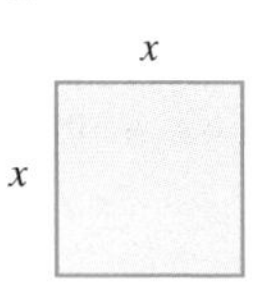

iii

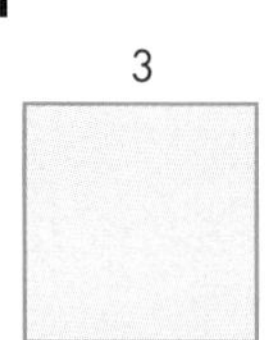

iv

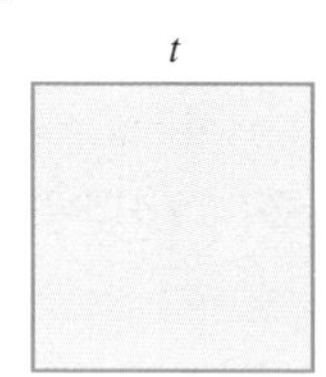

a Work out the area and perimeter of each square.

b Copy and complete these rules.

i The perimeter, P, of a square of side s centimetres is $P = \ldots\ldots$

ii The area, A, of a square of side s centimetres is $A = \ldots\ldots$

3 Asha, Bernice and Charu are three sisters. Bernice is x years old. Asha is three years older than Bernice. Charu is four years younger than Bernice.

a How old is Asha?

b How old is Charu?

4 An approximation method of converting from degrees Celsius to degrees Fahrenheit is given by this rule:

Multiply by 2 and add 30.

Using C to stand for degrees Celsius and F to stand for degrees Fahrenheit, complete this formula.

$F = \ldots\ldots$

5 Cows have four legs. Which of these formulae connects the number of legs (L) and the number of cows (C)?

a $C = 4L$

b $L = C + 4$

c $L = 4C$

d $L + C = 4$

6 There are 3 feet in a yard. The rule $F = 3Y$ connects the number of feet (F) and the number of yards (Y). Write down rules, using the letters shown, to connect:

a the number of centimetres (C) in metres (M)

b the number of inches (N) in feet (F)

c the number of wheels (W) on cars (C)

d the number of heads (H) on people (P).

Check your formula with a numerical example. In 4 yards there are 12 feet, so, if $F = 3Y$ is correct, then $12 = 3 \times 4$, which is true.

7 **a** Anne has three bags of marbles. Each bag contains n marbles. How many marbles does she have altogether?

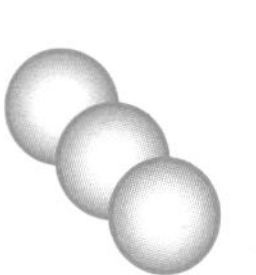

b Bea gives her another three marbles. How many marbles does Anne have now?

c Anne puts one of her new marbles in each bag. How many marbles are there now in each bag?

d Anne takes two marbles out of each bag. How many marbles are there now in each bag?

8 Simon has n cubes.

- Rob has twice as many cubes as Simon.
- Tom has two more than Simon.
- Vic has three fewer than Simon.
- Will has three more than Rob.

How many cubes does each person have?

HINTS AND TIPS

Remember that you do not have to write down a multiplication sign between numbers and letters, or letters and letters.

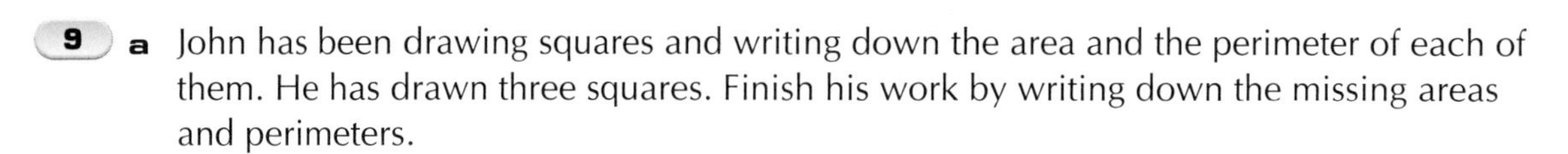

9 **a** John has been drawing squares and writing down the area and the perimeter of each of them. He has drawn three squares. Finish his work by writing down the missing areas and perimeters.

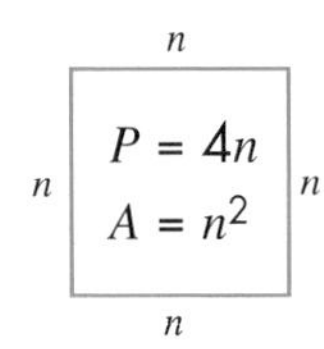

$2n$

$P = \ldots$
$A = 4n^2$

$3n$

$P = 12n$
$A = \ldots$

b Write down the area and the perimeter of this partly-covered square.

10 **a** I go shopping with £10 and spend £6. How much do I have left?

b I go shopping with £10 and spend £x. How much do I have left?

c I go shopping with £y and spend £x. How much do I have left?

d I go shopping with £$3x$ and spend £x. How much do I have left?

11 Give the total cost of:

a five pens at 15p each

b x pens at 15p each

c four pens at Ap each

d y pens at Ap each.

12 A boy went shopping with £A. He spent £B. How much did he have left?

13 Five ties cost £A. What is the cost of one tie?

PS **14** **a** My dad is 72 and I am T years old. How old shall we each be in x years' time?

b My mum is 64 years old. In two years' time she will be twice as old as I am. What age am I now?

15 I am twice as old as my son. I am T years old.

a How old is my son?

b How old will my son be in four years' time?

c How old was I x years ago?

16 What is the perimeter of each of these figures?

a

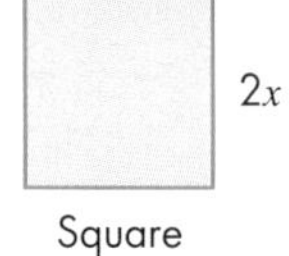

Square

b

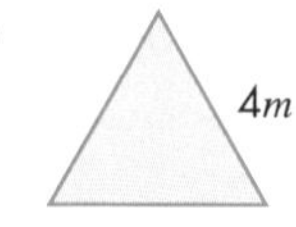

Equilateral triangle

c

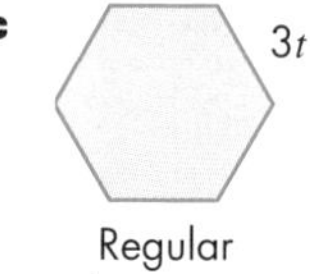

Regular hexagon

E

17 Write down the number of marbles each student ends up with.

Student	Action	Marbles
Andrea	Start with three bags each containing n marbles and give away one marble from each bag	
Bert	Start with three bags each containing n marbles and give away one marble from one bag	
Colin	Start with three bags each containing n marbles and give away two marbles from each bag	
Davina	Start with three bags each containing n marbles and give away n marbles from each bag	
Emma	Start with three bags each containing n marbles and give away n marbles from one bag	
Florinda	Start with three bags each containing n marbles and give away m marbles from each bag	

AU 18 The answer to $3 \times 4m$ is $12m$.

Write down two **different** expressions for which the answer is $12m$.

AU 19 Three expressions for the perimeter of a rectangle with length l and width w are:

$P = l + w + l + w$

$P = 2l + 2w$

$P = 2(l + w)$

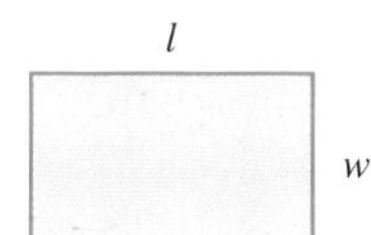

Show, using a numerical example, that they all give the same result.

HINTS AND TIPS

Just pick some easy numbers for l and w.

PS 20 My sister is three years older than I am.

The sum of our ages is 29.

How old am I?

HINTS AND TIPS

If I am x years old, work out how old my sister is in terms of x and use this to set up a simple equation.

PS 21 Ali has 65p and Heidi has 95p. How much should Heidi give to Ali so they both have the same amount?

22 Say if the following are expressions (E), equations (Q) or formula (F).

a $2x - 5$

b $s = \sqrt{A}$

c $2x - 3 = 1$

8.2 Simplifying expressions

This section will show you how to:

- simplify algebraic expressions by multiplying terms
- simplify algebraic expressions by collecting like terms

Key words
coefficient
like terms
simplify

Simplifying an algebraic expression means making it neater and, usually, shorter by combining its terms where possible.

Multiplying expressions

When you multiply algebraic expressions, first you multiply the numbers, then the letters.

EXAMPLE 4

Simplify:

a $2 \times t$ **b** $m \times t$ **c** $2t \times 5$ **d** $3y \times 2m$

The convention is to write the number first then the letters, but if there is no number just put the letters in alphabetical order. The number in front of the letter is called the **coefficient**.

a $2 \times t = 2t$ **b** $m \times t = mt$ **c** $2t \times 5 = 10t$ **d** $3y \times 2m = 6my$

In an examination you will not be penalised for writing $2ba$ instead of $2ab$, but you will be penalised if you write $ab2$ as this can be confused with powers, so *always* write the number first.

EXAMPLE 5

Simplify:

a $t \times t$ **b** $3t \times 2t$ **c** $3t^2 \times 4t$ **d** $2t^3 \times 4t^2$

Multiply the same variables, using powers. The indices are added together.

a $t \times t = t^2$ (Remember: $t = t^1$) **b** $3t \times 2t = 6t^2$

c $3t^2 \times 4t = 12t^3$ **d** $2t^3 \times 4t^2 = 8t^5$

EXERCISE 8B

> **HINTS AND TIPS**
> Remember to multiply numbers and add indices.

E

1 Simplify the following expressions.

a $2 \times 3t$ **b** $5y \times 3$ **c** $2w \times 4$

d $5b \times b$ **e** $2w \times w$ **f** $4p \times 2p$

g $3t \times 2t$ **h** $5t \times 3t$ **i** $m \times 2t$

j $5t \times q$ **k** $n \times 6m$ **l** $3t \times 2q$

m $5h \times 2k$ **n** $3p \times 7r$

D

AU 2 **a** Which of the following expressions are equivalent?

$2m \times 6n$ $4m \times 3n$ $2m \times 6m$ $3m \times 4n$

b The expressions $2x$ and x^2 are the same for only two values of x. What are these values?

PS 3 A square and a rectangle have the same area.

The rectangle has sides $2x$ cm and $8x$ cm.

What is the length of a side of the square?

C

4 Simplify the following expressions.

a $y^2 \times y$ **b** $3m \times m^2$ **c** $4t^2 \times t$

d $3n \times 2n^2$ **e** $t^2 \times t^2$ **f** $h^3 \times h^2$

g $3n^2 \times 4n^3$ **h** $3a^4 \times 2a^3$ **i** $k^5 \times 4k^2$

j $-t^2 \times -t$ **k** $-4d^2 \times -3d$ **l** $-3p^4 \times -5p^2$

m $3mp \times p$ **n** $3mn \times 2m$ **o** $4mp \times 2mp$

FM 5 There are 2000 students at Highville school. One student starts a rumour by telling it to two other students. The next day, those two students each tell the rumour to two other students who have not heard it already. The next day, those four students each tell the rumour to two other students who have not heard it before, and so on. How many days will it be before the whole school has heard the rumour?

> **HINTS AND TIPS**
> Fill in a table like this.

Day	1	2	3	4
Number told	2	4	8	16
Total who know	3	7	15	31

> In an examination there will always be space to the right of the table to draw in two columns – one for the mid-point and the other for the mid-point frequency.

Collecting like terms

Like terms are those that are multiples of the same variable or of the same combination of variables. For example, a, $3a$, $9a$, $\frac{1}{4}a$ and $-5a$ are all like terms.

So are $2xy$, $7xy$ and $-5xy$, and so are $6x^2$, x^2 and $-3x^2$.

Collecting like terms generally involves two steps.

- Collect like terms into groups.
- Then combine the like terms in each group.

Only like terms can be added or subtracted to simplify an expression. For example,

$a + 3a + 9a - 5a$	simplifies to	$8a$
$2xy + 7xy - 5xy$	simplifies to	$4xy$

Note that the variable does not change. All you have to do is combine the coefficients.

For example,

$$6x^2 + x^2 - 3x^2 = (6 + 1 - 3)x^2 = 4x^2$$

But an expression such as $4p + 8t + 5x - 9$ cannot be simplified, because $4p$, $8t$, $5x$ and 9 are *not like terms*, which *cannot* be combined.

EXAMPLE 6

Simplify the expression:

$7x^2 + 3y - 6z + 2x^2 + 3z - y + w + 9$

Write out the expression: $7x^2 + 3y - 6z + 2x^2 + 3z - y + w + 9$

Then collect like terms: $\boxed{7x^2 + 2x^2}\ \boxed{+3y - y}\ \boxed{-6z + 3z}\ \boxed{+w}\ \boxed{+9}$

Then combine them: $9x^2 \quad + \quad 2y \quad - \quad 3z \quad + w \quad + 9$

So, the expression in its simplest form is:

$9x^2 + 2y - 3z + w + 9$

EXERCISE 8C

1 Joseph is given £t, John has £3 more than Joseph, and Joy has £$2t$.

a How much more money has Joy than Joseph?

b How much do the three of them have altogether?

2 Write down an expression for the perimeter of each of these shapes.

a

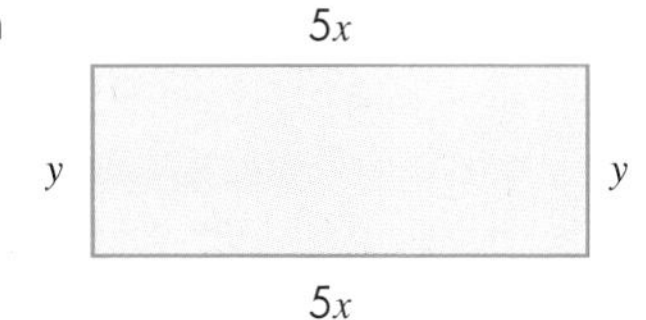

b

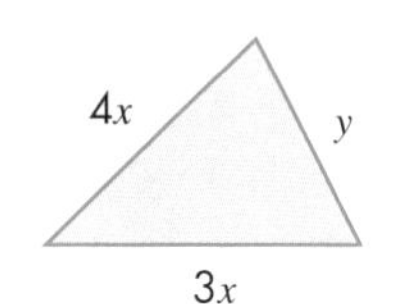

c

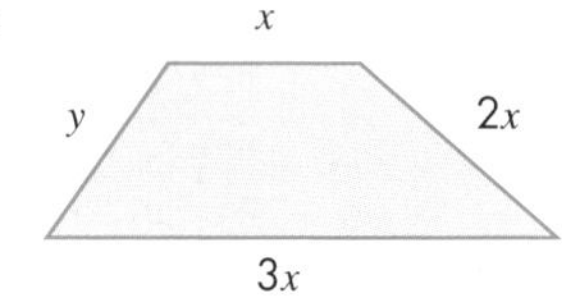

3 Write each of these expressions in a shorter form.

a $a + a + a + a + a$ **b** $c + c + c + c + c + c$

c $4e + 5e$ **d** $f + 2f + 3f$

e $5j + j - 2j$ **f** $9q - 3q - 3q$

g $3r - 3r$ **h** $2w + 4w - 7w$

i $5x^2 + 6x^2 - 7x^2 + 2x^2$ **j** $8y^2 + 5y^2 - 7y^2 - y^2$

k $2z^2 - 2z^2 + 3z^2 - 3z^2$

HINTS AND TIPS

The term a has a coefficient of 1, i.e. $a = 1a$, but you do not need to write the 1.

4 Simplify each of the following expressions.

a $3x + 4x$ **b** $5t - 2t$

c $-2x - 3x$ **d** $-k - 4k$

e $m^2 + 2m^2 - m^2$ **f** $2y^2 + 3y^2 - 5y^2$

HINTS AND TIPS

Remember that only **like** terms can be added or subtracted.
If all the terms cancel out, just write 0 rather than $0x^2$, for example.

5 Simplify each of the following expressions.

a $5x + 8 + 2x - 3$ **b** $7 - 2x - 1 + 7x$

c $4p + 2t + p - 2t$ **d** $8 + x + 4x - 2$

e $3 + 2t + p - t + 2 + 4p$ **f** $5w - 2k - 2w - 3k + 5w$

g $a + b + c + d - a - b - d$ **h** $9k - y - 5y - k + 10$

6 Simplify these expressions. (Be careful – two of them will not simplify.)

a $c + d + d + d + c$ **b** $2d + 2e + 3d$

c $f + 3g + 4h$ **d** $5u - 4v + u + v$

e $4m - 5n + 3m - 2n$ **f** $3k + 2m + 5p$

g $2v - 5w + 5w$ **h** $2w + 4y - 7y$

i $5x^2 + 6x^2 - 7y + 2y$ **j** $8y^2 + 5z - 7z - 9y^2$

k $2z^2 - 2x^2 + 3x^2 - 3z^2$

7 Find the perimeter of each of these shapes, giving it in its simplest form.

a

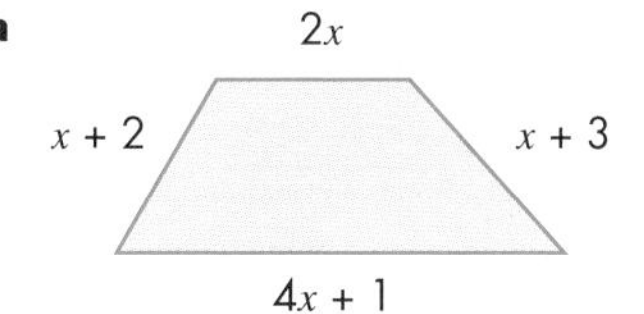

b

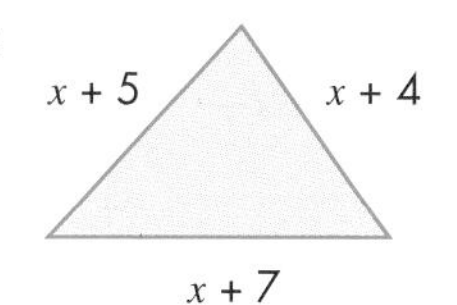

c

$x + 3$

$y + 1$

E

AU 8 $3x + 5y + 2x - y = 5x + 4y$

Write down two other **different** expressions which are equal to $5x + 4y$.

D

AU 9 Find the missing terms to make these equations true.

a $4x + 5y + \ldots\ldots\ldots - \ldots\ldots\ldots = 6x + 3y$

b $3a - 6b - \ldots\ldots\ldots + \ldots\ldots\ldots = 2a + b$

PS 10 ABCDEF is an L-shape.

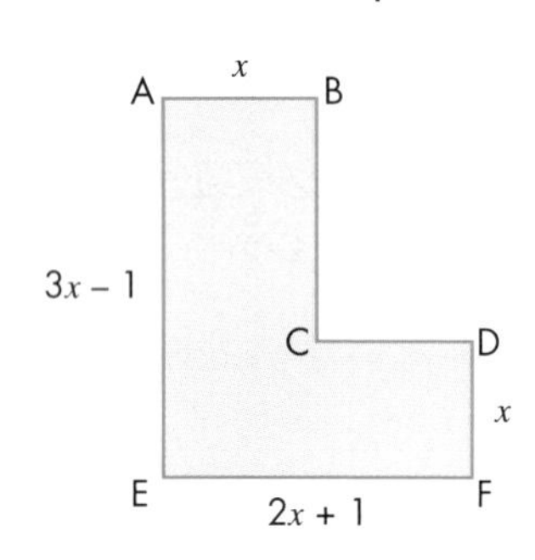

$AB = DF = x$

$AE = 3x - 1$ and
$EF = 2x + 1$

HINTS AND TIPS

Make sure your explanation uses expressions. Do not try to explain in words alone.

a Explain why the length $BC = 2x - 1$.

b Find the perimeter of the shape in terms of x.

c If $x = 2.5$ cm, what is the perimeter of the shape?

C

FM 11 Sean wants to measure the size of a rectangular lawn but he does not have a tape measure. Instead he measures the length and width using his pace and his shoe length. He finds that the length is four paces plus a shoe length and the width is two paces and two shoe lengths. He writes this as $4p + s$ and $2p + 2s$.

$4p + s$

$2p + 2s$

HINTS AND TIPS

Convert 1 m to 100 cm, then convert the answer back to metres.

a Work out the perimeter in terms of p and s.

b Later he finds that his pace is 1 m and his shoe length is 25 cm. What is the actual perimeter of the lawn?Give your answer in metres.

AU 12 A teacher asks her class to work out the perimeter of this L shape.

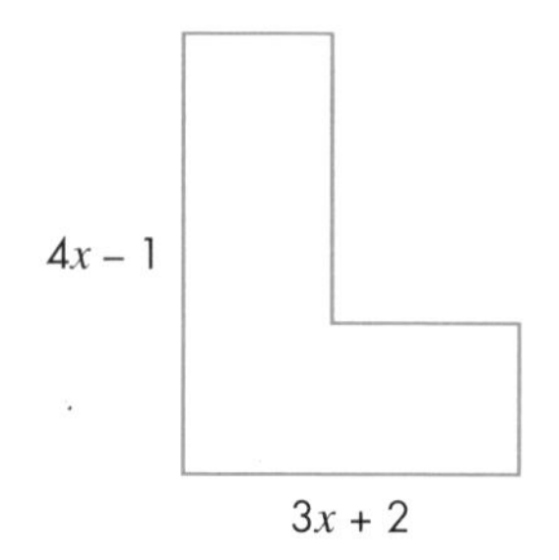

Tia says: 'There is information missing so you cannot work out the perimeter.'

Maria says: 'The perimeter is $4x - 1 + 4x - 1 + 3x + 2 + 3x + 2$.'

Who is correct?

Explain your answer.

8.3 Expanding brackets

This section will show you how to:

- expand brackets such as $2(x - 3)$
- expand and simplify brackets

Key words
expand
multiply out
simplify

Expanding

In mathematics, the term '**expand**' usually means '**multiply out**'. For example, expressions such as $3(y + 2)$ and $4y^2(2y + 3)$ can be expanded by multiplying them out.

Remember that there is an invisible multiplication sign between the outside number and the opening bracket. So $3(y + 2)$ is really $3 \times (y + 2)$, and $4y^2(2y + 3)$ is really $4y^2 \times (2y + 3)$.

You expand by multiplying *everything inside* the brackets by what is outside the brackets.

EXAMPLE 7

Expand $3(y + 2)$.

$$3(y + 2) = 3 \times (y + 2) = 3 \times y + 3 \times 2 = 3y + 6$$

EXAMPLE 8

Expand $4y^2(2y + 3)$.

$$4y^2(2y + 3) = 4y^2 \times (2y + 3) = 4y^2 \times 2y + 4y^2 \times 3 = 8y^3 + 12y^2$$

Look at these next examples of expansion, which show how each term inside the brackets has been multiplied by the term outside the brackets.

$2(m + 3) = 2m + 6$

$3(2t + 5) = 6t + 15$

$m(p + 7) = mp + 7m$

$x(x - 6) = x^2 - 6x$

$4t(t + 2) = 4t^2 + 8t$

$y(y^2 - 4x) = y^3 - 4xy$

$3x^2(4x + 5) = 12x^3 + 15x^2$

$3(2 + 3x) = 6 + 9x$

$2x(3 - 4x) = 6x - 8x^2$

$3t(2 + 5t - p) = 6t + 15t^2 - 3pt$

EXERCISE 8D

D

1 Expand these expressions.

a $2(3 + m)$ **b** $5(2 + l)$ **c** $3(4 - y)$ **d** $4(5 + 2k)$

e $4(3d - 2n)$ **f** $t(t + 3)$ **g** $m(m + 5)$ **h** $k(k - 3)$

i $g(3g + 2)$ **j** $y(5y - 1)$ **k** $p(5 - 3p)$ **l** $3m(m + 4)$

m $3t(5 - 4t)$ **n** $3d(2d + 4e)$ **o** $2y(3y + 4k)$ **p** $5m(3m - 2p)$

C

2 Expand these expressions.

a $y(y^2 + 5)$ **b** $h(h^3 + 7)$ **c** $k(k^2 - 5)$ **d** $3t(t^2 + 4)$

e $3d(5d^2 - d^3)$ **f** $3w(2w^2 + t)$ **g** $5a(3a^2 - 2b)$ **h** $3p(4p^3 - 5m)$

i $m^2(5 + 4m)$ **j** $t^3(t + 2t^2)$ **k** $g^2(5t - 4g^2)$ **l** $3t^2(5t + m)$

FM 3 The local supermarket is offering £1 off a large tin of biscuits. Morris wants five tins.

a If the original price of one tin is £t, which of the expressions below represents how much it will cost Morris to buy five tins?

$5(t - 1)$ $5t - 1$ $t - 5$ $5t - 5$

b Morris has £20 to spend. If each tin is £4.50, will he have enough money for five tins? Show your working to justify your answer.

AU 4 Dylan wrote the following.

$3(5x - 4) = 8x - 4$

Dylan has made two mistakes.

Explain the mistakes that Dylan has made.

HINTS AND TIPS

It is not enough just to give the right answer. You must try to explain, for example, why Dylan wrote $8x$ and what he should really have written if this is wrong.

PS 5 The expansion $2(x + 3) = 2x + 6$ can be shown by this diagram.

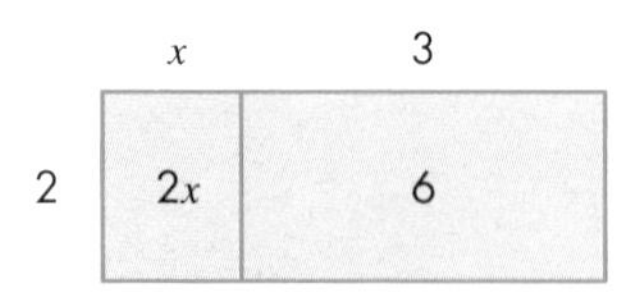

a What expansion is shown by this diagram?

3	$6y$	9

b Write down an expansion that can be shown by this diagram.

$12z$	8

Expand and simplify

This usually means that you need to expand more than one set of brackets and **simplify** the resulting expressions.

You will often be asked to expand and simplify expressions.

EXAMPLE 9

Expand and simplify $3(4 + m) + 2(5 + 2m)$.

$$3(4 + m) + 2(5 + 2m) = 12 + 3m + 10 + 4m = 22 + 7m$$

EXAMPLE 10

Expand and simplify $3t(5t + 4) - 2t(3t - 5)$.

$$3t(5t + 4) - 2t(3t - 5) = 15t^2 + 12t - 6t^2 + 10t = 9t^2 + 22t$$

Notice that multiplying $-2t$ by -5 gives an answer of $+10t$.

EXAMPLE 11

Expand and simplify $4a(2b - 3f) - 3b(a + 2f)$.

$$4a(2b - 3f) - 3b(a + 2f) = 8ab - 12af - 3ab - 6bf = 5ab - 12af - 6bf$$

EXERCISE 8E

E

1 Simplify these expressions.

a $4t + 3t$ **b** $2y + y$ **c** $3d + 2d + 4d$

d $5e - 2e$ **e** $4p - p$ **f** $3t - t$

g $2t^2 + 3t^2$ **h** $3ab + 2ab$ **i** $7a^2d - 4a^2d$

C

2 Expand and simplify these expressions.

a $3(4 + t) + 2(5 + t)$ **b** $5(3 + 2k) + 3(2 + 3k)$

c $4(1 + 3m) + 2(3 + 2m)$ **d** $2(5 + 4y) + 3(2 + 3y)$

e $4(3 + 2f) + 2(5 - 3f)$ **f** $5(1 + 3g) + 3(3 - 4g)$

g $3(2 + 5t) + 4(1 - t)$ **h** $4(3 + 3w) + 2(5 - 4w)$

HINTS AND TIPS

Expand the expression before trying to collect like terms. If you try to expand and collect at the same time you will probably make a mistake.

C

3 Expand and simplify these expressions.

a $4(3 + 2h) - 2(5 + 3h)$ **b** $5(3g + 4) - 3(2g + 5)$

c $3(4y + 5) - 2(3y + 2)$ **d** $3(5t + 2) - 2(4t + 5)$

e $5(5k + 2) - 2(4k - 3)$ **f** $4(4e + 3) - 2(5e - 4)$

g $3(5m - 2) - 2(4m - 5)$ **h** $2(6t - 1) - 3(3t - 4)$

4 Expand and simplify these expressions.

a $m(4 + p) + p(3 + m)$ **b** $k(3 + 2h) + h(4 + 3k)$

c $t(2 + 3n) + n(3 + 4t)$ **d** $p(2q + 3) + q(4p + 7)$

e $3h(2 + 3j) + 2j(2h + 3)$ **f** $2y(3t + 4) + 3t(2 + 5y)$

g $4r(3 + 4p) + 3p(8 - r)$ **h** $5k(3m + 4) - 2m(3 - 2k)$

HINTS AND TIPS

Be careful with minus signs. They are causes of the most common errors students make in examinations. Remember $-2 \times -4 = 8$ but $-2 \times 5 = -10$. You will learn more about multiplying and dividing with negative numbers in Chapter 3.

FM 5 A two-carriage train has f first-class seats and $2s$ standard-class seats.

A three-carriage train has $2f$ first-class seats and $3s$ standard-class seats.

On a weekday, five two-carriage trains and two three-carriage trains travel from Hull to Liverpool.

a Write down an expression for the total number of first-class and standard-class seats available during the day.

b On average on any day, half of the first-class seats are used. Each first-class seat costs £60.

On average on any day, three-quarters of the standard-class seats are used. Each standard-class seat costs £40.

How much money does the rail company earn in an average day on this route? Give your answer in terms of f and s.

c f =15 and s = 80. It costs the rail company £30 000 per day to operate this route. How much profit do they make on an average day?

AU 6 Fill in whole-number values so that the following expansion is true.

$3(\ldots\ldots x + \ldots\ldots y) + 2(\ldots\ldots x + \ldots\ldots y) = 11x + 17y$

HINTS AND TIPS

There is more than one answer. You don't have to give them all.

PS 7 A rectangle with sides 5 and $3x + 2$ has a smaller rectangle with sides 3 and $2x - 1$ cut from it.

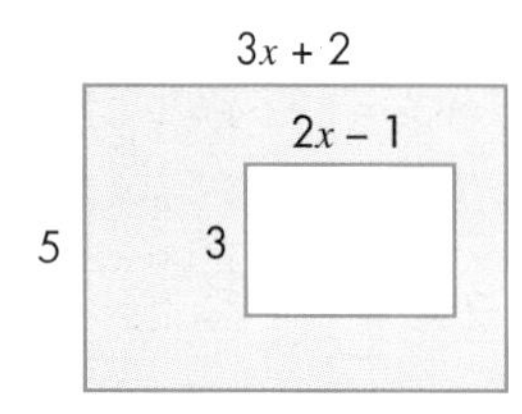

Work out the remaining area.

HINTS AND TIPS

Write out the expression for the difference of the two answers and then work it out.

8.4 Factorisation

This section will show you how to:

- 'reverse' the process of expanding brackets by taking out a common factor from each term in an expression

Key words
factor
factorisation

Factorisation is the opposite of expansion. It puts an expression into brackets.

To factorise an expression, look for the common **factors** in every term of the expression. Follow through the examples below to see how this works.

EXAMPLE 12

Factorise each expression.

a $6t + 9m$ **b** $6my + 4py$
c $8kp + 4k - 12km$ **d** $8kp + 4kt - 12km$

a The common factor is 3, so $6t + 9m = 3(2t + 3m)$

b The common factor is $2y$, so $6my + 4py = 2y(3m + 2p)$

c The common factor is $4k$, so $8kp + 4k - 12km = 4k(2p + 1 - 3m)$

d The common factor is $4k$, so $8kp + 4kt - 12km = 4k(2p + t - 3m)$

Notice that if you multiply out each answer you will get the expressions you started with.

This diagram may help you to see the difference and the connection between expansion and factorisation.

Note: When the whole term is the common factor, as in part **c**, then you are left with 1, not 0, inside the brackets.

Expanding
$3(2t + 3m) = 6t + 9m$
Factorising

EXERCISE 8F

1 Factorise the following expressions. The first three have been started for you.

a $6m + 12t = 6(\quad)$ **b** $9t + 3p = 3(\quad)$ **c** $8m + 12k = 4(\quad)$
d $4r + 8t$ **e** $mn + 3m$ **f** $5g^2 + 3g$
g $4w - 6t$ **h** $8p - 6k$ **i** $16h - 10k$
j $2mp + 2mk$ **k** $4bc + 2bk$ **l** $6ab + 4ac$
m $3y^2 + 2y$ **n** $4t^2 - 3t$ **o** $4d^2 - 2d$
p $3m^2 - 3mp$

HINTS AND TIPS

First look for a common factor of the numbers and then look for common factors of the letters.

D

C

2 Factorise the following expressions.

a $6p^2 + 9pt$ **b** $8pt + 6mp$ **c** $8ab - 4bc$

d $12a^2 - 8ab$ **e** $9mt - 6pt$ **f** $16at^2 + 12at$

g $5b^2c - 10bc$ **h** $8abc + 6bed$ **i** $4a^2 + 6a + 8$

j $6ab + 9bc + 3bd$ **k** $5t^2 + 4t + at$ **l** $6mt^2 - 3mt + 9m^2t$

m $8ab^2 + 2ab - 4a^2b$ **n** $10pt^2 + 15pt + 5p^2t$

3 Factorise the following expressions where possible. List those that cannot be factorised.

a $7m - 6t$ **b** $5m + 2mp$ **c** $t^2 - 7t$

d $8pt + 5ab$ **e** $4m^2 - 6mp$ **f** $a^2 + b$

g $4a^2 - 5ab$ **h** $3ab + 4cd$ **i** $5ab - 3b^2c$

FM 4 Three friends have a meal together. They each have a main meal costing £6.75 and a dessert costing £3.25.

Chris says that the bill in pounds, will be $3 \times 6.75 + 3 \times 3.25$.

Mary says that she has an easier way to work out the bill as $3 \times (6.75 + 3.25)$.

a Explain why Chris and Mary's methods both give the correct answer.

b Explain why Mary's method is better.

c What is the total bill?

AU 5 Three students are asked to factorise the expression $12m - 8$.

These are their answers.

Aidan	Bella	Craig
$2(6m - 4)$	$4(3m - 2)$	$4m\left(3 - \frac{2}{m}\right)$

All the answers are factorised correctly, but only one is the normally accepted answer.

a Which student gave the answer that is normally accepted as correct?

b Explain why the other two students' answers are not normally accepted as correct.

PS 6 Explain why $5m + 6p$ cannot be factorised.

ACTIVITY

Algebra dominoes

This is an activity for two people.

You need some card to make a set of algebra dominoes like those below.

$4 \times n$	t^2	$2b$	$2 - t$	$\frac{12n}{2}$	$0.5n$	$5w$	$b + b$
$3t - 2$	$3 \times 2y$	$5 + y$	$n + 2 + n + 3$	$\frac{4t + 2n}{2}$	$t \times t$	$5b - 3b$	$6n$
$6y$	$t - 2$	$2a + 2$	$2t + 2 - 3t$	$y + 5$	$7n - n$	$3n + 3n$	$4t - 2 - t$
b^2	$\frac{1}{2}n$	$t + 3 - 2$	$2n - 1$	$t + 5$	$b \times 2$	$10w \div 2$	$n + n + n + n$
$2t + n$	$4n$	$n + 2 + n - 3$	$2n + 5$	$\frac{n}{2}$	$2(a + 1)$	$n \div 2$	$n \times 4$

Turn the dominoes over and shuffle them. Deal five dominoes to each player.

One player starts by putting down a domino.

The other player may put down a domino that matches either end of the domino on the table. For example, if one player has put down the domino with $6n$ at one end, the other player could put down a domino that has an expression equal in value, such as the domino with $3n + 3n$ because $3n + 3n = 6n$. Otherwise, this player must pick up a domino from the spares.

The first player follows, playing a domino or picking one up, and so on, in turn.

The winner is the first player who has no dominoes.

Make up your own set of algebra dominoes.

8.5 Substitution

This section will show you how to:
- substitute numbers for letters in formulae and evaluate the resulting numerical expression
- use a calculator to evaluate numerical expressions

Key words
brackets
calculator
formula
substitution

One of the most important features of algebra is the use of expressions and **formulae**, and the **substitution** of real numbers into them.

The value of an expression, such as $3x + 2$, changes when different values of x are substituted into it. For example, the expression $3x + 2$ has the value:

5 when $x = 1$ 14 when $x = 4$

and so on. A formula expresses the value of one variable as the others in the formula change. For example, the formula for the area, A, of a triangle of base b and height h is:

$$A = \frac{b \times h}{2}$$

When $b = 4$ and $h = 8$:

$$A = \frac{4 \times 8}{2} = 16$$

EXAMPLE 13

The formula for the area of a trapezium is:

$$A = \frac{(a + b)h}{2}$$

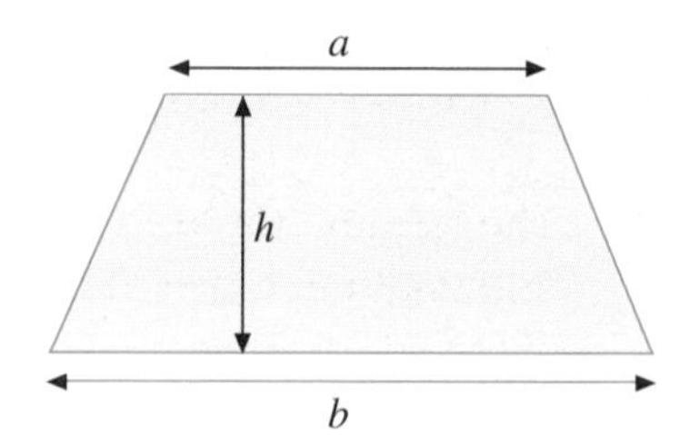

Find the area of the trapezium when $a = 5$, $b = 9$ and $h = 3$.

$$A = \frac{(5 + 9) \times 3}{2} = \frac{14 \times 3}{2} = 21$$

Always substitute the numbers for the letters before trying to work out the value of the expression. You are less likely to make a mistake this way. It is also useful to write **brackets** around each number, especially with negative numbers.

EXERCISE 8G

1 Find the value of $3x + 2$ when:

a $x = 2$ **b** $x = 5$ **c** $x = 10$

> **HINTS AND TIPS**
>
> It helps to put the numbers in brackets.
> $3(2) + 2 = 6 + 2 = 8$
> $3(5) + 2 = 15 + 2 = 17$
> etc …

2 Find the value of $4k - 1$ when:

a $k = 1$ **b** $k = 3$ **c** $k = 11$

3 Find the value of $5 + 2t$ when:

a $t = 2$ **b** $t = 5$ **c** $t = 12$

4 Evaluate $15 - 2f$ when: **a** $f = 3$ **b** $f = 5$ **c** $f = 8$

5 Evaluate $5m + 3$ when: **a** $m = 2$ **b** $m = 6$ **c** $m = 15$

6 Evaluate $3d - 2$ when: **a** $d = 4$ **b** $d = 5$ **c** $d = 20$

FM 7 A taxi company uses the following rule to calculate their fares.

Fare = £2.50 plus 50p per kilometre.

a How much is the fare for a journey of 3 km?

b Farook pays £9.00 for a taxi ride. How far was the journey?

c Maisy knows that her house is 5 miles from town. She has £5.50 left in her purse after a night out. Has she got enough for a taxi ride home?

AU 8 Kaz knows that x, y and z have the values 2, 8 and 11, but she does not know which variable has which value.

a What is the maximum value that the expression $2x + 6y - 3z$ could be?

b What is the minimum value that the expression $5x - 2y + 3z$ could be?

> **HINTS AND TIPS**
>
> You could just try all combinations, but if you think for a moment you will find that the $6y$ term must give the largest number. This will give you a clue to the other terms.

PS 9 The formula for the area, A, of a rectangle with length l and width w is $A = lw$.

The formula for the area, T, of a triangle with base b and height h is $T = \frac{1}{2}bh$.

Find values of l, w, b and h so that $A = T$.

10 Find the value of $\frac{8 \times 4h}{5}$ when: **a** $h = 5$ **b** $h = 10$ **c** $h = 25$

11 Find the value of $\frac{25 - 3p}{2}$ when: **a** $p = 4$ **b** $p = 8$ **c** $p = 10$

12 Evaluate $\frac{x}{3}$ when: **a** $x = 6$ **b** $x = 24$ **c** $x = -30$

E

D

D

13 Evaluate $\frac{A}{4}$ when: **a** $A = 12$ **b** $A = 10$ **c** $A = -20$

14 Find the value of $\frac{12}{y}$ when: **a** $y = 2$ **b** $y = 4$ **c** $y = 6$

15 Find the value of $\frac{24}{x}$ when: **a** $x = 2$ **b** $x = 3$ **c** $x = 16$

FM 16 A holiday cottage costs £150 per day to rent.

A group of friends decide to rent the cottage for seven days.

a Which formula represents the cost of the rental for each person if there are n people in the group? Assume that they share the cost equally.

$\frac{150}{n}$ $\frac{150}{7n}$ $\frac{1050}{n}$ $\frac{150n}{n}$

HINTS AND TIPS

To check your choice in part **a**, make up some numbers and try them in the formula. For example, take $n = 5$.

b Eventually 10 people go on the holiday. When they get the bill, they find that there is a discount for a seven-day rental.

After the discount, they each find it cost them £12.50 less than they expected.

How much does a 7-day rental cost?

C

AU 17 **a** p is an odd number and q is an even number.

Say if each of these expressions is odd or even.

i $p + q$ **ii** $p^2 + q$ **iii** $2p + q$ **iv** $p^2 + q^2$

b x, y and z are all odd numbers.

Write an expression, using x, y and z, so that the value of the expression is always even.

HINTS AND TIPS

There are many answers for **b** and **a** should give you a clue.

PS 18 A formula for the cost of delivery, in pounds, of orders from a do-it-yourself warehouse is:

$$D = 2M - \frac{C}{5}$$

where D is the cost of the delivery, M is the distance in miles from the store and C is the cost of the goods to be delivered.

HINTS AND TIPS

Note: a rebate is a refund of some of the money that someone has already paid for goods or services.

a How much does the delivery cost when $M = 30$ and $C = 200$?

b Bob buys goods worth £300 and lives 10 miles from the store.

i The formula gives the cost of delivery as a negative value. What is this value?

ii Explain why Bob will not get a rebate from the store.

c Maya buys goods worth £400. She calculates that her cost of delivery will be zero. What is the greatest distance that Maya could live from the store?

GRADE BOOSTER

- **F** You can use a formula expressed in words
- **F** You can substitute numbers into expressions and use letters to write a simple algebraic expression
- **E** You can simplify expressions by collecting like terms
- **D** You can use letters to write more complicated expressions, expand expressions with brackets and factorise simple expressions
- **C** You can expand and simplify expressions with brackets and factorise expressions

What you should know now

- How to simplify a variety of algebraic expressions by multiplying, collecting like terms and expanding brackets
- How to factorise expressions by removing common factors
- How to substitute into expressions, using positive or negative whole numbers and decimals

1 Simplify each expression.

a $p + 5p - 2p$ **b** $3q \times 4r$

c $7t - 10t$

2 Simplify $d + d + d + d + d$ (1)

(Total 1 mark)

Edexcel, March 2008, Paper 9 Foundation, Question 2

3 James packs books into boxes.

He packs 20 books into each box.

James packs x boxes of books.

Write an expression, in terms of x, for the number of books he packs. (1)

(Total 1 mark)

Edexcel, November 2007, Paper 10 Section B Foundation, Question 7

4 **a** Simplify $5a + 2b - a + 5b$

b Expand $5(p + 2q - 3r)$

5 Using the formula $v = 4u - 3t$, calculate the value of v when $u = 12.1$ and $t = 7.2$

6 **a** Simplify $2x + 4y - x + 4y$

b Find the value of $3p + 5q$ when $p = 2$ and $q = -1$

c Find the value of $u^2 + v^2$ when $u = 4$ and $v = -3$

7 **a** Matt buys 10 boxes of apple juice at 24 pence each.

i Calculate the total cost.

ii He pays with a £10 note.
How much change will he receive?

b Aisha buys c oranges at 20 pence each.

i Write down an expression for the total cost in terms of c.

ii She now buys d apples at 15 pence each.
Write down an expression for the total cost of the apples and oranges.

8 Graham is y years old.

Harriet is 5 years older than Graham.

a Write down an expression for Harriet's age.

b Jane is half as old as Harriet.
Write down an expression for Jane's age.

9 **a** Simplify $4e \times 3f$ (1)

b Expand $4(2x + 5)$ (1)

c Simplify $4r - 2t + 3r - 7t$ (2)

(Total 4 marks)

Edexcel, March 2007, Paper 10 Section B Foundation, Question 7

10 Simplify

a $a + a + a + a$ (1)

b $2 \times p \times q$ (1)

c $3a + 4b + a - 2b$ (1)

(Total 3 marks)

Edexcel, June 2007, Paper 10 Section B Foundation, Question 6

11 **a** Simplify $g + g + g$ (1)

b Simplify $5 \times h \times k$ (1)

(Total 2 marks)

Edexcel, November 2007, Paper 10 Section A Foundation, Question 8

12 Simplify $3a + 5b - a - 2b$ (2)

(Total 2 marks)

Edexcel, March 2008, Paper 10 Section B Foundation, Question 7a

13 A can of cola costs x pence.

An ice cream costs y pence.

Atif buys 3 cans of cola and 2 ice creams.

Write down an expression, in terms of x and y, for the total cost, in pence. (2)

(Total 2 marks)

Edexcel, March 2008, Paper 9 Foundation, Question 9

14 **a** Find the value of a^3 when $a = 4$

b Find the value of $5x + 3y$ when $x = -2$ and $y = 4$

c There are p seats in a standard class coach and q seats in a first class coach. A train has five standard class coaches and two first class coaches.

Write down an expression in terms of p and q for the total number of seats in the train.

C D E

15 $d = 3e + 2h^2$

Calculate the value of d when $e = 3.7$ and $h = 2$

16 **a** Expand and simplify this expression.

$2(x + 3) + 5(x + 2)$

b Expand and simplify this expression.

$(4x + y) - (2x - y)$

17 **a** Simplify this expression.

$3x + 4y + 6x - 3y - 5x$

b Factorise this expression.

$6c + 9$

c Factorise this expression.

$z^2 + 6z$

18 **a** Factorise $5x + 10$ (1)

b Factorise $x^2 - 8x$ (1)

(Total 2 marks)

Edexcel, March 2008, Paper 10 Section B Foundation, Question 7c and d

19 **a** **i** Multiply out and simplify this expression.
$3(x - 3) + 2(x + 2)$

ii Multiply out and simplify this expression.
$(n - 1)^2$

b Factorise completely the following expressions.

i $6a^2 + a$

ii $6x^2y^3 - 4xy^2$

20 **a** Expand and simplify this expression.
$3(x - 1) + 2(3x - 5)$

b Expand and simplify this expression.
$(x - 3)(x - 2)$

Worked Examination Questions

1 Factorise completely:

$4x^2 - 8xy$

$$4x^2 - 8xy = 4x \times x - 4x \times 2y$$
$$= 4x(x - 2y)$$

Total: 2 marks

Note the words 'Factorise completely'. This is a clue that there is more than one common factor. Look for a common factor of 4 and 8, e.g. 4. Look for a common factor of x^2 and xy, e.g. x.

Split up the terms, using the common factors.

Write as a factorised expression.

The correct answer gets 2 marks (1 for accuracy and 1 for method). A partial factorisation such as $4(x^2 - 2xy)$ or $2x(2x - 4y)$ would get 1 mark.

2 Expand and simplify:

$3(x - 2) + 2(3x + 5)$

$$3(x - 2) + 2(3x + 5) = 3x - 6 + 6x + 10$$
$$= 3x + 6x - 6 + 10$$
$$= 9x + 4$$

Total: 3 marks

'Expand and simplify' means expand the brackets then collect like terms together. Expand each bracket. This would get 1 method mark for the attempt to expand and 1 accuracy mark if both brackets were expanded correctly.

Rearrange terms to get like terms together.

Collect like terms by adding or subtracting the coefficients of x and the numbers. The final answer gets 1 mark.

Worked Examination Questions

FM **3** Two plumbers use different rules for working out how much to charge for a job.

Dwayne Pipes: £60 + £25 per hour
Walter Bucket: £45 + £30 per hour

Louise has a pipe leak under the floor. Both plumbers estimate that fixing the leak will take between two hours and five hours.

Which plumber should Louise give the job, to and why?

Time (hours)	1	2	3	4	5
Dwayne	£85	£110	£135	£160	£185
Walter	£70	£105	£135	£165	£195

This question is called an 'open-ended' question and there are many ways to answer it, but the marks will be awarded for doing certain things.

Work out how much each plumber will charge for different times.
You could set your results out in a table.

Recognise that they both charge the same at three hours. You will need to write this down to make sure the examiner knows you have spotted this.

Louise should use Dwayne because, even though he is more expensive for a job under three hours, he will be cheaper for jobs over three hours. So, for the estimated time, Dwayne is more expensive for the first two hours, but he will be cheaper if the job takes longer than three hours, so he is the better choice.

There will be 2 marks for your conclusion based on the other facts in the question.

Total: 4 marks

EQ **4** A rectangle has a length of $2x + 4$ and width of $x + 2$.

a Show that the perimeter can be written as $6(x + 2)$.

b Mark says that the perimeter must always be an even number.

Find a value of x that proves that Mark is wrong.

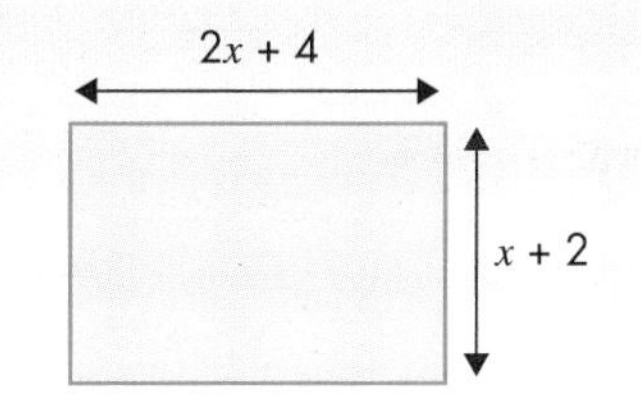

a Perimeter $= 2 \times (2x + 4) + 2 \times (x + 2)$

$= 2 \times 2(x + 2) + 2(x + 2)$

$= 4(x + 2) + 2(x + 2)$

$= 6(x + 2)$

Write down an expression for the perimeter. This will get you 1 method mark.

Take out the common factor.

Collect the like terms. These final two steps will get you 1 mark.

An alternative is

$2x + 4 + 2x + 4 + x + 2 + x + 2 = 6x + 12$

$6x + 12 = 6(x + 2)$

2 marks

b $6(x + 2) = 15$

$x + 2 = 2.5$

$x = 0.5$

Pick an odd value for the perimeter, say 15. Any value that works will be worth 1 mark.

1 mark

There are many answers. Any value that makes the value inside the brackets 'something and a half', such as $x = 1.5$ or $x = 2.5$ will work.

Total: 3 marks

8

Functional Maths
Walking using Naismith's rule

Many people go walking each weekend. It is good exercise and can be a very enjoyable pastime.

When walkers set out they often try to estimate the length of time the walk will take. There are many factors that could influence this, but one rule that can help in estimating how long the walk will take is Naismith's Rule.

Naismith's rule

Naismith's rule is a rule of thumb that you can use when planning a walk, by calculating how long it will take. The rule was devised by William Naismith, a Scottish mountaineer, in 1892.

The basic rule is:

Allow 1 hour for every 3 miles (5 km) forward, plus $\frac{1}{2}$ hour for every 1000 feet (300 m) of ascent.

Getting started

Before you begin your main task, you may find it useful to fill in the following table to practise using Naismith's Rule.

Can you use algebra to display the rule?

Day	Distance (km)	Height (m)	Time (m)
1	16	250	
2	18	0	
3	11	340	
4	13	100	
5	14	120	

Now, in small groups think about:

- What kind of things influence the speed at which you walk?
- Do different types of routes make people walk at different rates?
- If there is a large group of people will they all walk at the same rate?

Use all the ideas you have just discussed as you move on to your main task.

Your task

You are going to compare data to see if Naismith's rule is still a useful way to work out how much time to allow for different walks.

The table on the right shows the actual times taken by a school group as they did five different walks in five days. Use this information to work out the following:

1 If the group had started at the same times and had the same breaks, how long would the group have taken each day, according to Naismith's rule?

2 Do you think Naismith's rule is still valid today? Explain your reasons.

3 If your friend was going to climb Ben Nevis, setting out at 11.30 am, would you advise them to do the walk? You will need to research the distance and climb details of the pathway up Ben Nevis, in order to advise them fully.

Day	Distance (km)	Height (m)	Time (minutes)	Time (hours/ minutes)	Start	Breaks	Finish
1	16	250	255	4 h 15 m	10.00 am	2 h	4.15 pm
2	18	0	270	4 h 30 m	10.00 am	1 h 30 m	4.00 pm
3	11	340	199	3 h 19 m	09.30 am	2 h 30 m	3.19 pm
4	13	100	195	3 h 15 m	10.30 am	2 h 30 m	4.15 pm
5	14	120	222	3 h 42 m	10.30 am	2 h 30 m	4.42 pm

Why this chapter matters

Line graphs are used in many media, including newspapers and the textbooks of most of the subjects that you learn in school.

Graphs show the relationship between two variables. Often one of these variables is time and the graph shows how the other variable changes over time.

For example, the graph top right shows how the exchange rate between the dollar and the pound changed over five months in 2009.

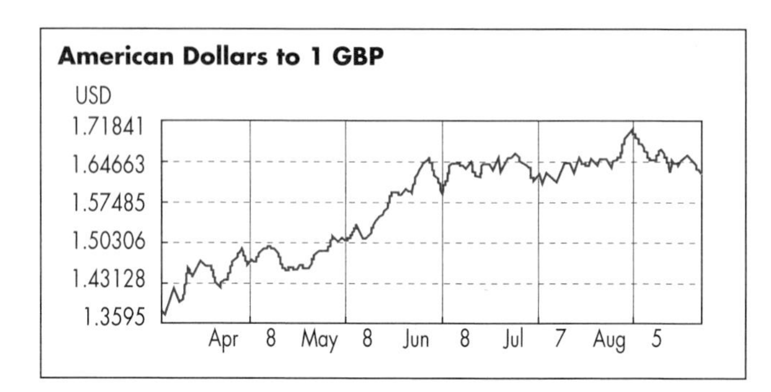

The earliest line graphs, such as the one shown on the right, appeared in the book *A Commercial and Political Atlas*, written in 1786 by William Playfair, who also used bar charts and pie charts for the first time. Playfair argued that charts and graphs communicated information to an audience better than tables of data. Do you think this is true?

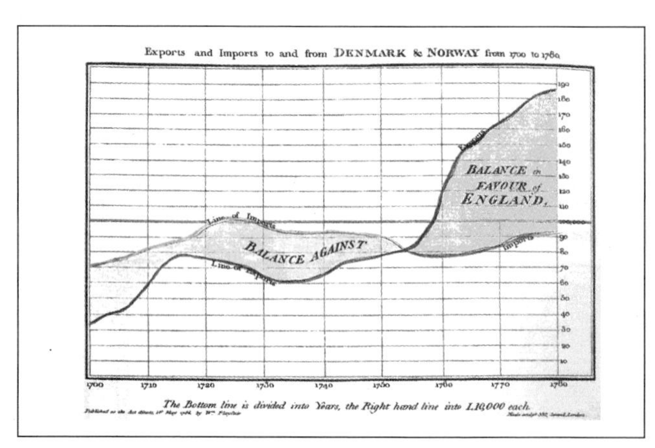

A line graph showing imports and exports to and from Denmark and Norway by William Playfair.

The graph below right shows all the data from a racing car going round a circuit. Engineers can use this to fine-tune parts of the car to give the best performance. It perfectly illustrates that graphs give a visual representation of how variables change and can be used to compare data in a way that looking at lists of data cannot.

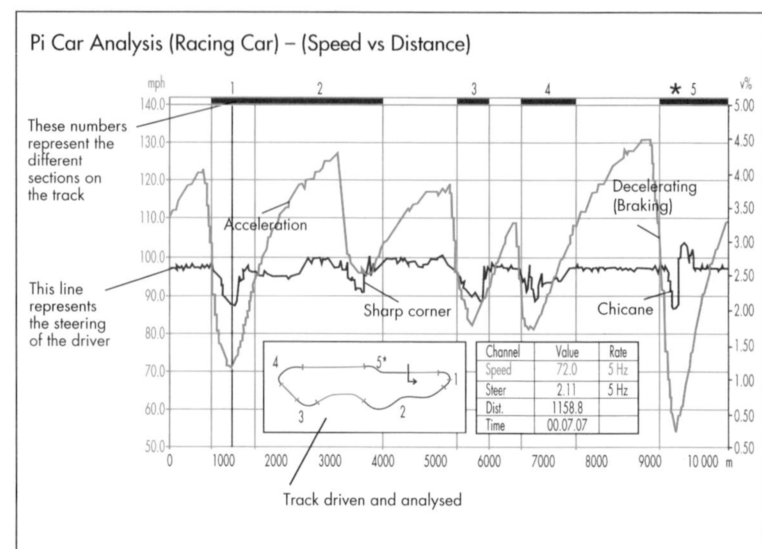

Think about instances in school and everyday life where a line graph would help you to communicate information more effectively.

Chapter

Algebra: Graphs

The grades given in this chapter are target grades.

This chapter will show you ...

F how to read information from a conversion graph

E to D how to draw a straight-line graph from its equation

D how to read information from a travel graph

Visual overview

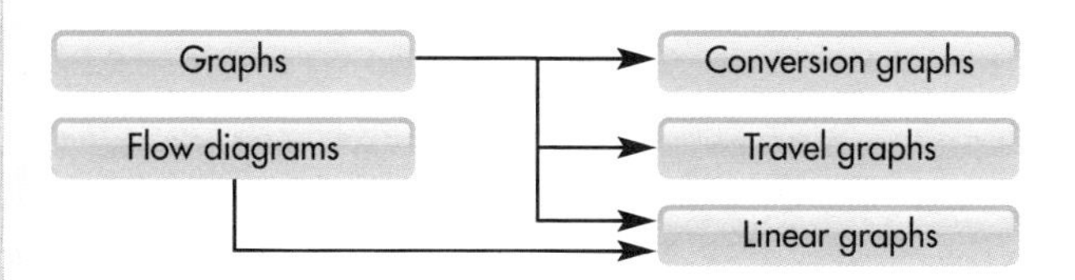

What you should already know

- How to plot coordinates in the first quadrant **(KS3 level 3, GCSE grade G)**
- How speed, distance and time are related (from Chapter 9) **(KS3 level 6, GCSE grade D)**
- How to substitute numbers into a formula (from Chapter 7) **(KS3 level 5, GCSE grade E)**
- How to use a flow diagram to set up an expression (from Chapter 13) **(KS3 level 5, GCSE grade E)**
- How to read and estimate from scales **(KS3 level 5, GCSE grade E)**

Quick check

Write down the coordinates of the following points.

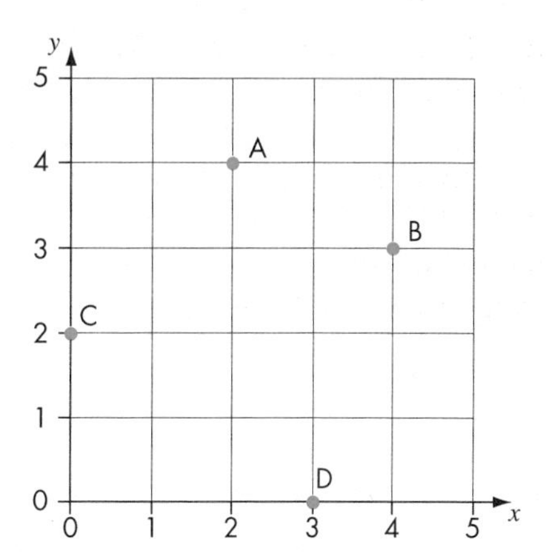

9.1 Conversion graphs

This section will show you how to:
- convert from one unit to another unit by using a graph

Key words
conversion graph
estimate
scales

Look at Examples 1 and 2, and make sure that you can understand the conversions. You need to be able to read these types of graph by finding a value on one axis and following it through to the other axis. Make sure you understand the **scales** on the axes to help you **estimate** the answers.

EXAMPLE 1

This is a **conversion graph** between litres and gallons.

a How many litres are there in 5 gallons?

b How many gallons are there in 15 litres?

From the graph you can see that:

a 5 gallons are approximately equivalent to 23 litres.

b 15 litres are approximately equivalent to $3\frac{1}{4}$ gallons.

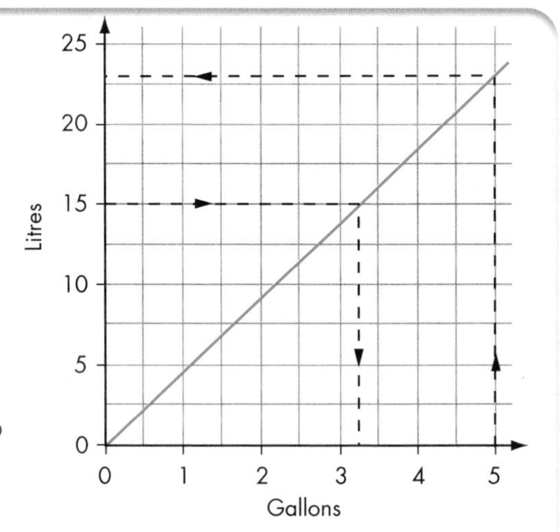

EXAMPLE 2

This is a graph of the charges made for units of electricity used in the home.

a How much will a customer who uses 500 units of electricity be charged?

b How many units of electricity will a customer who is charged £20 have used?

From the graph you can see that:

a A customer who uses 500 units of electricity will be charged £45.

b A customer who is charged £20 will have used about 150 units.

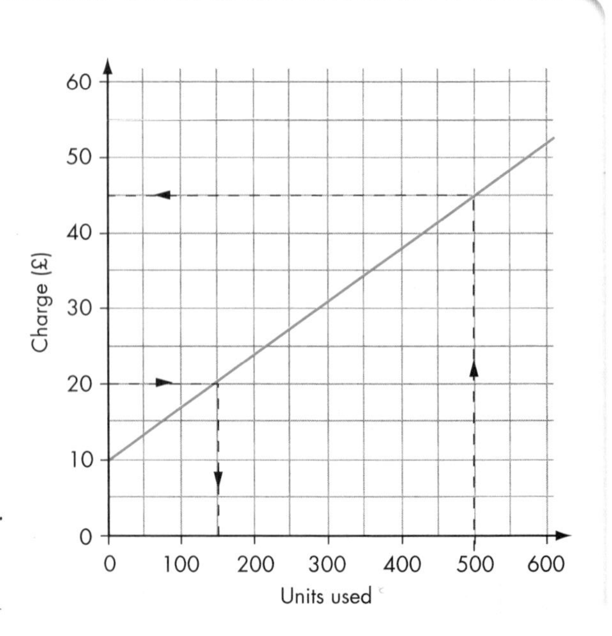

EXERCISE 9A

FM 1 This is a conversion graph between kilograms (kg) and pounds (lb).

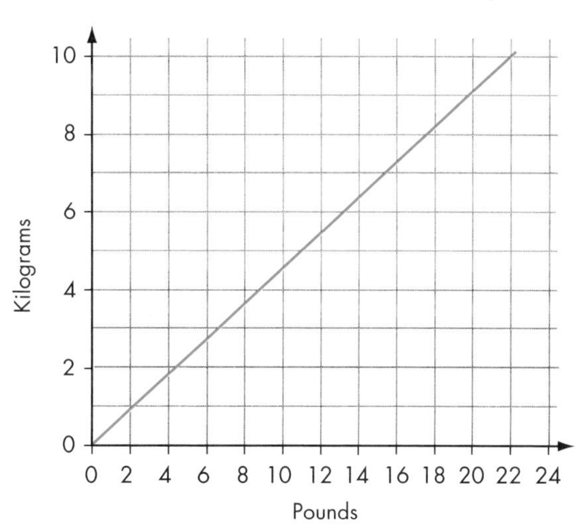

a Use the graph to make an approximate conversion of:

i 18 lb to kilograms

ii 5 lb to kilograms

iii 4 kg to pounds

iv 10 kg to pounds.

b Approximately how many pounds are equivalent to 1 kg?

c Explain how you could use the graph to convert 48 lb to kilograms.

FM 2 This is a conversion graph between inches (in) and centimetres (cm).

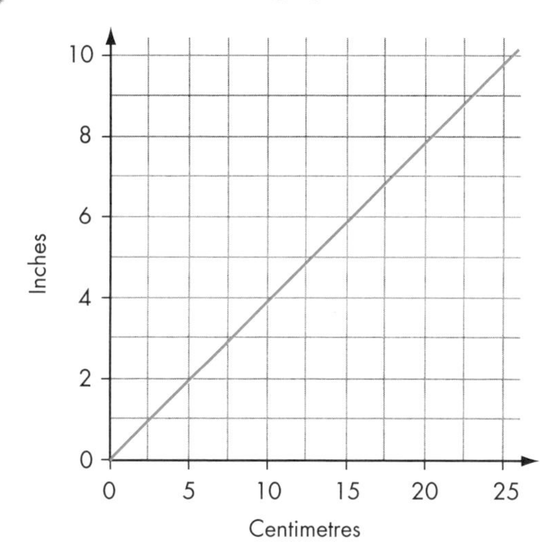

a Use the graph to make an approximate conversion of:

i 4 in to centimetres

ii 9 in to centimetres

iii 5 cm to inches

iv 22 cm to inches.

b Approximately how many centimetres are equivalent to 1 in?

c Explain how you could use the graph to convert 18 into centimetres.

FM 3 This graph was produced to show the approximate equivalence of the British pound (£) to the Singapore dollar ($).

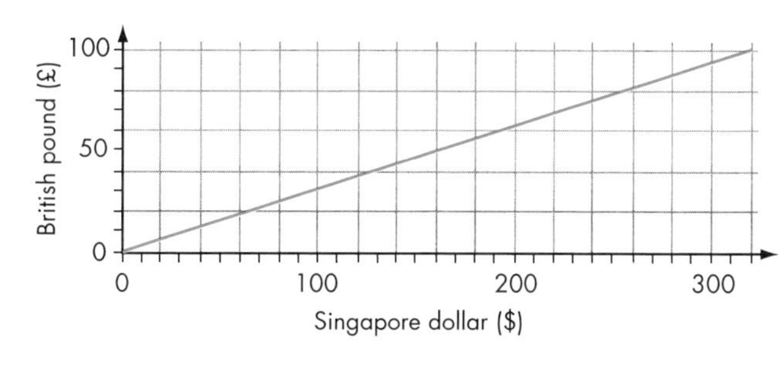

a Use the graph to make an approximate conversion of:

i £100 to Singapore dollars

ii £30 to Singapore dollars

iii $150 to British pounds

iv $250 to British pounds.

b Approximately how many Singapore dollars are equivalent to £1?

AU **c** What would happen to the conversion line on the graph if the pound became weaker against the Singapore dollar?

F

4 A hire firm hired out industrial blow heaters. They used the following graph to approximate what the charges would be.

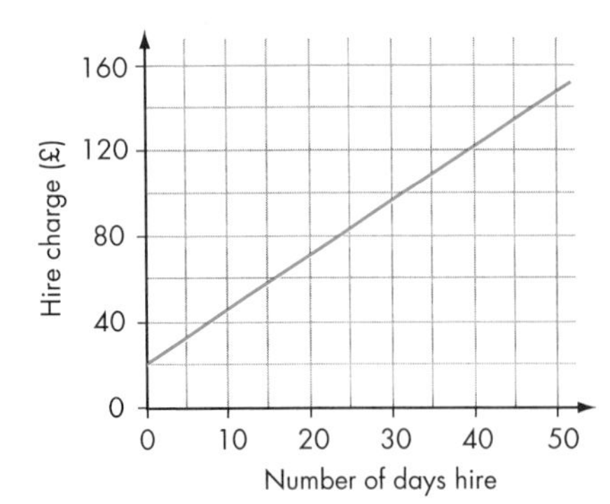

a Use the graph to find the approximate charge for hiring a heater for:

i 40 days

ii 25 days.

b Use the graph to find out how many days' hire you would get for a cost of:

i £100

ii £140.

5 A conference centre had the following chart on the office wall so that the staff could see the approximate cost of a conference, based on the number of people attending it.

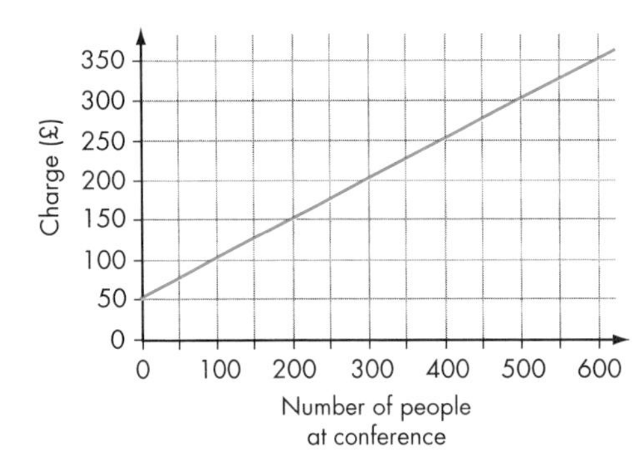

a Use the graph to find the approximate charge for:

i 100 people

ii 550 people.

b Use the graph to estimate how many people can attend a conference at the centre for a cost of:

i £300

ii £175.

6 At a small shop, the manager marked all goods at the pre-VAT prices and the sales assistant had to use the following chart to convert these marked prices to selling prices.

a Use the chart to find the selling price of goods marked:

i £60

ii £25.

b What was the marked price if you bought something for:

i £100

ii £45?

F

7 Granny McAllister still finds it hard to think in degrees Celsius. So she always uses the following conversion graph to help her to understand the weather forecast.

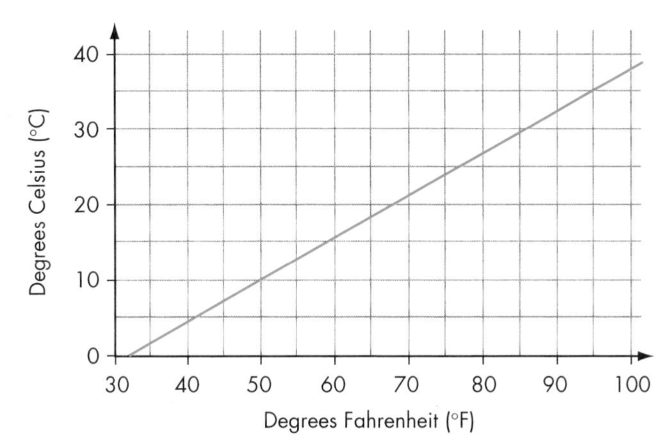

a Use the graph to make an approximate conversion of:

i 35 °C to Fahrenheit

ii 20 °C to Fahrenheit

iii 50 °F to Celsius

iv 90 °F to Celsius.

b Water freezes at 0 °C. What temperature is this in Fahrenheit?

E

FM 8 Tea is sold at a school fete between 1.00 pm and 2.30 pm. The numbers of cups of tea that had been sold were noted at half-hour intervals.

Time	1.00	1.30	2.00	2.30	3.00	3.30
No. of cups of tea sold	0	24	48	72	96	120

a Draw a graph to illustrate this information. Use a scale from 1 to 4 hours on the horizontal time axis, and from 1 to 120 on the vertical axis for numbers of cups of tea sold.

b Use your graph to estimate when the 60th cup of tea was sold.

FM 9 I lost my fuel bill, but while talking to my friends I found out that:

Bill, who had used 850 units, was charged £57.50
Wendy, who had used 320 units, was charged £31
Rhanni, who had used 540 units, was charged £42.

a Plot the given information and draw a straight-line graph. Use a scale from 0 to 900 on the horizontal units axis, and from £0 to £60 on the vertical cost axis.

b Use your graph to find what I will be charged for 700 units.

AU 10 The graph shows the number of passengers arriving in Exeter each day on a particular train that is due at 0815 in December 2009.

The first of December was a Tuesday. Sundays are marked with red lines.

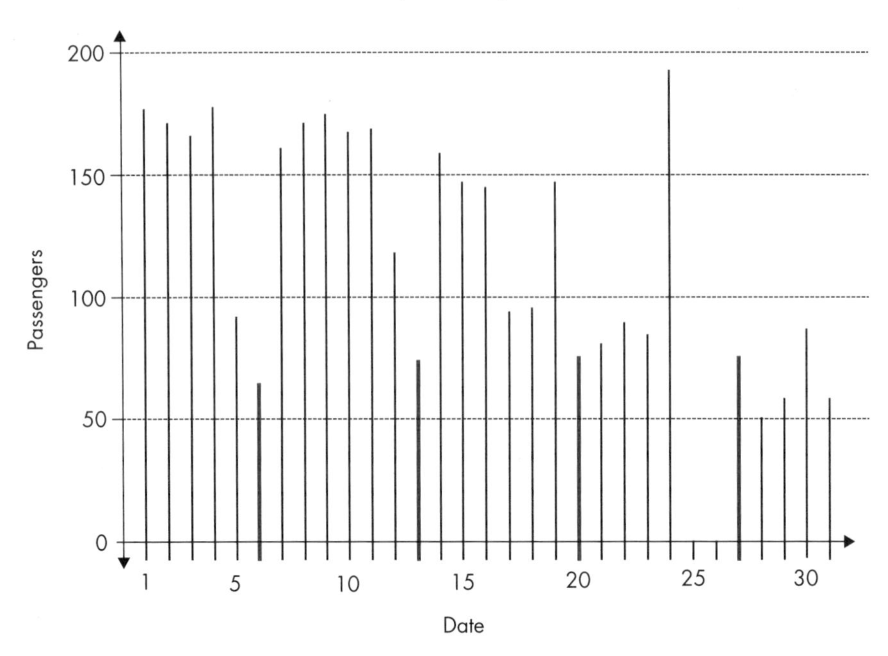

a No passengers used the train on the 25th and 26th. Why was this?

b One of the schools in Exeter closed for Christmas on 16th December. What evidence is there on the graph to support this?

c There was a steady increase in passengers using the train on Saturday through the month and a large number using the train on the 24th. What reason can you give for this?

PS AU 11 Leon is travelling from Paris to Calais.

When he sets off he has enough fuel to drive about 100 miles.

Sometime into the journey he sees this sign.

Calais 125 km ⇨

⇦ Paris 75 km

Does he have enough fuel to get to Calais or will he have to visit a petrol station?

Justify your answer.

PS 12 Two taxi companies use these rules for calculating fares:

CabCo: £2.50 basic charge and £0.75 per kilometre

YellaCabs: £2.00 basic charge and £0.80 per kilometre

This map shows the distances, in kilometres, that three friends, Anya (A), Bettina (B) and Calista (C) live from a restaurant (R) and from each other.

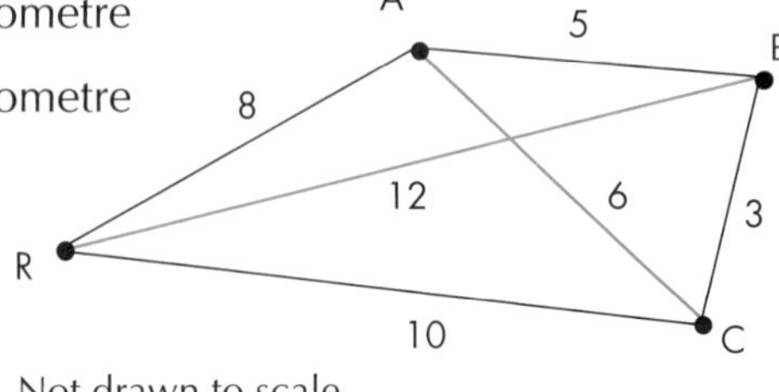

Not drawn to scale

You may find a copy of the grid below useful in answering the following question.

a If they each take an individual cab home, which company should they each choose?

b Work out the cheapest way they can travel home if two, or all three, share a cab.

HINTS AND TIPS

Draw a graph for both companies on the grid. Use this to work out the costs of the journeys.

C

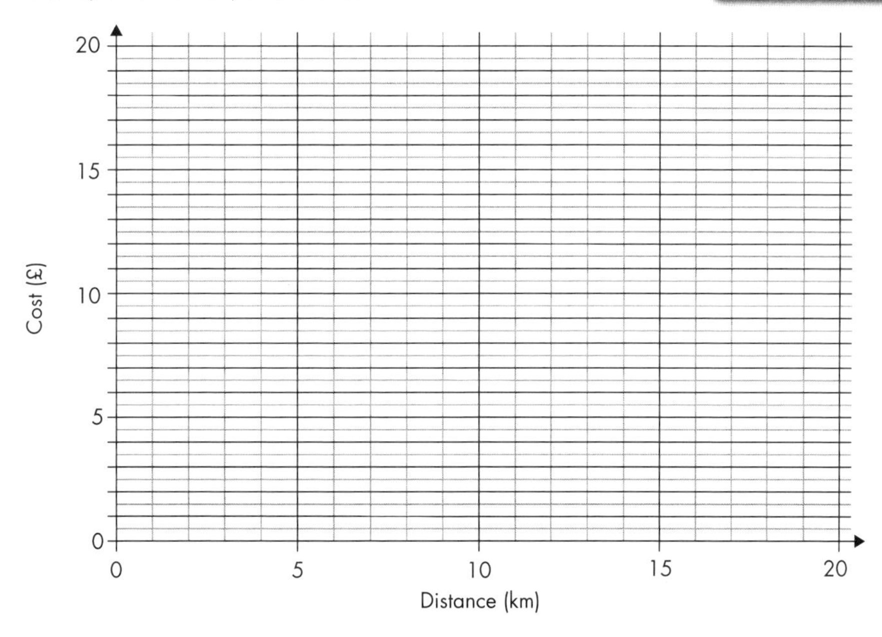

9.2 Travel graphs

This section will show you how to:
- read information from a travel graph
- find an average speed from a travel graph

Key words
average speed
distance–time graph
travel graph

As the name suggests, a **travel graph** gives information about how someone or something has travelled over a given time period. It is also called a **distance–time graph**.

A travel graph is read in a similar way to the conversion graphs you have just done. But you can also find the **average speed** from a distance–time graph by using the formula:

$$\text{average speed} = \frac{\text{total distance travelled}}{\text{total time taken}}$$

EXAMPLE 3

The distance–time graph below represents a car journey from Barnsley to Nottingham, a distance of 50 km, and back again.

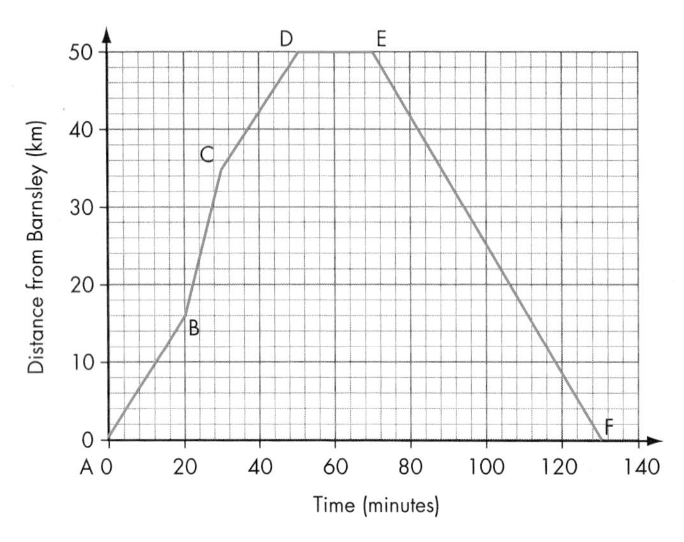

a What can you say about points B, C and D?

b What can you say about the journey from D to F?

c Work out the average speed for each of the five stages of the journey.

From the graph:

a B: After 20 minutes the car was 16 km away from Barnsley.

C: After 30 minutes the car was 35 km away from Barnsley.

D: After 50 minutes the car was 50 km away from Barnsley, so at Nottingham.

b D–F: The car stayed at Nottingham for 20 minutes, and then took 60 minutes for the return journey.

c The average speeds over the five stages of the journey are worked out as follows.

A to B represents 16 km in 20 minutes.

20 minutes is $\frac{1}{3}$ of an hour, so we need to multiply by 3 to give distance/hour.
Multiplying both numbers by 3 gives 48 km in 60 minutes, which is 48 km/h.

B to C represents 19 km in 10 minutes.

Multiplying both numbers by 6 gives 114 km in 60 minutes, which is 114 km/h.

C to D represents 15 km in 20 minutes.

Multiplying both numbers by 3 gives 45 km in 60 minutes, which is 45 km/h.

D to E represents a stop: no further distance travelled.

E to F represents the return journey of 50 km in 60 minutes, which is 50 km/h.

So, the return journey was at an average speed of 50 km/h.

You always work out the distance travelled in 1 hour to get the speed in kilometres per hour (km/h) or miles per hour (mph or miles/h).

EXERCISE 9B

FM **1** Paul was travelling in his car to a meeting. This distance–time graph illustrates his journey.

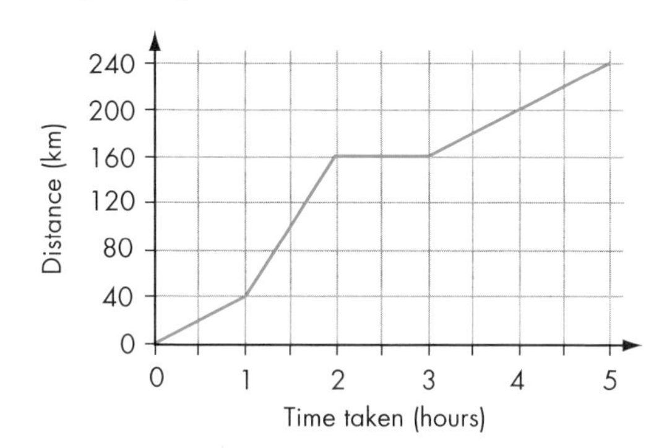

HINTS AND TIPS

Read the question carefully. Paul set off at 7 o'clock in the morning and the graph shows the time after this.

a How long after he set off did he

i stop for his break

ii set off after his break

iii get to his meeting place?

b At what average speed was he travelling:

i over the first hour

ii over the second hour

iii for the last part of his journey?

HINTS AND TIPS

If part of a journey takes 30 minutes, for example, just double the distance to get the average speed per hour.

c The meeting was scheduled to start at 10.30 am.

What is the latest time he should have left home?

2 A small bus set off from Leeds to pick up Mike and his family. It then went on to pick up Mike's parents and grandparents. It then travelled further, dropping them all off at a hotel. The bus then went on a further 10 km to pick up another party and took them back to Leeds. This distance–time graph illustrates the journey.

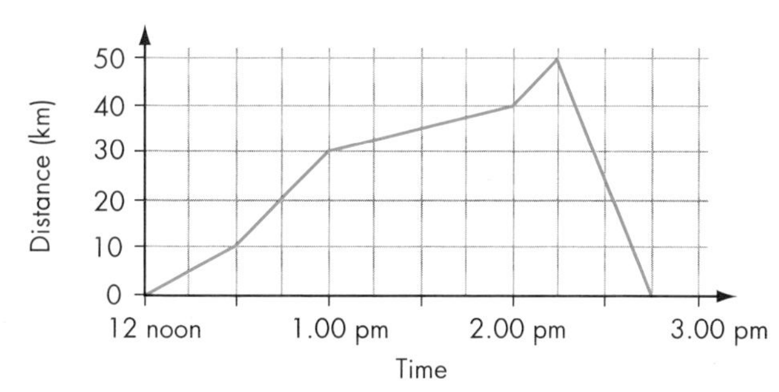

a How far from Leeds did Mike's parents and grandparents live?

b How far from Leeds is the hotel at which they all stayed?

c What was the average speed of the bus on its way back to Leeds?

D

3 James was travelling to Cornwall on his holidays. This distance–time graph illustrates his journey.

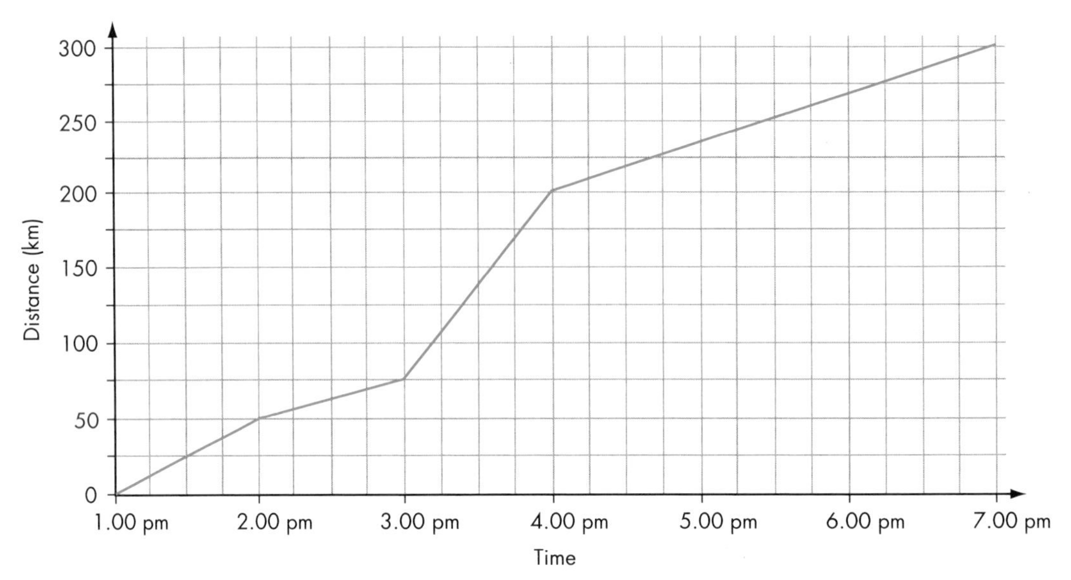

a His greatest speed was on the motorway.

i How far did he travel along the motorway?

ii What was his average speed on the motorway?

b **i** When did he travel most slowly?

ii What was his lowest average speed?

HINTS AND TIPS

Remember that the graph is made up of straight lines, as it shows average speed for each section of the journey. In reality, speed is rarely constant – except sometimes on motorways.

PS **4** Azam and Jafar were having a race. The distance–time graph below illustrates the distances covered.

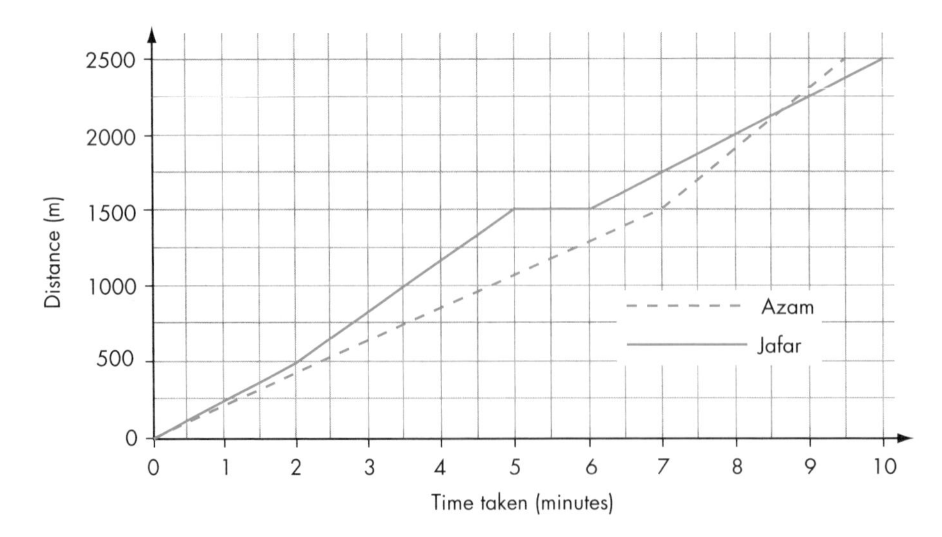

Write a commentary to describe the race.

D

FM **5** Three friends, Patrick, Araf and Sean, ran a 1000 metres race. The race is illustrated on the distance–time graph below.

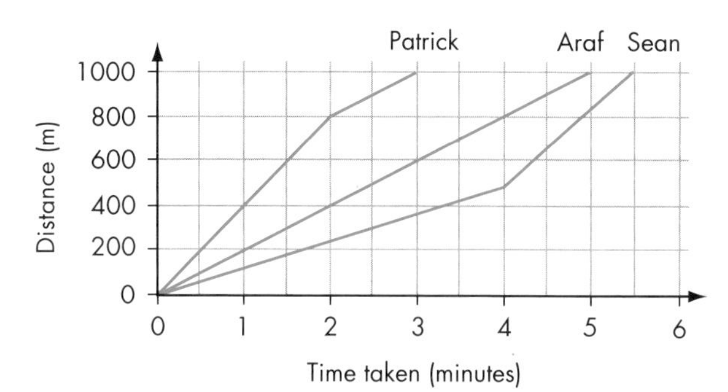

The school newspaper gave the following report of Patrick's race:

'Patrick took an early lead, running the first 800 metres in 2 minutes. He then slowed down a lot and ran the last 200 metres in 1 minute, to finish first in a total time of 3 minutes.'

a Describe the races of Araf and Sean in a similar way.

b **i** What is the average speed of Patrick in kilometres per hour?

ii What is the average speed of Araf in kilometres per hour?

iii What is the average speed of Sean in kilometres per hour?

AU **6** A walker sets off at 9.00 am from point P to walk along a trail at a steady pace of 6 km per hour.

90 minutes later, a cyclist sets off from P on the same trail at a steady pace of 15 km per hour.

At what time did the cyclist overtake the walker?

You may use a graph to help you solve this question.

HINTS AND TIPS

This question can be done by many methods, but drawing a distance–time graph is the easiest.

D

PS 7 Three school friends all set off from school at the same time, 3.45 pm. They all lived 12 km away from the school. The distance–time graph below illustrates their journeys.

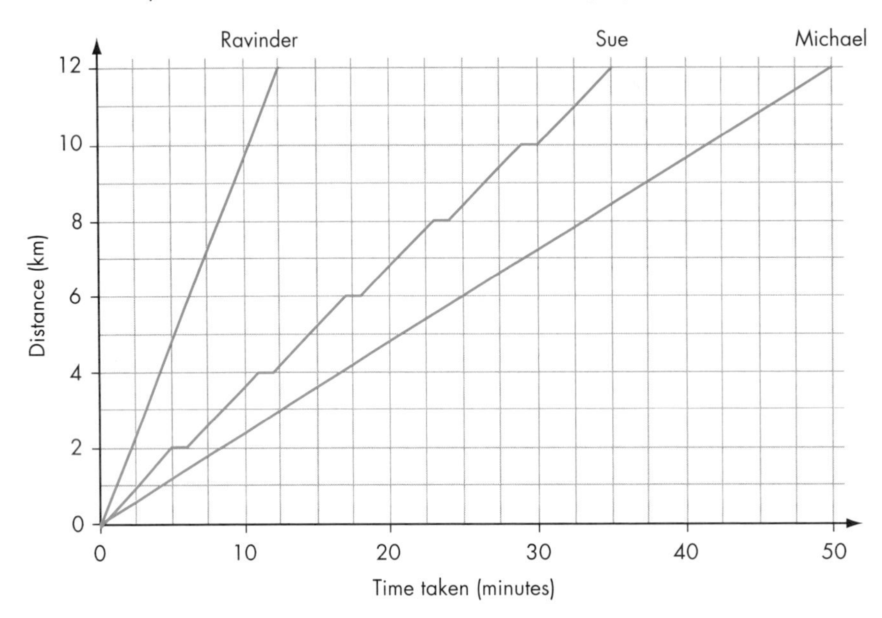

One of them went by bus, one cycled and one was taken by car.

a **i** Explain how you know that Sue used the bus.

ii Who went by car?

b At what time did each friend get home?

c **i** When the bus was moving, it covered 2 kilometres in 5 minutes. What is this speed in kilometres per hour?

ii Overall, the bus covered 12 kilometres in 35 minutes. What is this speed in kilometres per hour?

iii How many stops did the bus make before Sue got home?

9.3 Flow diagrams and graphs

This section will show you how to:
- find the equations of horizontal and vertical lines
- use flow diagrams to draw graphs

Key words
equation of a line
flow diagram
function
input value
line segment
negative coordinates
output value
x-value
y-value

Plotting negative coordinates

A set of axes can form four sectors called quadrants, but so far, all the points you have read or plotted on graphs have been coordinates in the first quadrant. The grid below shows you how to read and plot coordinates in all four quadrants and how to find the equations of vertical and horizontal lines. This involves using **negative coordinates**.

The coordinates of a point are given in the form (x, y), where x is the number along the x-axis and y is the number up the y-axis.

The coordinates of the four points on the grid are:

A(2, 3) B(–1, 2) C(–3, –4) D(1, –3)

The x-coordinate of all the points on line X are 3. So you can say the **equation of line** X is $x = 3$.

The y-coordinate of all the points on line Y are –2. So you can say the equation of line Y is $y = -2$.

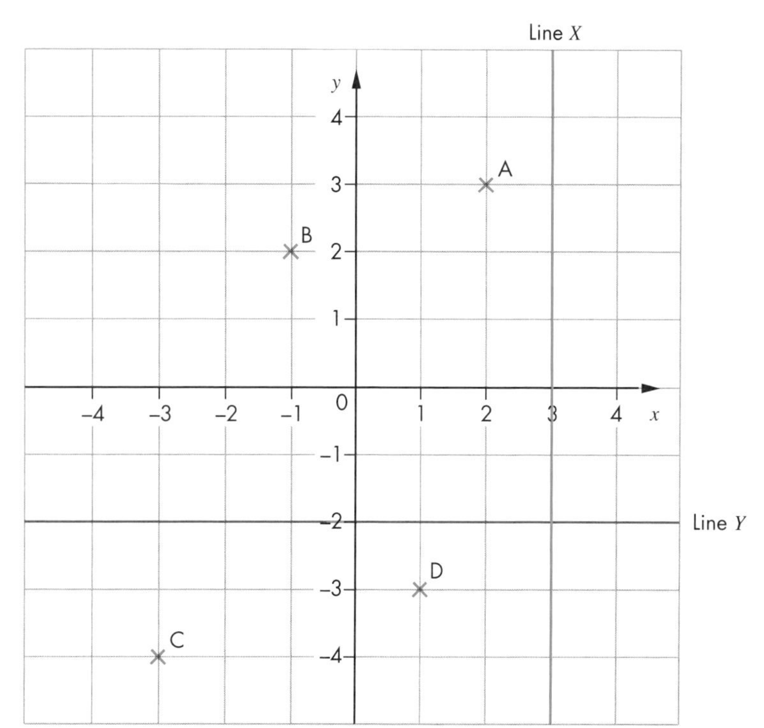

Note: The equation of the x-axis is $y = 0$ and the equation of the y-axis is $x = 0$.

Flow diagrams

One way of drawing a graph is to obtain a set of coordinates from an equation by means of a **flow diagram**. These coordinates are then plotted and the graph is drawn.

In its simplest form, a flow diagram consists of a single box, which may be thought of as containing a mathematical operation, called a **function**. A set of numbers fed into one side of the box is changed by the operation into another set, which comes out from the opposite side of the box. For example, the box shown below represents the operation of multiplying by 3.

Input: 0, 1, 2, 3, 4 → $\times 3$ → Output: 0, 3, 6, 9, 12

The numbers that are fed into the box are called **input values** and the numbers that come out are called **output values**.

The input and output values can be arranged in a table.

x	0	1	2	3	4
y	0	3	6	9	12

The input values are called ***x*-values** and the output values are called ***y*-values**. These form a set of coordinates that can be *plotted on a graph*. In this case, the coordinates are (0, 0), (1, 3), (2, 6), (3, 9) and (4, 12).

Most functions consist of more than one operation, so the flow diagrams consist of more than one box. In such cases, you need to match the *first* input values to the *last* output values. The values produced in the middle operations are just working numbers and can be missed out.

0, 1, 2, 3, 4 → $\times 2$ → 0, 2, 4, 6, 8 → $+3$ → 3, 5, 7, 9, 11

So, for the two-box flow diagram the table looks like this.

x	0	1	2	3	4
y	3	5	7	9	11

This gives the coordinates (0, 3), (1, 5), (2, 7), (3, 9) and (4, 11).

The two flow diagrams above represent respectively the equation $y = 3x$ and the equation $y = 2x + 3$, as shown below.

x → $\times 3$ → $3x$

$y = 3x$

x → $\times 2$ → $2x$ → $+ 3$ → $2x + 3$

$y = 2x + 3$

It is now an easy step to plot the coordinates for each equation on a set of axes, to produce the graphs of $y = 3x$ and $y = 2x + 3$, as shown below.

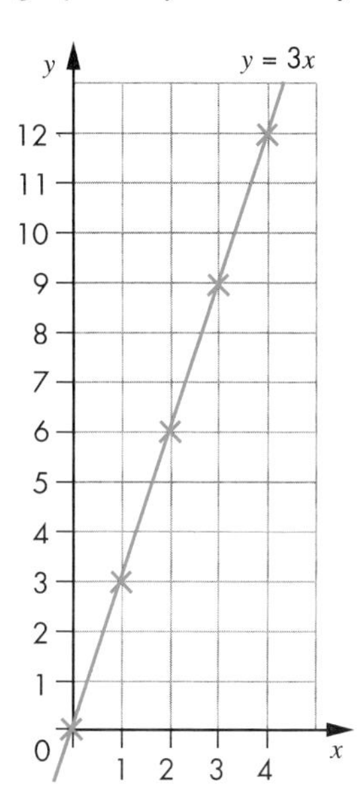

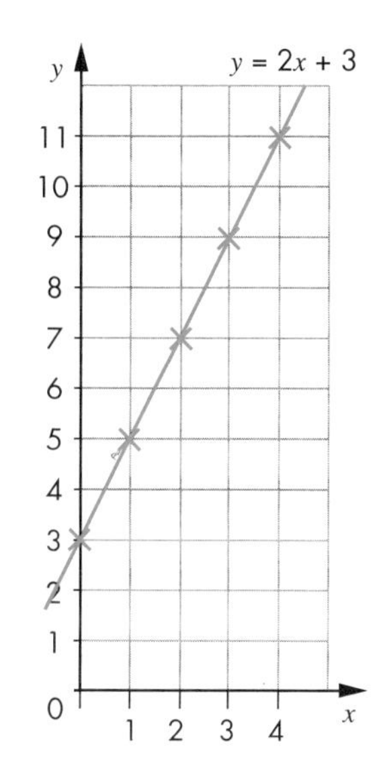

Remember:
Always label graphs.

Note: The line drawn is a **line segment**, as it is only part of an infinitely long line. You will not be penalised if you extend the line beyond the given range of values.

One of the practical problems in graph work is deciding the range of values for the axes. In examinations this is not usually a problem as the axes are drawn for you. Throughout this section, diagrams like the one below will show you the range for your axes for each question. These diagrams are not necessarily drawn to scale.

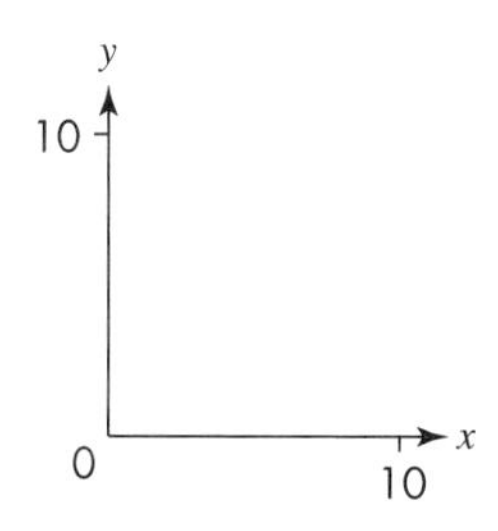

This particular diagram means draw the x-axis (horizontal axis) from 0 to 10 and the y-axis (vertical axis) from 0 to 10. You can use any type of graph or squared paper to draw your axes.

Note that the *scale* on each axis need *not always be the same.*

EXAMPLE 4

Use the flow diagram below to draw the graph of $y = 4x - 1$.

x —— 0, 1, 2, 3, 4 ——> [$\times 4$] —— ?, ?, ?, ?, ? ——> [-1] —— ?, ?, ?, ?, ? ——> y

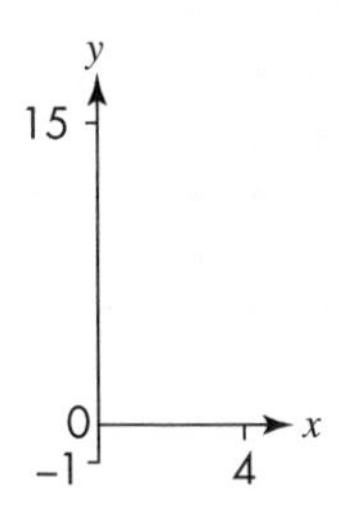

Now enter the values in a table.

x	0	1	2	3	4
y					

The table becomes:

x	0	1	2	3	4
y	–1	3	7	11	15

So, the coordinates are:

(0, –1), (1, 3), (2, 7), (3, 11), (4, 15)

Plot these points and join them up to obtain the graph shown on the right.

This is the graph of $y = 4x - 1$.

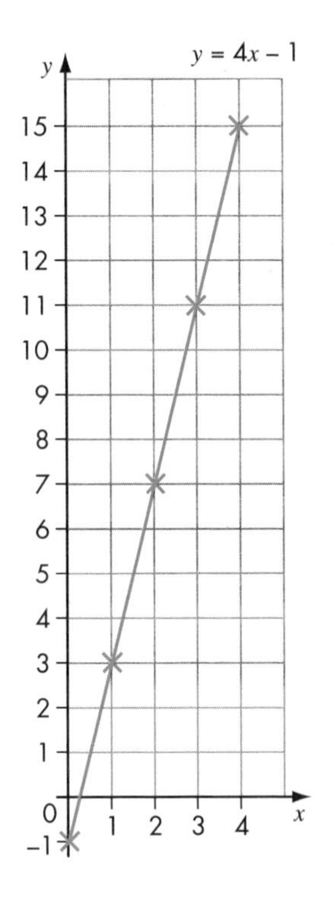

Always label your graphs. In an examination, you may need to draw more than one graph on the same axes. If you do not label your graphs you may lose marks.

EXERCISE 9C

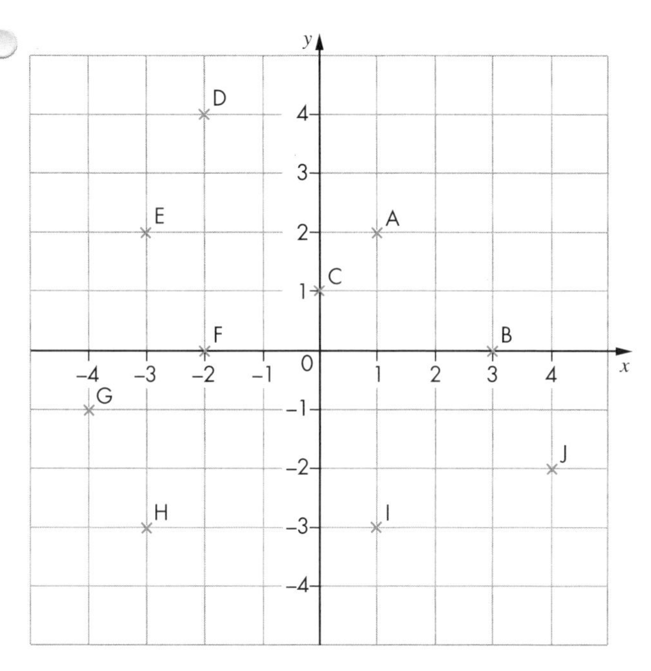

a Write down the coordinates of all the points A to J on the grid.

b Write down the coordinates of the midpoint of the line joining:

i A and B **ii** H and I **iii** D and J.

c Write down the equations of the lines labelled 1 to 4 on the grid.

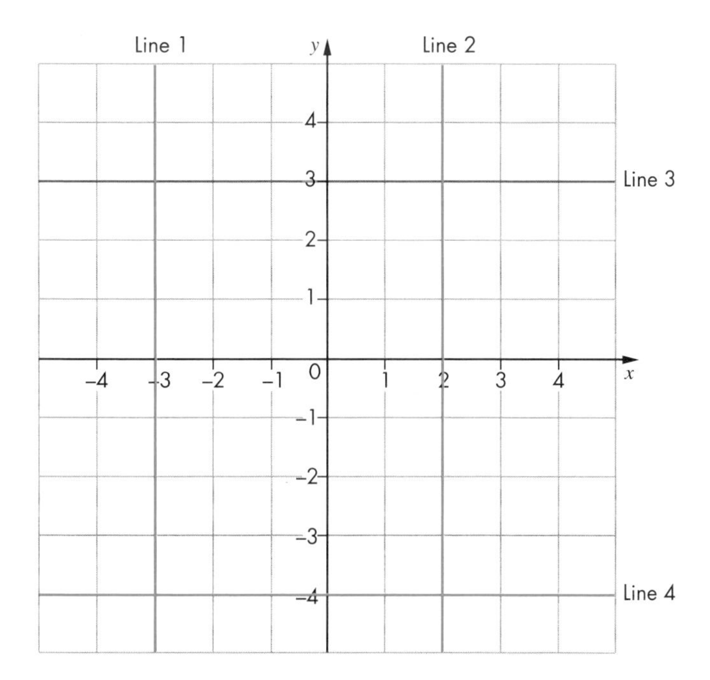

d Write down the equation of the line that is exactly halfway between:

i line 1 and line 2 **ii** line 3 and line 4.

2 Draw the graph of $y = x + 2$.

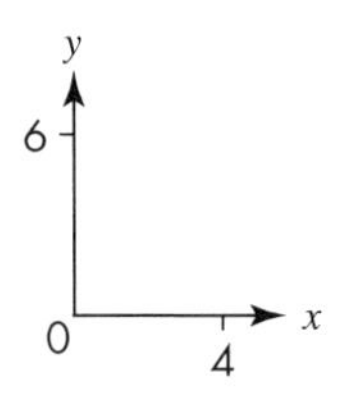

x	0	1	2	3	4
y					

3 Draw the graph of $y = 2x - 2$.

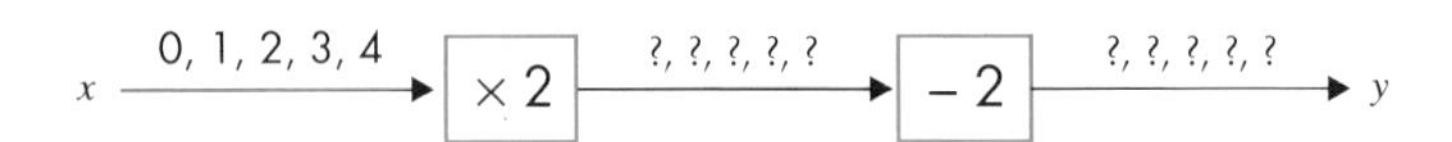

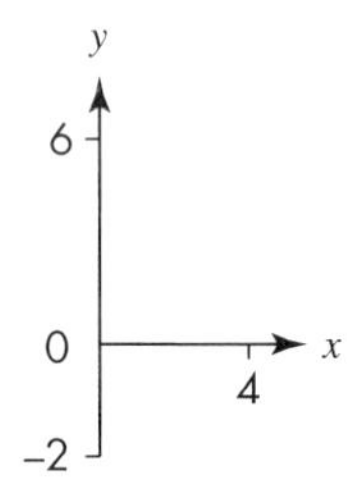

x	0	1	2	3	4
y					

4 Draw the graph of $y = \frac{x}{3} + 1$.

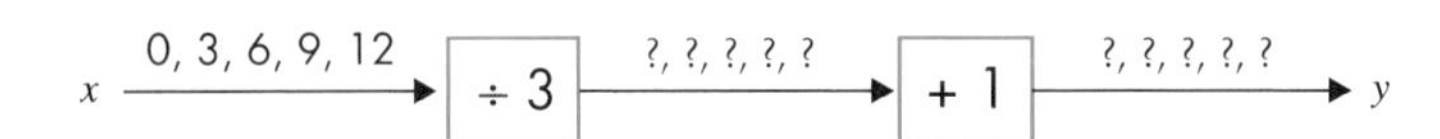

y
5
0
12
x

x	0	3	6	9	12
y					

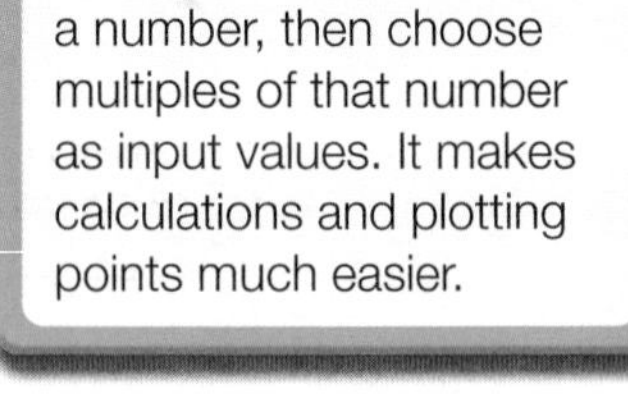

5 Draw the graph of $y = \frac{x}{2} - 4$.

x
0, 2, 4, 6, 8
÷ 2
?, ?, ?, ?, ?
– 4
?, ?, ?, ?, ?
y

x	0	2	4	6	8
y					

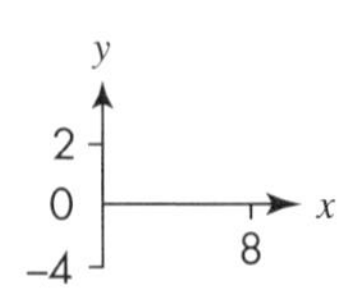

6 **a** Draw the graphs of $y = 2x$ and $y = x + 6$ on the same grid.

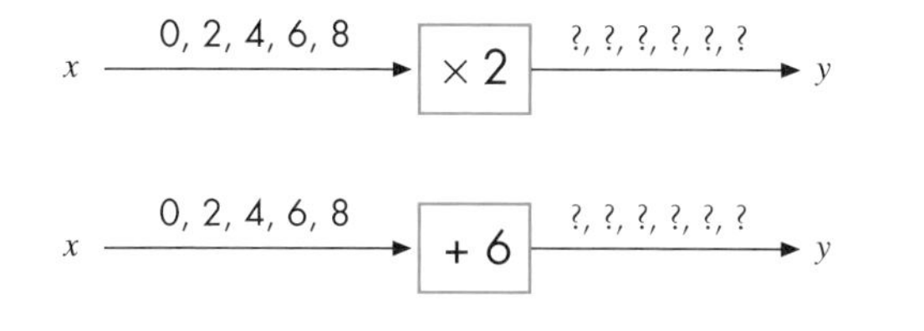

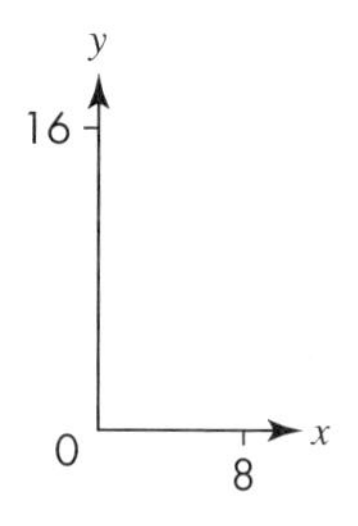

b At which point do the lines intersect?

7 **a** Draw the graphs of $y = x - 3$ and $y = 2x - 6$ on the same grid.

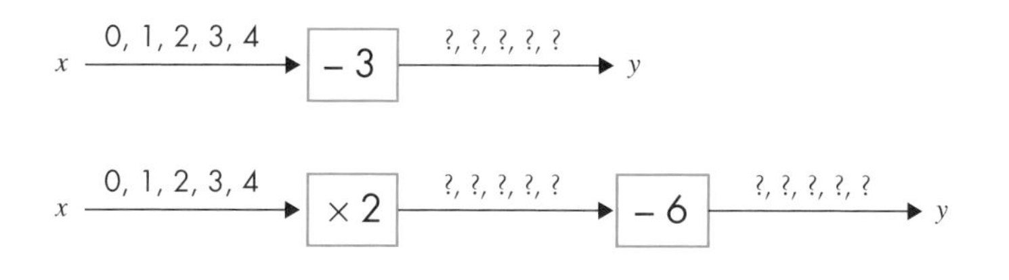

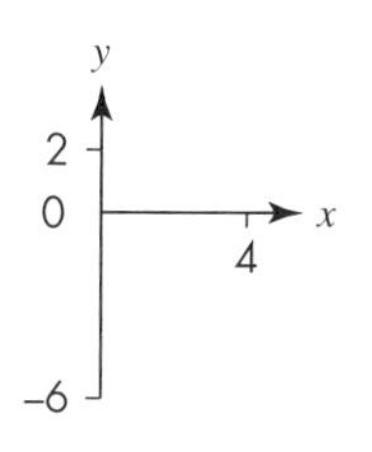

b At which point do the lines intersect?

8 Draw the graph of $y = 5x - 1$. Choose your own inputs and axes.

FM 9 A tea shop sells two types of afternoon tea: a cream tea which costs £3.50 and a high tea which costs £5.00.

To work out the cost of different combinations of teas, they use a wall chart or a flow diagram.

a Use the flow chart to work out the cost of three cream teas and two high teas.

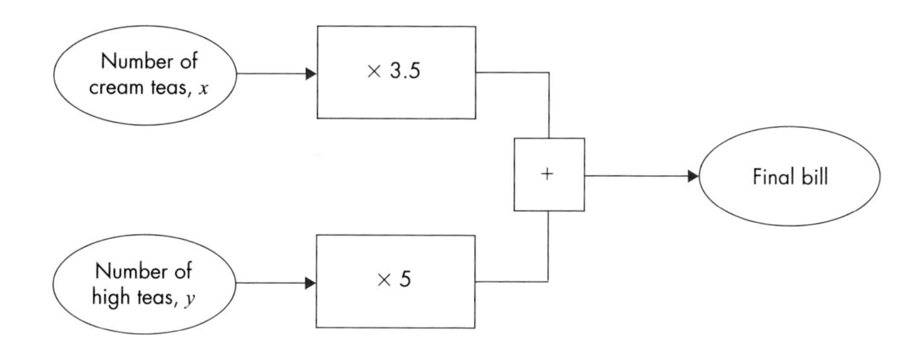

D

b The wall chart is partially filled in.

Use the flow diagram to complete the chart.

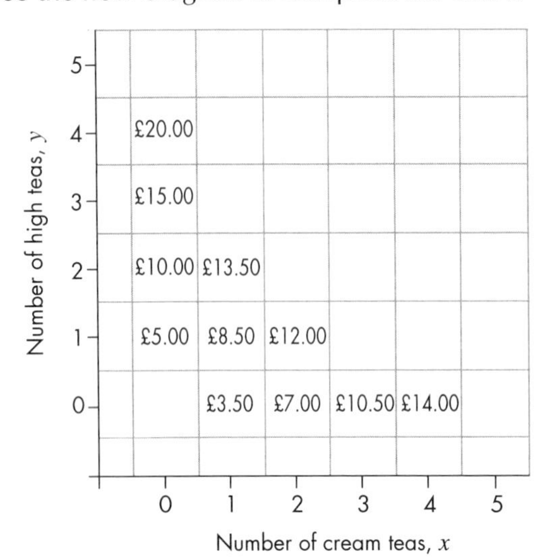

c A party paid £30.50. Can you say for sure what their order was?

AU 10 A teacher reads out the following 'think of a number' problem:

'I am thinking of a number: I multiply it by 3 and add 1.'

a Represent this using a flow diagram.

b If the input is x and the output is y, write down a relationship between x and y.

c Draw a graph for x-values from 0 to 5.

d Explain how you could use the graph to find the number the teacher thought of if the final answer was 13.

PS 11 This flow diagram connects two variables X and Y.

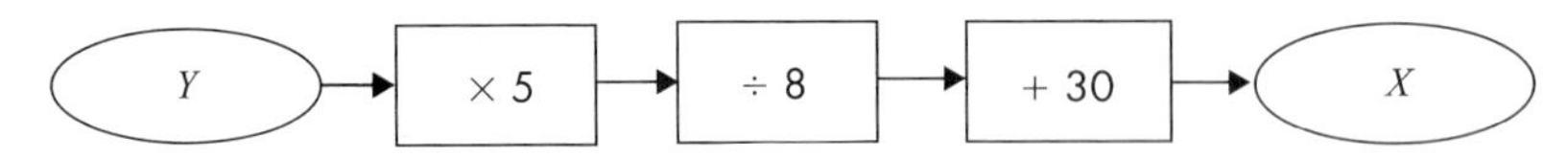

This graph connects the variable Y and X.

Fill in the missing values on the Y-axis.

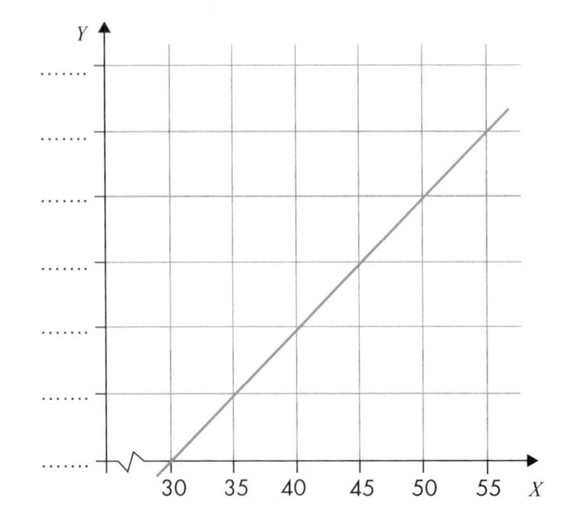

9.4 Linear graphs

This section will show you how to:
- draw linear graphs without using flow diagrams
- find the gradient of a straight line
- use the gradient to draw a straight line

Key words
gradient
linear graphs
slope

This chapter is concerned with drawing straight-line graphs. These graphs are usually referred to as **linear graphs**.

The minimum number of points needed to draw a linear graph is two but it is better to plot three or more because that gives at least one point to act as a check. There is no rule about how many points to plot but here are some tips for drawing graphs.

- Use a sharp pencil and mark each point with an accurate cross.
- Position your eyes directly over the graph. If you look from the side, you will not be able to line up your ruler accurately.

Drawing graphs by finding points

This method is a bit quicker and does not need flow diagrams. However, if you prefer flow diagrams, use them.

Follow through Example 5 to see how this method works.

EXAMPLE 5

Draw the graph of $y = 4x - 5$ for values of x from 0 to 5. This is usually written as $0 \leqslant x \leqslant 5$.

Choose three values for x: these should be the highest and lowest x-values and one in between.

Work out the y-values by substituting the x-values into the equation.

Keep a record of your calculations in a table, as shown below.

x	0	3	5
y			

When $x = 0$, $y = 4(0) - 5 = -5$
This gives the point (0, –5).

When $x = 3$, $y = 4(3) - 5 = 7$
This gives the point (3, 7).

When $x = 5$, $y = 4(5) - 5 = 15$
This gives the point (5, 15).

Hence your table is:

x	0	3	5
y	–5	7	15

You now have to decide the extent (range) of the axes. You can find this out by looking at the coordinates that you have so far.

The smallest x-value is 0, the largest is 5.
The smallest y-value is –5, the largest is 15.

Now draw the axes, plot the points and complete the graph.

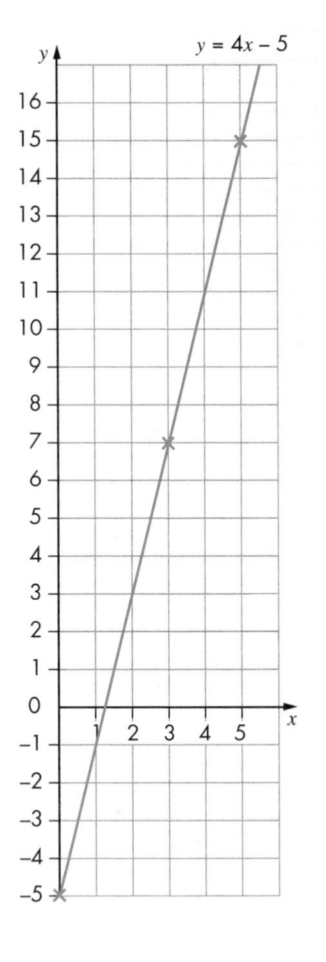

It is nearly always a good idea to choose 0 as one of the x-values. In an examination, the range for the x-values will usually be given and the axes will already be drawn.

EXERCISE 9D

Read through these hints before drawing the following linear graphs.

- Use the highest and lowest values of x given in the range.
- Do not pick x-values that are too close together, such as 1 and 2. Try to space them out so that you can draw a more accurate graph.
- Always label your graph with its equation. This is particularly important when you are drawing two graphs on the same set of axes.
- If you want to use a flow diagram, use one.
- Create a table of values. You will often have to complete these in your examinations.

D

1 Draw the graph of $y = 3x + 4$ for x-values from 0 to 5 ($0 \leq x \leq 5$).

2 Draw the graph of $y = 2x - 5$ for $0 \leq x \leq 5$.

3 Draw the graph of $y = \frac{x}{2} - 3$ for $0 \leq x \leq 10$.

4 Draw the graph of $y = 3x + 5$ for $-3 \leq x \leq 3$.

5 Draw the graph of $y = \frac{x}{3} + 4$ for $-6 \leq x \leq 6$.

HINTS AND TIPS

Complete the table of values first, then you will know the extent of the y-axis.

6 **a** On the same set of axes, draw the graphs of $y = 3x - 2$ and $y = 2x + 1$ for $0 \leq x \leq 5$.

b At which point do the two lines intersect?

7 **a** On the same axes, draw the graphs of $y = 4x - 5$ and $y = 2x + 3$ for $0 \leq x \leq 5$.

b At which point do the two lines intersect?

8 **a** On the same axes, draw the graphs of $y = \frac{x}{3} - 1$ and $y = \frac{x}{2} - 2$ for $0 \leq x \leq 12$.

b At which point do the two lines intersect?

9 **a** On the same axes, draw the graphs of $y = 3x + 1$ and $y = 3x - 2$ for $0 \leq x \leq 4$.

b Do the two lines intersect? If not, why not?

10 **a** Copy and complete the table to draw the graph of $x + y = 5$ for $0 \leq x \leq 5$.

x	0	1	2	3	4	5
y	5		3		1	

b Now draw the graph of $x + y = 7$ for $0 \leq x \leq 7$.

11 Ian the electrician used this formula to work out how much to charge for a job:

$C = 25 + 30H$

where C is the charge and H is how long the job takes.

John the electrician uses this formula:

$C = 35 + 27.5H$

a On a copy of the grid, draw lines to represent these formulae.

FM **b** For what length of job do Ian and John charge the same amount?

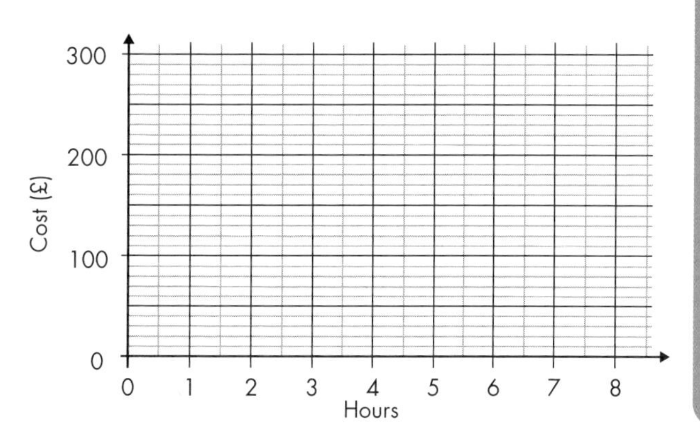

AU **12** **a** Draw the graphs $y = 4$, $y = x$ and $x = 1$ on a copy of the grid on the right.

b What is the area of the triangle formed by the three lines?

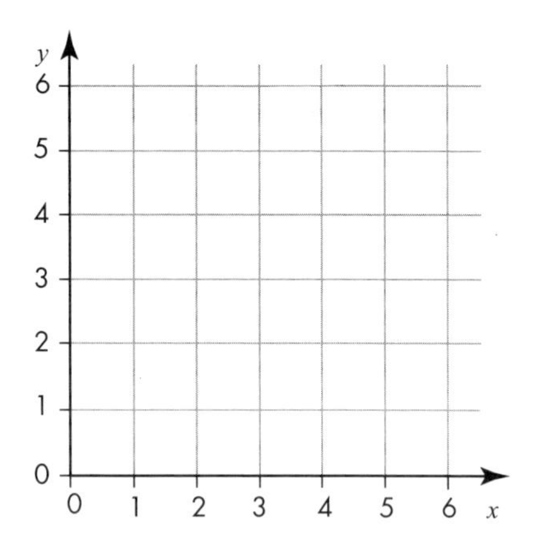

PS **13** The two graphs below show y against x and y against z.

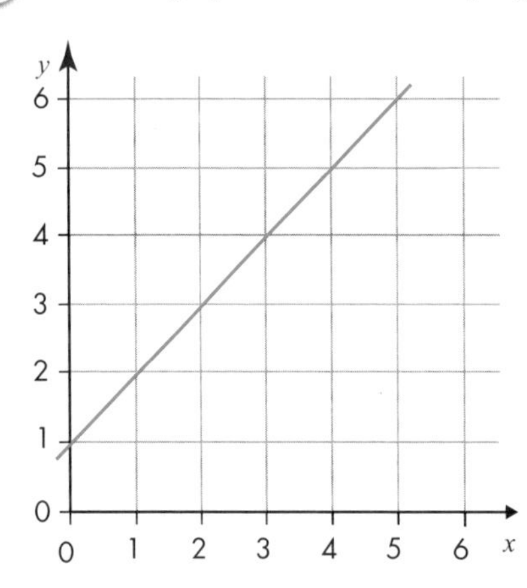

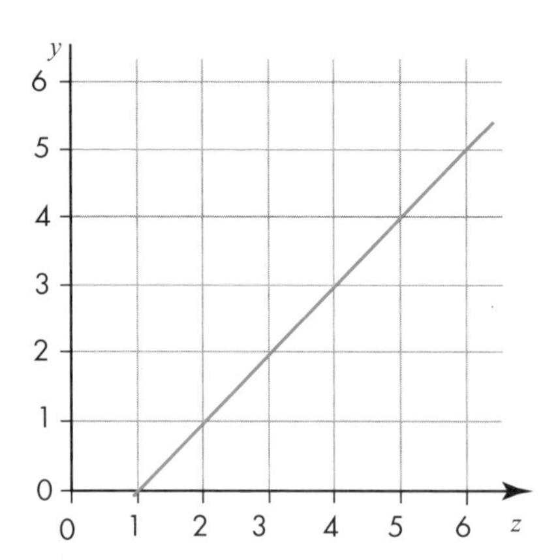

On a copy of the blank grid, show the graph of x against z.

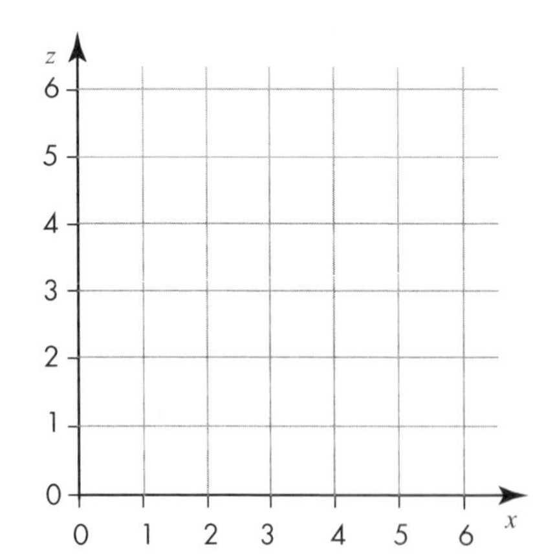

HINTS AND TIPS

When $x = 1$, $y = 2$ and when $y = 2$, $z = 3$.
So, plot coordinates for $x = 1$, $z = 3$.

Gradient

The **slope** of a line is called its **gradient**. The steeper the slope of the line, the larger the value of the gradient.

The gradient of the line shown here can be measured by drawing, as large as possible, a right-angled triangle which has part of the line as its hypotenuse (sloping side). The gradient is then given by:

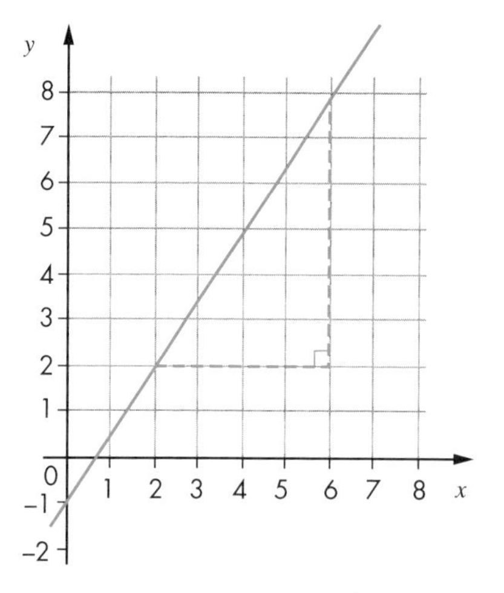

$$\text{gradient} = \frac{\text{distance measured up}}{\text{distance measured along}}$$

$$= \frac{\text{difference on } y\text{-axis}}{\text{difference on } x\text{-axis}}$$

For example, to measure the steepness of the line in the first diagram, below, you first draw a right-angled triangle that has part of this line as its hypotenuse. It does not matter where you draw the triangle but it makes the calculations much easier if you choose a sensible place. This usually means using existing grid lines, so that you avoid fractional values. See the second and third diagrams below.

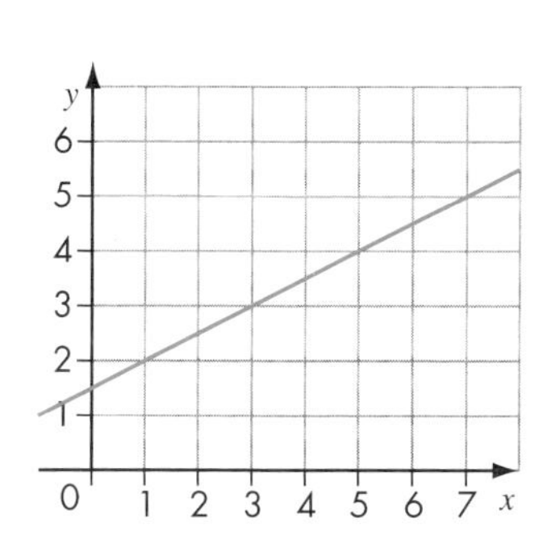

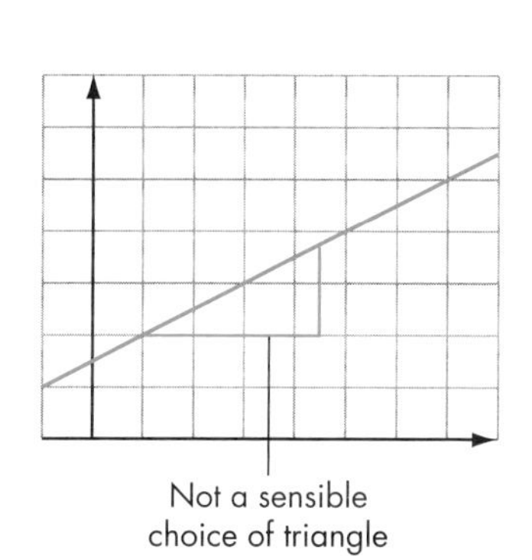

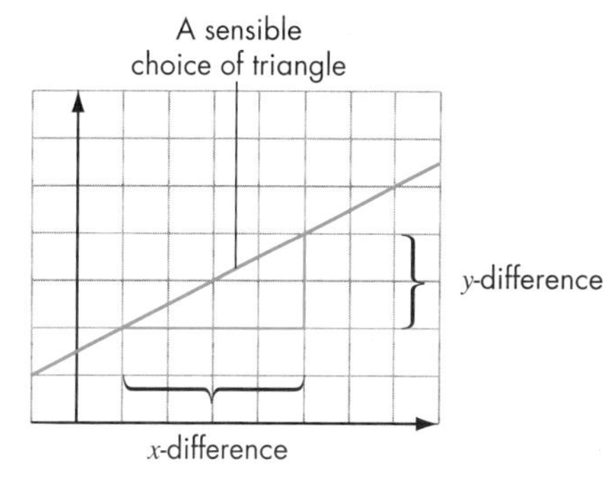

After you have drawn the triangle, measure (or count) how many squares there are on the vertical side. This is the difference between the y-coordinates. In this case, it is 2.

Then measure (or count) how many squares there are on the horizontal side. This is the difference between the x-coordinates. In the case above, this is 4.

To work out the gradient, you make the following calculation.

$$\text{gradient} = \frac{\text{difference of the } y\text{-coordinates}}{\text{difference of the } x\text{-coordinates}}$$

$$= \frac{2}{4} = \frac{1}{2} \text{ or } 0.5$$

Note that the value of the gradient is not affected by where the triangle is drawn. As you are calculating the ratio of two sides of the triangle, the gradient will always be the same wherever you draw the triangle.

Remember: Take care when finding the differences between the coordinates of the two points. Choose one point as the first and the other as the second, and subtract in the *same order* each time to find the difference. When a line slopes *down from right to left* (/) the gradient is always positive, but when a line slopes *down from left to right* (\) the gradient is always negative, so you must make sure there is a minus sign in front of the fraction.

EXAMPLE 6

Find the gradient of each of these lines.

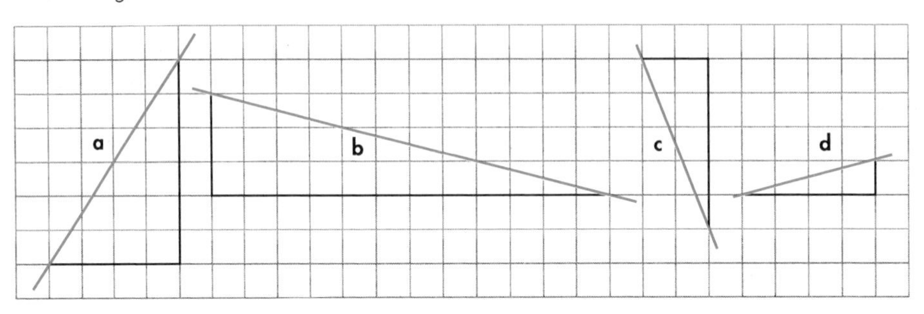

In each case, a sensible choice of triangle has already been made.

a y-difference = 6, x-difference = 4 Gradient $= 6 \div 4 = \frac{3}{2} = 1.5$

b y-difference = 3, x-difference = 12 Line slopes down from left to right,

so gradient $= -(3 \div 12) = -\frac{1}{4} = -0.25$

c y-difference = 5, x-difference = 2 Line slopes down from left to right,

so gradient $= -(5 \div 2) = -\frac{5}{2} = -2.5$

d y-difference = 1, x-difference = 4 Gradient $= 1 \div 4 = \frac{1}{4} = 0.25$

Drawing a line with a certain gradient

To draw a line with a certain gradient, you need to 'reverse' the process described above. Use the given gradient to draw the right-angled triangle first. For example, take a gradient of 2.

Start at a convenient point (A in the diagrams opposite). A gradient of 2 means for an x-step of 1 the y-step must be 2 (because 2 is the fraction $\frac{2}{1}$). So, move one square across and two squares up, and mark a dot.

Repeat this as many times as you like and draw the line. You can also move one square back and two squares down, which gives the same gradient, as the third diagram shows.

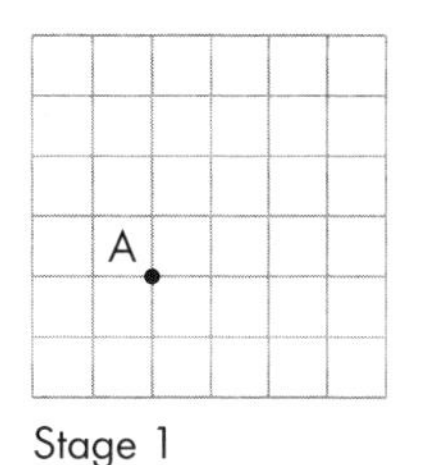

Stage 1

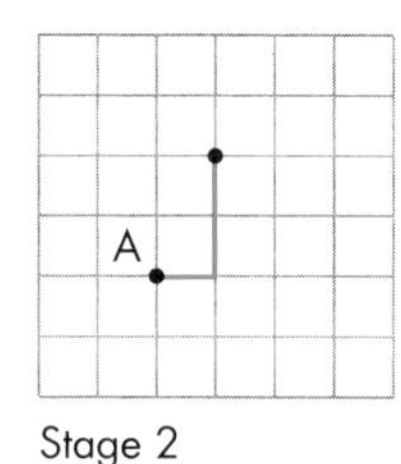

Stage 2

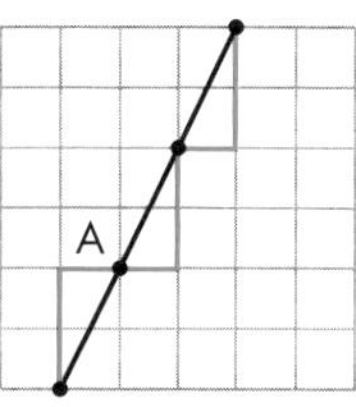

Stage 3

Remember: For a positive gradient you move across (left to right) and then *up*. For a negative gradient you move across (left to right) and then *down*.

EXAMPLE 7

Draw lines with these gradients. **a** $\frac{1}{3}$ **b** -3 **c** $-\frac{1}{4}$

a This is a fractional gradient which has a y-step of 1 and an x-step of 3. Move three squares across and one square up every time.

b This is a negative gradient, so for every one square across, move three squares down.

c This is also a negative gradient and it is a fraction. So for every four squares across, move one square down.

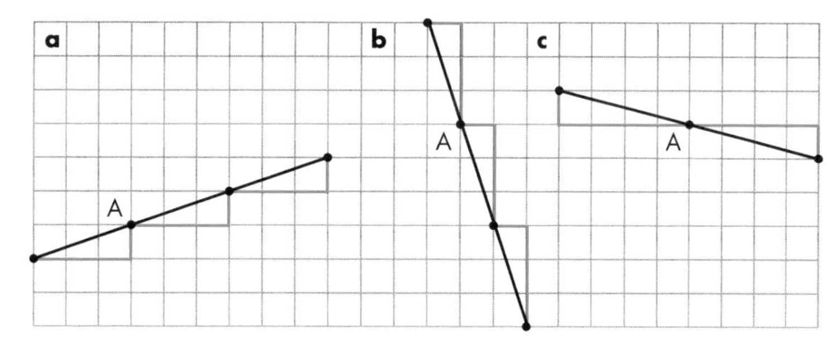

EXERCISE 9E

D

FM **1** Ravi was ill in hospital.
This is his temperature chart for the two weeks he was in hospital.

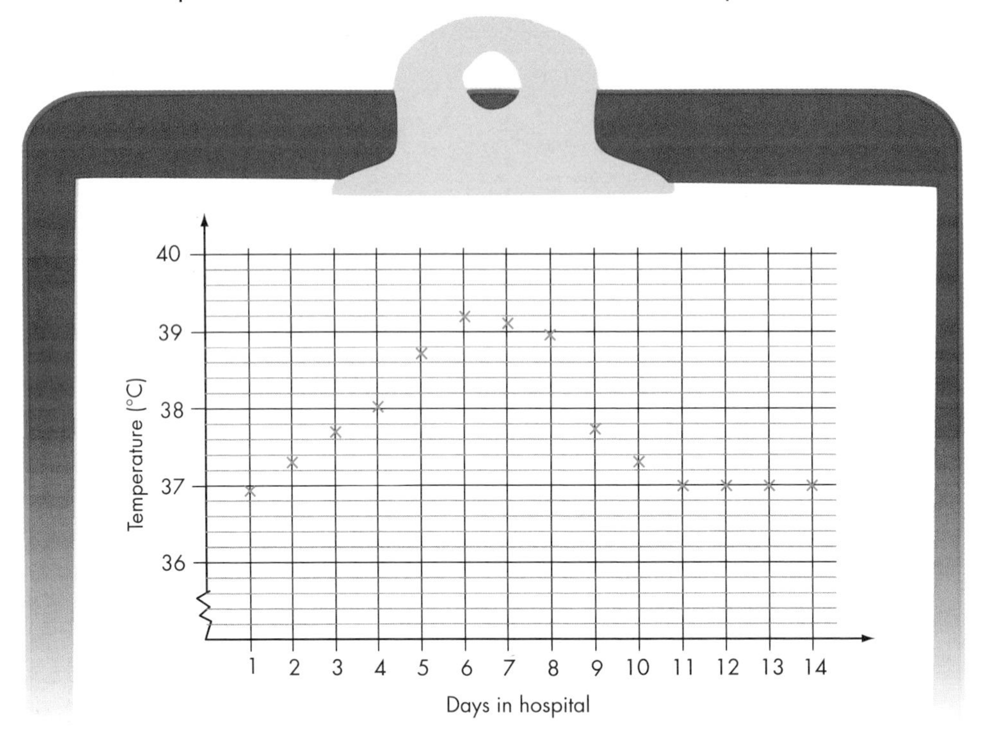

a What was Ravi's highest temperature?

b Between which days did Ravi's temperature increase the most? Explain how you can tell.

c Between which days did Ravi's temperature fall the most? Explain how you can tell.

d When Ravi's temperature went over 38.5 °C he was put on an antibiotic drip.

  i On what day did Ravi go on the drip?

  ii How many days did it take for the antibiotics to work before Ravi's temperature started to come down?

e Once a patient's temperature returns to normal for four days, they are allowed home. What is the normal body temperature?

C

2 Find the gradient of each of these lines.

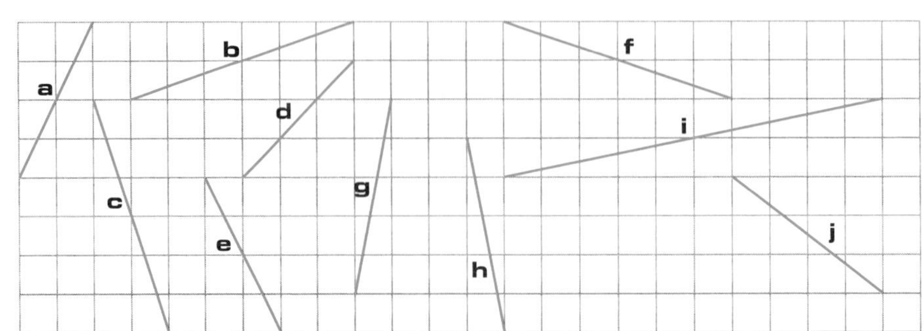

C

3 Find the gradient of each of these lines. What is special about these lines?

a

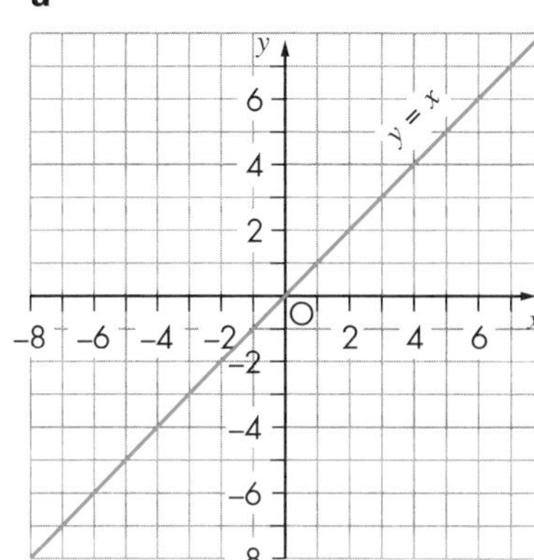

b

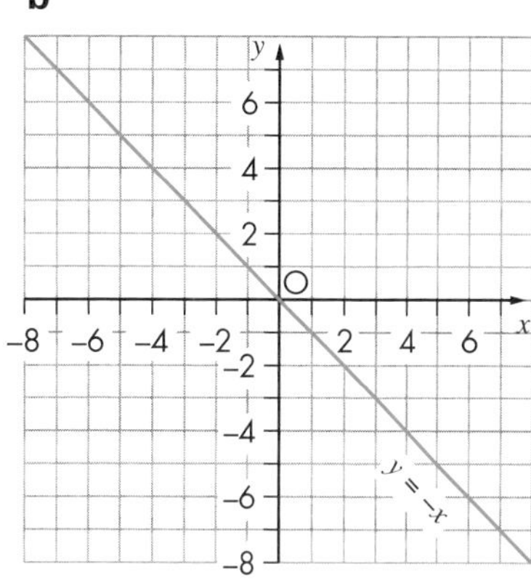

4 Draw lines with these gradients.

a 4 **b** $\frac{2}{3}$ **c** −2 **d** $-\frac{4}{5}$ **e** 6 **f** −6

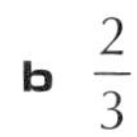

AU 5 Students in a class were asked to predict the y-value for an x-value of 10 for this line.

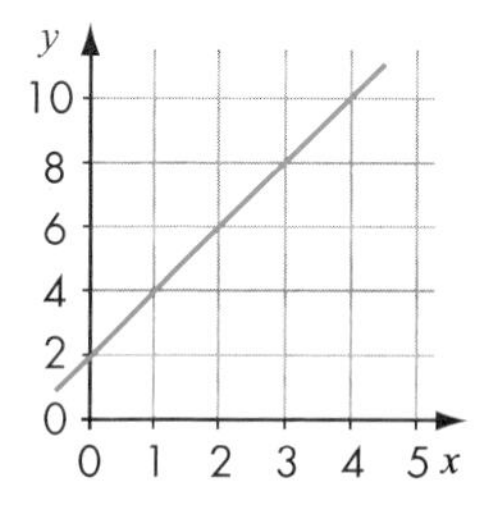

Rob says 'The gradient is 1, so the line is $y = x + 2$. When $x = 10$, $y = 12$.'

Rob is wrong.

Explain why and work out the correct y-value.

FM PS 6 The Health and Safety regulations for vent pipes from gas appliances state that the minimum height depends on the pitch (gradient) of the roof.

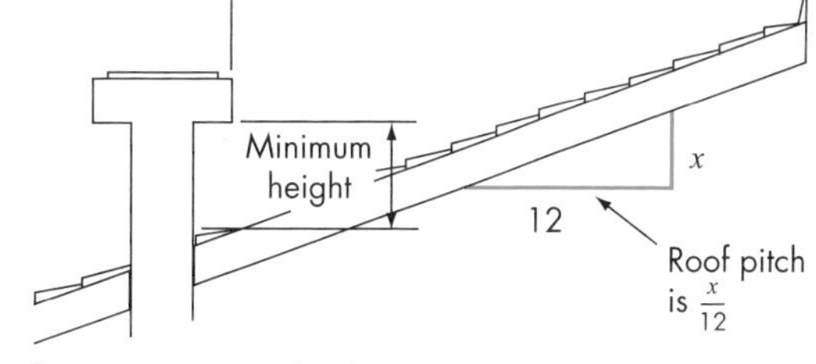

This is the rule:

Minimum height = 1 metre or twice the pitch in metres, whichever is greatest

a What is the minimum height of a roof with a pitch of 2?

b What is the minimum height of a roof with a pitch of 0.5?

c What is the minimum height for these two roofs?

i

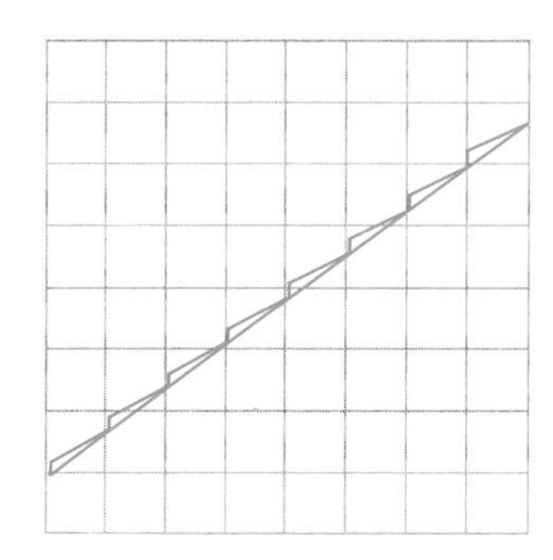

ii

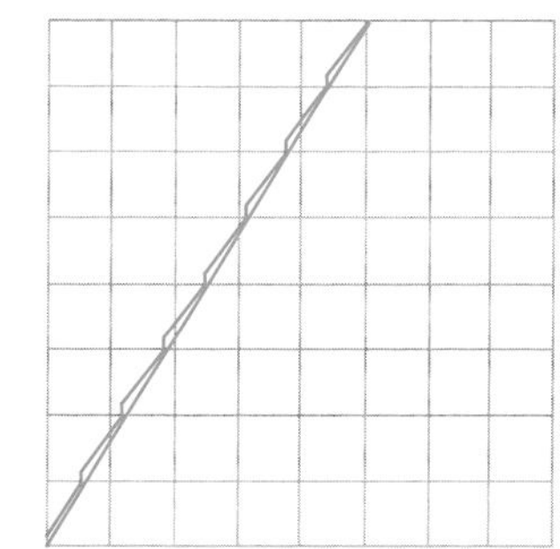

GRADE BOOSTER

- **F** You can read off values from a conversion graph
- **E** You can plot points in all four quadrants
- **E** You can draw a linear graph given a table of values to complete
- **D** You can find an average speed from a travel graph
- **D** You can draw a linear graph without being given a table of values
- **D** You can read off distances and times from a travel graph

What you should know now

- How to use conversion graphs
- How to use travel graphs to find distances, times and speeds
- How to draw a linear graph

1 This conversion graph can be used to change between metres and feet.

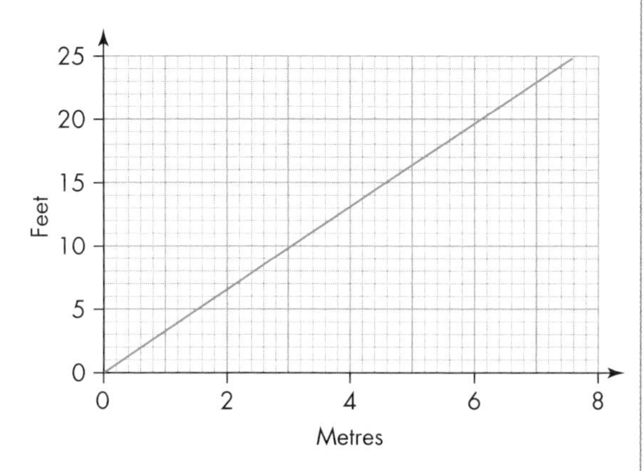

a Use the conversion graph to change 6 metres to feet. (1)

b Use the conversion graph to change 8 feet to metres. (1)

Robert jumps 4 metres.

James jumps 12 feet.

c **i** Who jumps furthest, Robert or James?

ii How did you get your answer? (2)

(Total 4 marks)

Edexcel, November 2008, Paper 13 Foundation, Question 10

2

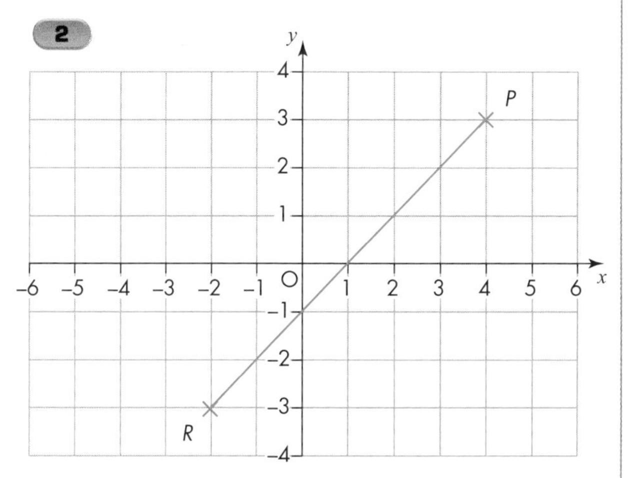

a Write down the coordinates of the point *P*. (1)

b On the grid, mark the point (–3, 1) with a cross (x). (1)

c Write down the coordinates of the midpoint of the line *PR*. (2)

(Total 4 marks)

Edexcel, June 2007, Paper 10 Foundation, Question 3

3

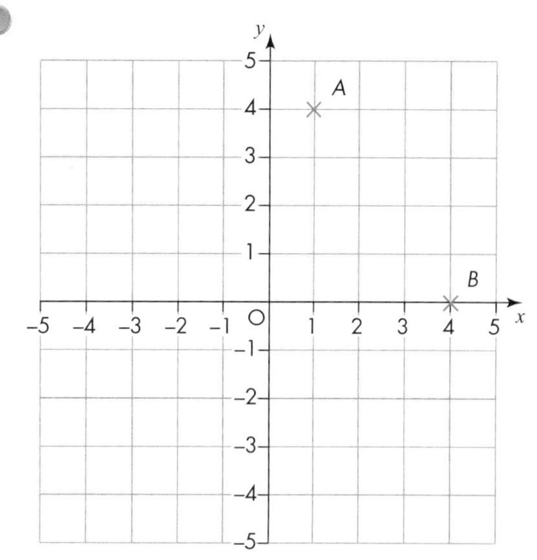

a **i** Write down the coordinates of the point *A*.

ii Write down the coordinates of the point *B*. (2)

b **i** On the grid, plot the point (3, 2).

Label this point *P*.

ii On the grid, plot the point (–4, 3).

Label this point *Q*. (2)

(Total 4 marks)

Edexcel, May 2008, Paper 1 Foundation, Question 7

4 **a** Complete the table of values for $y = 2x - 3$

x	–1	0	1	2	3
y	–5		–1		3

b On a grid, draw the graph of $y = 2x - 3$ for values of x from –1 to +3. Take the x-axis from –1 to +3 and the y-axis from –5 to 3

c Find the coordinates of the point where the line $y = 2x - 3$ crosses the line $y = -2$

5 **a** Complete the table of values for $y = 3x - 1$

x	−2	−1	0	1	2
y		−4		2	

(2)

b On the grid, draw the graph of $y = 3x - 1$

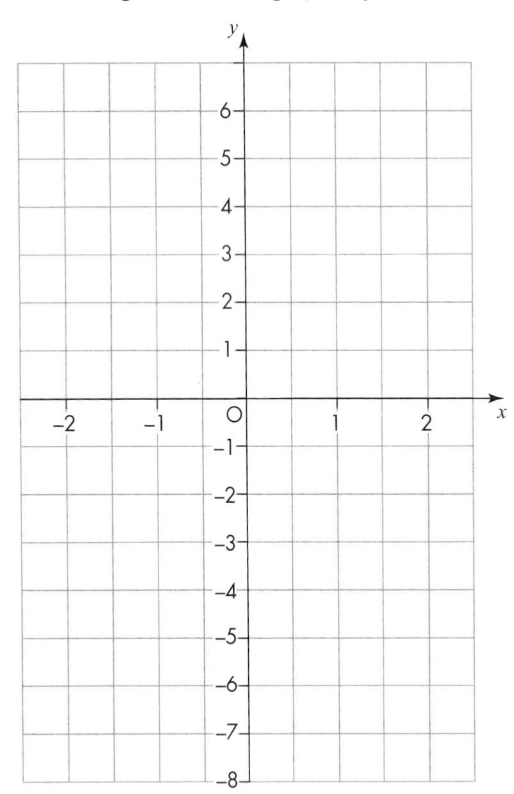

(2)

(Total 4 marks)

Edexcel, November 2008, Paper 10 Foundation, Question 8

6 **a** Complete the table of values for $y = 3x + 1$

x	−3	−2	−1	0	1	2
y	−8		−2			

(2)

b On the grid, draw the graph of $y = 3x + 1$

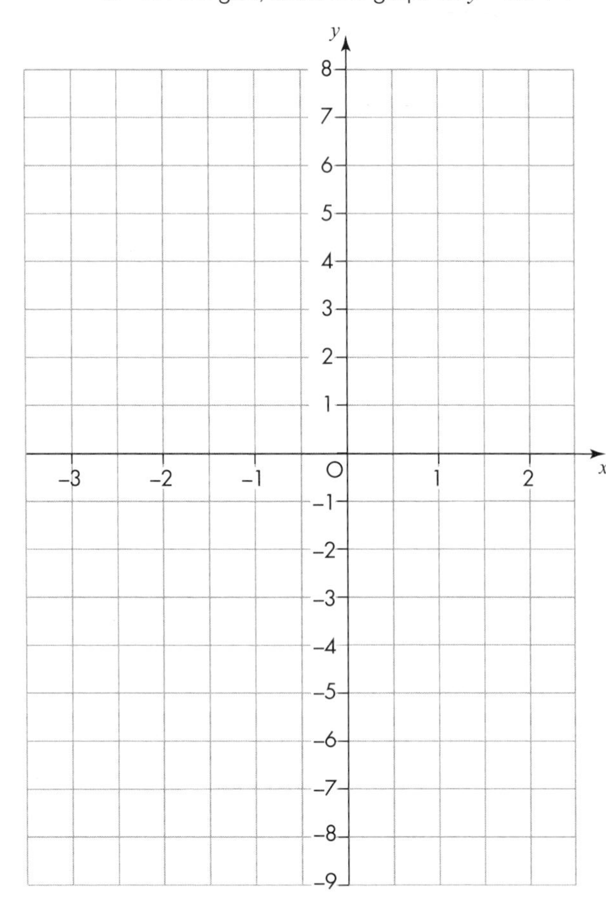

(2)

(Total 4 marks)

Edexcel, May 2008, Paper 1 Foundation, Question 18

7 **a** Draw a set of axes on a grid and label the x-axis from −4 to +4 and the y-axis from −8 to +10. On this grid, draw and label the lines $y = -5$ and $y = 2x + 1$

b Write down the coordinates of the point where the lines $y = -5$ and $y = 2x + 1$ cross.

8 Here is a travel graph of Siân's journey from her house to the library and back to her house.

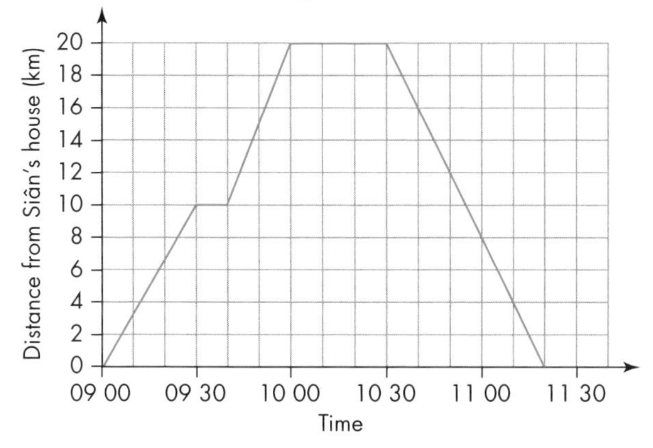

a How far is Siân from her house at 09 30? (1)

The library is 20 km from Siân's house.

b i At what time did Siân arrive at the library?

ii How long did Siân spend at the library? (2)

Siân left the library at 10 30 to travel back to her house.

c At what time did Siân arrive back at her house? (1)

(Total 4 marks)

Edexcel, May 2008, Paper 12 Foundation, Question 10

9 Pete visited his friend and then returned home.

The travel graph shows some information about Pete's journey.

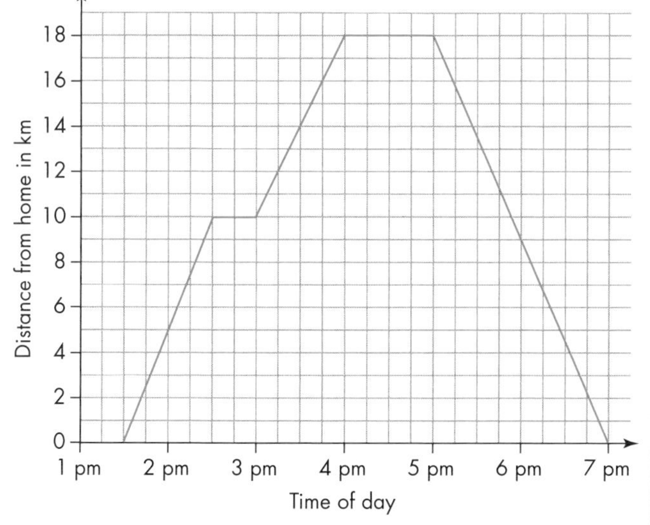

a Write down the time that Pete started his journey. (1)

At 2.30 pm Pete stopped for a rest.

b i Find his distance from home when he stopped for this rest.

ii How many minutes was this rest? (2)

Pete stayed with his friend for one hour. He then returned home.

c Work out the total distance travelled by Pete on this journey. (2)

(Total 5 marks)

Edexcel, November 2008, Paper 12 Foundation, Question 8

10 Here are six temperature/time graphs.

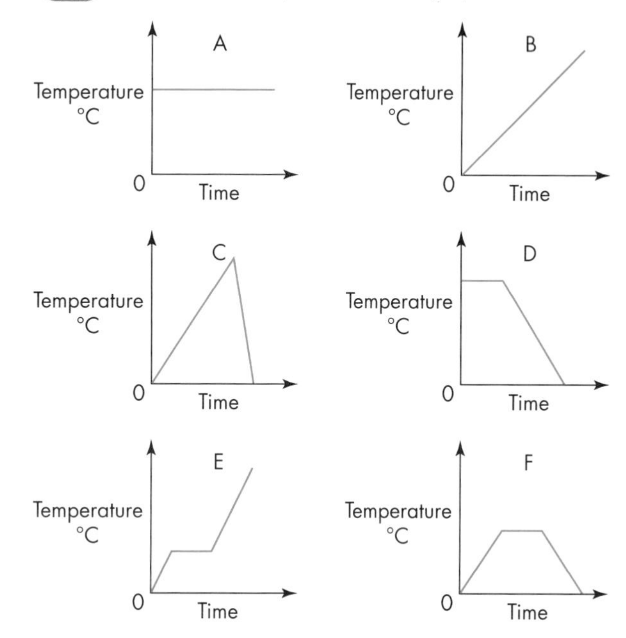

Each sentence in the table describes one of the graphs. Write the letter of the correct graph next to each sentence.

The first one has been done for you. (3)

The temperature starts at 0°C and keeps rising.	**B**
The temperature stays the same for a time and then falls.	
The temperature rises and then falls quickly.	
The temperature is always the same.	
The temperature rises, stays the same for a time and then falls.	
The temperature rises, stays the same for a time and then rises again.	

(Total 3 marks)

Edexcel, November 2008, Paper 12 Foundation, Question 11

Worked Examination Questions

FM **1** The distance–time graph shows the journey of a train between two stations. The stations are 6 kilometres apart.

a During the journey the train stopped at a signal. For how long was the train stopped?

b What was the average speed of the train for the whole journey? Give your answer in kilometres per hour.

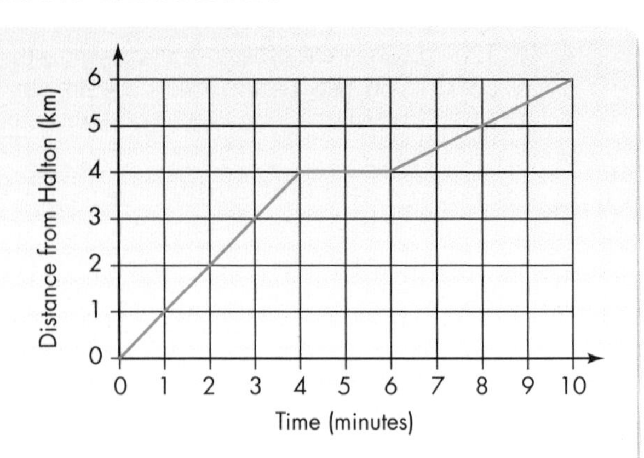

a The train stopped for 2 minutes (where the line is horizontal).

You get 1 mark for the correct answer.

b The train travels 6 km in 10 minutes. This is 36 km in 60 minutes (multiply both numbers by 6). So the average speed is 36 km/h.

You will get 1 method mark for writing down any distance and an equivalent time and 1 accuracy mark for the answer.

Total: 3 marks

FM **2** The graph shows the increase in rail fares as a percentage of the fares in 1997 since the railways were privatised in 1997.

a In what year did first class fares double in price from 1997?

b Approximately how much would a first class fare that cost £100 in 1997 cost in 2010?

c During which period did first class fares rise the most? How can you tell?

d An unregulated standard class fare cost £35 in 2008. Approximately how much would this fare have been in 1997?

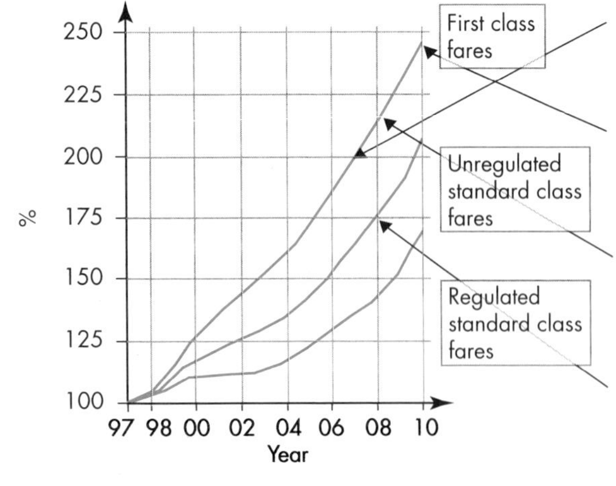

a 2007.

Read when the fares were 200%. This is worth 1 mark.

b £223.

Read the percentage increase in 2010. The correct answer is worth 1 mark.

c 2005–2010:

This is the steepest part of the graph. The correct answer is worth 1 mark.

d £20.

Unregulated fares have gone up by 75%. 75% of £20 is £15. The correct answer is worth 1 mark.

Total: 5 marks

Worked Examination Questions

PS **3** This graph shows the conversion between two variables x and y.

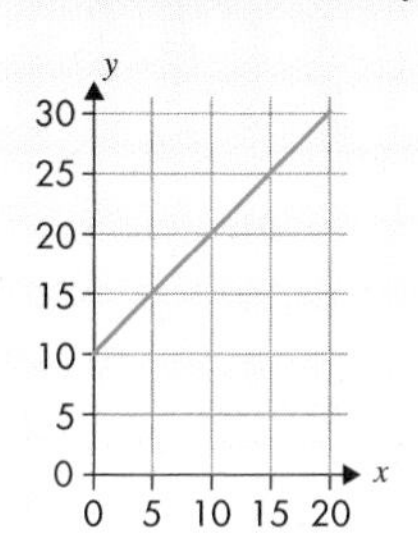

This graph shows the conversions between two variables y and z.

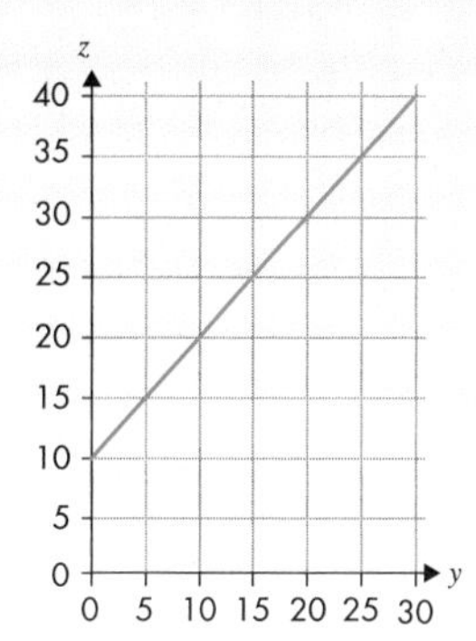

On the graph below, draw the conversion between x and z.

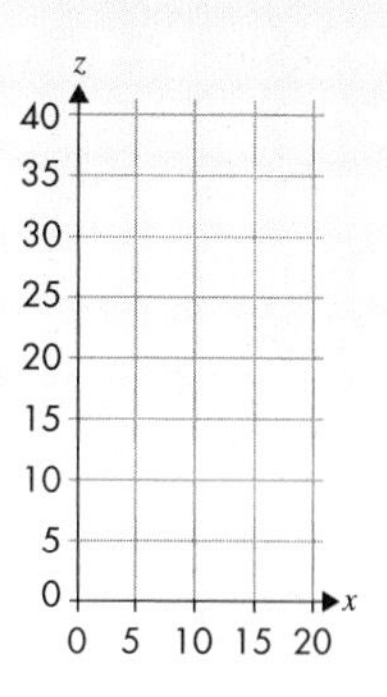

When $x = 0$, $y = 10$ and when $y = 10$, $z = 20$

When $x = 20$, $y = 30$ and when $y = 30$, $z = 40$

Connect at least two x-values to y-values. The obvious choices are 0 and 20 as these are the limits for x. This is worth 1 method mark and 1 accuracy mark.

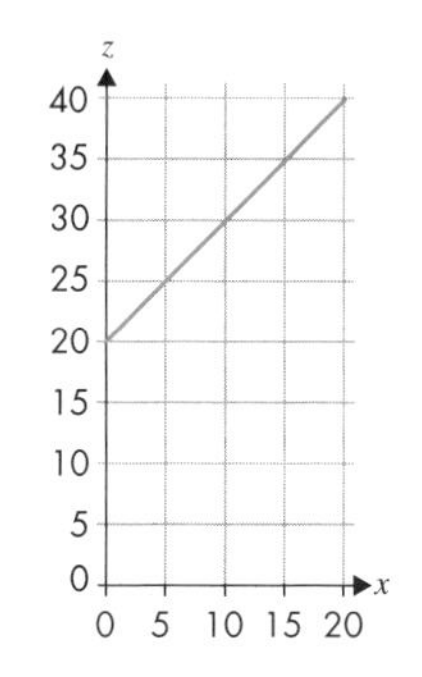

Draw a line joining the pairs of x- and z-values, i.e. (0, 20) to (20, 40). This is worth 1 accuracy mark.

Total: 3 marks

AU **4** Three lines are shown:

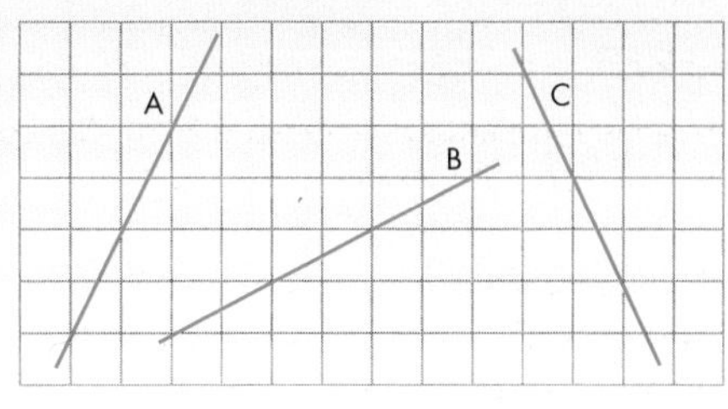

a Write down one thing that is the same about the gradients of line A and B.

b Write down one thing that is the same about the gradients of line A and C.

c Write down one thing that is different about the gradients of line B and C.

Gradient of A is 2, B is $\frac{1}{2}$ and C is -2

First work out gradients for 1 mark.

a They are both positive.

b They both have values of 2.

c One is positive and one is negative.

Write down something obvious. This question is testing your understanding of gradients so you could have said anything valid. You get 1 mark for a statement about line B and 1 mark for a statement about line C.

Total: 4 marks

Functional Maths
Planning a motorbike trip to France

A group of friends are going on a holiday to France. They have decided to go on motorbikes and have asked you to join them as a pillion passenger. The ferry will take your group to Boulogne and the destination is Perpignan.
Planning your motorbike trip will involve a range of mathematics, much of which can be represented on graphs.

Your task

The following task will require you to work in groups of 2–3.

Using all the information that you gather from these pages and your own knowledge, investigate the key mathematical elements of your motorbike trip.

You must draw at least one conversion graph and try to use as many different mathematical methods as possible.

Getting started

Start by thinking about the mathematics that you use when you go on holiday. Here are a few questions to get you going.

- What information do you need before you travel?
- How many euros (€) are there in one British pound (£)?
- What differences might you find when you travel abroad?
- What are the differences between metric and imperial units of measure? (It may help to list the conversion facts that you know.)
- After approximately how long would you need to stop to rest when travelling?

Handy hints

There are a number of measures and units in France that will need converting when you get there. Two of the most noticeable are:

- **Currency:** in France the currency is in euros
- **Distances**: in France, distances are measured in kilometres, and hence speeds are in kilometres per hour.

1 Euro = £0.90

1 gallon ≈ 4 litres

1 mile ≈ 1.6 km

A motorbike fuel tank holds 3 gallons and travels about 45 miles per gallon.

The cost of petrol in France is 1.2 per litre.

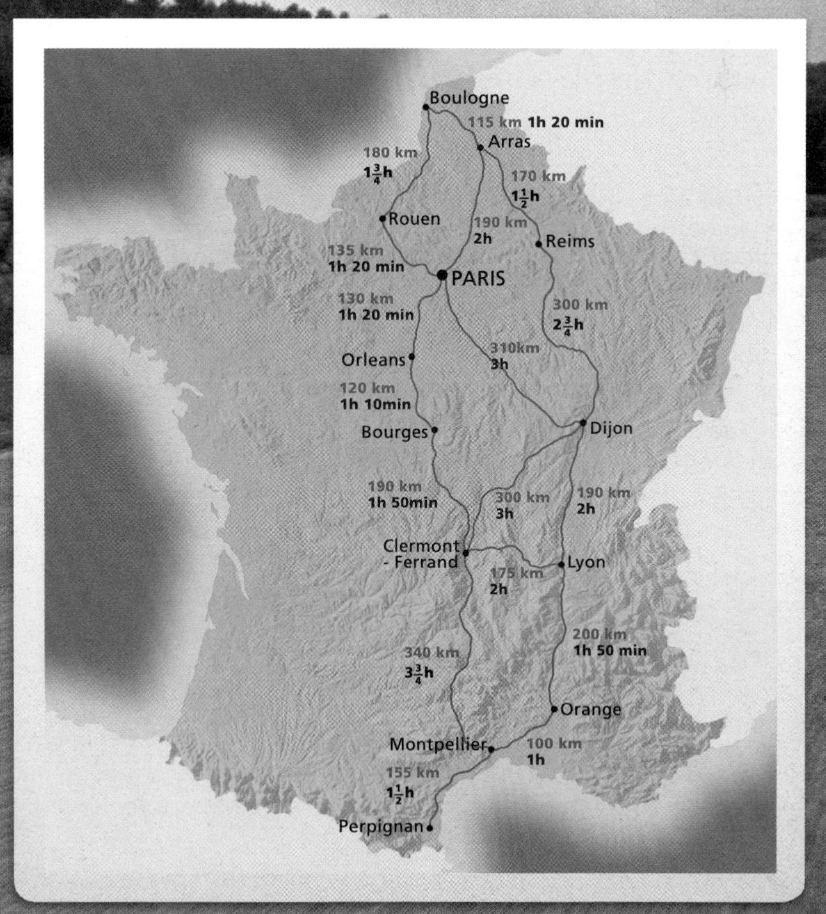

Why this chapter matters

For centuries statistical graphs such as bar charts, line graphs and pictograms have been used in many different areas, from science to politics. They provide a way to represent, analyse and interpret information.

Developing statistical analysis

The development of statistical graphs was spurred on by:

- the need in the 17th and 18th centuries to base policies on demographic and economic data and for this information to be shared with a large number of people
- greater understanding of measures and numbers in the 19th century
- increasing interest in analysing and understanding social conditions in the 19th and 20th centuries.

Since the later part of the 20th century, statistical graphs have become an important way of analysing information, and computer-generated statistical graphs are seen every day on TV, in newspapers and in magazines.

Who invented statistical diagrams?

William Playfair was a Scottish engineer, who is thought to be the founder of representing statistics in a graphical way.

He invented three types of diagrams: in 1786 the line graph and bar chart, then in 1801 the pie chart (which you will come across in later chapters).

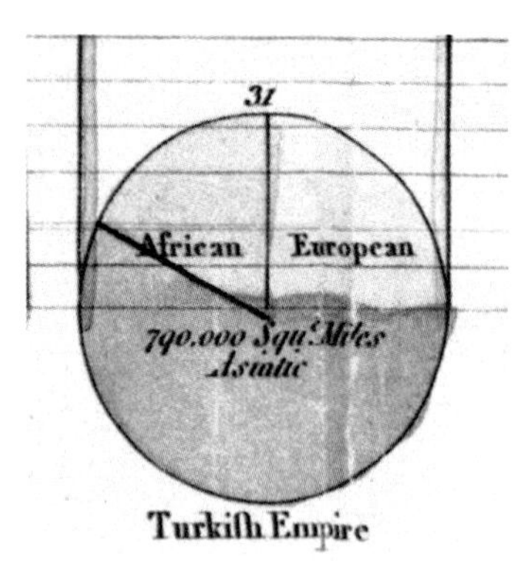

William Playfair pioneered charts and graphs such as this pie chart.

Florence Nightingale was born in 1820. She was very good at mathematics from an early age, becoming a pioneer in presenting information visually. She developed a form of the pie chart now known as the 'polar area diagram' or the 'Nightingale rose diagram', which was like a modern circular bar chart. It illustrated monthly patient deaths in military field hospitals. She called these diagrams 'coxcombs' and used them a great deal to present reports on the conditions of medical care in the Crimean War to Parliament and to civil servants who may not have fully understood traditional statistical reports.

Florence Nightingale was a nurse and hospital reformer. She used charts and graphs in her work.

In 1859, Florence Nightingale was elected the first female member of the Royal Statistical Society.

This chapter introduces you to some of the most common forms of statistical representation. They fall into two groups: graphical diagrams, such as pictograms and bar charts; and quantitive diagrams, such as frequency tables and stem-and-leaf diagrams.

Chapter

Statistics: Statistical representation

The grades given in this chapter are target grades.

This chapter will show you ...

- **F** how to collect and organise data, and how to represent data on various types of diagram
- **G** to **D** how to draw conclusions from statistical diagrams
- **F** to **D** how to draw diagrams for data, including line graphs for time series and frequency diagrams
- **E** to **D** how to draw diagrams for discrete data, including stem-and-leaf diagrams

Visual overview

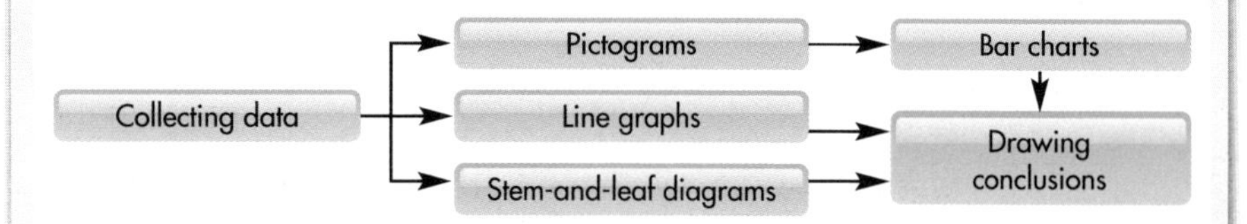

What you should already know

- How to use a tally for recording data **(KS3 level 3, GCSE grade G)**
- How to read information from charts and tables **(KS3 level 3, GCSE grade G)**

Quick check

Zoe works in a dress shop. She recorded the sizes of all the dresses sold during a week. The table shows the results.

Day	Size of dresses sold									
Monday	12	8	10	8	14	8	12	8	8	
Tuesday	10	10	8	12	14	16	8	12	14	16
Wednesday	16	8	12	10						
Thursday	12	8	8	10	12	14	16	12	8	
Friday	10	10	8	10	12	14	14	12	10	8
Saturday	10	8	8	12	10	12	8	10		

a Use a tallying method to make a table showing how many dresses of each size were sold in the week.

b Which dress size had the most sales?

10.1 Frequency diagrams

This section will show you how to:

- collect and represent discrete and grouped data using tally charts and frequency tables

Key words
class
class interval
data collection sheet
experiment
frequency
frequency table
grouped data
grouped frequency table
observation
sample
tally chart

Statistics is concerned with the collection and organisation of data, the representation of data on diagrams and the interpretation of data.

When you are collecting data for simple surveys, it is usual to use a **data collection sheet**, also called a **tally chart**. For example, data collection sheets are used to gather information on how people travel to work, how students spend their free time and the amount of time people spend watching TV.

It is easy to record the data by using tally marks, as shown in Example 1. Counting up the tally marks in each row of the chart gives the **frequency** of each category. By listing the frequencies in a column on the right-hand side of the chart, you can make a **frequency table** (see Example 1). Frequency tables are an important part of making statistical calculations, as you will see in Chapter 11.

EXAMPLE 1

Sandra wanted to find out about the ways in which students travelled to school. She carried out a survey. Her frequency table looked like this:

Method of travel	Tally	Frequency
Walk	𝍸 𝍸 𝍸 𝍸 𝍸 III	28
Car	𝍸 𝍸 II	12
Bus	𝍸 𝍸 𝍸 𝍸 III	23
Bicycle	𝍸	5
Taxi	II	2

By adding together all the frequencies, you can see that 70 students took part in the survey. The frequencies also show you that more students travelled to school on foot than by any other method of transport.

Three methods are used to collect data.

- **Taking a sample:** For example, to find out which 'soaps' students watch, you would need to take a sample from the whole school population by asking at random an equal number of boys and girls from each year group. In this case, a good sample size would be 50.
- **Observation:** For example, to find how many vehicles a day use a certain road, you would need to count and record the number of vehicles passing a point at different times of the day.
- **Experiment:** For example, to find out how often a six occurs when you throw a dice, you would need to throw the dice 50 times or more and record each score.

EXAMPLE 2

Andrew wanted to find out the most likely outcome when two coins are tossed. He carried out an experiment by tossing two coins 50 times. His frequency table looked like this.

Number of heads	Tally	Frequency
0	~~IIII~~ ~~IIII~~ II	12
1	~~IIII~~ ~~IIII~~ ~~IIII~~ ~~IIII~~ ~~IIII~~ II	27
2	~~IIII~~ ~~IIII~~ I	11

From Andrew's table, you can see that a single head appeared the highest number of times.

Grouped data

Many surveys produce a lot of data that covers a wide range of values. In these cases, it is sensible to put the data into groups before attempting to compile a frequency table. These groups of data are called **classes** or **class intervals**.

Once the data has been grouped into classes, a **grouped frequency table** can be completed. The method is shown in Example 3.

EXAMPLE 3

These marks are for 36 students in a Year 10 mathematics examination.

31	49	52	79	40	29	66	71	73	19	51	47
81	67	40	52	20	84	65	73	60	54	60	59
25	89	21	91	84	77	18	37	55	41	72	38

a Construct a frequency table, using classes of 1–20, 21–40 and so on.

b What was the most frequent interval of marks?

a Draw the grid of the table shown below and put in the headings.

Next, list the classes, in order, in the column headed 'Marks'.

Using tally marks, indicate each student's score against the class to which it belongs. For example, 81, 84, 89 and 91 belong to the class 81–100, giving five tally marks, as shown below.

Finally, count the tally marks for each class and enter the result in the column headed 'Frequency'. The table is now complete.

Marks	Tally	Frequency
1–20	III	3
21–40	~~IIII~~ III	8
41–60	~~IIII~~ ~~IIII~~ I	11
61–80	~~IIII~~ IIII	9
81–100	~~IIII~~	5

b From the grouped frequency table, you can see that the highest number of students obtained a mark in the 41–60 interval.

EXERCISE 10A

G

1 Philip kept a record of the number of goals scored by Burnley Rangers in the last 20 matches. These are his results:

0 1 1 0 2 0 1 3 2 1

0 1 0 3 2 1 0 2 1 1

a Draw a frequency table for his data.

b Which was the most frequent score?

c How many goals were scored in total for the 20 matches?

F

FM **2** Monica was doing a geography project on the weather. As part of her work, she kept a record of the daily midday temperatures in June.

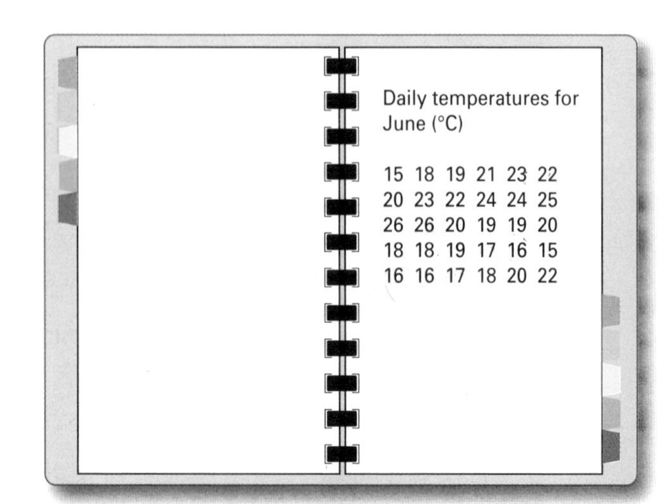

F

a Copy and complete the grouped frequency table for her data.

Temperature (°C)	Tally	Frequency
14–16		
17–19		
20–22		
23–25		
26–28		

b In which interval do the most temperatures lie?

c Describe what the weather was probably like throughout the month.

E

3 For the following surveys, decide whether the data should be collected by:

i sampling **ii** observation **iii** experiment.

a The number of people using a new superstore.

b How people will vote in a forthcoming election.

c The number of times a person scores double top in a game of darts.

d Where people go for their summer holidays.

e The frequency of a bus service on a particular route.

f The number of times a drawing pin lands point up when dropped.

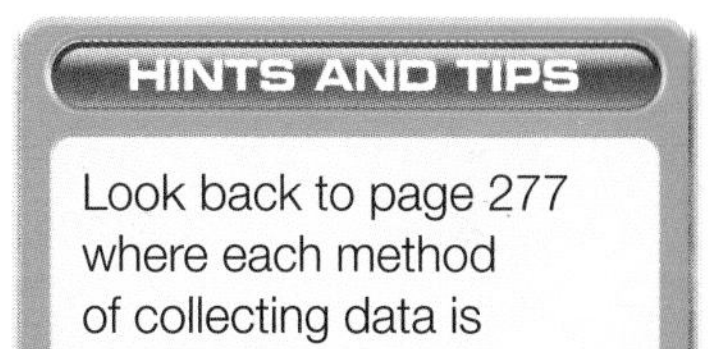
HINTS AND TIPS

Look back to page 277 where each method of collecting data is discussed.

4 In a game, Mitesh used a six-sided dice. He decided to keep a record of his scores to see whether the dice was fair. His scores were:

2 4 2 6 1 5 4 3 3 2 3 6 2 1 3

5 4 3 4 2 1 6 5 1 6 4 1 2 3 4

a Draw a frequency table for his data.

b How many throws did Mitesh have during the game?

c Do you think the dice was a fair one? Explain why.

5 The data shows the heights, in centimetres, of a sample of 32 Year 10 students.

172 158 160 175 180 167 159 180

167 166 178 184 179 156 165 166

184 175 170 165 164 172 154 186

167 172 170 181 157 165 152 164

a Draw a grouped frequency table for the data, using class intervals 151–155, 156–160, …

b In which interval do the most heights lie?

c Does this agree with a survey of the students in your class?

PS 6 Kathy used a stopwatch to time how long it took her rabbit to find food left in its hutch.

The following is her record in seconds.

7	30	14	27	8	31	8	28	10	41	51	37	15	21	37	16	38
23	20	9	11	55	9	33	8	35	45	35	25	25	49	23	43	55
45	8	13	9	39	12	57	16	37	26	32	19	48	29	37		

Find the best way to put this data into a frequency chart to illustrate the length of time it took the rabbit to find the food.

AU 7 David was doing a survey to find the ages of people at a football competition.

He said that he would make a frequency table with the regions 15–20, 20–25, 25–30.

Explain what difficulty David could have with these class divisions.

8 Conduct some surveys of your own choice and draw frequency tables for your data.

Double dice

This is an activity for two or more players. Each player needs two six-sided dice.

Each player throws their two dice together 100 times. For each throw, add together the two scores to get a total score.

What is the lowest total score anyone can get? What is the highest total score?

Everyone keeps a record of their 100 throws in a frequency table.

Compare your frequency table with someone else's and comment on what you notice. For example: Which scores appear the most often? What about 'doubles'?

How might this information be useful in games that use two dice?

Repeat the activity in one or more of the following ways.

- For each throw, multiply the score on one dice by the score on the other.
- Use two four-sided dice (tetrahedral dice), adding or multiplying the scores.
- Use two different-sided dice, adding or multiplying the scores.
- Use three or more dice, adding and/or multiplying the scores.

10.2 Statistical diagrams

This section will show you how to:

- show collected data as pictograms

Key words
key
pictograms
symbol

Data collected from a survey can be presented in pictorial or diagrammatic form to help people to understand it more quickly. You see plenty of examples of this in newspapers and magazines and on TV, where every type of visual aid is used to communicate statistical information.

Pictograms

A **pictogram** is a frequency table in which frequency is represented by a repeated **symbol**. The symbol itself usually represents a number of items, as Example 5 shows. However, sometimes it is more sensible to let a symbol represent just a single unit, as in Example 4. The **key** tells you how many items are represented by a symbol.

EXAMPLE 4

The pictogram shows the number of phone calls made by Mandy from her mobile phone during a week.

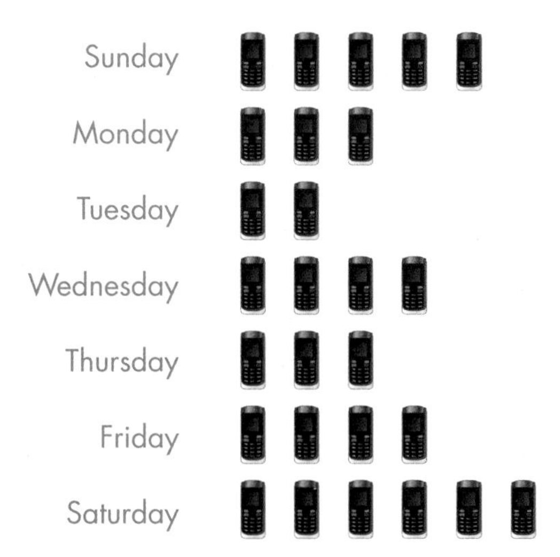

How many calls did Mandy make in the week?

From the pictogram, you can see that Mandy made a total of 27 calls.

Although pictograms can have great visual impact (particularly as used in advertising) and are easy to understand, they have a serious drawback. Apart from a half, fractions of a symbol cannot usually be drawn accurately and so frequencies are often represented only approximately by symbols.

Examples 5 and 6 highlight this difficulty.

EXAMPLE 5

The pictogram shows the number of Year 10 students who were late for school during a week.

Monday

Tuesday

Wednesday

Thursday

Friday

Key represents 5 students

How many students were late on:

a Monday **b** Thursday?

Precisely how many students were late on Monday and Thursday respectively?

If you assume that each 'limb' of the symbol represents one student and its 'body' also represents one student, then the answers are:

a 19 students were late on Monday. **b** 13 on Thursday.

EXAMPLE 6

This pictogram is used to show how many trains ran late in the course of one weekend.

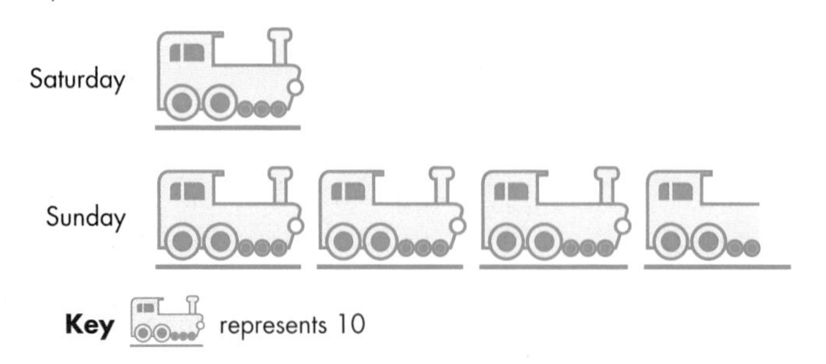

Give a reason why the pictogram is difficult to read.

The last train of those representing the number of trains late on Sunday is drawn as a fraction of 10. However, we cannot easily make out what this fraction is specifically.

EXERCISE 10B

1 The frequency table shows the numbers of cars parked in a supermarket's car park at various times of the day. Draw a pictogram to illustrate the data. Use a key of 1 symbol = 5 cars.

Time	9 am	11 am	1 pm	3 pm	5 pm
Frequency	40	50	70	65	45

2 Mr Weeks, a milkman, kept a record of how many pints of milk he delivered to 10 flats on a particular morning. Draw a pictogram for the data. Use a key of 1 symbol = 1 pint.

Flat 1	Flat 2	Flat 3	Flat 4	Flat 5	Flat 6	Flat 7	Flat 8	Flat 9	Flat 10
2	3	1	2	4	3	2	1	5	1

3 The pictogram, taken from a Suntours brochure, shows the average daily hours of sunshine for five months in Tenerife.

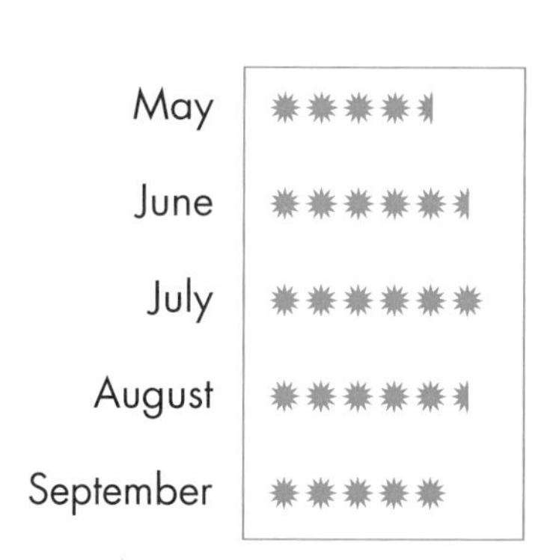

a Write down the average daily hours of sunshine for each month.

b Which month had the most sunshine?

c Give a reason why pictograms are useful in holiday brochures.

4 The pictogram shows the amounts of money collected by six students after they had completed a sponsored walk for charity.

Key £ represents £5

a Who raised the most money?

b How much money was raised altogether by the six students?

c Robert also took part in the walk and raised £32. Why would it be difficult to include him on the pictogram?

FM 5 A newspaper showed the following pictogram about one of its team member's family and the number of emails they each received during one Sunday.

Key ✉ represents 4 emails

		Frequency
Dad	✉ ✉ ✉	
Mum	✉ ◸	
Teenage son	✉ ✉ ✉ ▷	
Teenage daughter		23
Young son		9

F

a How many emails did:

i Dad receive

ii Mum receive

iii the teenage son receive?

b Copy and complete the pictogram.

c How many emails were received altogether?

PS 6 A survey was taken on the types of books read by students of a school.

	Frequency
Thriller	51
Romance	119
Science fiction	187
Historical	136

This information is to be put into a pictogram.

Design a pictogram to show this information with as few symbols as possible.

AU 7 A pictogram is to be made from this frequency table which shows the ways students travel to school.

Car	342
Bus	336
Walk	524

Explain why a key of four students to a symbol is not a good idea.

8 Draw pictograms of your own to show the following data.

a The number of hours for which you watched TV every evening last week.

b The magazines that students in your class read.

c The favourite colours of students in your class.

AU 9 A children's charity produces an advert to show how much money they have raised over the last two years.

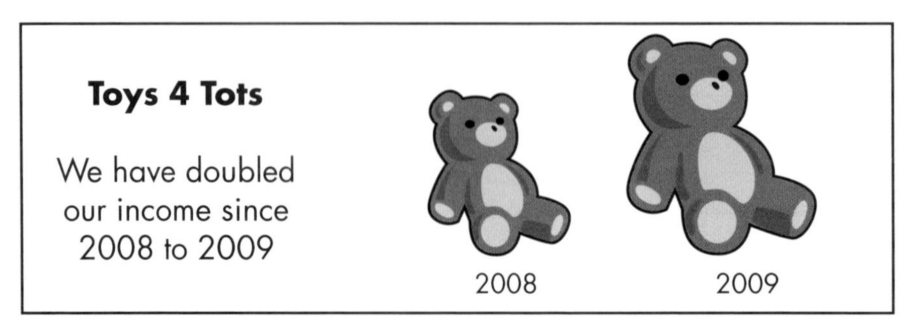

Why is the advert misleading?

10.3 Bar charts

This section will show you how to:
- draw bar charts to represent statistical data

Key words
axis
bar chart
class interval
dual bar chart

A **bar chart** consists of a series of bars or blocks of the *same* width, drawn either vertically or horizontally from an **axis**.

The heights or lengths of the bars always represent *frequencies*.

Sometimes, the bars are separated by narrow gaps of equal width, which makes the chart easier to read.

EXAMPLE 7

The grouped frequency table below shows the marks of 24 students in a test.
Draw a bar chart for the data.

Marks	1–10	11–20	21–30	31–40	41–50
Frequency	2	3	5	8	6

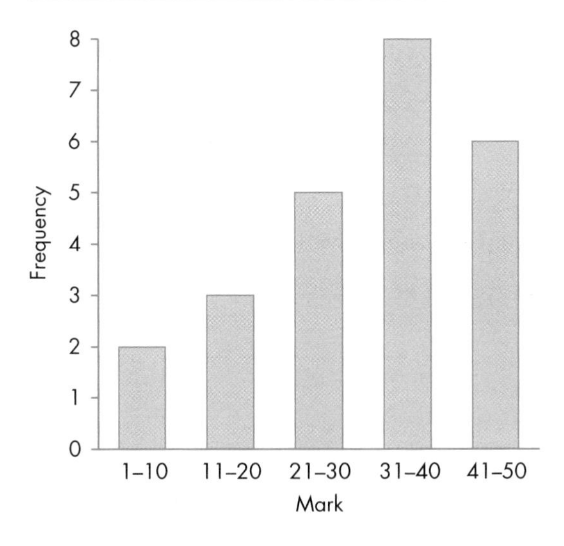

Note:

- Both axes are labelled.
- The **class intervals** are written under the middle of each bar.
- The bars are separated by equal spaces.

By using a **dual bar chart**, it is easy to compare two sets of related data, as Example 8 shows.

EXAMPLE 8

This dual bar chart shows the average daily maximum temperatures for England and Turkey over a five-month period.

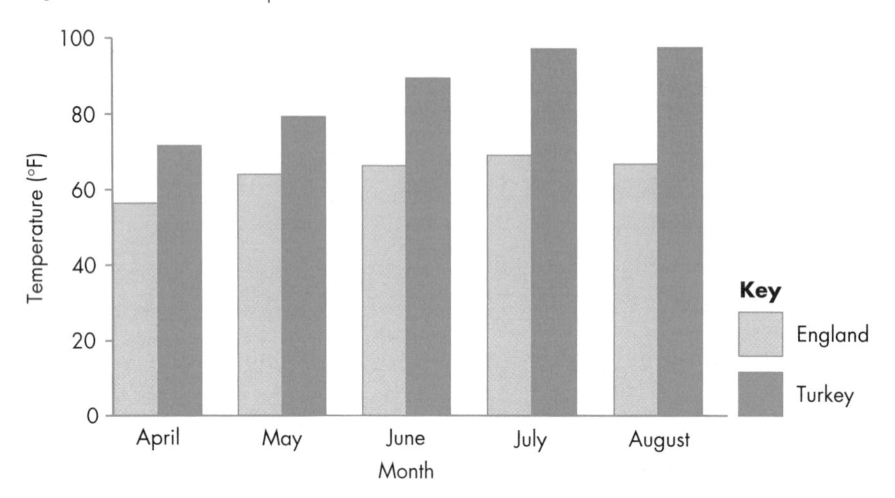

In which month was the difference between temperatures in England and Turkey the greatest?

The largest difference can be seen in August.

Note: You must always include a key to identify the two different sets of data.

EXERCISE 10C

F

1 For her survey on fitness, Maureen asked a sample of people, as they left a sports centre, which activity they had taken part in. She then drew a bar chart to show her data.

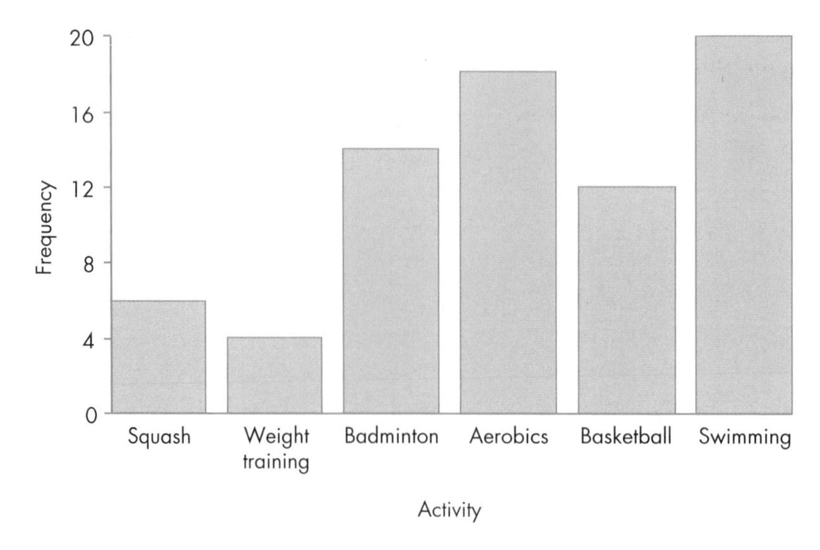

F

a Which was the most popular activity?

b How many people took part in Maureen's survey?

c Give a probable reason why fewer people took part in weight training than in any other activity.

d Is a sports centre a good place in which to do a survey on fitness? Explain why.

2 The frequency table below shows the levels achieved by 100 Year 10 students in their mock GCSE examinations.

Grade	F	E	D	C	B	A
Frequency	12	22	24	25	15	2

a Draw a suitable bar chart to illustrate the data.

b What fraction of the students achieved a grade C or grade B?

c Give one advantage of drawing a bar chart rather than a pictogram for this data.

3 This table shows the number of points Richard and Derek were each awarded in eight rounds of a general knowledge quiz.

Round	**1**	**2**	**3**	**4**	**5**	**6**	**7**	**8**
Richard	7	8	7	6	8	6	9	4
Derek	6	7	6	9	6	8	5	6

a Draw a dual bar chart to illustrate the data.

b Comment on how well each of them did in the quiz.

4 Kay did a survey on the time it took students in her form to get to school on a particular morning. She wrote down their times to the nearest minute.

15 23 36 45 8 20 34 15 27 49

10 60 5 48 30 18 21 2 12 56

49 33 17 44 50 35 46 24 11 34

a Draw a grouped frequency table for Kay's data, using class intervals 1–10, 11–20, …

b Draw a bar chart to illustrate the data.

c What conclusions can Kay draw from the bar chart?

5 This table shows the number of accidents at a dangerous crossroads over a six-year period.

Year	2000	2001	2002	2003	2004	2005
No. of accidents	6	8	7	9	6	4

a Draw a pictogram for the data.

b Draw a bar chart for the data.

AU **c** Which diagram would you use if you were going to write to your local council to suggest that traffic lights should be installed at the crossroads? Explain why.

F

AU 6 The diagram below shows the minimum and maximum temperatures, in degrees Celsius, for one day in August in five cities.

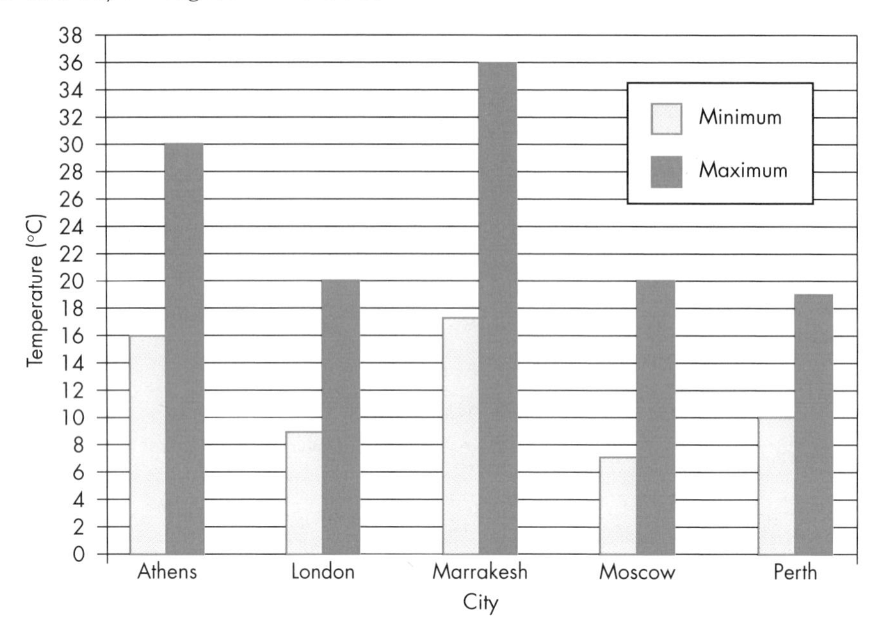

Chris says that the minimum temperature is always about half the maximum temperature for most cities.

Is Chris correct?

Give reasons to justify your answer.

AU 7 The bar chart shows the average daily temperatures, in degrees Celsius, in England and Scotland.

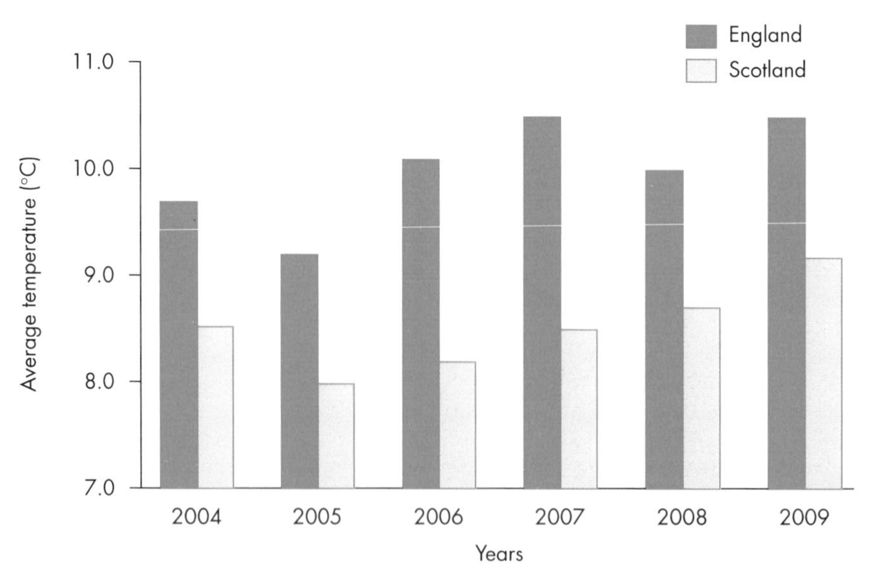

Derek says that the graph shows that the temperature in England is always more than double the temperature in Scotland.

Explain why Derek is wrong.

FM **8** Conduct a survey to find the colours of cars that pass your school or your home.

a Draw pictograms and bar charts to illustrate your data.

b Compare your results with someone else's in your class and comment on any conclusions you can draw concerning the colours of cars in your area.

FM **9** Choose a broadsheet newspaper, such as *The Times* or the *Guardian* and a tabloid newspaper, such as the *Sun* or the *Mirror*. Take a fairly long article from both papers, preferably on the same topic. Count the number of words in the first 50 sentences of each article.

a For each article, draw a grouped frequency table for the number of words in each of the first 50 sentences.

b Draw a dual bar chart for your data.

c Do your results support the hypothesis that

'Sentences in broadsheet newspapers are longer than the sentences in tabloid newspapers'?

FM AU **10** The first bar chart shows the stock of crisps in the school tuck shop at the start of break.

The second bar chart shows how many of each flavour were left at the end of break.

40 bags of prawn cocktail crisps were sold.

Each bag of crisps was 30p.

How much money did the tuck shop make selling crisps?

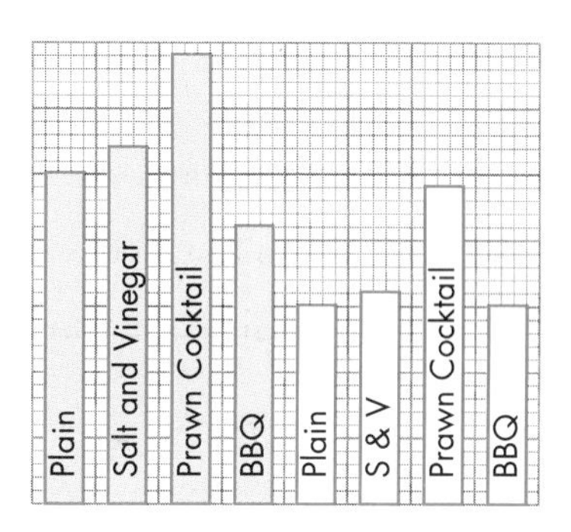

F

E

10.4 Line graphs

This section will show you how to:
- draw a line graph to show trends in data

Key words
line graphs
trends

Line graphs are usually used in statistics to show how data changes over a period of time. One such use is to indicate **trends**, for example, whether the Earth's temperature is increasing as the concentration of carbon dioxide builds up in the atmosphere, or whether a firm's profit margin is falling year-on-year.

Line graphs are best drawn on graph paper.

EXAMPLE 9

This line graph shows the outside temperature at a weather station, taken at hourly intervals. Estimate the temperature at 3.30 pm.

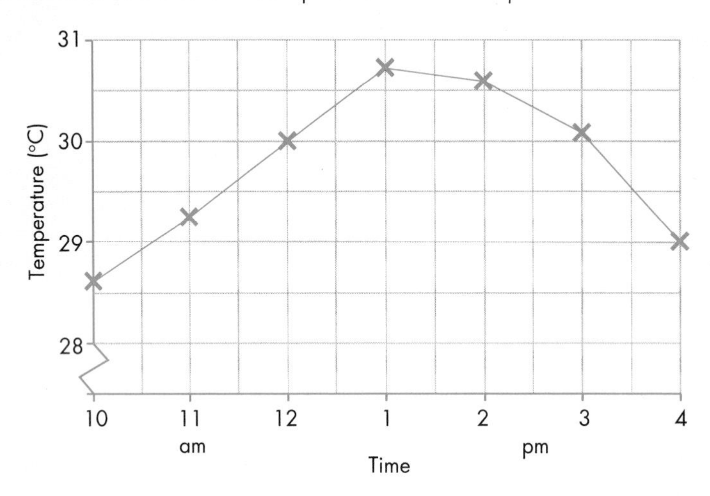

At 3.30 the temperature is approximately 29.5 °C.

Note: The temperature axis starts at 28 °C rather than 0 °C. This allows the use of a scale which makes it easy to plot the points and then to read the graph. The points are joined with lines so that the intermediate temperatures can be estimated for other times of the day.

EXAMPLE 10

This line graph shows the profit made each year by a company over a six-year period. Between which years did the company have the greatest increase in profits?

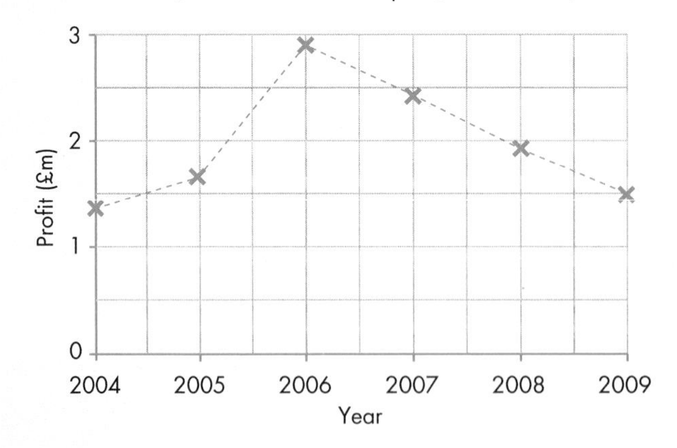

The greatest increase in profits was between 2005 and 2006.

For this graph, the values between the plotted points have no meaning because the profit of the company would have been calculated at the end of every year. In cases like this, the lines are often dashed. Although the trend appears to be that profits have fallen after 2006, it would not be sensible to predict what would happen after 2009.

EXERCISE 10D

1 This line graph shows the value of Spevadon shares on seven consecutive trading days.

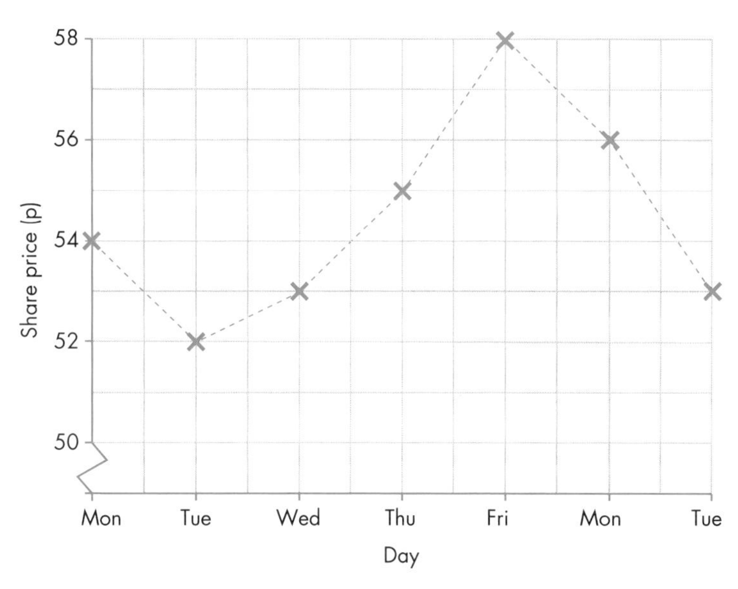

a On which day did the share price have its lowest value and what was that value?

b By how much did the share price rise from Wednesday to Thursday?

c Which day had the greatest rise in the share price from the previous day?

FM **d** Mr Hardy sold 500 shares on Friday. How much profit did he make if he originally bought the shares at 40p each?

2 The table shows the population of a town, rounded to the nearest thousand, after each census.

Year	1941	1951	1961	1971	1981	1991	2001
Population (1000s)	12	14	15	18	21	25	23

a Draw a line graph for the data.

b From your graph estimate the population in 1966.

c Between which two consecutive censuses did the population increase the most?

AU **d** Can you predict the population for 2011? Give a reason for your answer.

D

3 The number of ants in an ants' nest are counted at the end of each week.

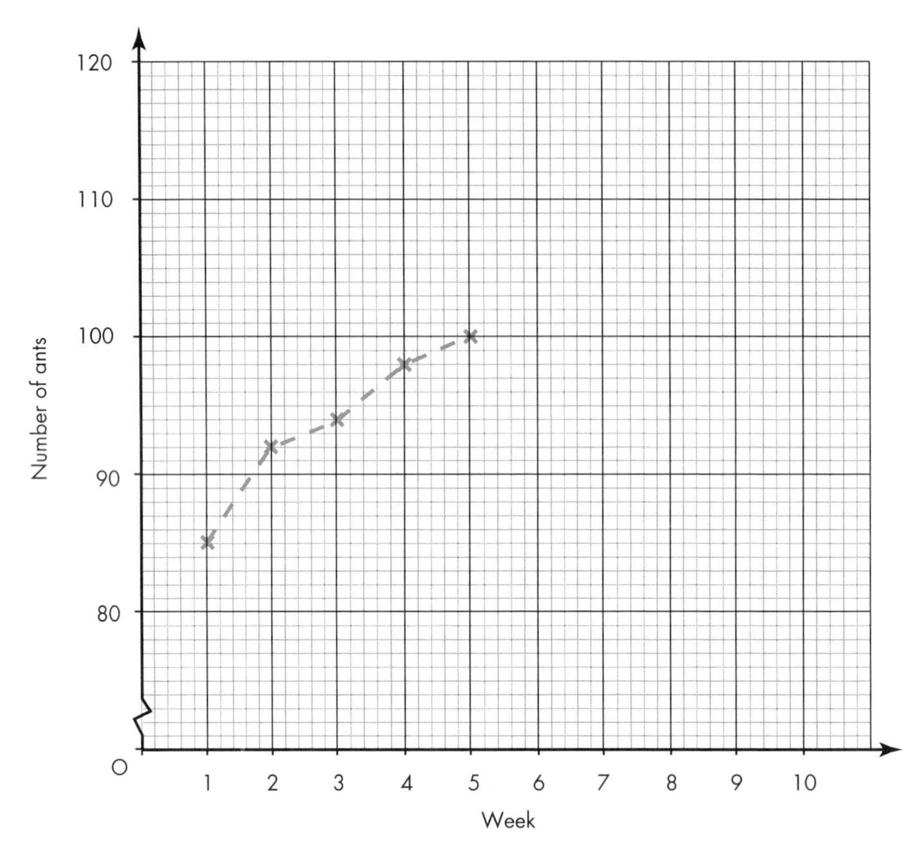

The graph shows the number of ants. At the end of week 5 the number is 100.

a At the end of week 6 the number is 104. At the end of week 10 the number is 120.

i Copy the graph and plot these points on the graph.

ii Complete the graph with straight lines.

b Use your graph to estimate the number of ants at the end of week 8.

FM 4 The table shows the estimated number of tourists worldwide.

Year	1970	1975	1980	1985	1990	1995	2000	2005
No. of tourists (millions)	100	150	220	280	290	320	340	380

a Draw a line graph for the data.

b Use your graph to estimate the number of tourists in 2010.

c Between which two consecutive years did tourism increase the most?

d Explain the trend in tourism. What reasons can you give to explain this trend?

D

5 The table shows the maximum and minimum daily temperatures for London over a week.

Day	Sunday	Monday	Tuesday	Wednesday	Thursday	Friday	Saturday
Maximum (°C)	12	14	16	15	16	14	10
Minimum (°C)	4	5	7	8	7	4	3

a Draw line graphs on the *same* axes to show the maximum and minimum temperatures.

b Find the smallest and greatest differences between the maximum and minimum temperatures.

PS 6 A puppy is weighed at the end of each week as shown in the table.

Week	1	2	3	4	5
Weight (g)	850	920	940	980	1000

Estimate how much the puppy would weigh after eight weeks.

AU 7 When plotting a graph to show the summer midday temperatures in Spain, Abbass decided to start his graph at the temperature 20 °C.

Explain why he might have done that.

Diagrams from the press

This is an activity for a group of two or more people. You will need a large selection of recent newspapers and magazines.

In a group, look through the newspapers and magazines.

Cut out any statistical diagrams and stick them on large sheets of coloured paper.

Underneath each diagram, explain what the diagram shows and how useful the diagram is in showing that information.

If any of the diagrams appears to be misleading, explain why.

You now have a lot of up-to-date statistics to display in your classroom.

10.5 Stem-and-leaf diagrams

This section will show you how to:
- draw and read information from an ordered stem-and-leaf diagram

Key words
discrete data
ordered data
raw data
unordered data

Raw data

If you were recording the ages of the first 20 people who line up at a bus stop in the morning, the **raw data** might look like this.

23, 13, 34, 44, 26, 12, 41, 31, 20, 18, 19, 31, 48, 32, 45, 14, 12, 27, 31, 19

This data is **unordered** and is difficult to read and analyse. When the data is **ordered**, it will look like this.

12, 12, 13, 14, 18, 19, 19, 20, 23, 26, 27, 31, 31, 31, 32, 34, 41, 44, 45, 48

This is easier to read and analyse.

Another method for displaying **discrete data** is a stem-and-leaf diagram. The tens digits will be the 'stem' and the units digits will be the 'leaves'.

Key 1 | 2 represents 12

1	2	2	3	4	8	9	9
2	0	3	6	7			
3	1	1	1	2	4		
4	1	4	5	8			

This is called an ordered stem-and-leaf diagram and gives a better idea of how the data is distributed.

A stem-and-leaf diagram should always have a key.

EXAMPLE 11

Put the following data into an ordered stem-and-leaf diagram.

45, 62, 58, 58, 61, 49, 61, 47, 52, 58, 48, 56, 65, 46, 54

a What is the largest value?

b What is the most common value?

c What is the difference between the largest and smallest values?

First decide on the stem and the leaf.

In this case, the tens digit will be the stem and the units digit will be the leaf.

Key 4 | 5 represents 45

Stem	Leaf					
4	5	6	7	8	9	
5	2	4	6	8	8	8
6	1	1	2	5		

a The largest value is 65.

b The most common value is 58 which occurs three times.

c The difference between the largest and the smallest is $65 - 45 = 20$.

EXERCISE 10E

1 The following stem-and-leaf diagram shows the times taken for 15 students to complete a mathematical puzzle.

Key 1 | 7 represents 17 seconds

Stem	Leaf					
1	7	8	8	9		
2	2	2	2	5	6	9
3	3	4	5	5	8	

a What is the shortest time to complete the puzzle?

b What is the most common time to complete the puzzle?

c What is the difference between the longest time and the shortest time to complete the puzzle?

E

E

FM 2 This stem-and-leaf diagram shows the marks for the boys and girls in form 10E in a maths test.

Key Boys: 2 | 4 means 42 marks

Girls: 3 | 5 means 35 marks

HINTS AND TIPS

Read the boys' marks from right to left.

Boys					Girls					
6	4	2	3	3	5	7	9			
9	9	6	2	4	2	2	3	8	8	8
7	6	6	6	5	1	1	5			

a What was the highest mark for the boys?

b What was the highest mark for the girls?

c What was the most common mark for the boys?

d What was the most common mark for the girls?

e What overall conclusions can you draw from this data?

D

3 The heights of 15 sunflowers were measured.

43 cm, 39 cm, 41 cm, 29 cm, 36 cm,

34 cm, 43 cm, 48 cm, 38 cm, 35 cm,

41 cm, 38 cm, 43 cm, 28 cm, 48 cm

a Show the results in an ordered stem-and-leaf diagram, using this key:

Key 4 | 3 represents 43 cm

b What was the largest height measured?

c What was the most common height measured?

d What is the difference between the largest and smallest heights measured?

4 A student records the number of text messages she receives each day for two weeks.

12, 18, 21, 9, 17, 23, 8, 2, 20, 13, 17, 22, 9, 9

a Show the results in an ordered stem-and-leaf diagram, using this key:

Key 1 | 2 represents 12 messages

b What was the largest number of text messages received in a day?

c What is the most common number of text messages received in a day?

PS 5 The stem-and-leaf diagram shows some heights of boys and girls in the same form.

6	15	3 5 5 7 7 9
2 5 5 6 8 9 9	16	1 1 5 6 6 7 8 9
3 4 5 5 7	17	4
2 3 3	18	

Explain what the diagram is telling you about the students in the form.

AU 6 The number of matches in a set of boxes were each counted with the following results.

50, 52, 51, 53, 52, 51, 51, 53, 54, 55, 54, 52, 52, 51, 50, 53

Explain why a stem-and-leaf diagram is not a good way to represent this information.

PUZZLE

Data problems

1. The weather

The chart represents the weather in Yorkshire over one month.

A quarter of the days were sunny.
There were more cloudy days than rainy days.
There were more mixed weather days than any other.

Copy and label the chart with Sunny, Rainy, Cloudy and Mixed.

2. Who has all the CDs?

Helen, Tom, Evie and Gran all have some CDs.
The bar chart represents how many each has.

Tom has twice as many as Helen.
Evie has the least number of CDs

Copy the chart and label each bar with either Helen, Evie, Tom or Gran.

3. Holidays

A class talked about their summer holidays.

Twice as many went to England than to Wales.
Two more went to Wales than to Scotland.
The same number went to Wales as went to Ireland and Scotland together.
Two more went to England than France.

Copy and complete the pictogram representing the students' holidays.

Ireland	☺
Wales	☺
Scotland	☺
England	☺
France	☺

Key ☺ = 2 students

GRADE BOOSTER

- F You can draw and read information from bar charts, dual bar charts and pictograms
- F You can work out the total frequency from a frequency table and compare data in bar charts
- E You can read information from a stem-and-leaf diagram
- E You can read information from a line graph
- C You can draw an ordered stem-and-leaf diagram

What you should know now

- How to draw frequency tables for grouped and ungrouped data
- How to draw and interpret pictograms, bar charts and line graphs
- How to read information from statistical diagrams, including stem-and-leaf diagrams

1 Mary threw a dice 24 times.

Here are the 24 scores.

3	5	3	4	1	2	4	5
6	2	3	4	3	1	4	3
2	3	5	5	3	4	2	1

Complete the frequency table. (3)

Score	Tally	Frequency
1		
2		
3		
4		
5		
6		

(Total 3 marks)

Edexcel, March 2007, Paper 8 Foundation, Unit 2 Test, Question 1

2 The bar chart shows the number of TVs sold by a shop six days last week.

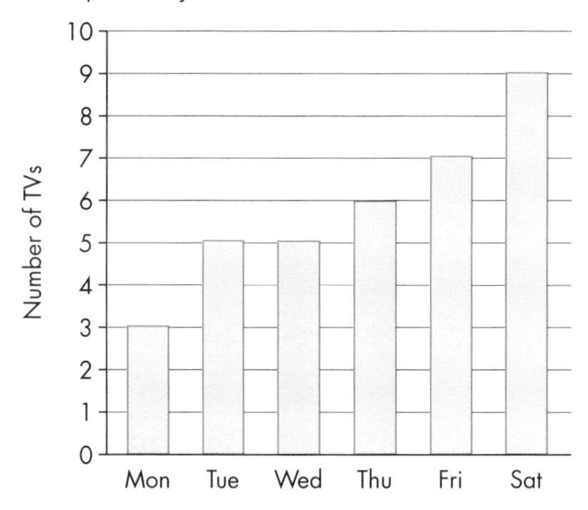

a How many TVs were sold on Friday? (1)

b On which day was the **least** number of TVs sold? (1)

c On which two days were the same number of TVs sold? (1)

(Total 3 marks)

Edexcel, March 2009, Paper 5 Foundation, Unit 1 Test, Question 1

3 The pictogram shows the number of plates sold by a shop on Monday, Tuesday, Wednesday and Thursday of one week.

Monday	○ ○
Tuesday	○ ◖
Wednesday	○ ○ ○
Thursday	○
Friday	
Saturday	

Key: ○ represents 10 plates

a Work out the number of plates sold on Monday. (1)

b Work out the number of plates sold on Tuesday. (1)

The shop sold 40 plates on Friday.

The shop sold 25 plates on Saturday.

c Use this information to complete the pictogram. (2)

(Total 4 marks)

Edexcel, November 2008, Paper 2 Foundation, Question 1

4 Steve asked his friends to tell him their favourite colour.

Here are his results.

Favourite colour	Tally	Frequency
Red	~~IIII~~ I	6
Blue	~~IIII~~ III	8
Green	~~IIII~~	5
Yellow	III	3

a Complete the bar chart to show his results. (2)

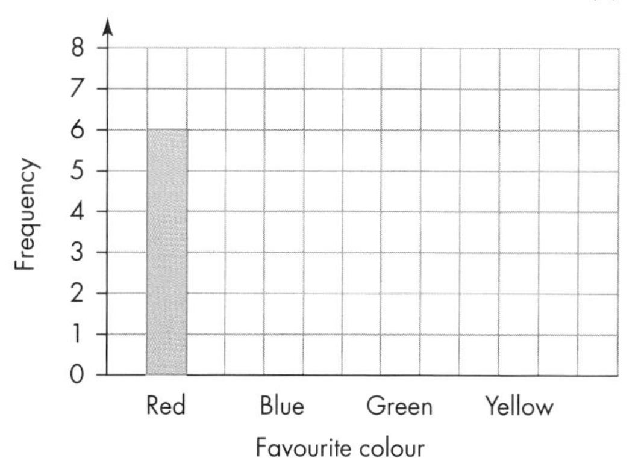

b Which colour did most of his friends say? (1)

(Total 3 marks)

Edexcel, May 2008, Paper 1 Foundation, Question 8

5 Here is a pictogram.

It shows the number of books read by Asad, by Betty, and by Chris.

Asad	
Betty	
Chris	
Diana	
Erikas	

Key: represents 4 books

a Write down the number of books read by

i Asad **ii** Chris (2)

Diana read 12 books.
Erikas read 9 books.

b Show this information on the pictogram. (2)

(Total 4 marks)

Edexcel, March 2009, Paper 5 Foundation, Unit 1 Test, Question 1

6 The bar chart shows information about the amount of time, in minutes, that Andrew and Karen spent watching television on four days last week.

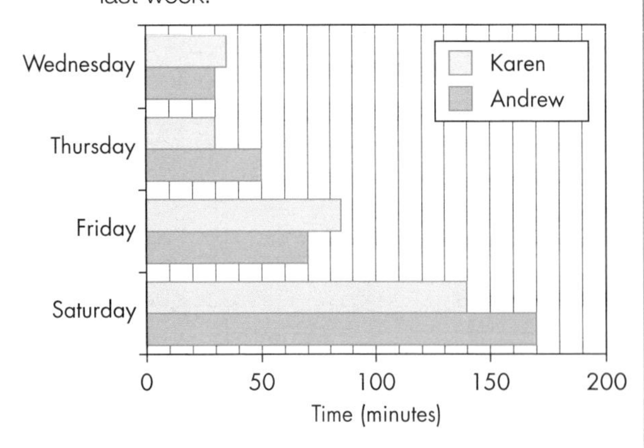

Karen spent more time watching television than Andrew on two of these four days.

a Write down these two days. (2)

b Work out the total amount of time Andrew spent watching television on these four days. (2)

(Total 4 marks)

Edexcel, June 2008, Paper 5 Foundation, Unit 1 Test, Question 2

7 The bar chart shows the number of buses going to different towns from Speedville each day.

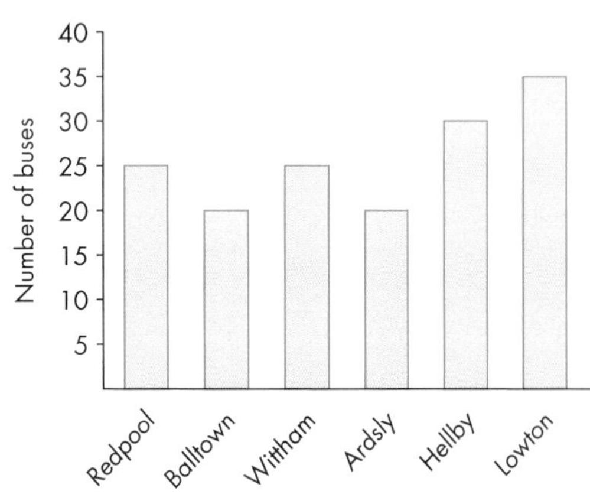

The bar chart was redrawn as a pictogram for the newspaper, as shown below.

Town		Frequency
Redpool		25
Balltown		20
Witham		27
Ardsley		20
Hellby		30
Lowton		35

Explain what is wrong with the pictogram.

8 The stem-and-leaf diagram shows the test scores of 13 students.

Key 2 | 5 is a score of 25

1	9
2	2 2 3 5 7
3	3 4 4 8 8 9
4	0

How many students scored less than 20?

9 Here are the amounts, in £, spent by some shoppers at a supermarket.

37	56	23	40
38	56	31	48
25	49	32	46

Draw an ordered stem-and-leaf diagram for these amounts.

You must include a key. (3)

2	
3	
4	
5	

Key:

(Total 3 marks)

Edexcel, March 2008, Paper 5 Foundation, Unit 1 Test, Question 4

10 Here are the ages, in years, of some members of a swimming club.

9	12	18	10	9	7	21	30	23	16
19	32	17	28	15	8	10	15	21	10

Draw an ordered stem-and-leaf diagram for these ages.

You must include a key. (3)

0	
1	
2	
3	

Key:

(Total 3 marks)

Edexcel, March 2007, Paper 8 Foundation, Unit 2 Test, Question 3

11 Tom and Barbara grew tomatoes. They compared their tomatoes by selecting 100 of each one weekend. The table shows the mean weight of Tom's tomatoes.

Weight, w (grams)	Tom's tomatoes
$50 \leqslant w < 100$	21
$100 \leqslant w < 150$	28
$150 \leqslant w < 200$	26
$200 \leqslant w < 250$	14
$250 \leqslant w < 300$	9
$300 \leqslant w < 350$	2

a Which class interval contains the median weight for Tom's tomatoes?

b The graphs for Barbara's tomatoes is drawn on the following grid. Copy it on to graph paper. On the same grid draw the graph for Tom's tomatoes.

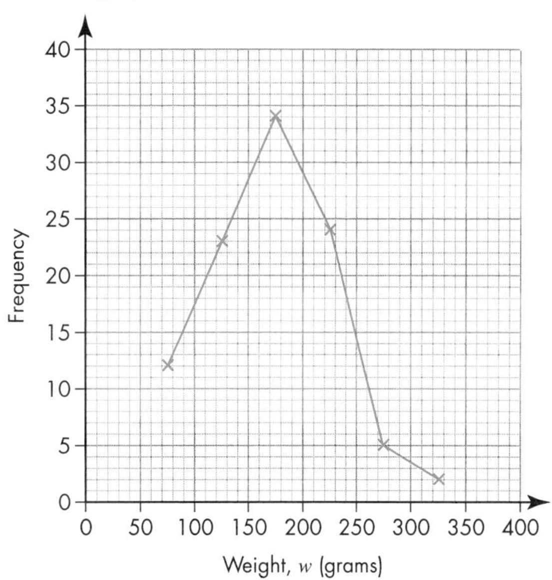

c Use the graphs to write down one comparison between Tom and Barbara's tomatoes.

Worked Examination Questions

1 The temperature is recorded in 20 towns on one day.

8	14	21	15	2	10	11	17	7	24
23	18	5	11	4	20	18	23	4	19

Draw a **stem-and-leaf diagram** to represent these data.

Key 1 | 4 represents 14

Draw a basic stem-and-leaf diagram when the stem is the tens digits and the leaves are the units digits. Create a key, using any value for 1 method mark.

Stem	Leaves
0	2 4 4 5 7 9
1	0 1 1 4 5 7 8 8 9
2	0 1 3 3 4

Now complete the stem-and-leaf diagram, keeping the data in order. You get 2 marks for accuracy and 1 mark if you make only one error.

Total: 3 marks

2 a A supermarket has some special offers.

What is the saving on the orange juice?

b Here are some more special offers.

i Which item is now half price?

ii On which item do you save the most money?

Item	Was	Now
Orange juice	£1.54	99p
Marmalade	84p	64p
Oven chips	£1.50	80p
Cat food	54p	27p
Cornflakes	£1.30	80p
Tin of salmon	£2	£1.85

a £1.54 – 99p = 55p

Taking the new price from the original price is worth 1 mark for accuracy.

1 mark

b i Cat food.

Double each new price to see if it becomes the old price. This is only true for cat food. This is worth 1 accuracy mark.

ii Orange juice difference is 55p
Marmalade difference is 20p
Oven chips difference is 70p
Cat food difference is 27p
Cornflakes difference is 50p
Tin of salmon difference is 15p
The greatest difference is
Oven chips, 70p saved

Realising that the difference between the two prices shows the saving is worth 1 method mark.

You get 1 mark for accuracy if the 70p is also shown.
Note that without the 70p shown, the correct answer will only get 1 mark.

3 marks

Worked Examination Questions

3 c The prices of three of these items are shown on this bar chart. The full length of the bar shows the original price. The darker shaded part of the bar shows the saving.

i Copy and complete the bar chart.

ii Which item has a saving of about one-quarter of the original price?

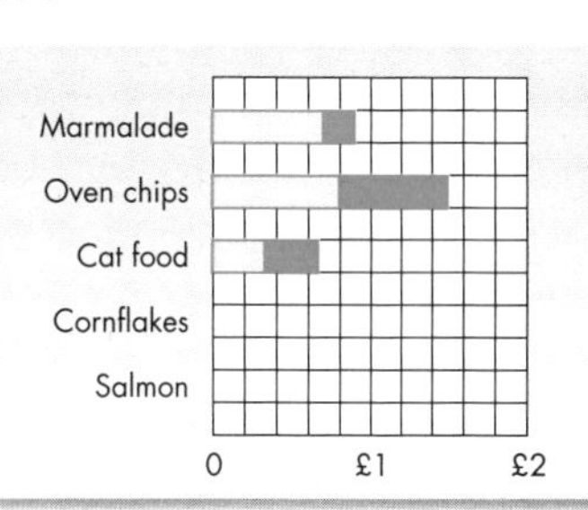

c i

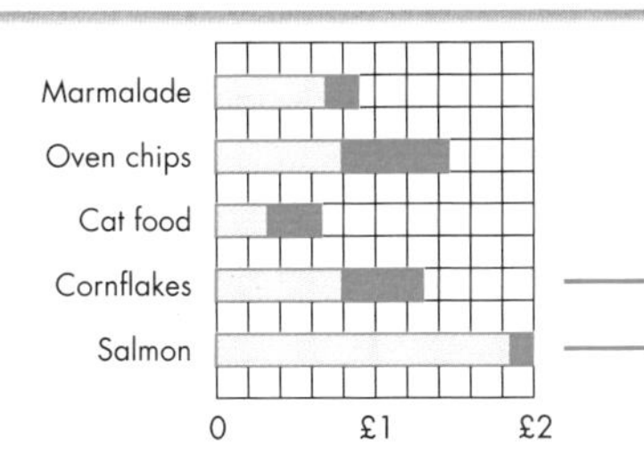

You get 1 mark for showing the saving to be 80p
You get 1 mark for showing the original price to be £1.30
You get 1 mark for showing the saving to be £1.85
You get 1 mark for showing the original price to be £2

ii Marmalade, $\frac{1}{4}$ of 84p = 21p
The saving is 20p
This is almost $\frac{1}{4}$

This is a question in which you could be assessed on your quality of written communication, so make sure you show that a quarter of the price and the saving are about the same. You will get 2 marks – one for identifying 20p and 21p, and one for showing that these amounts are about the same.

6 marks **Total:** 10 marks

PS 4 Andrew was born on 27 March and had his weight recorded regularly as shown in the table.

Day	1	5	9	13	17
Weight (g)	4100	3800	4000	4500	4900

Estimate how much Andrew would weigh after 3 weeks.

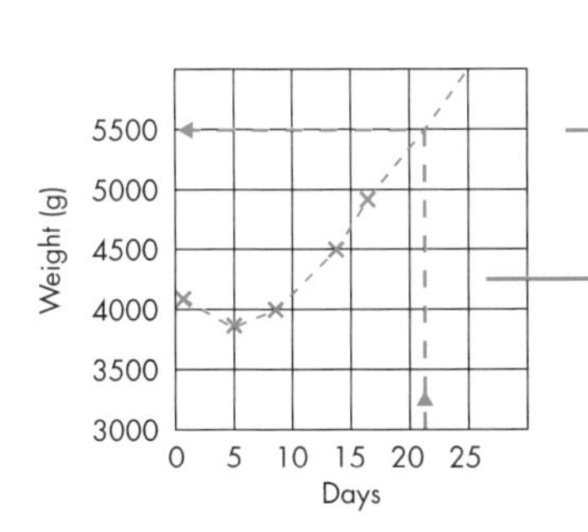

Use the table to draw the given points and plot a line graph.

Plot the points on a graph so that a prediction can be made by extending the graph. You get 1 mark for accuracy for this.

You get 1 method mark for showing 21 days (3 weeks) being read vertically to the extended graph.

You get 1 method mark for showing a horizontal reading from the graph to the weight.

Estimated weight is 5500 g

You get 1 mark for accuracy for an answer close to 5500 g.

Total: 4 marks

AU 5 When drawing a bar chart to show the summer midday temperatures in various capital cities of Europe, Joy decided to start her graph at the temperature 10 °C.

Explain why she might have done that.

The lowest temperature to be shown would be about 10 degrees and she wanted to emphasise the differences between the cities.

State a lowest temperature of 10 °C or just higher. This is worth 1 mark.

Suggest that she wanted to emphasise the differences. This is worth 1 mark.

Total: 2 marks

10 Functional Maths
Reporting the weather

The weather often appears in the news headlines and a weather report is given regularly on the television. In this activity you are required to look at the data supplied in a weather report and interpret its meaning.

Your task

The type of information given on the map and in the table is found in most newspapers in the UK each day. The map gives a forecast of the weather for the day in several large towns, while the table summarises the weather on the previous day.

It is your task to use appropriate statistical diagrams and measures to summarise the data given in the map and table.

Then, you must write a report to describe fully the weather in the UK on Friday 21st April 2010. You should use your statistical analysis to support your descriptions.

Getting started

Look at the data provided in the table.

- Which cities had the most sun, which had the most rain and which were the warmest on this particular day?
- Is there any other information you can add?

Friday 21st April 2010

	Sun (hrs)	Rain (mm)	Max (°C)	Min (°C)	Daytime weather
Aberdeen	5	3	10	5	rain
Barnstaple	5	0	12	8	sunny
Belfast	5	0	10	4	cloudy
Birmingham	5	0	11	5	sunny
Bournemouth	5	0	12	6	sunny
Bradford	5	1	10	5	rain
Brighton	5	0	11	5	mixed
Bristol	5	2	11	6	rain
Cardiff	5	2	11	6	rain
Carlisle	5	3	10	5	rain
Chester	5	3	10	4	rain
Eastbourne	5	0	9	5	windy
Edinburgh	5	0	9	5	cloudy
Falmouth	5	0	12	8	sunny
Glasgow	5	0	9	4	cloudy
Grimsby	5	1	10	4	mixed
Holyhead	5	0	8	3	windy
Ipswich	5	1	11	6	mixed
Isle of Man	5	0	9	3	windy
Isle of Wight	5	0	13	6	sunny

	Sun (hrs)	Rain (mm)	Max (°C)	Min (°C)	Daytime weather
Leeds	5	0	11	6	cloudy
Lincoln	5	3	10	4	rain
Liverpool	5	1	12	6	mixed
London	5	2	13	6	rain
Manchester	5	0	13	7	sunny
Middlesbrough	5	1	11	6	rain
Newcastle	5	0	11	6	cloudy
Newquay	5	0	12	8	sunny
Nottingham	5	0	12	6	cloudy
Oxford	5	2	12	5	rain
Plymouth	6	0	12	9	sunny
Rhyl	5	1	9	4	mixed
Scunthorpe	5	1	10	4	mixed
Sheffield	5	0	12	7	sunny
Shrewsbury	5	0	9	4	windy
Southampton	5	0	12	6	sunny
Swindon	5	1	11	5	mixed
Weymouth	5	0	12	6	sunny
Windermere	5	2	10	4	mixed
York	5	0	11	5	cloudy

Why this chapter matters

Today, people always seem to be asking "Am I average?" But, how do we answer this question? The idea of an 'average' is not a modern one. History shows us that people have always been concerned with averages as the examples below demonstrate.

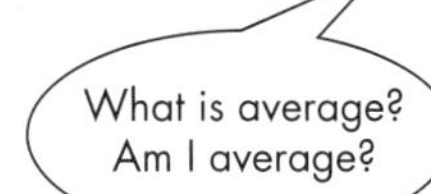

Mean in Ancient India

There is a story about Rtuparna who was born in India around 5000BC. He wanted to estimate the amount of fruit on a single tree. He counted how much fruit was on one branch, then estimated the number of branches on the tree. He multiplied the estimated number of branches by the counted fruit on one branch and was amazed that the total was very close to the actual counted number of fruit when it was picked.

This is seen as a first attempt at an arithmetic **mean**, because the one branch he chose would have been an average one representing all the branches. So the number of fruit on that branch would have been in the middle of the smallest and largest number of fruit on other branches on the tree.

Rtuparna may have been estimating the amount of fruit on a mango tree. These trees are common in India.

Mode in Ancient Greece

The Athenian army would have needed regular-sized bricks, as on this wall, to make their calculations accurate and useful.

This story comes from the Peloponnesian War in Ancient Greece (431–404BC). It is about a battle between the Peloponnesian League (led by Sparta) and the Delian League (led by Athens).

The Athenians had to get over the Peloponnesian Wall so they needed to work out the height of the wall. They did this by looking at the wall and counting the layers of bricks. This was done by hundreds of soldiers at the same time because many of them would get it wrong – but the majority would get it about right.

They then had to guess the height of one brick and so calculate the total height of the wall. They could then make ladders long enough to reach the top of the wall.

This is seen as an early use of the **mode**: the number of layers that occurred the most was deemed as the one most likely to be correct.

The other average that we use is the **median**, and there is no record of any use of this (which finds the middle value) being used until the early 17th century.

These ancient examples demonstrate that we do not always work out the average in the same way – we must choose a method that is appropriate to the situation. Bearing this in mind, how will you seek to answer the question "Am I average?"?

Chapter

Statistics: Averages

The grades given in this chapter are target grades.

This chapter will show you ...

- **F** how to calculate the mode, median, mean and range of small sets of discrete data
- **E** how to decide which is the best average for different types of data
- **D** how to calculate the mode, median, mean and range from frequency tables of discrete data
- **D** how to draw frequency polygons
- **C** how to use and recognise the modal class and calculate an estimate of the mean from frequency tables of grouped data

Visual overview

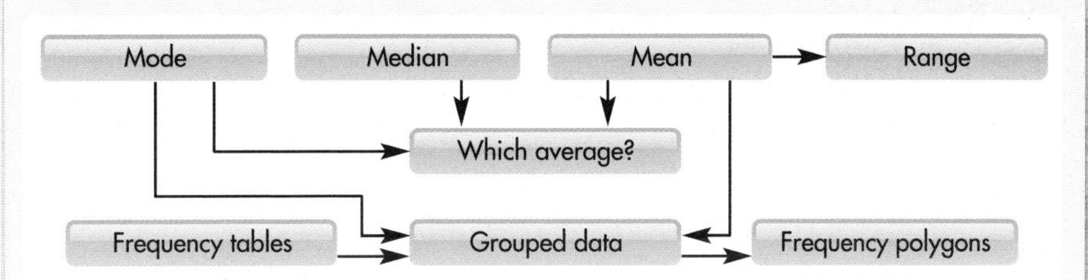

What you should already know

- How to collect and organise data **(KS3 level 4, GCSE grade F)**
- How to draw frequency tables **(KS3 level 4, GCSE grade F)**
- How to extract information from tables and diagrams **(KS3 level 4, GCSE grade F)**

Quick check

The marks for 15 students in a mathematics test are:

2, 3, 4, 5, 5, 6, 6, 6, 7, 7, 7, 7, 7, 8, 10

a What is the most common mark?

b What is the middle value in the list?

c Find the difference between the highest mark and the lowest mark.

d Find the total of all 15 marks.

Average is a term often used when describing or comparing sets of data, for example, the average rainfall in Britain, the average score of a batsman, an average weekly wage or the average mark in an examination.

In each of the above examples, you are representing the whole set of many values by just a single, 'typical' value, which is called the average.

The idea of an average is extremely useful, because it enables you to compare one set of data with another set by comparing just two values – their averages.

There are several ways of expressing an average, but the most commonly used averages are the **mode**, the **median** and the **mean**.

11.1 The mode

This section will show you how to:

- find the mode from lists of data and from frequency tables

Key words
frequency
modal class
modal value
mode

The **mode** is the value that occurs the most in a set of data. That is, it is the value with the highest **frequency**.

The mode is a useful average because it is very easy to find and it can be applied to non-numerical data (qualitative data). For example, you could find the modal style of skirts sold in a particular month.

EXAMPLE 1

Suhail scored the following number of goals in 12 school football matches:

1 2 1 0 1 0 0 1 2 1 0 2

What is the mode of his scores?

The number which occurs most often in this list is 1. So, the mode is 1.

You can also say that the modal score or **modal value** is 1.

EXAMPLE 2

Barbara asked her friends how many books they had each taken out of the school library during the previous month. Their responses were:

2 1 3 4 6 4 1 3 0 2 6 0

Find the mode.

Here, there is *no* mode, because no number occurs more than any of the others.

EXERCISE 11A

> **HINTS AND TIPS**
>
> It helps to put the data in order or group all the same things together.

1 Find the mode for each set of data.

a 3, 4, 7, 3, 2, 4, 5, 3, 4, 6, 8, 4, 2, 7

b 47, 49, 45, 50, 47, 48, 51, 48, 51, 48, 52, 48

c –1, 1, 0, –1, 2, –2, –2, –1, 0, 1, –1, 1, 0, –1, 2, –1, 2

d $\frac{1}{2}, \frac{1}{4}, 1, \frac{1}{2}, \frac{3}{4}, \frac{1}{4}, 0, 1, \frac{3}{4}, \frac{1}{4}, 1, \frac{1}{4}, \frac{3}{4}, \frac{1}{4}, \frac{1}{2}$

e 100, 10, 1000, 10, 100, 1000, 10, 1000, 100, 1000, 100, 10

f 1.23, 3.21, 2.31, 3.21, 1.23, 3.12, 2.31, 1.32, 3.21, 2.31, 3.21

2 Find the modal category for each set of data.

a red, green, red, amber, green, red, amber, green, red, amber

b rain, sun, cloud, sun, rain, fog, snow, rain, fog, sun, snow, sun

c α, γ, α, β, γ, α, α, γ, β, α, β, γ, β, β, α, β, γ, β

d ❋, ☆, ★, ★, ☆, ❋, ★, ✰, ★, ✰, ✭, ❋, ✪, ✰, ★, ✭, ✰

FM 3 Joan did a survey to find the shoe sizes of students in her class. The bar chart illustrates her data.

a How many students are in Joan's class?

b What is the modal shoe size?

c Can you tell from the bar chart which are the boys or which are the girls in her class?

d Joan then decided to draw a bar chart to show the shoe sizes of the boys and the girls separately. Do you think that the mode for the boys and the mode for the girls will be the same as the mode for the whole class? Explain your answer.

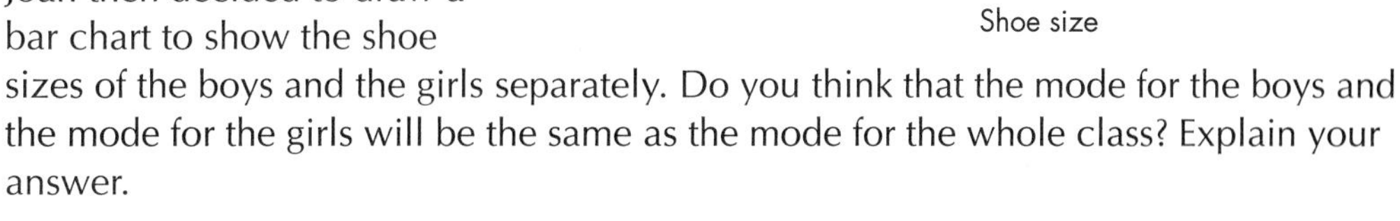

4 The frequency table shows the marks that Form 10MP obtained in a spelling test.

Mark	3	4	5	6	7	8	9	10
Frequency	1	2	6	5	5	4	3	4

a Write down the mode for their marks.

AU b Do you think this is a typical mark for the form? Explain your answer.

F

5 The grouped frequency table shows the number of e-mails each household in Orchard Street received during one day.

No. of e-mails	0–4	5–9	10–14	15–19	20–24	25–29	30–34	35–39
Frequency	9	12	14	11	10	8	4	2

a Draw a bar chart to illustrate the data.

b How many households are there in Orchard Street?

c How many households received 20 or more e-mails?

AU **d** How many households did not receive any e-mails during the week? Explain your answer.

e Write down the modal class for the data in the table.

HINTS AND TIPS

You cannot find the mode of the data in a grouped frequency table. So, instead, you need to find the **modal class**, which is the class interval with the highest frequency.

D

6 Explain why the mode is often referred to as the 'shopkeeper's average'.

7 This table shows the colours of eyes of the students in form 11P.

	Blue	Brown	Green
Boys	4	8	1
Girls	8	5	2

a How many students are in form 11P?

b What is the modal eye colour for:

i boys **ii** girls **iii** the whole form?

AU **c** After two students join the form the modal eye colour for the whole form is blue. Which of the following statements is true?

- Both students had green eyes.
- Both students had brown eyes.
- Both students had blue eyes.
- You cannot tell what their eye colours were.

AU **8** Here is a large set of raw data.

5 6 8 2 4 8 9 8 1 3 4 2 7 2 4 6 7 5 3 8

9 1 3 1 5 6 2 5 7 9 4 1 4 3 3 5 6 8 6 9

8 4 8 9 3 4 6 7 7 4 5 4 2 3 4 6 7 6 5 5

a What problems may occur if you attempted to find the mode by counting individual numbers?

b Explain a method that would make finding the mode more efficient and accurate.

c Use your method to find the mode of the data.

11.2 The median

This section will show you how to:
- find the median from a list of data, a table of data and a stem-and-leaf diagram

Key words
median
middle value

The **median** is the **middle value** of a list of values when they are put in *order* of size, from lowest to highest.

The advantage of using the median as an average is that half the data-values are below the median value and half are above it. Therefore, the average is only slightly affected by the presence of any particularly high or low values that are not typical of the data as a whole.

EXAMPLE 3

Find the median for the following list of numbers:

2, 3, 5, 6, 1, 2, 3, 4, 5, 4, 6

Putting the list in numerical order gives:

1, 2, 2, 3, 3, **4**, 4, 5, 5, 6, 6

There are 11 numbers in the list, so the middle of the list is the 6th number. Therefore, the median is 4.

EXAMPLE 4

Find the median of the data shown in the frequency table.

Value	2	3	4	5	6	7
Frequency	2	4	6	7	8	3

First, add up the frequencies to find out how many pieces of data there are.

The total is 30 so the median value will be between the 15th and 16th values.

Now, add up the frequencies to give a running total, to find out where the 15th and 16th values are.

Value	2	3	4	5	6	7
Frequency	2	4	6	7	8	3
Total frequency	2	6	12	19	27	30

There are 12 data-values up to the value 4 and 19 up to the value 5.

Both the 15th and 16th values are 5, so the median is 5.

To find the median in a list of n values, written in order, use the rule:

$$\text{median} = \frac{n+1}{2}\text{th value}$$

For a set of data that has a lot of values, it is sometimes more convenient and quicker to draw a stem-and-leaf diagram. Example 5 shows you how to do this.

EXAMPLE 5

The ages of 20 people attending a conference were as follows:

28, 34, 46, 23, 28, 34, 52, 61, 45, 34, 39, 50, 26, 44, 60, 53, 31, 25, 37, 48

Find the modal age and median age of the group.

Taking the tens to be the 'stem' and the units to be the 'leaves', draw the stem-and-leaf diagram as shown below.

Stem	Leaves
2	3 5 6 8 8
3	1 4 4 4 7 9
4	4 5 6 8
5	0 2 3
6	0 1

Key 2 | 3 represents 23 people

The most common value is 34, so the mode is 34.

There is an even number of values in this list, so the middle of the list is between the two central values, which are the 10th and 11th values. To find the central values count *up* 10 from the lowest value, 23, 25, 26, 28, 28, 31 … or *down* 10 from the highest value 61, 60, 53, 52, 50, 48 …

Therefore, the median is exactly midway between 37 and 39.

Hence, the median is 38.

EXERCISE 11B

G

1 Find the median for each set of data.

a 7, 6, 2, 3, 1, 9, 5, 4, 8

b 26, 34, 45, 28, 27, 38, 40, 24, 27, 33, 32, 41, 38

c 4, 12, 7, 6, 10, 5, 11, 8, 14, 3, 2, 9

d 12, 16, 12, 32, 28, 24, 20, 28, 24, 32, 36, 16

e 10, 6, 0, 5, 7, 13, 11, 14, 6, 13, 15, 1, 4, 15

f –1, –8, 5, –3, 0, 1, –2, 4, 0, 2, –4, –3, 2

g 5.5, 5.05, 5.15, 5.2, 5.3, 5.35, 5.08, 5.9, 5.25

HINTS AND TIPS

Remember to put the data in order before finding the median.

HINTS AND TIPS

If there is an even number of pieces of data, the median will be halfway between the two middle values.

2 A group of 15 sixth-formers had lunch in the school's cafeteria. Given below are the amounts that they spent.

£2.30, £2.20, £2, £2.50, £2.20, £3.50, £2.20, £2.25, £2.20, £2.30, £2.40, £2.20, £2.30, £2, £2.35

a Find the mode for the data.

b Find the median for the data.

AU **c** Which is the better average to use? Explain your answer.

3 **a** Find the median of 7, 4, 3, 8, 2, 6, 5, 2, 9, 8, 3.

b Without putting them in numerical order, write down the median for each of these sets.

i 17, 14, 13, 18, 12, 16, 15, 12, 19, 18, 13

ii 217, 214, 213, 218, 212, 216, 215, 212, 219, 218, 213

iii 12, 9, 8, 13, 7, 11, 10, 7, 14, 13, 8

iv 14, 8, 6, 16, 4, 12, 10, 4, 18, 16, 6

HINTS AND TIPS

Look for a connection between the original data and the new data. For example, in **i**, the numbers are each 10 more than those in part **a**.

4 Given below are the age, height and weight of each of the seven players in a netball team.

	Ella	Linda	Pat	Marion	Amina	Martha	Elisa
Age (yr)	13	15	12	11	11	15	14
Height (cm)	161	165	162	158	154	168	169
Weight (kg)	41	42	37	32	35	42	40

a Find the median age of the team. Which player has the median age?

b Find the median height of the team. Which player has the median height?

c Find the median weight of the team. Which player has the median weight?

AU **d** Who would you choose as the average player in the team? Give a reason for your answer.

F

5 The table shows the number of sandwiches sold in a corner shop over 25 days.

Sandwiches sold	10	11	12	13	14	15	16
Frequency	2	3	6	4	3	4	3

a What is the modal number of sandwiches sold?

b What is the median number of sandwiches sold?

6 The bar chart shows the marks that Mrs Woodhead gave her students for their first Functional Mathematics task.

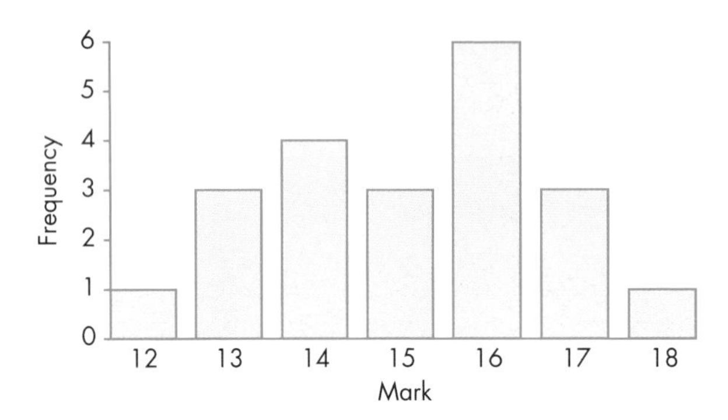

a How many students are there in Mrs Woodhead's class?

b What is the modal mark?

c Copy and complete this frequency table.

Mark	12	13	14	15	16	17	18
Frequency	1	3					

d What is the median mark?

E

PS 7 **a** Write down a list of nine numbers that has a median of 12.

b Write down a list of 10 numbers that has a median of 12.

c Write down a list of nine numbers that has a median of 12 and a mode of 8.

d Write down a list of 10 numbers that has a median of 12 and a mode of 8.

8 The following stem-and-leaf diagram shows the times taken for 15 students to complete a mathematical puzzle.

Key 1 | 7 represents 17 seconds

1	7 8 8 9
2	2 2 2 5 6 9
3	3 4 5 5 8

a What is the modal time taken to complete the puzzle?

b What is the median time taken to complete the puzzle?

9 The stem-and-leaf diagram shows the marks for 13 boys and 12 girls in form 7E in a science test.

Key 2 | 3 represents 32 marks for boys
3 | 5 represents 35 marks for girls

Boys		Girls
6 4 2	3	5 7 9
9 9 6 2	4	2 2 3 8 8 8
7 6 6 6 5 3	5	1 1 5

a What was the modal mark for the boys?

b What was the modal mark for the girls?

c What was the median mark for the boys?

d What was the median mark for the girls?

FM **e** Who did better in the test, the boys or the girls? Give a reason for your answer.

HINTS AND TIPS

Read the boys' marks from right to left.

HINTS AND TIPS

To find the middle value of two numbers, add them together and divide the result by 2. For example, for 43 and 48, 43 + 48 = 91, 91 ÷ 2 = 45.5.

PS 10 A list contains seven even numbers. The largest number is 24. The smallest number is half the largest. The mode is 14 and the median is 16. Two of the numbers add up to 42. What are the seven numbers?

11 The marks of 25 students in an English examination were as follows.

55, 63, 24, 47, 60, 45, 50, 89, 39, 47, 38, 42, 69, 73, 38, 47, 53, 64, 58, 71, 41, 48, 68, 64, 75

Draw a stem-and-leaf diagram to find the median.

PS 12 Look at this list of numbers.

4, 4, 5, 8, 10, 11, 12, 15, 15, 16, 20

a Add four numbers to make the median 12.

b Add six numbers to make the median 12.

c What is the least number of numbers to add that will make the median 4?

AU 13 Here are five payments.

£3, £5, £8, £100, £3000

Explain why the median is not a good average to use in this set of payments.

E

D

11.3 The mean

This section will show you how to:
- calculate the mean of a set of data

Key words
average
mean

The **mean** of a set of data is the sum of all the values in the set divided by the total number of values in the set. That is:

$$\text{mean} = \frac{\text{sum of all values}}{\text{total number of values}}$$

This is what most people mean when they use the term '**average**'.

Another name for this average is the arithmetic **mean**.

The advantage of using the mean as an average is that it takes into account all the values in the set of data.

EXAMPLE 6

Find the mean of 4, 8, 7, 5, 9, 4, 8, 3.

Sum of all the values = 4 + 8 + 7 + 5 + 9 + 4 + 8 + 3 = 48

Total number of values = 8

Therefore, mean $= \frac{48}{8} = 6$

EXAMPLE 7

The ages of 11 players in a football squad are:

21, 23, 20, 27, 25, 24, 25, 30, 21, 22, 28

What is the mean age of the squad?

Sum of all the ages = 266

Total number in squad = 11

Therefore, mean age $= \frac{266}{11} = 24.1818\ldots = 24.2$ (1 decimal place)

When the answer is not exact, it is usual to round the mean to 1 decimal place.

Using a calculator

If your calculator has a statistical mode, you can use it to find the mean of a set of numbers. Put your calculator into 1-VAR statistical mode and then key in each number from your set of data followed by [=] after each number. You then find the mean by keying [AC] [SHIFT] [1] (stat) [5] (var) [2] ($\bar{x}$) [=].

On some calculators, the statistical mode is represented by [SD].

EXAMPLE 8

The mean weight of eight members of a rowing crew is 89 kg. When the cox is included, the mean weight is 85 kg.

What is the weight of the cox?

The eight crew members have a total weight of $8 \times 89 = 712$ kg.

With the cox the total weight is $9 \times 85 = 765$ kg.

So the cox weighs $765 - 712 = 53$ kg.

EXAMPLE 9

Find the mean of 12, 16 and 17.

Your calculator may use the following process for finding the mean.

Set up into statistics mode by keying:

[MODE] [2] (stat) [1] (1-var)

Enter the data 12, 16, 17 by keying:

[1] [2] [=] [1] [6] [=] [1] [7]

Calculate the mean by keying:

[AC] [SHIFT] [1] (stat) [5] (var) [2] ($\bar{x}$) [=]

This gives: mean = 15

HINTS AND TIPS

It is generally more difficult to find the mean on a calculator by this method than simply adding up the data values and dividing. So don't use this method to find the mean unless you are very confident about using it.

On some calculators you may have to use the following key strokes.

Find the mean of 2, 3, 7, 8 and 10.

First put your calculator into statistical mode.

Then press the following keys:

[2] [DATA] [3] [DATA] [7] [DATA] [8] [DATA] [1] [0] [DATA] [$\bar{x}$]

You should find that the mean is given by [$\bar{x}$] = 6.

You can also find the number of data-values by pressing the [n] key.

EXERCISE 11C

1 Find, without the help of a calculator, the mean for each set of data.

a 7, 8, 3, 6, 7, 3, 8, 5, 4, 9

b 47, 3, 23, 19, 30, 22

c 42, 53, 47, 41, 37, 55, 40, 39, 44, 52

d 1.53, 1.51, 1.64, 1.55, 1.48, 1.62, 1.58, 1.65

e 1, 2, 0, 2, 5, 3, 1, 0, 1, 2, 3, 4

2 Calculate the mean for each set of data, giving your answer correct to 1 decimal place. You may use your calculator.

a 34, 56, 89, 34, 37, 56, 72, 60, 35, 66, 67

b 235, 256, 345, 267, 398, 456, 376, 307, 282

c 50, 70, 60, 50, 40, 80, 70, 60, 80, 40, 50, 40, 70

d 43.2, 56.5, 40.5, 37.9, 44.8, 49.7, 38.1, 41.6, 51.4

e 2, 3, 1, 0, 2, 5, 4, 3, 2, 0, 1, 3, 4, 5, 0, 3, 1, 2

FM **3** The table shows the marks that 10 students obtained in Mathematics, English and Science in their Year 10 examinations.

Student	Abigail	Brian	Chloe	David	Eric	Frances	Graham	Howard	Ingrid	Jane
Maths	45	56	47	77	82	39	78	32	92	62
English	54	55	59	69	66	49	60	56	88	44
Science	62	58	48	41	80	56	72	40	81	52

a Work out the mean mark for Mathematics.

b Work out the mean mark for English.

c Work out the mean mark for Science.

d Which student obtained marks closest to the mean in all three subjects?

e How many students were above the average mark in all three subjects?

4 Heather kept a record of the amount of time she spent on her homework over 10 days:

$\frac{1}{2}$ h, 20 min, 35 min, $\frac{1}{4}$ h, 1 h, $\frac{1}{2}$ h, $1\frac{1}{2}$ h, 40 min, $\frac{3}{4}$ h, 55 min

Calculate the mean time, in minutes, that Heather spent on her homework.

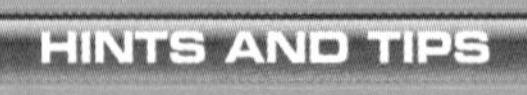

HINTS AND TIPS

Convert all times to minutes, for example, $\frac{1}{4}$h = 15 minutes.

5 The weekly wages of 10 people working in an office are:

£350 £200 £180 £200 £350 £200 £240 £480 £300 £280

a Find the modal wage.

b Find the median wage.

c Calculate the mean wage.

d Which of the three averages best represents the office staff's wages? Give a reason for your answer.

> **HINTS AND TIPS**
>
> Remember that the mean can be distorted by extreme values.

6 The ages of five people in a group of walkers are 38, 28, 30, 42 and 37.

a Calculate the mean age of the group.

b Steve, who is 41, joins the group. Calculate the new mean age of the group.

7 **a** Calculate the mean of 3, 7, 5, 8, 4, 6, 7, 8, 9 and 3.

b Calculate the mean of 13, 17, 15, 18, 14, 16, 17, 18, 19 and 13. What do you notice?

c Write down, without calculating, the mean for each of the following sets of data.

i 53, 57, 55, 58, 54, 56, 57, 58, 59, 53

ii 103, 107, 105, 108, 104, 106, 107, 108, 109, 103

iii 4, 8, 6, 9, 5, 7, 8, 9, 10, 4

> **HINTS AND TIPS**
>
> Look for a connection between the original data and the new data. For example in **i** the numbers are 50 more.

PS 8 Two families were in a competition.

Speed family		Roberts family	
Brian	aged 59	Frank	aged 64
Kath	aged 54	Marylin	aged 62
James	aged 34	David	aged 34
Helen	aged 34	James	aged 32
John	aged 30	Tom	aged 30
Joseph	aged 24	Helen	aged 30
Joy	aged 19	Evie	aged 16

Each family had to choose four members with a mean age of between 35 and 36.

Choose two teams, one from each family, that have this mean age between 35 and 36.

AU 9 Asif had an average batting score of 35 runs.
He had scored 315 runs in nine games of cricket.

What is the least number of runs he needs to score in the next match if he is to get a higher average score?

10 The mean age of a group of eight walkers is 42. Joanne joins the group and the mean age changes to 40. How old is Joanne?

D

C

11.4 The range

This section will show you how to:
- find the range of a set of data and compare different sets of data, using the mean and the range

Key words
consistency
range
spread

The **range** for a set of data is the highest value of the set minus the lowest value.

The range is *not* an average. It shows the **spread** of the data. It is, therefore, used when comparing two or more sets of similar data. You can also use it to comment on the **consistency** of two or more sets of data.

EXAMPLE 10

Rachel's marks in 10 mental arithmetic tests were 4, 4, 7, 6, 6, 5, 7, 6, 9 and 6.

Therefore, her mean mark is $60 \div 10 = 6$ and the range is $9 - 4 = 5$.

Adil's marks in the same tests were 6, 7, 6, 8, 5, 6, 5, 6, 5 and 6.

Therefore, his mean mark is $60 \div 10 = 6$ and the range is $8 - 5 = 3$.

Although the means are the same, Adil has a smaller range. This shows that Adil's results are more consistent.

G

1 Find the range for each set of data.

a 3, 8, 7, 4, 5, 9, 10, 6, 7, 4

b 62, 59, 81, 56, 70, 66, 82, 78, 62, 75

c 1, 0, 4, 5, 3, 2, 5, 4, 2, 1, 0, 1, 4, 4

d 3.5, 4.2, 5.5, 3.7, 3.2, 4.8, 5.6, 3.9, 5.5, 3.8

e 2, –1, 0, 3, –1, –2, 1, –4, 2, 3, 0, 2, –2, 0, –3

F

2 The table shows the maximum and minimum temperatures at midday for five cities in England during a week in August.

	Birmingham	Leeds	London	Newcastle	Sheffield
Maximum temperature (°C)	28	25	26	27	24
Minimum temperature (°C)	23	22	24	20	21

a Write down the range of the temperatures for each city.

b What do the ranges tell you about the weather for England during the week?

FM 3 Over a three-week period, the school tuck shop took the following amounts.

	Monday	Tuesday	Wednesday	Thursday	Friday
Week 1	£32	£29	£36	£30	£28
Week 2	£34	£33	£25	£28	£20
Week 3	£35	£34	£31	£33	£32

a Calculate the mean amount taken each week.

b Find the range for each week.

c What can you say about the total amounts taken for each of the three weeks?

E

4 In a womens' golf tournament, the club chairperson had to choose either Sheila or Fay to play in the first round. In the previous eight rounds, their scores were as follows.

Sheila's scores: 75, 92, 80, 73, 72, 88, 86, 90

Fay's scores: 80, 87, 85, 76, 85, 79, 84, 88

a Calculate the mean score for each golfer.

b Find the range for each golfer.

AU **c** Which golfer would you choose to play in the tournament? Explain why.

HINTS AND TIPS

The best person to choose may not be the one with the biggest mean but could be the most consistent player.

5 Dan has a choice of two buses to get to school: Number 50 or Number 63. Over a month, he kept a record of the number of minutes each bus was late when it set off from his home bus stop.

No. 50: 4, 2, 0, 6, 4, 8, 8, 6, 3, 9

No. 63: 3, 4, 0, 10, 3, 5, 13, 1, 0, 1

a For each bus, calculate the mean number of minutes late.

b Find the range for each bus.

AU **c** Which bus would you advise Dan to catch? Give a reason for your answer.

E

PS 6 The table gives the ages and heights of 10 children.

Name	Age (years)	Height (cm)
Billy	9	121
Isaac	4	73
Lilla	8	93
Lewis	10	118
Evie	3	66
Andrew	6	82
Oliver	4	78
Beatrice	2	69
Isambard	9	87
Chloe	7	82

a Chloe is having a party. She wants to invite as many children as possible but does not want the range of ages to be more than 5. Who will she invite?

b This is a sign at a theme park:

> You have to be taller than ... cm
>
> *and*
>
> shorter than ... cm to go on this ride

Isaac is the shortest person who can go on the ride and Isambard is the tallest.

What are the smallest and largest missing values on the sign?

AU 7 **a** The age range of a school quiz team is 20 years and the mean age is 34. Who would you expect to be in this team? Explain your answer.

b Another team has an average age of $15\frac{1}{2}$ and a range of 1. Who would you expect to be in this team? Explain your answer.

ACTIVITY

Your time is up

You are going to find out how good you are at estimating 1 minute.

You need a stopwatch and a calculator.

This is a group activity. One person in the group acts as a timekeeper, says 'Start' and starts the stopwatch.

When someone thinks 1 minute has passed, they say 'Stop', and the timekeeper writes down the actual time, in seconds, that has passed. The timekeeper should try to record everyone's estimate.

Repeat the activity, with every member of the group taking a turn as the timekeeper.

Collate all the times and, from the data, find the mean (to the nearest second) and the range.

- How close is the mean to 1 minute?
- Why is the range useful?
- What strategies did people use to estimate 1 minute?

Repeat the activity for estimating different times, for example, 30 seconds or 2 minutes.

Write a brief report on what you find out about people's ability to estimate time.

11.5 Which average to use

This section will show you how to:
- understand the advantages and disadvantages of each type of average and decide which one to use in different situations

Key words
appropriate
extreme values
representative

An average must be truly **representative** of a set of data. So, when you have to find an average, it is crucial to choose the **appropriate** type of average for this particular set of data.

If you use the wrong average, your results will be distorted and give misleading information.

This table, which compares the advantages and disadvantages of each type of average, will help you to make the correct decision.

	Mode	Median	Mean
Advantages	Very easy to find Not affected by **extreme values** Can be used for non-numerical data	Easy to find for ungrouped data Not affected by extreme values	Easy to find Uses all the values The total for a given number of values can be calculated from it
Disadvantages	Does not use all the values May not exist	Does not use all the values Often not understood	Extreme values can distort it Has to be calculated
Use for	Non-numerical data Finding the most likely value	Data with extreme values	Data with values that are spread in a balanced way

EXERCISE 11E

1 The ages of the members of a hockey team were:

29 26 21 24 26 28 35 23 29 28 29

a Give:

i the modal age **ii** the median age **iii** the mean age.

b What is the range of the ages?

2 **a** For each set of data, find the mode, the median and the mean.

i 6, 10, 3, 4, 3, 6, 2, 9, 3, 4 **ii** 6, 8, 6, 10, 6, 9, 6, 10, 6, 8

iii 7, 4, 5, 3, 28, 8, 2, 4, 10, 9

AU **b** For each set of data, decide which average is the best one to use and give a reason.

E

3 A newsagent sold the following numbers of copies of *The Evening Star* on 12 consecutive evenings during a promotion exercise organised by the newspaper's publisher.

65 73 75 86 90 112 92 87 77 73 68 62

a Find the mode, the median and the mean for the sales.

AU **b** The newsagent had to report the average sale to the publisher after the promotion. Which of the three averages would you advise the newsagent to use? Explain why.

4 The mean age of a group of 10 young people was 15.

a What do all their ages add up to?

b What will be their mean age in five years' time?

5 **a** Find the median of each list below.

i 2, 4, 6, 7, 9

ii 12, 14, 16, 17, 19

iii 22, 24, 26, 27, 29

iv 52, 54, 56, 57, 59

v 92, 94, 96, 97, 99

b What do you notice about the lists and your answers?

c Use your answer above to help find the medians of the following lists.

i 132, 134, 136, 137, 139

ii 577, 576, 572, 574, 579

iii 431, 438, 439, 432, 435

iv 855, 859, 856, 851, 857

d Find the mean of each of the sets of numbers in part **a**.

D

AU **6** Decide which average you would use for each of the following. Give a reason for your answer.

a The average mark in an examination

b The average pocket money for a group of 16-year-old students

c The average shoe size for all the girls in Year 10

d The average height for all the artistes on tour with a circus

e The average hair colour for students in your school

f The average weight of all newborn babies in a hospital's maternity ward.

7 A pack of matches consisted of 12 boxes. The contents of each box were counted as:

34 31 29 35 33 30 31 28 29 35 32 31

On the box it stated 'Average contents 32 matches'. Is this correct?

D

8 A firm showed the annual salaries for its employees as:

Chairman	£83 000
Managing director	£65 000
Floor manager	£34 000
Skilled worker 1	£28 000
Skilled worker 2	£28 000
Machinist	£20 000
Computer engineer	£20 000
Secretary	£20 000
Office junior	£8 000

a Give:

i the modal salary **ii** the median salary **iii** the mean salary.

b The management suggested a pay rise of 6% for all employees. The shopfloor workers suggested a pay rise of £1500 for all employees.

i One of the suggestions would cause problems for the firm. Which one is that and why?

ii What difference would each suggestion make to the modal, median and mean salaries?

AU 9 Mr Brennan, a caring maths teacher, told each student their test mark and only gave the test statistics to the whole class. He gave the class the modal mark, the median mark and the mean mark.

a Which average would tell a student whether they were in the top half or the bottom half of the class?

b Which average tells the students nothing really?

c Which average allows a student to gauge how well they have done compared with everyone else?

FM 10 Three players were hoping to be chosen for the basketball team.

The following table shows their scores in the last few games they played.

Tom	16, 10, 12, 10, 13, 8, 10
David	16, 8, 15, 25, 8
Mohaned	15, 2, 15, 3, 5

The teacher said they would be chosen by their best average score.

Which average would each boy want to be chosen by?

C

PS **11** **a** Find five numbers that have **both** the properties below:

- a range of 5
- a mean of 5.

b Find five numbers that have **all** the properties below:

- a range of 5
- a median of 5
- a mean of 5.

AU **12** What is the average pay at a factory with 10 employees?

The boss said: "£43 295"
A worker said: "£18 210"

They were both correct.
Explain how this can be.

PS **13** A list of nine numbers has a mean of 7.6. What number must be added to the list to give a new mean of 8?

PS **14** A dance group of 17 teenagers had a mean weight of 44.5 kg. To enter a competition there needed to be 18 teenagers with an average weight of 44.4 kg or less. What is the maximum weight that the eighteenth person must be?

11.6 Frequency tables

This section will show you how to:

- revise finding the mode and median from a frequency table
- learn how to calculate the mean from a frequency table

Key word
frequency table

When a lot of information has been gathered, it is often convenient to put it together in a **frequency table**. From this table you can then find the values of the three averages and the range.

EXAMPLE 11

A survey was done on the number of people in each car leaving the Meadowhall Shopping Centre, in Sheffield. The results are summarised in the table below.

Number of people in each car	1	2	3	4	5	6
Frequency	45	198	121	76	52	13

For the number of people in a car, calculate:

a the mode **b** the median **c** the mean.

a The modal number of people in a car is easy to spot. It is the number with the largest frequency, which is 198. Hence, the modal number of people in a car is 2.

b The median number of people in a car is found by working out where the middle of the set of numbers is located. First, add up frequencies to get the total number of cars surveyed, which comes to 505. Next, calculate the middle position.

$(505 + 1) \div 2 = 253$

Now add the frequencies across the table to find which group contains the 253rd item. The 243rd item is the end of the group with 2 in a car. Therefore, the 253rd item must be in the group with 3 in a car. Hence, the median number of people in a car is 3.

c To calculate the mean number of people in a car, multiply the number of people in the car by the frequency. This is best done in an extra column. Add these to find the total number of people and divide by the total frequency (the number of cars surveyed).

Number in car	Frequency	Number in these cars
1	45	$1 \times 45 = 45$
2	198	$2 \times 198 = 396$
3	121	$3 \times 121 = 363$
4	76	$4 \times 76 = 304$
5	52	$5 \times 52 = 260$
6	13	$6 \times 13 = 78$
Totals	505	1446

Hence, the mean number of people in a car is $1446 \div 505 = 2.9$ (to 1 decimal place).

Using your calculator

The previous example can also be done by using the statistical mode that is available on some calculators. However, not all calculators are the same, so you will have either to read your instruction manual or experiment with the statistical keys on your calculator.

You may find one labelled:

M+ or DATA or Σ+ or $\bar{x}$, where $\bar{x}$ is printed in blue.

Try the following key strokes:

1 × 4 5 M+ 2 × 1 9 8 M+ … 6 × 1 3 M+ $\bar{x}$

EXERCISE 11F

D

1 Find **i** the mode, **ii** the median and **iii** the mean from each frequency table below.

a A survey of the shoe sizes of all the Year 10 boys in a school gave these results.

Shoe size	4	5	6	7	8	9	10
Number of students	12	30	34	35	23	8	3

b A survey of the number of eggs laid by hens over a period of one week gave these results.

Number of eggs	0	1	2	3	4	5	6
Frequency	6	8	15	35	48	37	12

c This is a record of the number of babies born each week over one year in a small maternity unit.

Number of babies	0	1	2	3	4	5	6	7	8	9	10	11	12	13	14
Frequency	1	1	1	2	2	2	3	5	9	8	6	4	5	2	1

d A school did a survey on how many times in a week students arrived late at school. These are the findings.

Number of times late	0	1	2	3	4	5
Frequency	481	34	23	15	3	4

2 A survey of the number of children in each family of a school's intake gave these results.

Number of children	1	2	3	4	5
Frequency	214	328	97	26	3

a Assuming each child at the school is shown in the data, how many children are at the school?

b Calculate the mean number of children in a family.

c How many families have this mean number of children?

FM **d** How many families would consider themselves average from this survey?

FM **3** A dentist kept records of how many teeth he extracted from his patients.

In 1989 he extracted 598 teeth from 271 patients.

In 1999 he extracted 332 teeth from 196 patients.

In 2009 he extracted 374 teeth from 288 patients.

a Calculate the average number of teeth taken from each patient in each year.

AU **b** Explain why you think the average number of teeth extracted falls each year.

D

4 One hundred cases of apples delivered to a supermarket were inspected and the numbers of bad apples were recorded.

Bad apples	0	1	2	3	4	5	6	7	8	9
Frequency	52	29	9	3	2	1	3	0	0	1

Give:

a the modal number of bad apples per case

b the mean number of bad apples per case.

5 Two dice are thrown together 60 times. The sums of the scores are shown below.

Score	2	3	4	5	6	7	8	9	10	11	12
Frequency	1	2	6	9	12	15	6	5	2	1	1

Find: **a** the modal score **b** the median score **c** the mean score.

6 During a one-month period, the number of days off taken by 100 workers in a factory were noted as follows.

Number of days off	0	1	2	3	4
Number of workers	35	42	16	4	3

Calculate:

a the modal number of days off

b the median number of days off

c the mean number of days off.

7 Two friends often played golf together. They recorded their scores for each hole over the last five games to compare who was more consistent and who was the better player. Their results were summarised in the following table.

No. of shots to hole ball	1	2	3	4	5	6	7	8	9
Roger	0	0	0	14	37	27	12	0	0
Brian	5	12	15	18	14	8	8	8	2

a What is the modal score for each player?

b What is the range of scores for each player?

c What is the median score for each player?

d What is the mean score for each player?

AU **e** Which player is the more consistent and why?

AU **f** Who would you say is the better player and why?

D

PS 8 A tea stain on a newspaper removed four numbers from the following frequency table of goals scored in 40 league football matches one weekend.

Goals	0	1	2			5
Frequency	4	6	9			3

The mean number of goals scored is 2.4.

What could the missing four numbers be?

AU 9 Talera made day trips to Manchester frequently during a year.

The table shows how many days in a week she travelled.

Days	0	1	2	3	4	5
Frequency	17	2	4	13	15	1

Explain how you would find the median number of days Talera travelled in a week to Manchester.

11.7 Grouped data

This section will show you how to:

- identify the modal class
- calculate an estimate of the mean from a grouped table

Key words
estimated
grouped data
mean
modal class

Sometimes the information you are given is grouped in some way (called **grouped data**), as in Example 12, which shows the range of weekly pocket money given to Year 12 students in a particular class.

Normally, grouped tables use continuous data, which is data that can have any value within a range of values, for example, height, weight, time, area and capacity. In these situations, the **mean** can only be **estimated** as you do not have all the information.

Discrete data is data that consists of separate numbers, for example, goals scored, marks in a test, number of children and shoe sizes.

In both cases, when using a grouped table to estimate the mean, first find the midpoint of the interval by adding the two end-values and then dividing by two.

EXAMPLE 12

Pocket money, p (£)	$0 < p \leqslant 1$	$1 < p \leqslant 2$	$2 < p \leqslant 3$	$3 < p \leqslant 4$	$4 < p \leqslant 5$
No. of students	2	5	5	9	15

a Write down the **modal class**.

b Calculate an estimate of the mean weekly pocket money.

a The modal class is easy to pick out, since it is simply the one with the largest frequency. Here the modal class is £4 to £5.

b To estimate the mean, assume that each person in each class has the 'midpoint' amount, then build up the following table.

To find the midpoint value, the two end-values are added together and then divided by two.

Pocket money, p (£)	Frequency (f)	Midpoint (m)	$f \times m$
$0 < p \leqslant 1$	2	0.50	1.00
$1 < p \leqslant 2$	5	1.50	7.50
$2 < p \leqslant 3$	5	2.50	12.50
$3 < p \leqslant 4$	9	3.50	31.50
$4 < p \leqslant 5$	15	4.50	67.50
Totals	36		120

The estimated mean will be £120 ÷ 36 = £3.33 (rounded to the nearest penny).

Note the notation for the classes:

$0 < p \leqslant 1$ means any amount above 0p up to and including £1.

$1 < p \leqslant 2$ means any amount above £1 up to and including £2, and so on.

If you had written 0.01–1.00, 1.01–2.00 and so on for the groups, then the midpoints would have been 0.505, 1.505 and so on. This would not have had a significant effect on the final answer as it is only an estimate.

Note that you **cannot** find the **median** from a grouped table as you do not know the actual values.

You also **cannot** find the **range** but you can say what limits there are. In the table above the smallest possible value for pocket money in the first group is 1p (this is unlikely but it cannot be 0 as the range is $0 < p \leqslant 1$) and the largest is £1. In the last group the smallest possible value is £4.01 and the largest is £5. This means the range must be between £5 – 1p = £4.99 and £4.01 – £1 = £3.01.

EXERCISE 11G 

C

1 For each table of values given below, find:

i the modal group

ii an estimate for the mean.

> **HINTS AND TIPS**
>
> When you copy the tables, draw them vertically as in Example 12.

a

x	$0 < x \leqslant 10$	$10 < x \leqslant 20$	$20 < x \leqslant 30$	$30 < x \leqslant 40$	$40 < x \leqslant 50$
Frequency	4	6	11	17	9

b

y	$0<y\leqslant 100$	$100<y\leqslant 200$	$200<y\leqslant 300$	$300<y\leqslant 400$	$400<y\leqslant 500$	$500<x\leqslant 600$
Frequency	95	56	32	21	9	3

c

z	$0 < z \leqslant 5$	$5 < z \leqslant 10$	$10 < z \leqslant 15$	$15 < z \leqslant 20$
Frequency	16	27	19	13

d

Weeks	1–3	4–6	7–9	10–12	13–15
Frequency	5	8	14	10	7

2 Jason brought 100 pebbles back from the beach and weighed them all, recording each weight to the nearest gram. His results are summarised in the table below.

Weight, w (g)	$40<w\leqslant 60$	$60<w\leqslant 80$	$80<w\leqslant 100$
Frequency	5	9	22

Weight, w (g)	$100<w\leqslant 120$	$120<w\leqslant 140$	$140<w\leqslant 160$
Frequency	27	26	11

Find:

a the modal weight of the pebbles

b an estimate of the total weight of all the pebbles

c an estimate of the mean weight of the pebbles.

3 A gardener measured the heights of all his daffodils to the nearest centimetre and summarised his results as follows.

Height (cm)	10–14	15–18	19–22	23–26	27–40
Frequency	21	57	65	52	12

a How many daffodils did the gardener have?

b What is the modal height of the daffodils?

c What is the estimated mean height of the daffodils?

FM **4** A survey was created to see how quickly the AA attended calls that were not on a motorway. The following table summarises the results.

Time (min)	1–15	16–30	31–45	46–60	61–75	76–90	91–105
Frequency	2	23	48	31	27	18	11

a How many calls were used in the survey?

b Estimate the mean time taken per call.

c Which average would the AA use for the average call-out time?

d What percentage of calls do the AA get to within the hour?

5 One hundred light bulbs were tested by their manufacturer to see whether the average life-span of the manufacturer's bulbs was over 200 hours. The following table summarises the results.

Life span, h (hours)	$150 < h \leqslant 175$	$175 < h \leqslant 200$	$200 < h \leqslant 225$	$225 < h \leqslant 250$	$250 < h \leqslant 275$
Frequency	24	45	18	10	3

a What is the modal length of time a bulb lasts?

b What percentage of bulbs last longer than 200 hours?

c Estimate the mean life-span of the light bulbs.

d Do you think the test shows that the average life-span is over 200 hours? Fully explain your answer.

FM AU **6** Three supermarkets each claimed to have the lowest average price increase over the year. The following table summarises their price increases.

Price increase (p)	1–5	6–10	11–15	16–20	21–25	26–30	31–35
Soundbuy	4	10	14	23	19	8	2
Springfields	5	11	12	19	25	9	6
Setco	3	8	15	31	21	7	3

Using their average price increases, make a comparison of the supermarkets and write a report on which supermarket, in your opinion, has the lowest price increases over the year. Do not forget to justify your answers.

C

FM AU **7** The table shows the distances run, over a month, by an athlete who is training for a marathon.

Distance, d (miles)	$0 < d \leqslant 5$	$5 < d \leqslant 10$	$10 < d \leqslant 15$	$15 < d \leqslant 20$	$20 < d \leqslant 25$
Frequency	3	8	13	5	2

a A marathon is 26.2 miles. It is recommended that an athlete's daily average mileage should be at least a third of the distance of the race for which they are training. Is this athlete doing enough training?

b The athlete records the times of some runs and calculates that her average pace for all runs is $6\frac{1}{2}$ minutes to a mile. Explain why she is wrong to expect a finishing time for the marathon of $26.2 \times 6\frac{1}{2}$ minutes ≈ 170 minutes.

c The runner claims that the difference in length between her shortest and longest run is 21 miles. Could this be correct? Explain your answer.

PS **8** The table shows the points scored in a general-knowledge competition by all the players.

Points	0–9	10–19	20–29	30–39	40–49
Frequency	8	5	10	5	2

Helen noticed that two numbers were the wrong way round and that this made a difference of 1.7 to the arithmetic mean.

Which two numbers were the wrong way round?

AU **9** The profit made each week by a charity shop is shown in the table below.

Profit	£0–£500	£501–£1000	£1001–£1500	£1501–£2000
Frequency	15	26	8	3

Explain how you would estimate the mean profit made each week.

AU **10** The table shows the number of members of 100 football clubs.

Members	20–29	30–39	40–49	50–59	60–69
Frequency	16	34	27	18	5

a Roger claims that the median number of members is 39.5.

Is he correct? Explain your answer.

b He also says that the range of the number of members is 34.

Could he be correct? Explain your answer.

11.8 Frequency polygons

This section will show you how to:

- draw frequency polygons for discrete and continuous data

Key words
continuous data
discrete data
frequency polygon

To help people understand it, statistical information is often presented in pictorial or diagrammatic form, which includes the pie chart, the line graph, the bar chart and the stem-and-leaf diagram. These were covered in Chapter 10. Another method of showing data is by **frequency polygons**.

Frequency polygons can be used to represent both ungrouped data and grouped data, as shown in Example 13 and Example 14 respectively and are appropriate for both **discrete data** and **continuous data**.

Frequency polygons show the shapes of distributions and can be used to compare distributions.

EXAMPLE 13

No. of children	0	1	2	3	4	5
Frequency	12	23	36	28	16	11

This is the frequency polygon for the ungrouped data in the table.

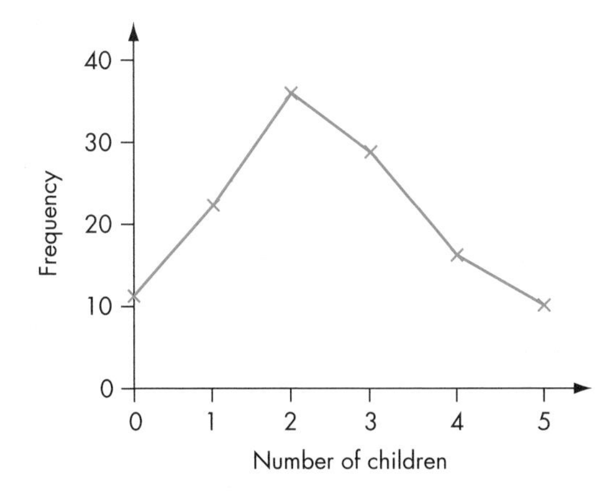

Note:

- The coordinates are plotted from the ordered pairs in the table.
- The polygon is completed by joining up the plotted points with straight lines.

EXAMPLE 14

Weight, w (kg)	$0 < w \leqslant 5$	$5 < w \leqslant 10$	$10 < w \leqslant 15$
Frequency	4	13	25

Weight, w (kg)	$15 < w \leqslant 20$	$20 < w \leqslant 25$	$25 < w \leqslant 30$
Frequency	32	17	9

This is the frequency polygon for the grouped data in the table.

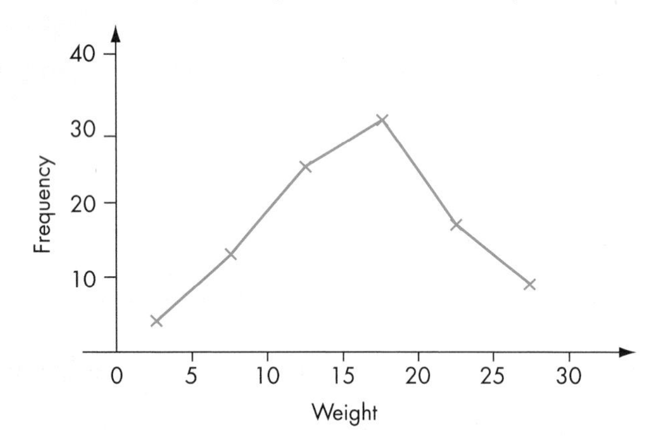

Note:

- The midpoint of each group is used, just as it was in estimating the mean.
- The ordered pairs of midpoints with frequency are plotted, namely:

  (2.5, 4), (7.5, 13), (12.5, 25), (17.5, 32), (22.5, 17), (27.5, 9)

- The polygon should be left like this. Any lines you draw before and after this have no meaning.

If you only have a frequency polygon you can work out the mean from the information on the graph.

EXAMPLE 15

The frequency polygon shows the lengths of 50 courgettes.

Work out the mean length.

The points are plotted at the midpoints so the mean is

$(7 \times 92.5 + 10 \times 97.5 + 16 \times 102.5 + 12 \times 107.5 + 5 \times 112.5) \div 50$

$= 102.3$ mm

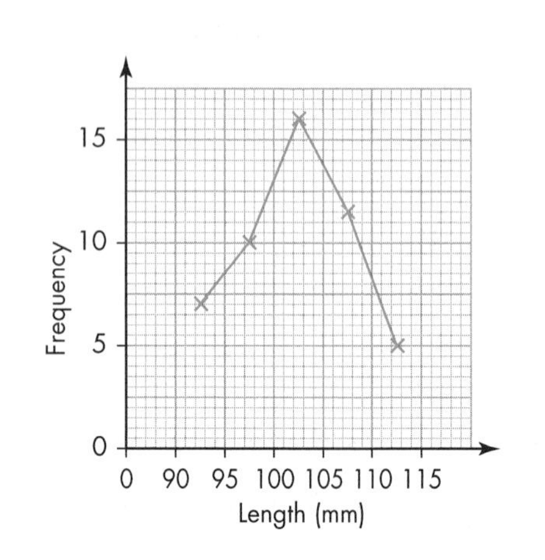

EXERCISE 11H

1 The following table shows how many students were absent from one particular class throughout the year.

Students absent	1	2	3	4	5
Frequency	48	32	12	3	1

a Draw a frequency polygon to illustrate the data.

b Estimate the mean number of absences each lesson.

2 The table below shows the number of goals scored by a hockey team in one season.

Goals	1	2	3	4	5
Frequency	3	9	7	5	2

a Draw the frequency polygon for this data.

b Estimate the mean number of goals scored per game this season.

3 After a spelling test, all the results were collated for girls and boys as below.

Number correct	1–4	5–8	9–12	13–16	17–20
Boys	3	7	21	26	15
Girls	4	8	17	23	20

a Draw frequency polygons to illustrate the differences between the boys' scores and the girls' scores.

b Estimate the mean score for boys and girls separately, and comment on the results.

HINTS AND TIPS

The highest point of the frequency polygon is the modal value.

4 A doctor was concerned at the length of time her patients had to wait to see her when they came to the morning surgery. The survey she did gave her the following results.

Time, m (min)	$0 < m \leqslant 10$	$10 < m \leqslant 20$	$20 < m \leqslant 30$
Monday	5	8	17
Tuesday	9	8	16
Wednesday	7	6	18

Time, m (min)	$30 < m \leqslant 40$	$40 < m \leqslant 50$	$50 < m \leqslant 60$
Monday	9	7	4
Tuesday	3	2	1
Wednesday	2	1	1

a Using the same pair of axes, draw a frequency polygon for each day.

b What is the average amount of time spent waiting each day?

c Why might the average times for each day be different?

D

C

C

5 The frequency polygon shows the amounts of money spent in a corner shop by the first 40 customers one morning.

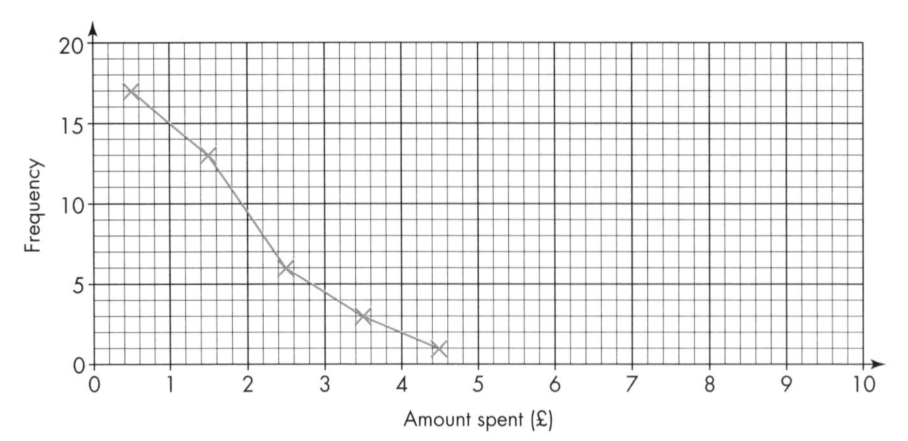

a **i** Use the frequency polygon to complete the table for the amounts spent by the first 40 customers.

Amount spent, *m* (£)	$0 < m \leqslant 1$	$1 < m \leqslant 2$	$2 < m \leqslant 3$	$3 < m \leqslant 4$	$4 < m \leqslant 5$
Frequency					

ii Work out the mean amount of money spent by these 40 customers.

b Mid-morning another 40 customers visit the shop and the shopkeeper records the amounts they spend. The table below shows the data.

Amount spent, *m* (£)	$0 < m \leqslant 2$	$2 < m \leqslant 4$	$4 < m \leqslant 6$	$6 < m \leqslant 8$	$8 < m \leqslant 10$
Frequency	3	5	18	10	4

i Copy the graph above and draw the frequency polygon to show this data.

ii Calculate the mean amount spent by the 40 mid-morning customers.

c Comment on the differences between the frequency polygons and the average amounts spent by the different groups of customers.

6 The frequency polygon shows the ages of 50 staff in a school.

a Draw up a grouped table to show the data.

b Calculate an estimate of the mean age.

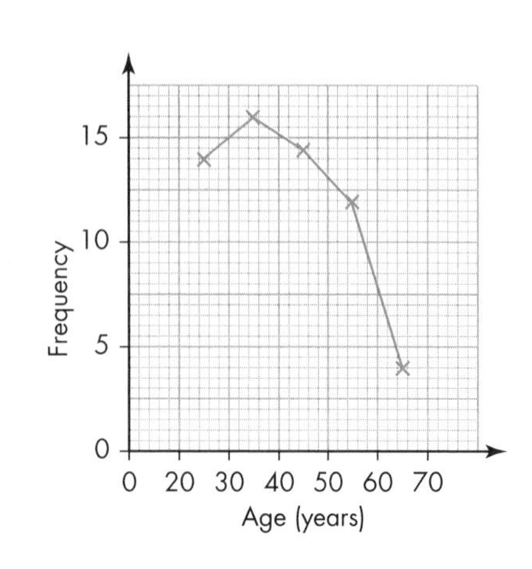

PS 7 The frequency polygon shows the lengths of time that students spent on homework one weekend.

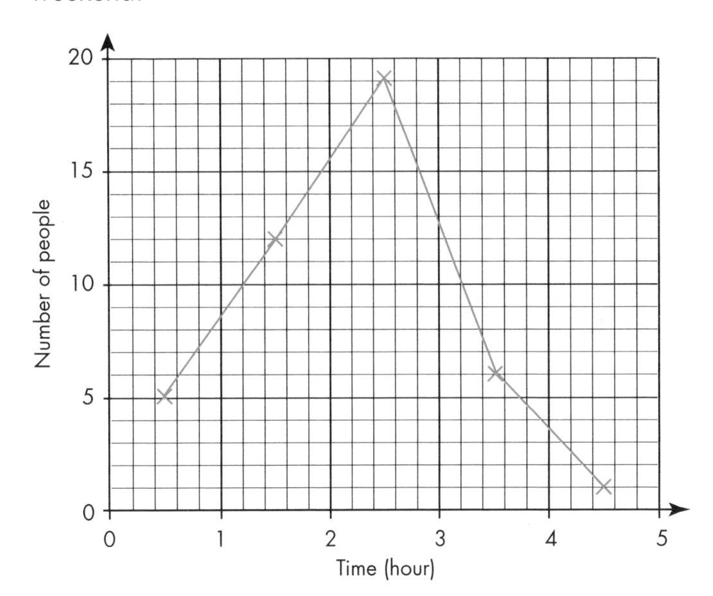

Calculate an estimate of the mean time spent on homework by the students.

AU 8 The frequency polygon shows the times that a number of people waited at a Post Office before being served one morning.

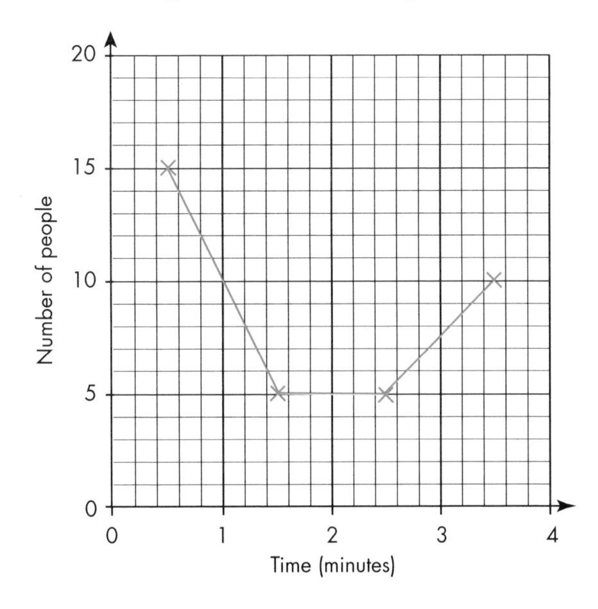

Julie said: "Most people spent 30 seconds waiting."

Explain why this might be wrong.

GRADE BOOSTER

G You can find the mode and median from a list of data

F You can find the range of a set of data and find the mean of a small set of data

E You can find the median and range from a stem-and-leaf diagram

D You can find the mean from a frequency table of discrete data and also draw a frequency polygon for such data

C You can find an estimate of the mean from a grouped table of continuous data and draw a frequency polygon for continuous data

What you should know now

- How to find the range, mode, median and mean of sets of discrete data
- Which average to use in different situations
- How to find the modal class and an estimated mean for continuous data
- How to draw frequency polygons for discrete and continuous data

1 Find **a** the mode **b** the median of

6, 6, 6, 8, 9, 10, 11, 12, 13

2 Here are ten numbers.

7 6 8 4 5 9 7 3 6 7

a Work out the range. (2)

b Work out the mean. (2)

(Total 4 marks)

Edexcel, November 2008, Paper 2 Foundation, Question 15

3 Here are the weights, in kg, of 8 people.

63 65 65 70 72 86 90 97

a Write down the mode of the 8 weights. (1)

b Work out the range of the weights. (2)

(Total 3 marks)

Edexcel, June 2007, Paper 8 Foundation, Unit 2 Test, Question 2

4 Leanne asked each of her friends which one country they would most like to visit.

Here are her results.

USA	France	Italy	USA	France
Australia	USA	Spain	France	Italy
Italy	USA	France	Italy	USA
USA	Spain	USA	Spain	Italy

a Complete the frequency table. (2)

Country	Tally	Frequency
Australia		
France		
Italy		
Spain		
USA		

b How many friends did Leanne ask? (1)

c Write down the mode. (1)

(Total 4 marks)

Edexcel, June 2008, Paper 5 Foundation, Unit 1 Test, Question 1

5 The table shows some information about five children.

Name	Gender	Age	Hair Colour
Aaron	Male	6	Black
Becky	Female	10	Brown
Kim	Female	6	Brown
Darren	Male	9	Blonde
Emily	Female	4	Red

a Write down the colour of Darren's hair. (1)

b Write down the name of the oldest child. (1)

c Work out the mean of the ages of the children. (2)

(Total 4 marks)

Edexcel, June 2008, Paper 5 Foundation, Unit 1 Test, Question 1

6 Fred wants to plant a conifer hedge. At the local garden centre he looks at 10 plants from two different varieties of conifer.

All the plants have been growing for six months.

The Sprucy Pine plants have a mean height of 74 cm and a range of 25 cm.

The Evergreen plants have a mean height of 52 cm and a range of 5 cm.

a Give one reason why Fred might decide to plant a hedge of Sprucy Pine trees.

b Give one reason why Fred might decide to plant a hedge of Evergreen trees.

7 The stem-and leaf-diagram shows the number of packages 15 drivers delivered.

Key: 3 | 5 means 35 packages

Stem	Leaf
3	5 7
4	1 3 8 8
5	0 2 5 6 7 9
6	6 9
7	2

a What is the range of the packets delivered?

b What is the median of the packets delivered?

c What is the mode of the packets delivered?

8 The weights, in kilograms, of each passenger in a minibus are:

86, 76, 84, 84, 81, 85, 80, 86, 33

a Calculate:

i their median weight

ii the range of their weights

iii their mean weight.

b Which of the two averages, mean or median, better describes the data above? Give a reason for your answer.

9 A company puts this advert in the local paper.

> **NCS Engineers**
> **Mechanic needed**
> Average wage over £500 per week

The following people work for the company.

Job	Wage per week (£)
Apprentice	210
Cleaner	210
Foreman	360
Manager	850
Mechanic	255
Parts Manager	650
Sales Manager	680

a What is the mode of these wages?

b What is the median wage?

c Calculate the mean wage.

d Explain why the advert is misleading.

10 The stem-and-leaf diagram shows information about the test marks of 17 students.

Key: 2 | 5 25 marks

Stem	Leaf
2	5 8
3	0 4 9
4	0 7 8 9
5	0 3 6 6 6 9
6	2 4

a Work out the range. (2)

b Find the median. (1)

(Total 3 marks)

Edexcel, June 2008, Paper 5 Foundation, Unit 1 Test, Question 3

11 Zoe recorded the weight of each of 15 people. She showed her results in a stem-and-leaf diagram.

Key: 4 | 6 means 46 kg

Stem	Leaf
4	6 8
5	1 2 8
6	0 3 4 6 8
7	4 7 8 9
8	7

a Write down the number of people with a weight of more than 70 kg. (1)

b Work out the range of the weights. (2)

(Total 3 marks)

Edexcel, March, 2009, Paper 5 Foundation, Unit 1 Test, Question 3

12 The stem-and-leaf diagram shows the test scores of 13 students.

Key: 2 | 5 is a score of 25

Stem	Leaf
1	9
2	2 2 3 5 7
3	3 4 4 8 8 9
4	0

How many students scored less than 20?

13 The numbers of people in 50 cars are recorded.

Number of people	Frequency
1	24
2	13
3	8
4	4
5	1

Calculate the mean number of people per car.

14 The table shows how many children there were in the family of each member of a class.

Number of children	Frequency
1	6
2	10
3	4
4	3
5	1

a How many children were in the class?

b What is the modal number of children per family?

c What is the median number of children per family?

d What is the mean number of children per family?

15 The table shows the distances travelled to work by 40 office workers.

Distance travelled, d (km)	Frequency
$0 < d \leqslant 2$	10
$2 < d \leqslant 4$	16
$4 < d \leqslant 6$	8
$6 < d \leqslant 8$	5
$8 < d \leqslant 10$	1

Calculate an estimate of the mean distance travelled to work by these office workers.

16 The mean weight of five rowers is 49.2 kg.

a Find the total weight of the rowers.

b The mean weight of the five rowers and the reserve is 50.5 kg.

Calculate the weight of the reserve.

17 The table shows information about the number of hours that 120 children used a computer last week.

Number of hours (h)	Frequency
$0 < h \leqslant 2$	10
$2 < h \leqslant 4$	15
$4 < h \leqslant 6$	30
$6 < h \leqslant 8$	35
$8 < h \leqslant 10$	25
$10 < h \leqslant 12$	5

Work out an estimate for the mean number of hours that the children used a computer. Give your answer correct to 2 decimal places.

Edexcel, Question 10, Paper 17 Intermediate, June 2005

Worked Examination Questions

1 A teacher asks all his class: "How many children are there in your family?" Their replies are given below.

Number of children in a family	Number of replies
1	7
2	12
3	5
4	2
5	0

a How many children are in the class?

b What is the modal number of children in a family?

c What is the median number in a family?

d What is the mean number in a family?

a $7 + 12 + 5 + 2 + 0 = 26$

The total number of children is 26.

Add up the frequencies. This is worth 1 mark.

1 mark

b The modal number of children is 2.

The largest frequency is 12 so the modal number is 2. This is worth 1 mark.

1 mark

c The median number of children is 2.

The median will be between the 13th and 14th values. Adding up the frequencies gives 7, 19, 24, 26, 26. So the required value is in the second row. This is worth 1 mark.

1 mark

d The mean number of children = $54 \div 26 = 2.1$

3 marks

Add an extra column to the table and multiply the number of children by the number of replies. This gives 7, 24, 15, 8, 0.

Add these to get 54. You get 1 mark for showing the total.

Divide 54 by 26. You get 1 mark for dividing the total by 26 and 1 mark for the answer.

Total: 6 marks

2 A teacher shows her class 25 objects on a tray. She leaves it in view for 1 minute.

She then covers the objects and asks the class to write down the names of as many objects as they can remember.

The results are shown in the table. What is the mean number of objects recalled by the class?

Number of objects recalled, x	Frequency, f
$0 < x \leqslant 5$	2
$5 < x \leqslant 10$	5
$10 < x \leqslant 15$	13
$15 < x \leqslant 20$	8
$20 < x \leqslant 25$	2
	30

Number of objects recalled, x	Frequency, f	Midpoint, m	$m \times f$
$0 < x \leqslant 5$	2	2.5	5
$5 < x \leqslant 10$	5	7.5	37.5
$10 < x \leqslant 15$	13	12.5	162.5
$15 < x \leqslant 20$	8	17.5	140
$20 < x \leqslant 25$	2	22.5	45
	30		390

First add a column for the midpoints. This is the two end-values added and divided by 2. This worth 1 mark.

Next, add a column for midpoint multiplied by frequency. This worth 1 mark.

Next, work out the totals for the frequency and the $m \times f$ columns. This worth 1 mark.

Mean $= 390 \div 30$

$= 13$

Finally, divide the total of the $m \times f$ column by the total frequency. This worth 1 mark.

You get 1 mark for the correct answer.

Total: 5 marks

Worked Examination Questions

3 The weights, in kilograms, of a rowing boat crew are:

91, 81, 89, 91, 91, 85, 89, 38

a Work out:

i the modal weight **ii** the median weight **iii** the range of the weights **iv** their mean weight.

AU **b** Which of the averages best describes the data? Explain your answer.

a i The modal weight is 91 kg.

Look for the weight that occurs most frequently. This is worth 1 mark.

ii The median weight is 89 kg.

Arrange the weights in order of size: 38, 81, 85, 89, 89, 91, 91, 91. The middle two are 89. This worth 1 mark.

iii The range is 53 kg.

Highest value – lowest value = 91 – 38. This is worth 1 mark.

iv The mean weight = $655 \div 8 = 81.875$

Add up all the weights and divide the total by the number of weights (8). You get 1 mark for method and 1 mark for accuracy.

5 marks

b The mean as it is the only average that takes into account all the weights.

The correct answer is worth 1 mark.

1 mark **Total:** 6 marks

4 The mean speed of each member of a cycling club over a long-distance race was recorded and a frequency polygon was drawn.

PS Work out an estimate of the mean speed for the whole club.

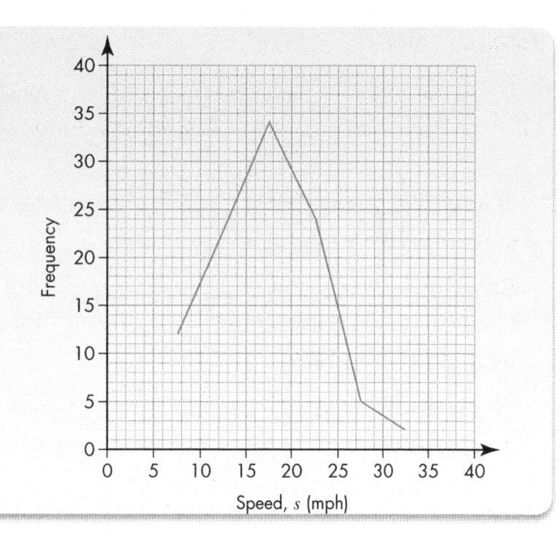

Create a grouped frequency table.

Speed, s (mph)	Frequency, f	Midpoint, m	$f \times m$
5–10	12	7.5	90
10–15	23	12.5	287.5
15–20	34	17.5	595
20–25	24	22.5	540
25–30	5	27.5	137.5
30–35	2	32.5	65
	100		1715

The table must include the midpoint values, the frequency and $f \times m$.

You get 1 mark for giving the frequencies and showing a total.

You get 1 mark for attempting to work out $f \times m$ and showing a total.

An estimate of the mean is $1715 \div 100 = 17.15$ mph

You get 1 mark for dividing the total $f \times m$ by the total f and 1 mark for the correct answer.

Total: 4 marks

11 Functional Maths
Fishing competition on the Avon

Averages are used to compare data and make statements about sets of data. They are used every day for a wide variety of purposes, from describing the weather to analysing the economy. In this activity averages will be applied to a sporting event.

A fishing competition

Kath's dad ran a fishing competition on the river Avon during the summer.

Kath kept an accurate record of the data collected during the competition.

This table shows a summary of Kath's records, collected from all of the anglers in the competition, for the first four weeks of July.

	Week 1	Week 2	Week 3	Week 4	Week 5
Mean number of fish caught	12.1	12.3	12.2	11.8	
Mean time spent fishing (hrs)	5.6	6.1	5.8	5.4	
Mean weight of fish caught (g)	1576	1728	1635	1437	
Mean length of longest fish caught (cm)	21.7	20.6	21.6	21.9	

In the last week of July, Kath again collected data from the anglers. She did not have time to add this data to her table.

The following tables show all the data that she collected in Week 5.

Number of fish caught	Frequency
0 – 5	6
6 – 10	11
11 – 15	8
16 – 20	5

Time spent fishing (hrs)	Frequency
0 – 4	2
4 hrs 1 min – 5	14
5 hrs 1 min – 6	6
6 hrs 1 min – 7	8

Weight of fish caught (g)	Frequency
0 – 500	1
501 – 1000	8
1001 – 1500	18
1501 – 2000	3

Longest fish caught (cm)	Frequency
0 – 10	2
11 – 15	6
16 – 20	12
21 – 25	10

Your task

Kath must do a presentation at the end of the month to summarise the fishing competition. Help her to write her presentation.

The presentation must include the following:

- The data for Week 5 inserted into the main table
- A graph to represent the data for all five weeks
- Statements to compare the five weeks of the competition
- A description of the 'average angler'.

Getting started

Start by thinking about how averages are calculated. What do the mean, median, mode and range show when applied to sets of data? How do they apply to frequency tables?

What are the mean, median, mode and range for this set of data?

3, 4, 8, 3, 2, 4, 5, 3, 4, 6, 8, 4, 2, 9, 1

Now think about the averages that you will need in your presentation and how these averages are best represented. Use your ideas in your presentation.

Why this chapter matters

This chapter extends the idea of statistical representation (which you first met in Chapter 10) by introducing pie charts and scatter diagrams.

Pie charts

The pie chart first appeared in 1801 in a publication called *The Statistical Breviary* by William Playfair. Do you remember him? He was one of the first to use bar charts, too.

William Playfair used graphical representations of quantitative data, such as bar charts and pie charts, because he believed that "making an appeal to the eye when trying to show data is the best and easiest method of giving any message that might be wanted to show through such diagrams".

He used circles in interesting ways to represent quantitative relationships, varying their sizes and subdividing them into slices to create circular charts – or 'pie' charts.

The term 'pie chart' was not actually used until years later and it is not the only food metaphor that has been used to describe it. The French referred to it as a camembert – a soft, round cheese!

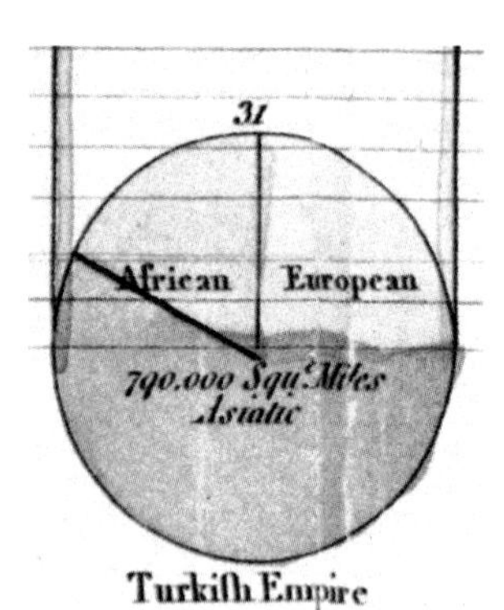

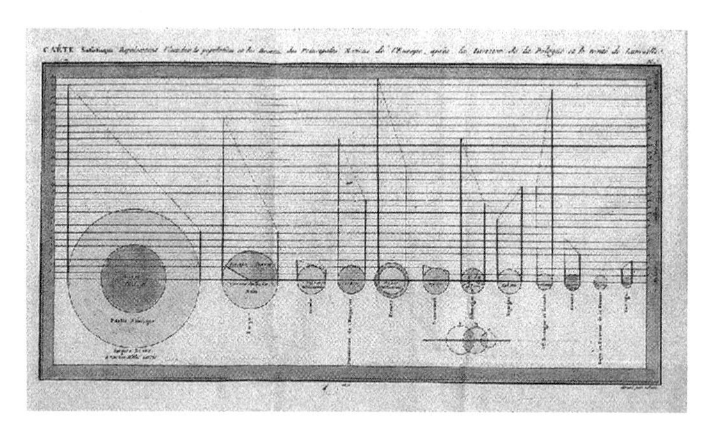

He was said to have been amused at the image of the pie chart (above left), showing a smile, and also liked the use of the pie chart above right, which is now seldom used as it's not clear which area is which.

Scatter diagrams

The word 'scatter' comes to us from Scandinavian influences in the 12th century, but we didn't see any scatter diagrams until one appeared in 1924 in a document from a university in what is now Pakistan. Apparently, they were one of the first establishments to use the technique of plotting points from two sources to see if any connections could be seen between the two.

Scatter diagrams were not used very much until the great energy debate in the late 1960s, when prices and sales of both gas and electricity were being studied. At that time, there was pressure for people to use more electricity, as it was thought to be an infinite power source, whereas gas would seemingly run out one day soon.

We now use both pie charts and scatter diagrams every day, for example, in business presentations, social studies and polls.

Chapter 12

Statistics: Pie charts, scatter diagrams and surveys

The grades given in this chapter are target grades.

This chapter will show you …

- E how to draw and interpret pie charts
- D how to design a survey sheet and questionnaire
- C how to draw scatter diagrams and lines of best fit
- C how to interpret scatter diagrams and the different types of correlation
- C how to describe the data-handling cycle
- C some of the common features of social statistics

Visual overview

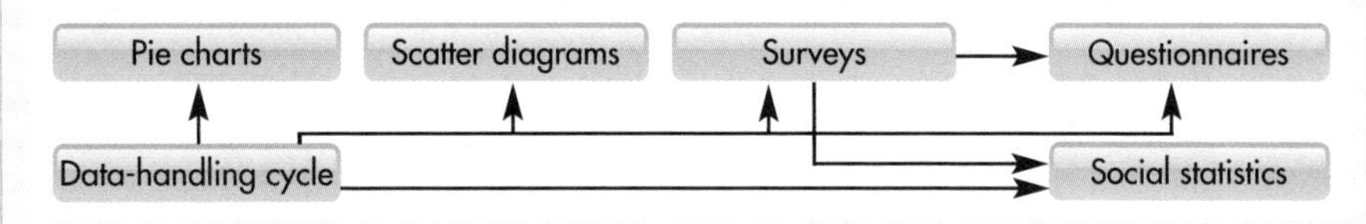

What you should already know

- How to draw and interpret pictograms, bar charts and line graphs **(KS3 level 3, GCSE grade G)**
- How to draw and measure angles **(KS3 level 4, GCSE grade F)**
- How to plot coordinates **(KS3 level 4, GCSE grade G)**

Quick check

1 The bar chart shows how many boys and girls are in five Year 7 forms.

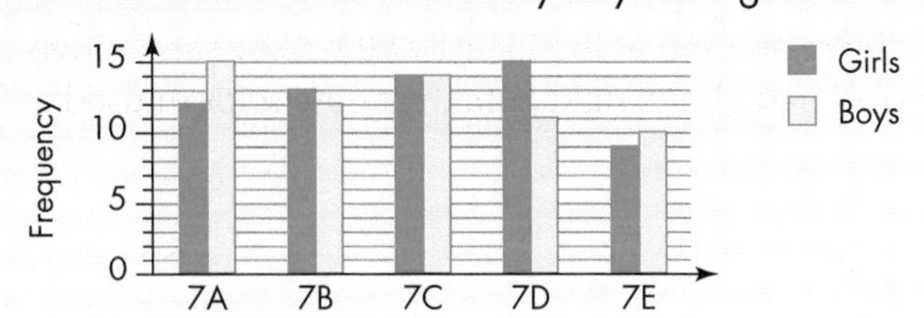

a How many pupils are in 7A?

b How many boys altogether are in the five forms?

2 Draw an angle of 72°.

3 Three points, A, B and C, are shown on the coordinate grid.

What are the coordinates of A, B and C?

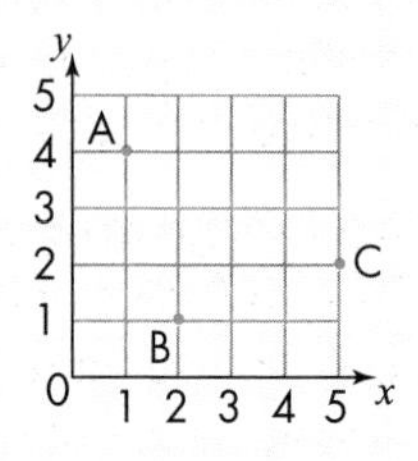

12.1 Pie charts

This section will show you how to:

- draw pie charts

Key words
angle
pie chart
sector

Pictograms, bar charts and line graphs (see Chapter 10) are easy to draw but they can be difficult to interpret when there is a big difference between the frequencies or there are only a few categories. In these cases, it is often more convenient to illustrate the data on a **pie chart**.

In a pie chart, the whole of the data is represented by a circle (the 'pie') and each category of it is represented by a **sector** of the circle (a 'slice of the pie'). The **angle** of each sector is proportional to the frequency of the category it represents.

So, a pie chart cannot show individual frequencies, like a bar chart can, for example. It can only show proportions.

Sometimes the pie chart will be marked off in equal sections rather than angles. In these cases, the numbers are always easy to work with.

EXAMPLE 1

20 people were surveyed about their preferred drink. Their replies are shown in the table.

Drink	Tea	Coffee	Milk	Pop
Frequency	6	7	4	3

Show the results on the pie chart given.

You can see that the pie chart has 10 equally-spaced divisions.

As there are 20 people, each division is worth two people. So the sector for tea will have three of these divisions. In the same way, coffee will have $3\frac{1}{2}$ divisions, milk will have 2 divisions and pop will have $1\frac{1}{2}$ divisions.

The finished pie chart will look like the one in the diagram.

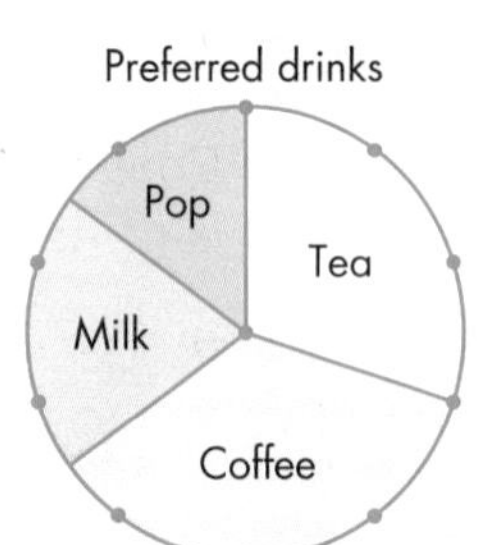

Note:

- You should always label the sectors of the chart (use shading and a separate key if there is not enough space to write on the chart).
- Give your chart a title.

EXAMPLE 2

In a survey on holidays, 120 people were asked to state which type of transport they used on their last holiday. This table shows the results of the survey. Draw a pie chart to illustrate the data.

Type of transport	Train	Coach	Car	Ship	Plane
Frequency	24	12	59	11	14

You need to find the angle for the fraction of 360° that represents each type of transport. This is usually done in a table, as shown below.

Type of transport	Frequency	Calculation	Angle
Train	24	$\frac{24}{120} \times 360° = 72°$	72°
Coach	12	$\frac{12}{120} \times 360° = 36°$	36°
Car	59	$\frac{59}{120} \times 360° = 177°$	177°
Ship	11	$\frac{11}{120} \times 360° = 33°$	33°
Plane	14	$\frac{14}{120} \times 360° = 42°$	42°
Totals	120		360°

Draw the pie chart, using the calculated angle for each sector.

Note:

- Use the frequency total (120 in this case) to calculate each fraction.
- Check that the sum of all the angles is 360°.
- Label each sector.
- The angles or frequencies do not have to be shown on the pie chart.

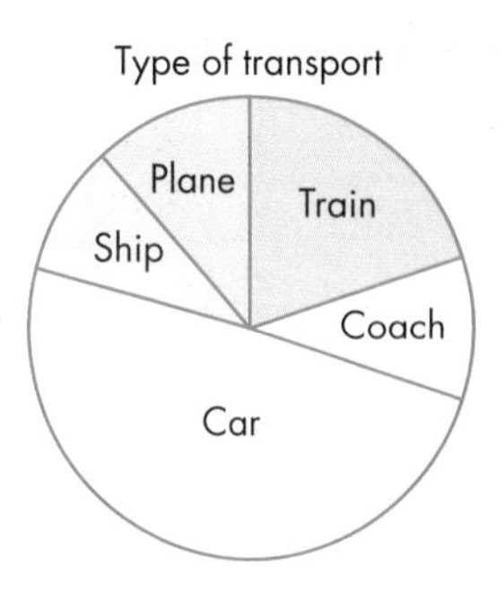

EXERCISE 12A

1 Copy the basic pie chart on the right and draw a pie chart to show each of the following sets of data.

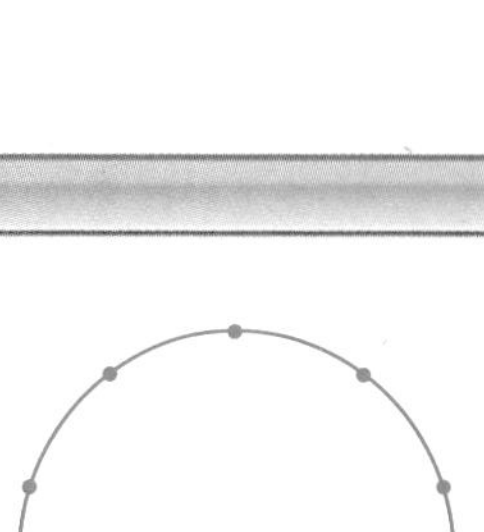

a The favourite pets of 10 children.

Pet	Dog	Cat	Rabbit
Frequency	4	5	1

b The makes of cars of 20 teachers.

Make of car	Ford	Toyota	Vauxhall	Nissan	Peugeot
Frequency	4	5	2	3	6

F

F

E

c The newspaper read by 40 office workers.

Newspaper	*Sun*	*Mirror*	*Guardian*	*The Times*
Frequency	14	8	6	12

2 Draw a pie chart to represent each of the following sets of data.

HINTS AND TIPS

Remember to complete a table as shown in the examples. Check that all angles add up to 360°.

a The number of children in 40 families.

No. of children	0	1	2	3	4
Frequency	4	10	14	9	3

b The favourite soap-opera of 60 students.

Programme	*Home and Away*	*Neighbours*	*Coronation Street*	*Eastenders*	*Emmerdale*
Frequency	15	18	10	13	4

c How 90 students get to school.

Journey to school	Walk	Car	Bus	Cycle
Frequency	42	13	25	10

3 Mariam asked 24 of her friends which sport they preferred to play. Her data is shown in this frequency table.

Sport	Rugby	Football	Tennis	Squash	Basketball
Frequency	4	11	3	1	5

Illustrate her data on a pie chart.

D

AU 4 Andy wrote down the number of lessons he had per week in each subject on his school timetable.

Mathematics 5 English 5 Science 8 Languages 6
Humanities 6 Arts 4 Games 2

a How many lessons did Andy have on his timetable?

b Draw a pie chart to show the data.

c Draw a bar chart to show the data.

d Which diagram better illustrates the data? Give a reason for your answer.

AU 5 In the run up to an election, 720 people were asked in a poll which political party they would vote for. The results are given in the table.

Conservative	248
Labour	264
Liberal-Democrat	152
Green Party	56

a Draw a pie chart to illustrate the data.

b Why do you think pie charts are used to show this sort of information during elections?

6 This pie chart shows the proportions of the different shoe sizes worn by 144 pupils in Year 11 in a London school.

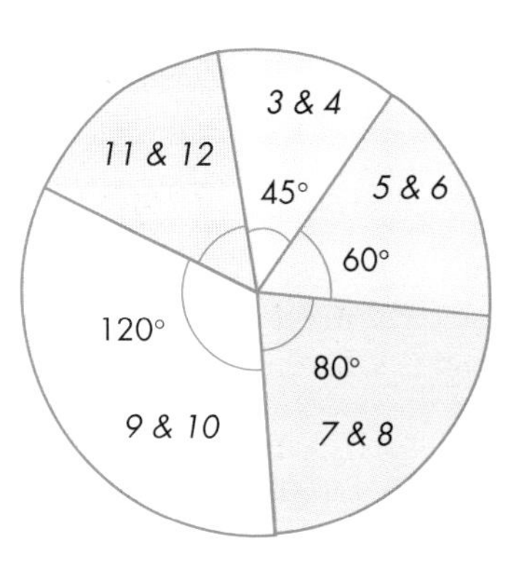

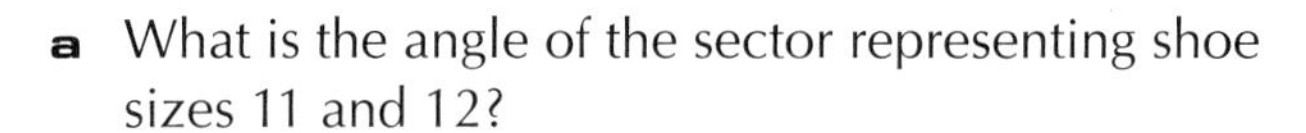

a What is the angle of the sector representing shoe sizes 11 and 12?

b How many pupils had a shoe size of 11 or 12?

c What percentage of pupils wore the modal size?

AU 7 The table below shows the numbers of candidates, at each grade, taking music examinations in Strings and Brass.

	Grades					
	3	**4**	**5**	**6**	**7**	**Total number of candidates**
Strings	300	980	1050	600	70	3000
Brass	250	360	300	120	70	1100

a Draw a pie chart to represent each of the two examinations.

b Compare the pie charts to decide which group of candidates, Strings or Brass, did better overall. Give reasons to justify your answer.

PS 8 In a survey, a rail company asked passengers whether their service had improved.

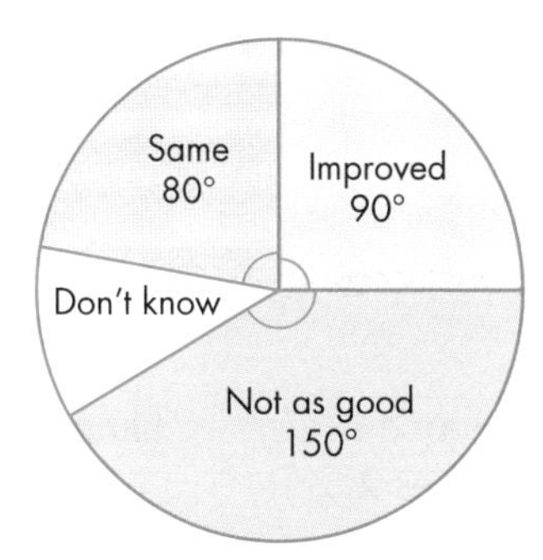

What is the probability that a person picked at random from this survey answered "Don't know"?

AU 9 You have been asked to draw a pie chart representing the different ways in which students come to school one morning.

What data would you collect to do this?

12.2 Scatter diagrams

This section will show you how to:
- draw, interpret and use scatter diagrams

Key words
correlation
line of best fit
negative correlation
no correlation
positive correlation
scatter diagram
variable

A **scatter diagram** (also called a scattergraph or scattergram) is a method of comparing two **variables** by plotting their corresponding values on a graph. These values are usually taken from a table.

In other words, the variables are treated just like a set of (x, y) coordinates. This is shown in the scatter diagram that follows, in which the marks scored in an English test are plotted against the marks scored in a mathematics test.

This graph shows **positive correlation**. This means that pupils who get high marks in mathematics tests also tend to get high marks in English tests.

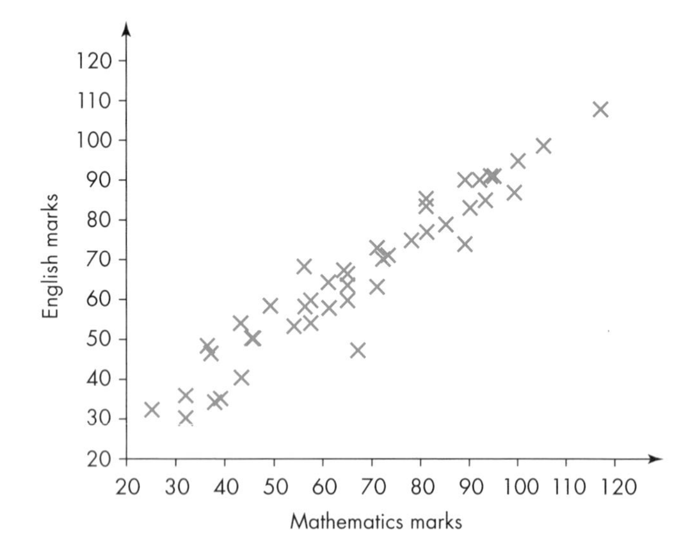

Correlation

There are different types of **correlation**. Here are three statements that may or may not be true.

- The taller people are, the wider their arm span is.
- The older a car is, the lower its value will be.
- The distance you live from your place of work will affect how much you earn.

These relationships could be tested by collecting data and plotting the data on a scatter diagram. For example, the first statement may give a scatter diagram like the first one below.

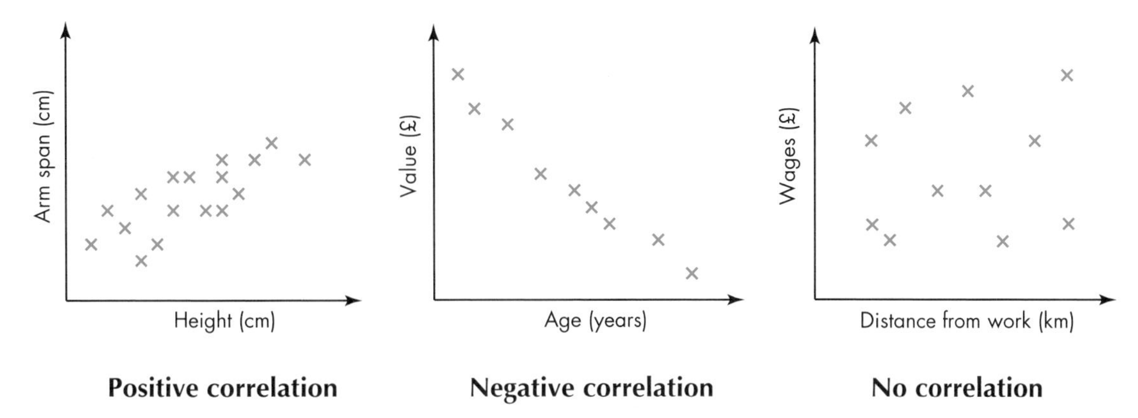

Positive correlation **Negative correlation** **No correlation**

This first diagram has **positive correlation** because, as one quantity increases, so does the other. From such a scatter diagram, you could say that the taller someone is, the wider their arm span.

Testing the second statement may give a scatter diagram like the middle one above. This has **negative correlation** because, as one quantity increases, the other quantity decreases. From such a scatter diagram, you could say that, as a car gets older, its value decreases.

Testing the third statement may give a scatter diagram like the one on the right, above. This scatter diagram has **no correlation**. There is no relationship between the distance a person lives from their work and how much they earn.

EXAMPLE 3

The graphs below show the relationship between the temperature and the amount of ice-cream sold, and that between the age of people and the amount of ice-cream they eat.

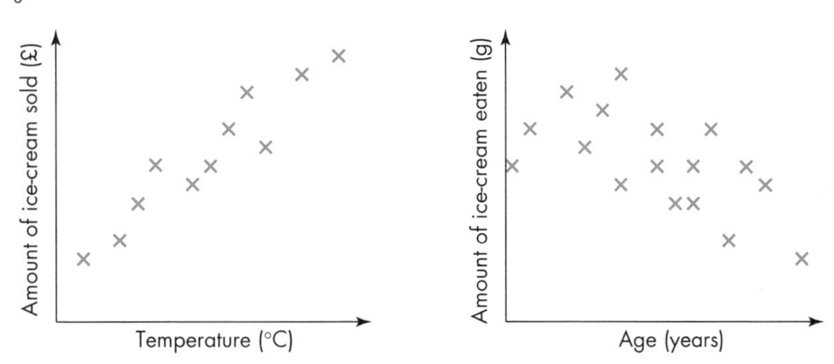

a Comment on the correlation of each graph.

b What does each graph tell you?

The first graph has positive correlation and tells us that, as the temperature increases, the amount of ice-cream sold increases.

The second graph has negative correlation and tells us that, as people get older, they eat less ice-cream.

Line of best fit

A **line of best fit** is a straight line that goes between all the points on a scatter diagram, passing as close as possible to all of them. You should try to have the same number of points on both sides of the line. Because you are drawing this line by eye, examiners make a generous allowance around the correct answer. The line of best fit for the scatter diagram at the start of this section is shown below, left.

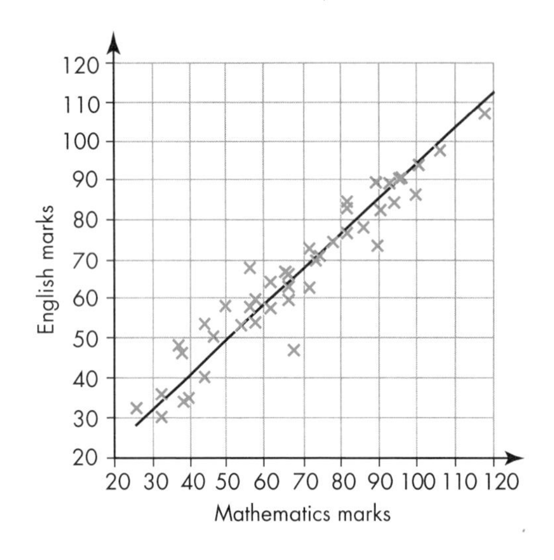

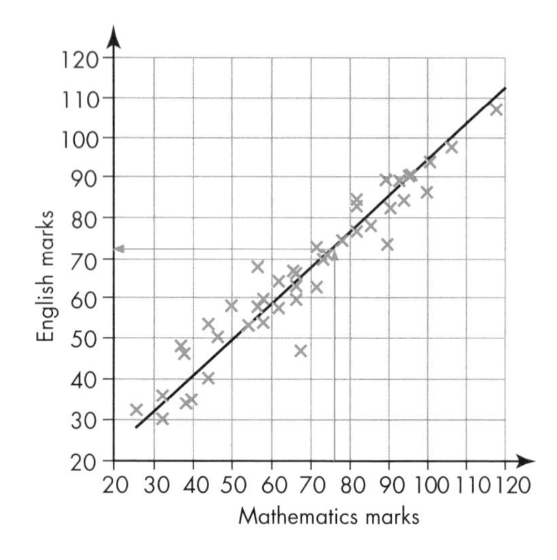

The line of best fit can be used to answer questions such as: "A girl took the mathematics test and scored 75 marks but was ill for the English test. How many marks was she likely to have scored?"

The answer is found by drawing a line up from 75 on the mathematics axis to the line of best fit and then drawing a line across to the English axis as shown in the graph above, right. This gives 73, which is the mark she is likely to have scored in the English test.

EXERCISE 12B

E

1 Describe the correlation of each of these four graphs.

a

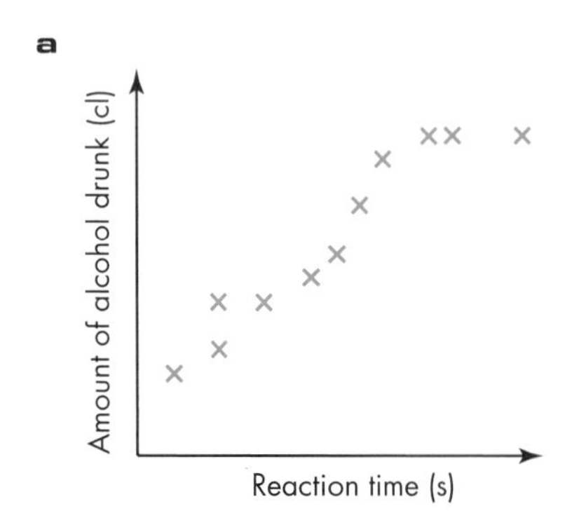

b

c

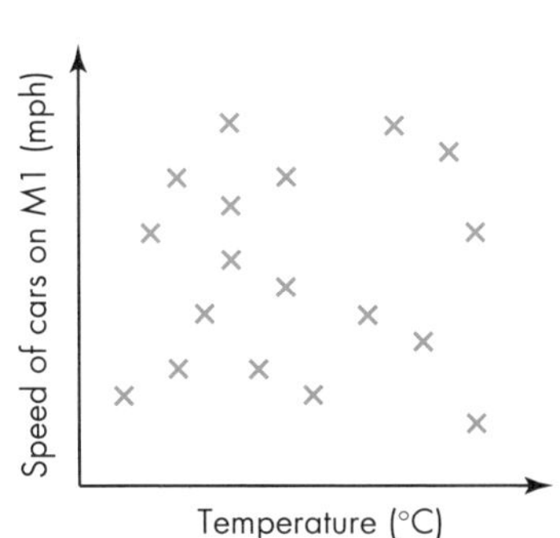

d

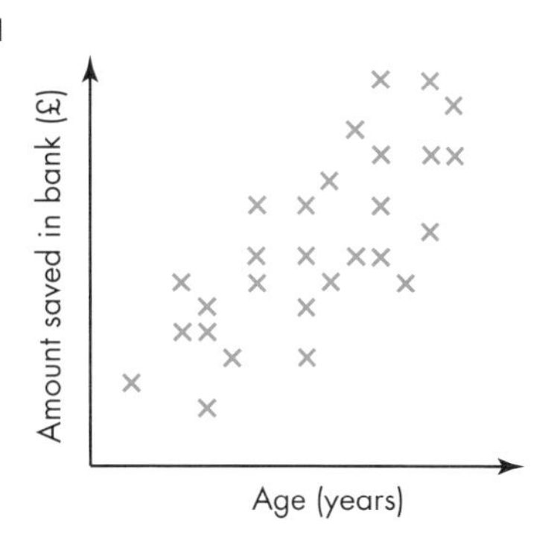

E

D

C

2 Write in words what each graph in question 1 tells you.

3 The table below shows the results of a science experiment in which a ball is rolled along a desk top. The speed of the ball is measured at various points.

Distance from start (cm)	10	20	30	40	50	60	70	80
Speed (cm/s)	18	16	13	10	7	5	3	0

a Plot the data on a scatter diagram.

b Draw the line of best fit.

c If the ball's speed had been measured at 5 cm from the start, what is it likely to have been?

d Estimate how far the ball was from the start when its speed was 12 cm/s.

HINTS AND TIPS

Often in exams axes are given and most, if not all, of the points are plotted.

4 The heights, in centimetres, of 20 mothers and their 15-year-old daughters were measured. These are the results.

Mother	153	162	147	183	174	169	152	164	186	178
Daughter	145	155	142	167	167	151	145	152	163	168
Mother	175	173	158	168	181	173	166	162	180	156
Daughter	172	167	160	154	170	164	156	150	160	152

a Plot these results on a scatter diagram. Take the x-axis for the mothers' heights from 140 to 200. Take the y-axis for the daughters' heights from 140 to 200.

b Is it true that the tall mothers have tall daughters?

C

FM 5 The table below shows the marks for ten students in their mathematics and geography examinations.

Student	Anna	Becky	Cath	Dema	Emma	Fatima	Greta	Hannah	Imogen	Sitara
Maths	57	65	34	87	42	35	59	61	25	35
Geog	45	61	30	78	41	36	35	57	23	34

a Plot the data on a scatter diagram. Take the x-axis for the mathematics scores and the y-axis for the geography scores.

b Draw the line of best fit.

c One of the students was ill when she took the geography examination. Which student was it most likely to be?

d If another student, Kate, was absent for the geography examination but scored 75 in mathematics, what mark would you expect her to have scored in geography?

e If another student, Lina, was absent for the mathematics examination but scored 65 in geography, what mark would you expect her to have scored in mathematics?

FM 6 A form teacher carried out a survey of 20 students from his class and asked them to say how many hours per week they spent playing sport and how many hours per week they spent watching TV. This table shows the results of the survey.

Student	1	2	3	4	5	6	7	8	9	10
Hours playing sport	12	3	5	15	11	0	9	7	6	12
Hours watching TV	18	26	24	16	19	27	12	13	17	14

Student	11	12	13	14	15	16	17	18	19	20
Hours playing sport	12	10	7	6	7	3	1	2	0	12
Hours watching TV	22	16	18	22	12	28	18	20	25	13

a Plot these results on a scatter diagram. Take the x-axis as the number of hours playing sport and the y-axis as the number of hours watching TV.

AU **b** If you knew that another student from the form watched 8 hours of TV a week, would you be able to predict how long they spent playing sport? Explain why.

FM 7 The table shows the times taken and distances travelled by a taxi driver in 10 journeys on one day.

Distance (km)	1.6	8.3	5.2	6.6	4.8	7.2	3.9	5.8	8.8	5.4
Time (minutes)	3	17	11	13	9	15	8	11	16	10

a Draw a scatter diagram of this information, with time on the horizontal axis.

b Draw a line of best fit on your diagram.

c If a taxi journey takes 5 minutes, how far, in kilometres, would you expect the journey to have been?

d How much time would you expect a journey of 4 km to take?

C

PS 8 Omar records the time taken, in hours, and the average speed, in miles per hour (mph), for several different journeys.

Time (h)	0.5	0.8	1.1	1.3	1.6	1.75	2	2.4	2.6
Speed (mph)	42	38	27	30	22	23	21	9	8

Estimate the average speed for a journey of 90 minutes.

AU 9 Describe what you would expect the scatter graph to look like if someone said that it showed negative correlation.

12.3 Surveys

This section will show you how to:

- conduct surveys
- ask good questions in order to collect reliable and valid data

Key words
data-collection sheet
leading question
questionnaire
response
survey

A **survey** is an organised way of asking a lot of people a few, well-constructed questions, or of making a lot of observations in an experiment, in order to reach a conclusion about something. Surveys are used to test out people's opinions or to test a hypothesis.

Simple data-collection sheet

If you need to collect some data to analyse, you will have to design a simple **data-collection sheet**.

Look at this example: "Where do you want to go for the Year 10 trip at the end of term – Blackpool, Alton Towers, The Great Western Show or London?"

You would put this question on the same day to a lot of Year 10 students and enter their answers straight onto a data-collection sheet, as below.

Place	Tally	Frequency
Blackpool	|||| |||| |||| |||| |||	23
Alton Towers	|||| |||| |||| |||| |||| |||| |||| |||| |||| |	46
The Great Western Show	|||| |||| ||||	14
London	|||| |||| |||| |||| ||	22

Notice how plenty of space is left for the tally marks and how the tallies are 'gated' in groups of five to make counting easier when the survey is complete.

This is a good, simple data-collection sheet because:

- only one question is asked ("Where do you want to go?")
- all the possible venues are listed
- the answer from each interviewee can be easily and quickly tallied, and the next interviewee questioned.

Notice too that, since the question listed specific places, they must all appear on the data collection sheet. You would lose marks in an examination if you just asked the open question: "Where do you want to go?"

Data sometimes needs to be collected to obtain **responses** for two different categories. The data-collection sheet is then in the form of a simple two-way table.

EXAMPLE 4

The head of a school wants to find out how long his students spend doing homework in a week. He carries out a survey on 60 students. He uses the two-way table to show the result.

	0–5 hours	0–10 hours	10–20 hours	More than 20 hours
Year 7				

This is not a good table as the categories overlap. A student who does 10 hours' work a week could tick any of two columns. Response categories should not overlap and there should be only one possible place to put a tick.

A better table would be:

	0 up to 5 hours	More than 5 and up to 10 hours	More than 10 and up to 15 hours	More than 15 hours
Year 7	~~IIII~~ II	~~IIII~~		
Year 8	~~IIII~~	~~IIII~~ II		
Year 9	III	~~IIII~~ II	II	
Year 10	III	~~IIII~~	III	I
Year 11	II	IIII	IIII	II

This gives a more accurate picture of the amount of homework done in each year group.

Using your computer

Once the data has been collected for your survey, it can be put into a computer database. This allows the data to be stored and amended or updated at a later date if necessary.

From the database, suitable statistical diagrams can easily be drawn within the software and averages calculated for you. Your results can then be published in, for example, the school magazine.

EXERCISE 12C

FM **1** "People like the supermarket to open on Sundays."

a To see whether this statement is true, design a data-collection sheet that will allow you to capture data while standing outside a supermarket.

b Does it matter on which day you collect data outside the supermarket?

2 The school tuck shop wanted to know which types of chocolate it should order to sell – plain, milk, fruit and nut, wholenut or white chocolate.

a Design a data-collection sheet that you could use to ask the pupils in your school which of these chocolate types are their favourite.

b Invent the first 30 entries on the chart.

> **HINTS AND TIPS**
> Include space for tallies.

3 What type of television programme do people in your age group watch the most? Is it crime, romance, comedy, documentary, sport or something else? Design a data-collection sheet to be used in a survey of your age group.

4 On what do people of your age tend to spend their money? Is it sport, magazines, clubs, cinema, sweets, clothes or something else? Design a data-collection sheet to be used in a survey of your age group.

5 Design two-way tables to show the following. Invent about 40 entries for each one.

a How students in different year groups travel to school in the morning.

b The type of programme that different age groups prefer to watch on TV.

c The favourite sport of boys and girls.

d How much time students in different year groups spend on the computer in the evening.

> **HINTS AND TIPS**
> Make sure all possible responses are covered.

FM **6** Carlos wanted to find out who eats healthy food.
He decided to investigate the hypothesis:

"Boys are less likely to eat healthy food than girls are."

a Design a data-capture form that Carlos could use to help him do this.

b Carlos records information from a sample of 40 boys and 25 girls. He finds that 17 boys and 15 girls eat healthy food. Based on this sample, is the hypothesis correct? Explain your answer.

AU **7** Show how you would find out what kind of tariffs your classmates use on their mobile phones.

AU **8** You have been asked to find out which shops the parents of the students at your school like to use. When creating a data-collection sheet, what two things must you include?

D

C

Questionnaires

When you are putting together a **questionnaire**, you must think very carefully about the sorts of question you are going to ask, to put together a clear, easy-to-use questionnaire.

Here are five rules that you should *always* follow.

- Never ask a **leading question** designed to get a particular response.
- Never ask a personal, irrelevant question.
- Keep each question as simple as possible.
- Include questions that will get a response from whomever is asked.
- Make sure the categories for the responses do not overlap and keep the number of choices to a reasonable number (six at the most).

The following questions are *badly constructed* and should *never* appear in any questionnaire.

✘ *What is your age?* This is personal. Many people will not want to answer. It is always better to give a range of ages such as:

☐ Under 15 ☐ 16–20 ☐ 21–30 ☐ 31–40 ☐ Over 40

✘ *Slaughtering animals for food is cruel to the poor defenceless animals. Don't you agree?* This is a leading question, designed to get a 'yes'. It is better ask an impersonal question such as:

Are you a vegetarian? ☐ Yes ☐ No

✘ *Do you go to discos when abroad?* This can be answered only by those who have been abroad. It is better to ask a starter question, with a follow-up question such as:

Have you been abroad for a holiday? ☐ Yes ☐ No

If 'Yes', did you go to a disco whilst you were away? ☐ Yes ☐ No

✘ *When you first get up in a morning and decide to have some sort of breakfast that might be made by somebody else, do you feel obliged to eat it all or not?* This question is too complicated. It is better to ask a series of shorter questions such as:

What time do you get up for school? ☐ Before 7 ☐ Between 7 and 8 ☐ After 8

Do you have breakfast every day? ☐ Yes ☐ No

If 'No', on how many schooldays do you have breakfast? ☐ 0 ☐ 1 ☐ 2 ☐ 3 ☐ 4 ☐ 5

A questionnaire is usually put together to test a hypothesis or a statement. For example: "People buy cheaper milk from the supermarket as they don't mind not getting it on their doorstep. They'd rather go out to buy it."

A questionnaire designed to test whether this statement is true or not should include these questions:

✓ *Do you have milk delivered to your doorstep?*
✓ *Do you buy cheaper milk from the supermarket?*
✓ *Would you buy your milk only from the supermarket?*

Once the data from these questions has been collected, it can be looked at to see whether or not the majority of people hold views that agree with the statement.

EXERCISE 12D

FM **1** These are questions from a questionnaire on healthy eating.

a *Fast food is bad for you. Don't you agree?*

☐ Strongly agree ☐ Agree ☐ Don't know

Give two criticisms of the question.

b *Do you eat fast food?* ☐ Yes ☐ No

If 'Yes', how many times on average per week do you eat fast food?

☐ Once or less ☐ 2 or 3 times ☐ 4 or 5 times ☐ More than 5 times

Give two reasons why this is a good question.

2 This is a question from a survey on pocket money:

How much pocket money do you get each week?

☐ £0–£2 ☐ £0–£5 ☐ £5–£10 ☐ £10 or more

a Give a reason why this is not a good question.

AU **b** Rewrite the question to make it a good question.

3 Design a questionnaire to test the following statement.

People under sixteen do not know what is meant by all the jargon used in the business news on TV, but the over-twenties do.

HINTS AND TIPS

Keep questions simple with clear response categories and no overlapping.

4 *The under-twenties feel quite at ease with computers, while the over-forties would rather not bother with them. The twenty-to-forties always try to look good with computers.*

Design a questionnaire to test this statement.

5 Design a questionnaire to test the following hypothesis.

The older you get, the less sleep you need.

AU **6** Carolina and André are doing a survey on the type of music people buy.

a This is one question from Carolina's survey.

> Folk music is just for country people.
>
> Don't you agree?
>
> Strongly agree ☐ Agree ☐ Don't know ☐

Give two criticisms of Carolina's question.

D

C

C

b This is a question from André's survey.

How many albums do you download each month?

2 or less ☐ 3 or 4 ☐ more than 4 ☐

Give two reasons why this is a good question.

c Make up another good question, with responses, that could be added to this survey.

PS **7** Design a questionnaire to test the hypothesis:

"People with back problems do not sit properly."

AU **8** For a survey, an assistant gave every customer leaving a store a questionnaire that included the following question.

Question: How much do you normally spend in this shop?

Response:	Less than £15 ☐	More than £25 ☐
	Less than £25 ☐	More than £50 ☐

Explain why the response section of this questionnaire is poor.

12.4 The data-handling cycle

This section will show you how:
- use the data-handling cycle to test a hypothesis

Key words
bias
hypothesis
population
primary data
sample
secondary data

The data-handling cycle

Testing out a **hypothesis** involves a cycle of planning, collecting data, evaluating the significance of the data and then interpreting the results, which may or may not show the hypothesis to be true. This cycle often leads to a refinement of the problem, which starts the cycle all over again.

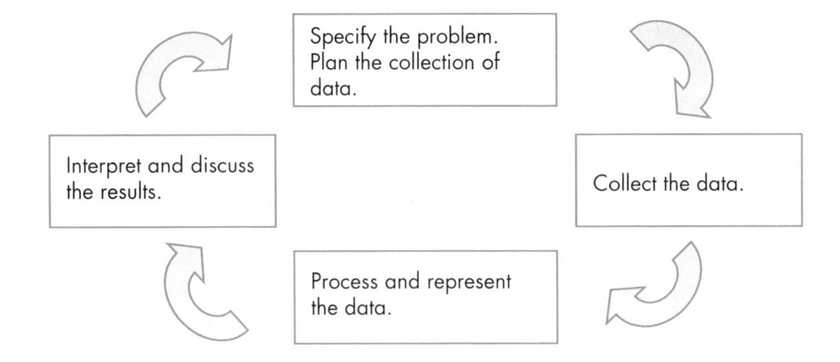

There are four parts to the data-handling cycle.

1. State the hypothesis, which is the idea being tested, outlining the problem and planning what needs to be done.
2. Plan the data collection and collect the data. Record the data collected clearly.
3. Choose the best way to process and represent the data. This will normally mean calculating averages (mean, median, mode) and measures of spread, then representing data in suitable diagrams.
4. Interpret the data and make conclusions.

Then the hypothesis can be refined or changes made to the data collected, for example a different type of data can be collected or the same data can be collected in a different way. In this way, the data-handling cycle helps to improve reliability in the collection and interpretation of data.

EXAMPLE 5

A gardener grows tomatoes, both in a greenhouse and outside.

He wants to investigate the following hypothesis:

> "Tomato plants grown inside the greenhouse produce more tomatoes than those grown outside."

Describe the data-handling cycle that may be applied to this problem.

Plan the data collection. Consider 10 tomato plants grown in the greenhouse, and 10 plants grown outside. Count the tomatoes on each plant.

Collect the data. Record the numbers of tomatoes collected from the plants between June and September. Only count those that are 'fit for purpose'.

Choose the best way to process and represent the data. Calculate the mean number collected per plant, as well as the range.

Interpret the data and make conclusions. Look at the statistics. What do they show? Is there a clear conclusion or do you need to alter the hypothesis in any way? Discuss the results, refine the method and continue the cycle.

As you see, in describing the data-handling cycle, you must refer to each of the four parts.

Data Collection

Data that you collect yourself is called **primary data**. You control it, in terms of accuracy and amount.

Data collected by someone else is called **secondary data**. Generally, there is a lot of this type of data available on the internet or in newspapers. This provides a huge volume of data but you have to rely on the sources being reliable, for accuracy.

EXERCISE 12E

Use the data-handling cycle to describe how you would test each of the following hypotheses. In each case state whether you would use primary or secondary data.

C

1. August is the hottest month of the year.

2. Boys are better than girls at estimating distances.

3. More men go to football matches than women.

4. Tennis is watched by more women than men.

5. The more revision you do, the better your exam results.

6. The older you are, the more likely you are to shop at a department store.

12.5 Other uses of statistics

This section will show you how:

- statistics are used in everyday life and what information the government needs about the population

Key words
margin of error
national census
polls
retail price index
social statistics
time series

This section will explain about **social statistics** and introduce some of the more common ones in daily use.

In daily life, many situations occur in which statistical techniques are used to produce data. The results of surveys appear in newspapers every day. There are many on-line **polls** and phone-ins that give people the chance to vote, such as in reality TV shows.

Results for these are usually given as a percentage with a **margin of error**, which is a measure of how accurate the information is.

Some common social statistics in daily use are briefly described below.

General index of retail prices

This is also know as the **retail price index** (RPI) and it measures how much the daily cost of living increases (or decreases). One year is chosen as the base year and given an index number, usually 100. The corresponding costs in subsequent years are compared to this and given a number proportional to the base year, such as 103.

Note: The numbers do not represent actual values but just compare current prices to those in the base year.

Time series

Like the RPI, a **time series** measures changes in a quantity over time. Unlike the RPI, though, the actual values of the quantity are used. A time series might track, for example, how the exchange rate between the pound and the dollar changes over time.

National census

A **national census** is a survey of all people and households in a country. Data about categories such as age, gender, religion and employment status is collected to enable governments to plan where to allocate future resources. In Britain a national census is taken every 10 years. The most recent census was in 2001.

D

EXERCISE 12F

1 In 2004 the cost of a litre of petrol was 78p. Using 2004 as a base year, the price index of petrol for each of the next five years is shown in this table.

Year	2004	2005	2006	2007	2008	2009
Index	100	103	108	109	112	120
Price	78p					

Work out the price of petrol in each subsequent year.

Give your answers to 1 decimal place.

2 The following is taken from the UK government statistics website.

In mid-2004 the UK was home to 59.8 million people, of which 50.1 million lived in England. The average age was 38.6 years, an increase on 1971 when it was 34.1 years. In mid-2004 approximately one in five people in the UK were aged under 16 and one in six people were aged 65 or over.

Use this extract to answer the following questions about the UK in 2004.

a How many of the population of the UK *did not* live in England?

b By how much had the average age increased since 1971?

c Approximately how many of the population were aged under 16?

d Approximately how many of the population were aged over 65?

3 The graph shows the exchange rate for the dollar against the pound for each month in one year.

a What was the exchange rate in January?

b Between which two consecutive months did the exchange rate fall most?

c Explain why you could not use the graph to predict the exchange rate in the January following this year.

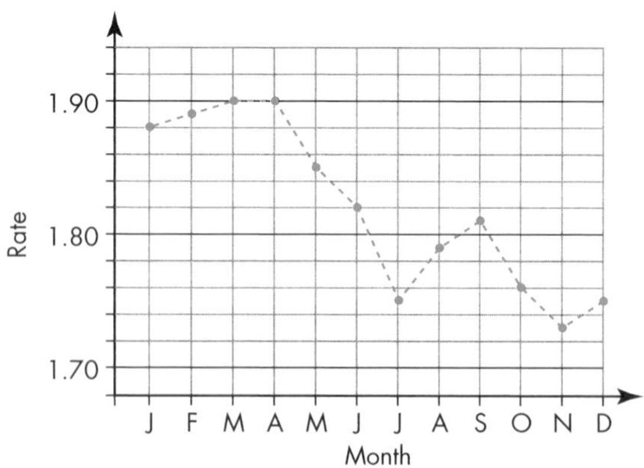

4 The general index of retail prices started in January 1987, when it was given a base number of 100. In January 2006 the index number was 194.1.

If the 'standard weekly shopping basket' cost £38.50 in January 1987, how much would it have cost in January 2006?

FM 5 The time series shows car production in Britain from November 2008 to November 2009.

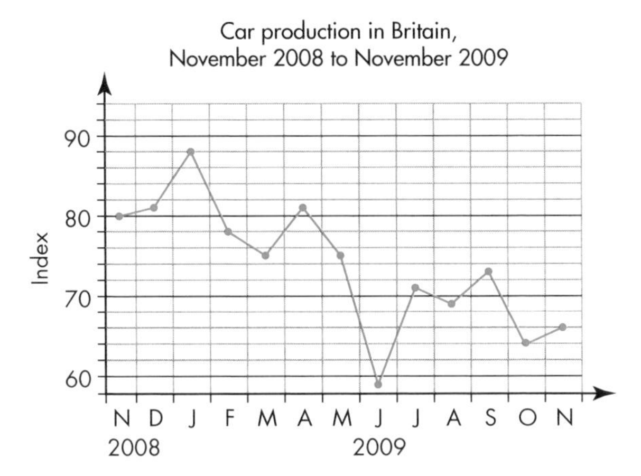

a Why was there a sharp drop in production in June 2009?

b The average production over the first three months shown was 172 000 cars.

i Work out an approximate value for the average production over the last three months shown.

ii The base month for the index is January 2005 when the index was 100. What was the approximate production in January 2005?

AU 6 The retail price index measures how much the daily cost of living increases or decreases. If 2008 is given a base index number of 100, then 2009 is given 98. What does this mean?

PS 7 On one day in 2009 Alex spent £51 in a supermarket.

She knew that the price index over the last three years was:

2007	2008	2009
100	103	102

How much would she have paid in the supermarket for the same goods in 2008?

C

GRADE BOOSTER

- **F** You can interpret a simple pie chart
- **E** You can draw a pie chart
- **E** You can recognise the different types of correlation
- **D** You can design a data-collection sheet
- **C** You can use a line of best fit and use it to make predictions
- **C** You can interpret a scatter diagram
- **C** You can design and criticise questions for questionnaires
- **C** You can describe the data-handling cycle

What you should know now

- How to read and draw pie charts
- How to plot scatter diagrams, recognise correlation, draw lines of best fit and use them to predict values
- How to design questionnaires and know how to ask suitable questions

1 The table gives information about the medals won by Austria in the 2002 Winter Olympic Games.

Draw an accurate pie chart to show this information.

Medal	Frequency
Gold	3
Silver	4
Bronze	11

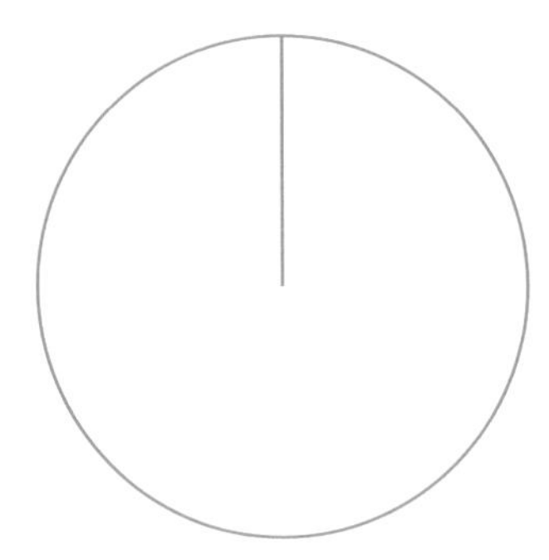

(Total 3 marks)

Edexcel, June 2005, Paper 4 Intermediate, Question 5

2 The table below shows how a number of men and women on a cruise ship rated their understanding of the rules of croquet.

	Totally understand	Understand	Understand some	Understand a little	Do not understand at all	Total
Men	160	520	560	320	40	1600
Women	160	240	200	80	40	720

The pie chart for men has been drawn for you.

a Copy and complete the pie chart for women.

b Which group, men or women, do you think had a better overall understanding of the rules of croquet? Give *one* reason to justify your answer.

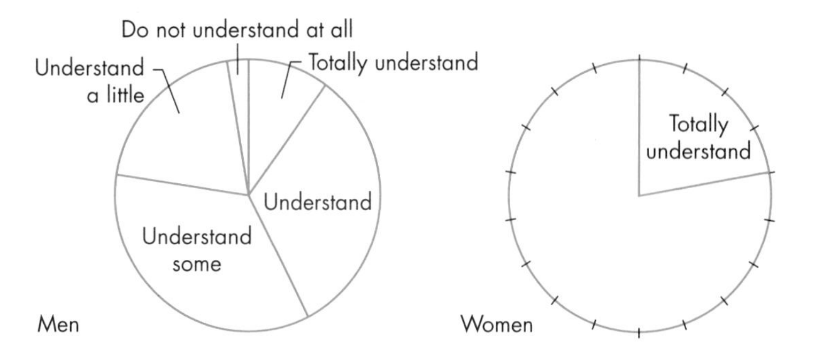

3 Some male students were asked to choose their favourite sport.

The pie chart shows information about the results.

The pie chart is drawn accurately.

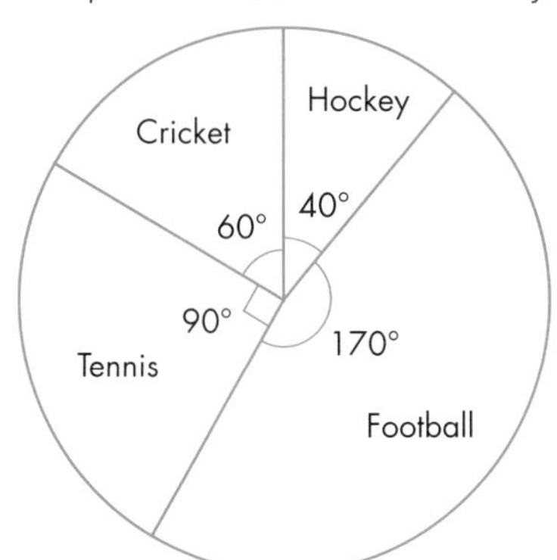

a 12 male students chose hockey.

Work out the number of male students who chose tennis. (3)

b A second pie is to be drawn for some female students.

There are 240 female students.

130 of the female students chose hockey.

Calculate the angle in the second pie chart for 130 female students. (2)

(Total 5 marks)

Edexcel, June 2009, IGCSE, Paper 2 Foundation, Question 13

4 The pie chart gives information about the mathematics exam grades of some students.

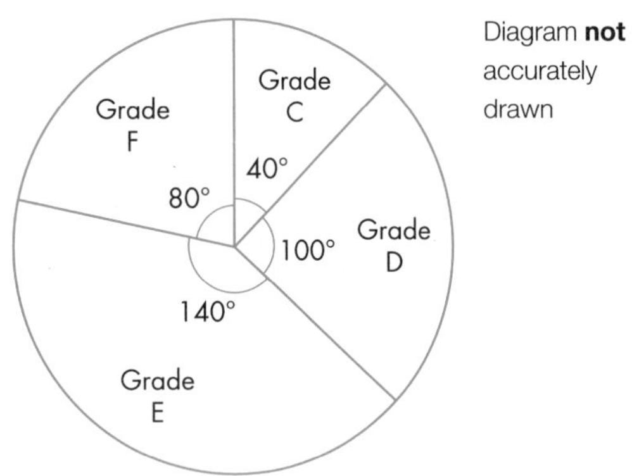

a What grade was the mode? (1)

b What fraction of the students got grade D? (1)

8 of the students got grade C.

c **i** How many of the students got grade F?

ii How many of the students took the exam? (3)

This accurate pie chart gives information about the English exam grades for a different set of students.

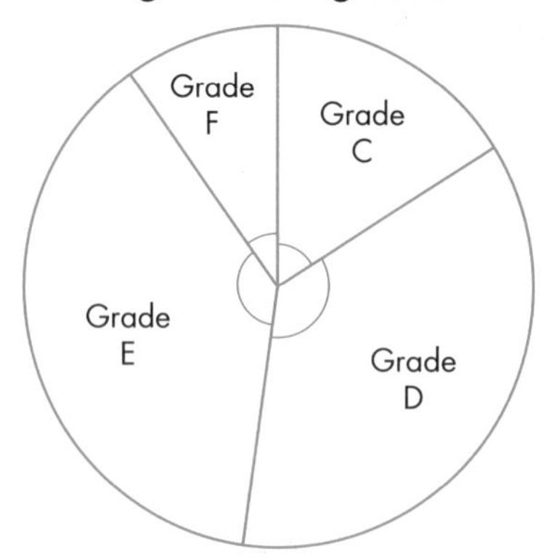

Sean says "More students got a grade D in English than in mathematics."

d Sean could be **wrong**.

Explain why. (1)

(Total 6 marks)

Edexcel, June 2008, Paper 2 Foundation, Question 16

5 The scatter diagrams below show the results of a survey on the average number of hours of sunshine in a week during the summer weeks in Bournemouth.

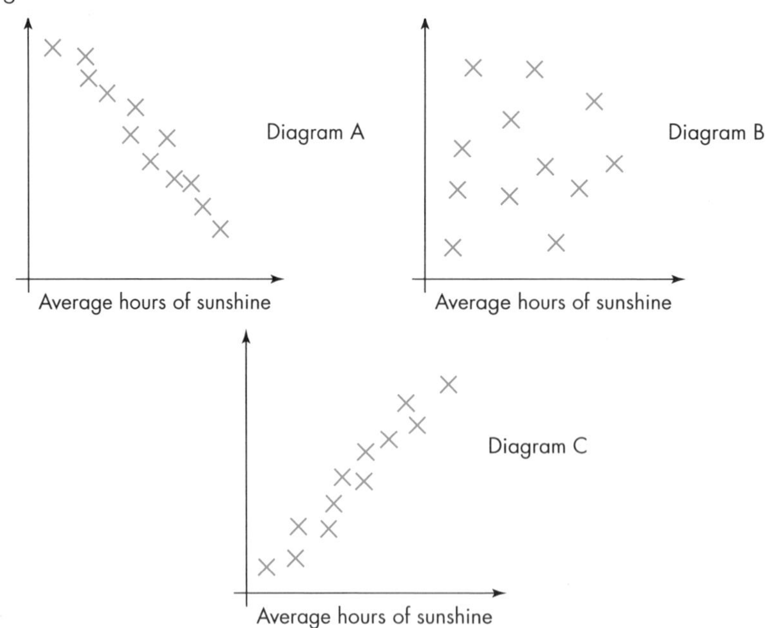

a Which scatter diagram shows the average hours of sunshine plotted against:

i the number of ice creams sold?

ii the number of umbrellas sold?

iii the number of births in the town?

b State which one of the diagrams shows a negative correlation.

6 Joy wants to find out who eats vegetarian food. She decides to investigate this hypothesis:

> Girls are more likely than boys to eat vegetarian food.

a Design a two-way table that Joy might use to help her do this.

b Joy records information from a sample of 40 boys and 30 girls. She finds that 18 boys and 16 girls eat vegetarian food. Based on this sample, is the hypothesis correct? Explain your answer.

7 Naomi wants to find out how often adults go to the cinema.

She uses this question on a questionnaire.

> "How many times do you go to the cinema?"
>
> Not very often ☐
>
> Sometimes ☐
>
> A lot ☐

a Write down **two** things wrong with this question. (2)

b Design a better question for her questionnaire to find out how often adults go to the cinema.

You should include some response boxes. (2)

(Total 4 marks)

Edexcel, November 2008, Paper 1 Foundation, Question 24

8 Mr Brown owns a café in the town centre.

He wants to find out what people think of the service in the café.

He uses this question on his questionnaire.

> What do you think of the service in the café?
>
> Excellent ☐
>
> Very good ☐
>
> Good ☐

a Write down one thing that is wrong with this question. (1)

Mr Brown wants to find out how often people visit the town centre.

b Design a suitable question for his questionnaire to find out how often people visit the town centre.

You must include some response boxes. (2)

(Total 3 marks)

Edexcel, March 2007, Paper 8 Foundation, Unit 2 Test, Question 4

9 James wants to find out how many text messages people send.

He uses this question on a questionnaire.

> "How many text messages do you send?"
>
> 1 to 10 ☐
>
> 11 to 20 ☐
>
> 21 to 30 ☐
>
> more than 30 ☐

a Write down **two** things wrong with this question. (2)

James asks 10 students in his class to complete his questionnaire.

b Give **one** reason why this may not be a suitable sample. (1)

(Total 3 marks)

Edexcel, March 2009, Paper 5 Foundation, Unit 1 Test, Question 4

10 The scatter graph shows some information about six newborn baby apes. For each baby ape, it shows the mother's leg length and the baby ape's birth weight.

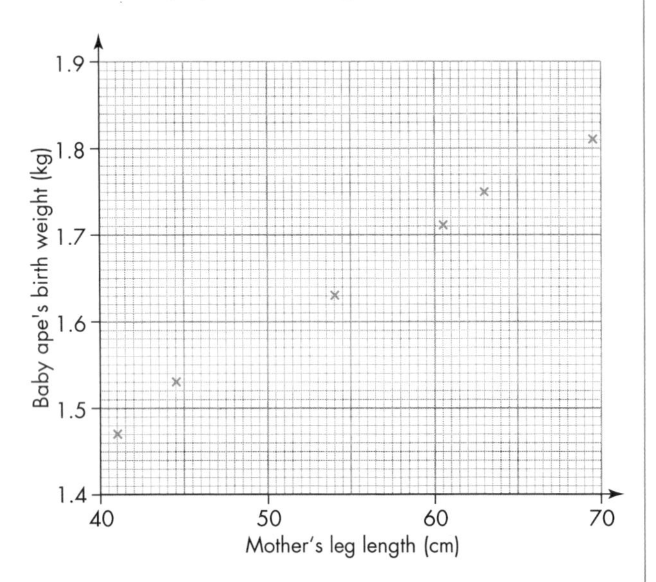

The table shows the mother's leg length and the birth weight of two more baby apes.

Mother's leg length (cm)	50	65
Baby ape's birth weight (kg)	1.6	1.75

a Copy the graph onto graph paper and plot the information from the table. (1)

b Describe the correlation between a mother's leg length and her baby ape's birth weight. (1)

c Draw a line of best fit on your graph. (1)

A mother's leg length is 55 cm.

d Use your line of best fit to estimate the birth weight of her baby ape. (1)

(Total 4 marks)

Edexcel, June 2005, Paper 16 Intermediate, Question 5

11 The table shows the time taken and distance travelled by a taxi driver for 10 journeys one day.

Time (min)	Distance (km)
3	1.7
17	8.3
11	5.1
13	6.7
9	4.7
15	7.3
8	3.8
11	5.7
16	8.7
10	5.3

a Plot on a grid, a scatter diagram with time, on the horizontal axis, from 0 to 20, and distance, on the vertical axis, from 0 to 10.

b Draw a line of best fit on your diagram.

c A taxi journey takes 4 minutes. How many kilometres is the journey?

d A taxi journey is 10 kilometres. How many minutes will it take?

12 Sanji goes fishing for pike.

The scatter graph shows information about the weights and lengths of some of the pike Sanji caught.

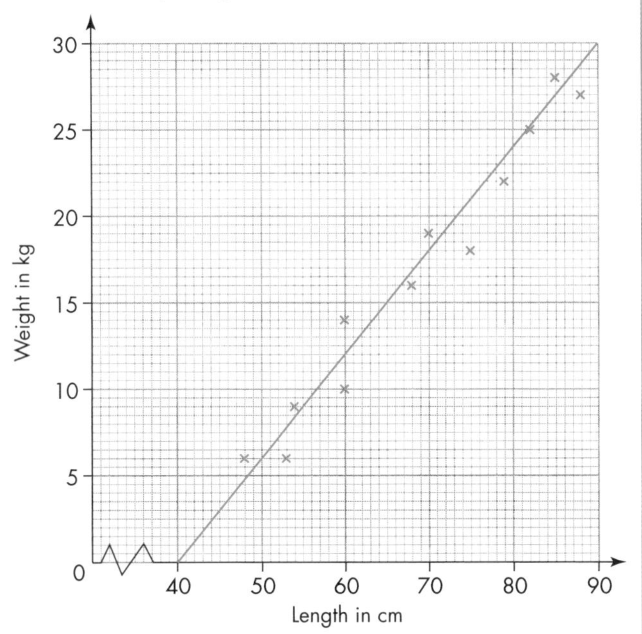

a Describe the relationship between the weight and the length of these pike. (1)

Sanji also caught a pike of weight 24 kg and length 78 cm.

b Show this information in the scatter graph. (1)

A pike has a length of 65 cm.

c Estimate the weight of this pike. (2)

(Total 4 marks)

Edexcel, June 2009, Paper 5 Foundation, Unit 1 Test, Question 4

13 The scatter graph shows information about eight sheep.

It shows the height and the length of each sheep.

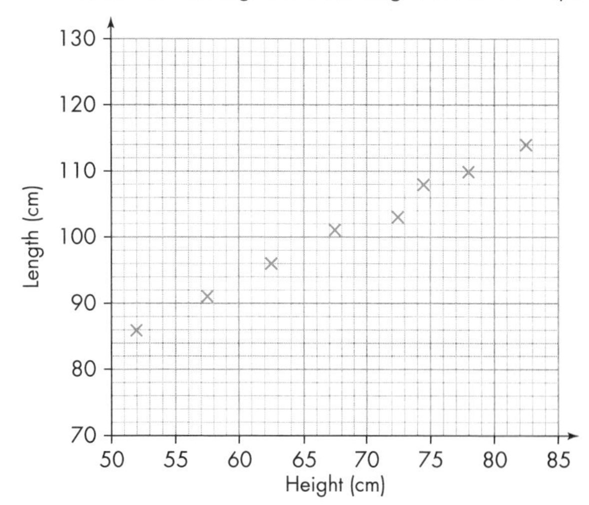

The table gives the height and the length of two more sheep.

Height (cm)	65	80
Length (cm)	100	110

a On the scatter graph, plot the information from the table. (1)

b Describe the relationship between the height and the length of these sheep. (1)

The height of a sheep is 76 cm.

c Estimate the length of this sheep. (2)

(Total 4 marks)

Edexcel, June 2009, Paper 2 Foundation, Question 21

Worked Examination Questions

1 The scatter diagram shows the relationship between the total mileage of a car and its value as a percentage of its original value.

Percentage of original value
100
A
D
B
C
0
0
Total mileage (thousands)
100

a Which of the four points, A, B, C or D, represents each of the statements below?

Alf: I have a rare car. It has done a lot of miles but it is still worth a lot of money.

Belinda: My car is quite new. It hasn't done many miles.

Charles: My car hasn't done many miles but it is really old and rusty.

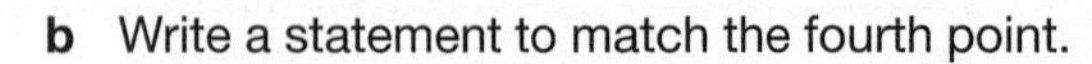

b Write a statement to match the fourth point.

c What does the graph tell you about the relationship between the mileage of a car and the percentage of its original value?

d Draw scatter diagrams to show the relationship between:

i the amount of petrol used and the distance driven

ii the value of a car and the age of the driver.

a *Alf is represented by point D.*
Belinda is represented by point A.
Charles is represented by point B.

3 marks

Read both axes. The horizontal axis is total mileage and the vertical axis is percentage of original value. So, Alf would be to the right of the horizontal axis and to the top of the vertical axis. Use similar reasoning for the other people. You get 1 mark for each correct person linked to a point.

b *'My car has done a lot of miles and isn't worth very much.'*

1 mark

The fourth point, C, is a car that has high mileage with low value.
Any statement that says something like this answer given would get the 1 mark available.

c *The more mileage a car has done, the less is its value.*

1 mark

The graph basically shows weak negative correlation, so as one variable increases, the other decreases.
Using this principle to give the correct answer earns you 1 mark.

d **i**

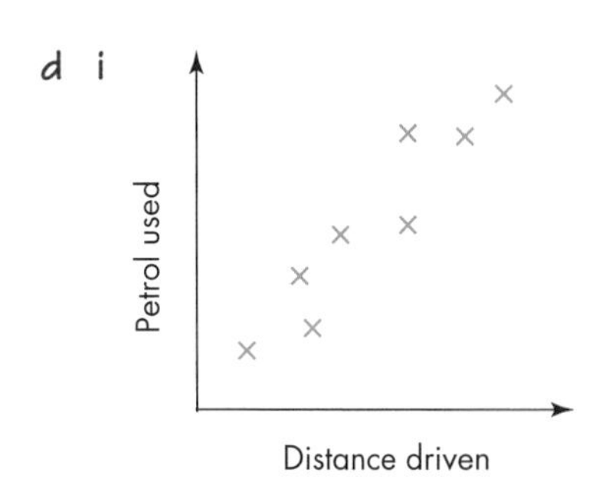

ii

Age of driver
Value of car

2 marks

The first diagram shows positive correlation, as the more miles are driven, the more petrol is used.
The second diagram shows no correlation as there is no relationship.
You earn 1 method mark for each correct diagram.

Total: 7 marks

Worked Examination Questions

AU **2** The table below shows the number of learners at each grade for two practice driving tests, Theory and Practical.

	Grades					
	Excellent	**Very good**	**Good**	**Pass**	**Fail**	**Total number of learners**
Theory	208	888	1032	696	56	2880
Practical	240	351	291	108	90	1080

a Represent each of the two practice tests in a pie chart.

b By comparing the pie charts, on which test, Theory or Practical, do you think learners did better overall? Give reasons to justify your answer.

a

	Theory		Practical	
Grade	Frequency	Angle	Frequency	Angle
Excellent	208	$360° \times 208 \div 2880 = 26°$	240	$360° \times 240 \div 1080 = 80°$
Very good	888	$360° \times 888 \div 2880 = 111°$	351	$360° \times 351 \div 1080 = 117°$
Good	1032	$360° \times 1032 \div 2880 = 129°$	291	$360° \times 291 \div 1080 = 97°$
Pass	696	$360° \times 696 \div 2880 = 87°$	108	$360° \times 108 \div 1080 = 36°$
Fail	56	$360° \times 56 \div 2880 = 7°$	90	$360° \times 90 \div 1080 = 30°$

You will earn 1 mark for showing in at least five places the correct process of 360° × frequency ÷ total frequency.

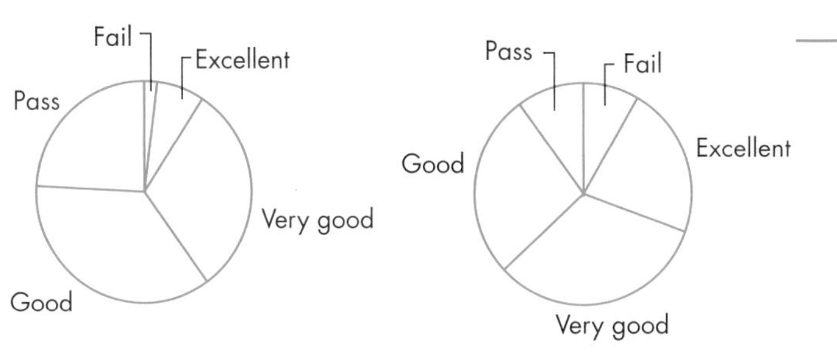

3 accuracy marks are available for the correct angles calculated. You will lose a mark for each incorrect one, to a minimum of 0.
Also available are: 1 method mark for drawing two separate pie charts.
1 method mark for correctly labelling both charts.
1 accuracy mark for each chart being correctly drawn.
Note that each angle can be up to 2° out for these marks to be given.

7 marks

b A greater proportion of learners got "Good" or better on the practical, but learners did better in the theory test as a much smaller proportion failed.

You earn 1 mark for stating one comparison correctly.

So, I would say the learners did better overall in the practical test.

You earn 1 mark for giving either of the tests as better overall but only with a justification as here.

2 marks

Total: 9 marks

Worked Examination Questions

PS **3**
- The older you are, the higher you score on the Speed test.
- The higher the score on the Speed test, the less TV you watch.
- The more TV you watch, the more hours you will sleep.

Suppose the above were all true. Sketch a scatter diagram to illustrate the relationship between age and hours slept.

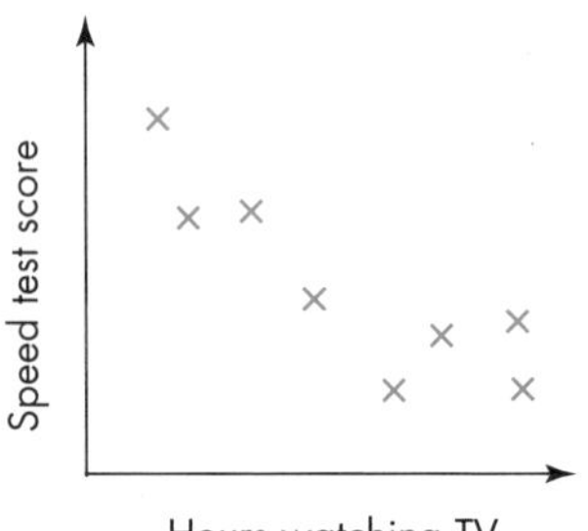

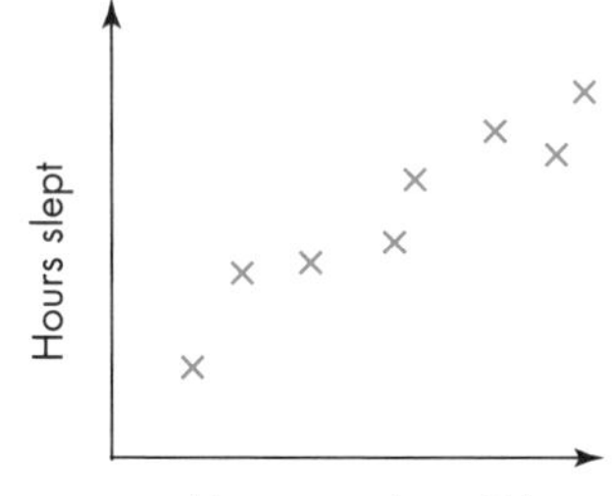

Sketch a scatter diagram for each bullet point given in the question. These three diagrams will help to show the relationships in the question.
You get 1 mark for showing some pictures similar to these.

Older person scores high on Speed test.
High score on Speed test means the person watches TV for a short time.

Imagine the extremes of age starting with the first diagram.
You get 2 independent marks for showing the link for extremes like this.

Watching TV for a short time gives short sleep time.
Hence "High age relates to small hours sleep".
Similarly, "Low age relates to long sleep".

You then get 1 mark for stating for each extreme.

So, the scatter diagram will show negative correlation and can be sketched as:

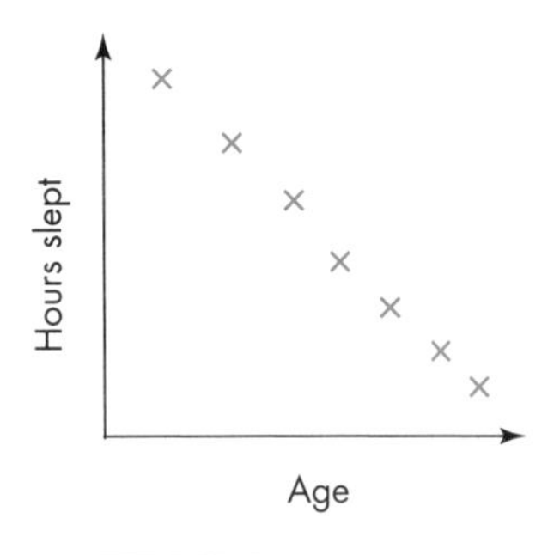

You get 1 mark for stating negative correlation, or suitable words.
You get 1 mark for sketching a diagram showing negative correlation.

Total: 7 marks

Worked Examination Question

AU **4** Godwin is asked to do a survey to investigate the hypothesis:

"As boys get older they lose interest in trains."

Explain the difficulties in trying to create a questionnaire to test out this hypothesis.

4 The difficulties are:

Measuring the level of boys' interest

How can you compare one person's interest with that of another?

> Measuring the level of interest will earn the first mark here.
> You get 1 mark for independent workings.

If you just ask the question to see what people's thoughts are on this then, again, you have not really tested the hypothesis other than asking peoples thoughts.

> You get 1 mark for independent workings will be available for another suitable comment about difficulties as seen.

Total: 2 marks

12 Functional Maths
Riding stables

Riding stables must carefully manage the welfare of their horses, including housing them in stables of the correct size and feeding them the correct amount of food for their weight and intended workload. Mr Owen owns a riding stable and wishes to understand his horses' needs, his customers and how he might go about gathering the right information to help him successfully expand his business.

Getting started

In pairs, consider the following points to help you in planning Mr Owen's expansion of his riding stables.

- There are several different elements to consider when running a riding stable: buying or breeding horses, looking after the horses (including feeding and cleaning), running and training the horses and running classes for different types of students.
- How would you find out if there is correlation between girth, weight and height? How could Mr Owen use this information to predict the body weight of horses of different girths? How useful would this information be to Mr Owen?
- How could you graphically represent the record that Mr Owen kept last summer of the different abilities of his customers? How could he use this information to help in planning his expansion?
- If Mr Owen's customers were asked what they wanted most from the stables, how could you collect, represent and interpret this information?

Your task

Mr Owen needs to work out how much feed to give his horses. He owns six horses, as detailed below. Working in pairs or individually, use the information on the six horses, the Bodyweight calculator and the Feed chart to find the best way to calculate the amount of feed each horse needs.

Feed chart

Body weight of horse (kg)	Weight of feed (kg) at different levels of work	
	Medium	Hard
300	2.4	3.0
350	2.8	3.5
400	3.2	4.0
450	3.6	4.5
500	4.0	5.0
Extra feed per 50 kg	300 g	400 g

Your task

Mr Owen also wants to understand the different abilities and numbers of riders who come to his riding stable, as he wishes to expand his business next year. He keeps a record of his students and the workload they place on the horses during one week of the summer holidays (see Riders table below). How would Mr Owen use this information to make sure he buys the right type of new horses to suit his riders?

Mr Owen is also aware that different riders look for different facilities at the stables. Help him to work out how he can decide which new facilities different riders would value the most.

Help Mr Owen to plan his expansion, using surveys and diagrams.

Bodyweight calculator

Girth (cm) Weight (kg) Length (cm)

Girth (cm): 111.8, 111.8, 120, 130, 140, 150, 160, 170, 180, 190, 200, 210, 220, 230, 240, 246.4

Weight (kg): 115.7, 125, 150, 175, 200, 225, 250, 275, 300, 325, 350, 400, 450, 500, 550, 600, 650, 700, 750, 800, 850, 900, 950, 1000, 1045.6

Length (cm): 71.75, 75, 80, 85, 90, 95, 100, 105, 110, 115, 120, 125, 130, 135, 140, 145, 150, 150.5

Using the bodyweight calculator

Place a ruler from your measurement for girth line to your measurement for the length line. The point at which the ruler crosses the weight line is the reading for the approximate weight of the horse. Note that the ruler will cross the weight line at an angle.

Riders	Ability/work	Total
Children	Medium	45
Female novice	Medium	20
Female experienced	Hard	61
Male novice	Medium	15
Male experienced	Hard	29

Horse:	Summer	Sally	Skip	Simon	Barney	Teddy
Girth:	220 cm	190 cm	200 cm	180 cm	160 cm	190 cm
Length:	142 cm	95 cm	114 cm	124 cm	110 cm	140 cm
Height: hands and inches	17 h 2 in	14 h 3 in	16 h 0 in	15 h 3 in	15 h 1 in	16 h 2 in
Work:	Medium	Hard	Hard	Medium	Medium	Hard

Why this chapter matters

Chance is a part of everyday life. Judgements are frequently made based on probability. For example,

- **there is an 80% chance that United will win the game tomorrow**
- **there is a 40% chance of rain tomorrow**
- **she has a 50–50 chance of having a baby girl**
- **there is a 10% chance of the bus being on time tonight.**

Certain probabilities are given to certain events, although two people might give different probabilities to those same events because of their different views. For example, one person might not agree that there is an 80% chance of United winning the game. They might well say that there is only a 70% chance of United winning tomorrow. A lot depends on what that person believes or has experienced.

Chance may be taken into account for lots of different events, from the lottery to tomorrow's weather. Probability is a branch of mathematics that describes the chance of outcomes.

Probability originated from the study of games of chance, such as tossing a dice or spinning a roulette wheel.

Mathematicians in the 16th and 17th centuries started to think about the mathematics of chance in games. Probability theory, as a branch of mathematics, developed in the 17th century when French gamblers asked mathematicians Blaise Pascal and Pierre de Fermat for help in their gambling.

Now, in the 21st century, probability theory is used to control the flow of traffic through road systems (above) or the running of telephone exchanges (right), and to look at patterns of the spread of infections.

There are many other everyday applications and as you work through this chapter you will see how frequently the language of probability is used.

Chapter

13 Probability: Probability of events

The grades given in this chapter are target grades.

1 Probability scale

2 Calculating probabilities

3 Probability that an outcome of an event will not happen

4 Addition rule for events

5 Experimental probability

6 Combined events

7 Expectation

8 Two-way tables

This chapter will show you ...

G how to use the the language of probability

F to **C** how to work out the probability of outcomes of events, using either theoretical models or experimental models

C how to predict outcomes, using theoretical models, and compare experimental and theoretical data

Visual overview

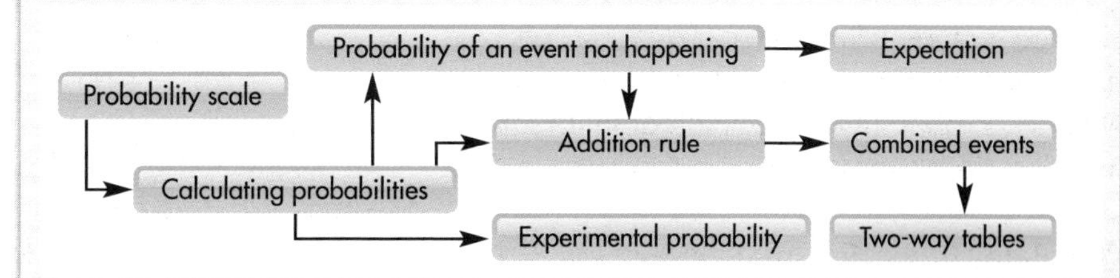

What you should already know

- How to add, subtract and cancel fractions **(KS3 level 4, GCSE grade F)**
- That outcomes of events cannot always be predicted and that the laws of chance apply to everyday events **(KS3 level 4, GCSE grade F)**
- How to list all the outcomes of an event in a systematic manner **(KS3 level 5, GCSE grade E)**

Quick check

1 Cancel the following fractions.

a $\frac{6}{8}$ **b** $\frac{6}{36}$ **c** $\frac{3}{12}$ **d** $\frac{8}{10}$ **e** $\frac{6}{9}$ **f** $\frac{5}{20}$

2 Calculate the following.

a $\frac{1}{8}+\frac{3}{8}$ **b** $\frac{5}{12}+\frac{3}{12}$ **c** $\frac{5}{36}+\frac{3}{36}$ **d** $\frac{2}{9}+\frac{1}{6}$ **e** $\frac{3}{5}+\frac{3}{20}$

3 Frank likes to wear brightly coloured hats and socks.

He has two hats, one is green and the other is yellow.
He has three pairs of socks, which are red, purple and pink.

Write down all the six possible combinations of hats and socks Frank could wear.

For example, he could wear a green hat and red socks.

13.1 Probability scale

This section will show you how to:

- use the probability scale and the basic language of probability

Key words
certain
chance
event
impossible
likely
outcome
probability
probability scale
unlikely

Almost daily, you hear somebody talking about the probability of whether this or that will happen. They usually use words such as '**chance**', 'likelihood' or 'risk' rather than 'probability'. For example:

"What is the likelihood of rain tomorrow?"
"What chance does she have of winning the 100 metre sprint"
"Is there a risk that his company will go bankrupt?"

You can give a value to the chance of any of these **outcomes** or **events** happening – and millions of others, as well. This value is called the **probability**.

It is true that some things are certain to happen and that some things cannot happen; that is, the chance of something happening can be anywhere between **impossible** and **certain**. This situation is represented on a sliding scale called the **probability scale**, as shown below.

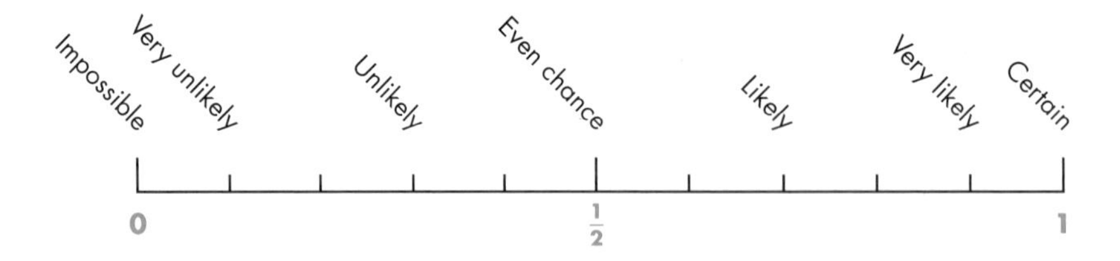

Note: All probabilities lie somewhere in the range of **0** to **1**.

An outcome or an event that cannot happen (is impossible) has a probability of 0.
For example, the probability that pigs will fly is 0.

An outcome or an event that is certain to happen has a probability of 1.
For example, the probability that the sun will rise tomorrow is 1.

EXAMPLE 1

Put arrows on the probability scale to show the probability of each of the outcomes of these events.

- **a** You will get a head when throwing a coin.
- **b** You will get a six when throwing a dice.
- **c** You will have maths homework this week.

a This outcome is an even chance. (Commonly described as a fifty-fifty chance.)

b This outcome is fairly **unlikely**.

c This outcome is **likely**.

The arrows show the approximate probabilities on the probability scale.

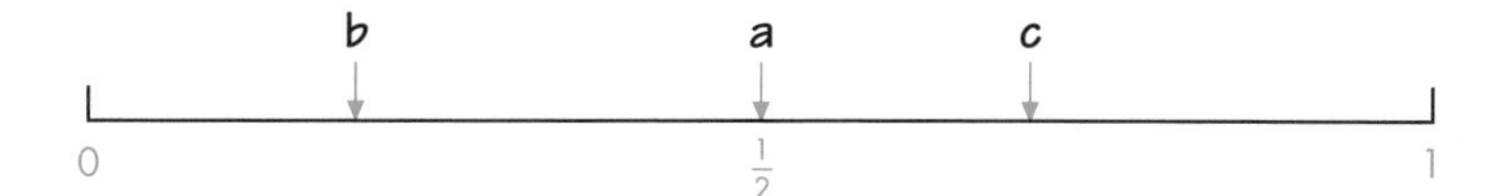

EXERCISE 13A

1 State whether each of the following events is impossible, very unlikely, unlikely, even chance, likely, very likely or certain.

- **a** Picking out a Heart from a well-shuffled pack of cards.
- **b** Christmas Day being on the 25th December.
- **c** Someone in your class is left-handed.
- **d** You will live to be 100.
- **e** A score of seven is obtained when throwing a dice.
- **f** You will watch some TV this evening.
- **g** A new-born baby will be a girl.

2 Draw a probability scale and put an arrow to show the approximate probability of each of the following events happening.

- **a** The next car you see will have been made in Europe.
- **b** A person in your class will have been born in the 20th century.
- **c** It will rain tomorrow.
- **d** In the next Olympic Games, someone will run the 1500 m race in 3 minutes.
- **e** During this week, you will have chips with a meal.

G

G

F

3 Draw a probability scale and mark an arrow to show the approximate probability of each event.

a The next person to come into the room will be male.

b The person sitting next to you in mathematics is over 16 years old.

c Someone in the class will have mobile phone.

4 Give two events of your own for which you think the probability of an outcome is as follows

a impossible
b very unlikely
c unlikely
d evens
e likely
f very likely
g certain.

5 In August, Janine and her family are going on holiday to Corsica, a French island off the south coast of France. She wonders about the chance of various events happening.

For each of the following, state whether the answer is likely to be:

impossible, very unlikely, unlikely, even chance, likely, very likely, certain.

a It being sunny most days.

b The aeroplane taking off smoothly.

c Her ears popping during the landing (they usually do).

d Hearing most people speak English on the plane.

e Hearing someone speak French in the resort they are going to.

f It raining in France when they arrive.

g Being able to wear her bikini while on holiday.

h The sea being warm when she swims.

i Her becoming the President of France.

j Her meeting her future husband while there.

k Her getting sunburnt if she doesn't put sun cream on but lies in the sun all day.

l Seeing the cost of meals in pounds in restaurants.

AU 6 "I have bought five lottery tickets this week, so I have a good chance of winning."

Explain what is wrong with this statement.

13.2 Calculating probabilities

This section will show you how to:

- calculate the probability of outcomes of events

Key words
equally likely
event
outcome
probability fraction
random

In question 2 of Exercise 13A, you may have had difficulty in knowing exactly where to put some of the arrows on the probability scale. It would have been easier for you if each result of the **event** could have been given a value, from 0 to 1, to represent the probability for that result.

For some events, this can be done by first finding all the possible results, or **outcomes**, for a particular event. For example, when you throw a coin there are two **equally likely** outcomes: heads or tails. If you want to calculate the probability of getting a head, there is only one outcome that is possible. So, you can say that there is a 1 in 2, or 1 out of 2, chance of getting a head. This is usually given as a **probability fraction**, namely $\frac{1}{2}$. So, you would write the event as:

$$P(\text{head}) = \frac{1}{2}$$

Probabilities can also be written as decimals or percentages, so that:

$$P(\text{head}) = \frac{1}{2} \text{ or } 0.5 \text{ or } 50\%$$

It is more usual to give probabilities as fractions in GCSE examinations but you will frequently come across probabilities given as percentages, for example, in the weather forecasts on TV.

The probability of an outcome is defined as:

$$P(\text{event}) = \frac{\text{number of ways the outcome can happen}}{\text{total number of possible outcomes}}$$

This definition always leads to a fraction, which should be cancelled to its simplest form.

Another probability term you will meet is at **random**. This means that the outcome cannot be predicted or affected by anyone.

EXAMPLE 2

A bag contains five red balls and three blue balls. A ball is taken out at random.

What is the probability that it is:

a red **b** blue **c** green?

Use the formula

$$P(\text{event}) = \frac{\text{number of ways the event can happen}}{\text{total number of possible outcomes}}$$

to work out these probabilities.

a There are five red balls out of a total of eight, so $P(\text{red}) = \frac{5}{8}$

b There are three blue balls out of a total of eight, so $P(\text{blue}) = \frac{3}{8}$

c There are no green balls, so this event is impossible: $P(\text{green}) = 0$

EXAMPLE 3

The spinner shown here is spun and the score on the side on which it lands is recorded.

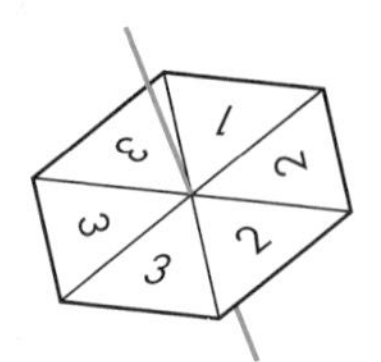

What is the probability that the score is:

a 2
b odd
c less than 5?

a There are two 2s out of six sides, so $P(2) = \frac{2}{6} = \frac{1}{3}$

b There are four odd numbers, so $P(\text{odd}) = \frac{4}{6} = \frac{2}{3}$

c All of the numbers are less than 5, so this is a certain event.

$P(\text{less than } 5) = 1$

EXAMPLE 4

Bernice is always early, just on time or late for work.

The probability that she is early is 0.1, the probability she is just on time is 0.5.

What is the probability that she is late?

As all the possibilities are covered – that is 'early', 'on time' and 'late' – the total probability is 1. So,

$P(\text{early}) + P(\text{on time}) = 0.1 + 0.5 = 0.6$

So, the probability of Bernice being late is $1 - 0.6 = 0.4$.

EXERCISE 13B

1 What is the probability of each of the following events?

- **a** Throwing a 2 with a fair, six-sided dice.
- **b** Throwing a 6 with a fair, six-sided dice.
- **c** Tossing a fair coin and getting a tail.
- **d** Drawing a Queen from a pack of cards.
- **e** Drawing a Heart from a pack of cards.
- **f** Drawing a black card from a pack of cards.
- **g** Throwing a 2 or a 6 with a fair, six-sided dice.
- **h** Drawing a black Queen from a pack of cards.
- **i** Drawing an Ace from a pack of cards.
- **j** Throwing a 7 with a fair, six-sided dice.

HINTS AND TIPS

If an event is impossible, just write the probability as 0, not as a fraction such as $\frac{0}{6}$. If it is certain, write the probability as 1, not as a fraction such as $\frac{6}{6}$.

HINTS AND TIPS

Remember to cancel the fractions if possible.

2 What is the probability of each of the following events?

- **a** Throwing an even number with a fair, six-sided dice.
- **b** Throwing a prime number with a fair, six-sided dice.
- **c** Getting a Heart or a Club from a pack of cards.
- **d** Drawing the King of Hearts from a pack of cards.
- **e** Drawing a picture card or an Ace from a pack of cards.
- **f** Drawing the seven of Diamonds from a pack of cards.

3 A bag contains only blue balls. If I take one out at random, what is the probability of each of these outcomes?

- **a** I get a black ball.
- **b** I get a blue ball.

4 Number cards with the numbers 1 to 10 inclusive are placed in a hat. Bob takes a number card out of the bag without looking. What is the probability that he draws:

- **a** the number 7
- **b** an even number
- **c** a number greater than 6
- **d** a number less than 3
- **e** a number between 3 and 8?

5 A bag contains one blue ball, one pink ball and one black ball. Craig takes a ball from the bag without looking. What is the probability that he takes out:

- **a** the blue ball
- **b** the pink ball
- **c** a ball that is not black?

HINTS AND TIPS

A ball that is not black must be pink or blue.

F

6 A pencil case contains six red pens and five blue pens. Geoff takes out a pen without looking at what it is. What is the probability that he takes out:

a a red pen **b** a blue pen **c** a pen that is not blue?

7 A bag contains 50 balls. 10 are green, 15 are red and the rest are white. Gemma takes a ball from the bag at random. What is the probability that she takes:

a a green ball **b** a white ball

c a ball that is not white **d** a ball that is green or white?

8 A box contains seven bags of cheese and onion crisps, two bags of beef crisps and six bags of plain crisps. Iklil takes out a bag of crisps at random. What is the probability that he gets:

a a bag of cheese and onion crisps **b** a bag of beef crisps

c a bag of crisps that are not cheese and onion **d** a bag of prawn cracker crisps

e a bag of crisps that is either plain or beef?

9 In a Christmas raffle, 2500 tickets are sold. One family has 50 tickets. What is the probability that that family wins the first prize?

E

10 Ashley, Bianca, Charles, Debbie and Eliza are in the same class. Their teacher wants two students to do a special job.

a Write down all the possible combinations of two people, for example, Ashley and Bianca, Ashley and Charles. (There are 10 combinations altogether).

HINTS AND TIPS

Try to be systematic when writing out all the pairs.

b How many pairs give two boys?

c What is the probability of choosing two boys?

d How many pairs give a boy and a girl?

e What is the probability of choosing a boy and a girl?

f What is the probability of choosing two girls?

11 In a sale at the supermarket, there is a box of 10 unlabelled tins. On the side it says: 4 tins of Creamed Rice and 6 tins of Chicken Soup. Mitesh buys this box. When he gets home he wants to have a lunch of chicken soup followed by creamed rice.

a What is the smallest number of tins he could open to get his lunch?

b What is the largest number of tins he could open to get his lunch?

c The first tin he opens is soup. What is the chance that the second tin he opens is:

i soup

ii rice?

E

12 What is the probability of each of the following events?

- **a** Drawing a Jack from a pack of cards.
- **b** Drawing a 10 from a pack of cards.
- **c** Drawing a red card from a pack of cards.
- **d** Drawing a 10 or a Jack from a pack of cards.
- **e** Drawing a Jack or a red card from a pack of cards.
- **f** Drawing a red Jack from a pack of cards.

13 A bag contains 25 coloured balls. 12 are red, 7 are blue and the rest are green. Martin takes a ball at random from the bag.

- **a** Find:
  - **i** P(he takes a red)
  - **ii** P(he takes a blue)
  - **iii** P(he takes a green).
- **b** Add together the three probabilities. What do you notice?
- **c** Explain your answer to part **b**.

14 The weather tomorrow will be sunny, cloudy or raining.

If P(sunny) = 40%, P(cloudy) = 25%, what is P(raining)?

15 At morning break, Priya has a choice of coffee, tea or hot chocolate.

If P(she chooses coffee) = 0.3 and P(she chooses hot chocolate) = 0.2, what is P(she chooses tea)?

PS 16 The following information is known about the classes at Bradway School.

Year	Y1		Y2		Y3		Y4		Y5		Y6	
Class	**P**	**Q**	**R**	**S**	**T**	**U**	**W**	**X**	**Y**	**Z**	**K**	**L**
Girls	7	8	8	10	10	10	9	11	8	12	14	15
Boys	9	10	9	10	12	13	11	12	10	8	16	17

A class representative is chosen at random from each class.

Which class has the best chance of choosing a boy as the representative?

AU 17 The teacher chooses, at random, a student to ring the school bell.

Tom says: "It's even chances that the teacher chooses a boy or a girl."

Explain why Tom might not be correct.

13.3 Probability that an outcome of an event will not happen

This section will show you how to:
- calculate the probability of an outcome of an event not happening when you know the probability of the outcome happening

Key word
outcome

In some questions in Exercise 13B, you were asked for the probability of something *not* happening. For example, in question 5 you were asked for the probability of picking a ball that is *not* black. You could answer this because you knew how many balls were in the bag. However, sometimes you do not have this type of information.

The probability of throwing a six on a fair, six-sided dice is $P(6) = \frac{1}{6}$.

There are five **outcomes** that are not sixes: 1, 2, 3, 4, 5.

So, the probability of *not* throwing a six on a dice is:

$$P(\text{not a } 6) = \frac{5}{6}$$

Notice that:

$$P(6) = \frac{1}{6} \quad \text{and} \quad P(\text{not a } 6) = \frac{5}{6}$$

So,

$$P(6) + P(\text{not a } 6) = 1$$

If you know that $P(6) = \frac{1}{6}$, then P(not a 6) is:

$$1 - \frac{1}{6} = \frac{5}{6}$$

So, if you know P(outcome happening), then:

P(outcome not happening) = 1 – P(outcome happening)

EXAMPLE 5

What is the probability of not picking an Ace from a pack of cards?

First, find the probability of picking an Ace:

P(picking an ace from a pack of cards) $= \frac{4}{52} = \frac{1}{13}$

Therefore:

P(not picking an ace from a pack of cards) $= 1 - \frac{1}{13} = \frac{12}{13}$

EXERCISE 13C

1 **a** The probability of winning a prize in a raffle is $\frac{1}{20}$. What is the probability of not winning a prize in the raffle?

b The probability that snow will fall during the Christmas holidays is 45%. What is the probability that it will not snow?

c The probability that Paddy wins a game of chess is 0.7 and the probability that he draws the game is 0.1. What is the probability that he loses the game?

2 Mary picks a card from a pack of well-shuffled playing cards.

Find the probability that she picks:

a **i** a picture card **ii** a card that is not a picture

b **i** a Club **ii** not a Club

c **i** an Ace or a King **ii** neither an Ace nor a King.

3 The following letter cards are put into a bag.

a Steve takes a letter card at random.

i What is the probability he takes a letter A?

ii What is the probability he does not take a letter A?

b Richard picks an M and keeps it. Sue now takes a letter from those remaining.

i What is the probability she takes a letter A?

ii What is the probability she does not take a letter A?

FM 4 The starting point in a board game is:

Start	Owens Park	Take a chance	Curry Mile	Oxford Road	Pay £500 in tax	Salford Quays	Exchange Quays	Station	Old Trafford	Rest area	The Lowry	Trafford Park

You roll a single dice and move, from the start, the number of places shown by the dice.

What is the probability of *not* landing on:

a a brown square

b the station

c a coloured square?

E

PS 5 Freddie and Taryn are playing a board game. On the next turn Freddie will go to jail if he rolls an even number. Taryn will go to jail if she rolls a 5 or a 6.

Who has the better chance of *not* going to jail on the next turn?

AU 6 Hamzah is told: "The chance of your winning this game is 0.3."

Hamzah says: "So I have a chance of 0.7 of losing."

Explain why Hamzah might be wrong.

13.4 Addition rule for events

This section will show you how to:

- work out the probability of two outcomes of events such as P(A) or P(B)

Key words
event
mutually exclusive

You have used this rule already but it has not yet been formally defined.

Mutually exclusive events are ones that cannot happen at the same time, such as throwing an odd number and an even number on a roll of a dice.

When two **events** are mutually exclusive, you can work out the probability of either of them occurring by adding up the separate probabilities. Mutually exclusive events are events for which, when one occurs, it does not have any effect on the probability of other events.

EXAMPLE 6

A bag contains 12 red balls, 8 green balls, 5 blue balls and 15 black balls. A ball is drawn at random. What is the probability that it is the following:

a red **b** black **c** red or black **d** not green?

a $P(\text{red}) = \frac{12}{40} = \frac{3}{10}$

b $P(\text{black}) = \frac{15}{40} = \frac{3}{8}$

c $P(\text{red or black}) = P(\text{red}) + P(\text{black}) = \frac{3}{10} + \frac{3}{8} = \frac{27}{40}$

d $P(\text{not green}) = \frac{32}{40} = \frac{4}{5}$

EXERCISE 13D

D

1 Iqbal throws an ordinary dice. What is the probability that he throws:

a a 2 **b** a 5 **c** a 2 or a 5?

2 Jennifer draws a card from a pack of cards. What is the probability that she draws:

a a Heart **b** a Club **c** a Heart or a Club?

3 A letter is chosen at random from the letters in the word PROBABILITY. What is the probability that the letter will be:

a B **b** a vowel **c** B or a vowel?

4 A bag contains 10 white balls, 12 black balls and 8 red balls. A ball is drawn at random from the bag. What is the probability that it will be:

a white **b** black

c black or white **d** not red

e not red or black?

HINTS AND TIPS

You can only add fractions with the same denominator.

C

5 At the local School Fayre the tombola stall gives out a prize if you draw from the drum a numbered ticket that ends in 0 or 5. There are 300 tickets in the drum altogether and the probability of getting a winning ticket is 0.4.

a What is the probability of getting a losing ticket?

b How many winning tickets are there in the drum?

6 John needs his calculator for his mathematics lesson. It is always in his pocket, bag or locker. The probability it is in his pocket is 0.35 and the probability it is in his bag is 0.45. What is the probability that:

a he will have the calculator for the lesson

b his calculator is in his locker?

7 Aneesa has 20 unlabelled CDs, 12 of which are rock, 5 are pop and 3 are classical. She picks a CD at random. What is the probability that it will be:

a rock or pop

b pop or classical

c not pop?

AU 8 The probability that it rains on Monday is 0.5. The probability that it rains on Tuesday is 0.5 and the probability that it rains on Wednesday is 0.5. Kelly argues that it is certain to rain on Monday, Tuesday or Wednesday because 0.5 + 0.5 + 0.5 = 1.5, which is bigger than 1 so it is a certain event. Explain why she is wrong.

FM 9 In a TV game show, contestants throw darts at the dartboard shown.

The angle at the centre of each black sector is 15°.

If a dart lands in a black sector the contestant loses.

Any dart missing the board is rethrown.

What is the probability that a contestant throwing a dart at random does not lose?

PS 10 There are 45 patients sitting in the hospital waiting room.

8 patients are waiting for Dr Speed.
12 patients are waiting for Dr Mayne.
9 patients are waiting for Dr Kildare.
10 patients are waiting for Dr Pattell.
6 patients are waiting for Dr Stone.

A patient suddenly has to go home.

What is the probability that the patient who left was due to see Dr Speed?

AU 11 The probability of it snowing on any one day in February is $\frac{1}{4}$.

One year, there was no snow for the first 14 days.

Ciara said: "The chance of it snowing on any day in the rest of February must now be $\frac{1}{2}$."

Explain why Ciara is wrong.

AU PS 12 At morning break, Pauline has a choice of coffee, tea or hot chocolate.
She also has a choice of a ginger biscuit, a rich tea biscuit or a doughnut.

The probabilities that she chooses each drink and snack are:

Drink	Coffee (C)	Tea (T)	Hot chocolate (H)
Probability	0.2	0.5	0.3

Snack	Ginger biscuit (G)	Rich tea biscuit (R)	Doughnut (D)
Probability	0.3	0.1	0.6

a Leon says that the probability that Pauline has coffee and a ginger biscuit is 0.5. Explain why Leon is wrong.

b There are nine possible combinations of drink and snack. Two of these are coffee and a ginger biscuit (C, G), coffee and rich tea biscuit (C, R).

Write down the other seven combinations.

c Leon now says 'I was wrong before. As there are nine possibilities the probability of Pauline having coffee and a ginger biscuit is $\frac{1}{9}$.'

Explain why Leon is wrong again.

d In fact the probability that Pauline chooses coffee and a ginger biscuit is $0.2 \times 0.3 = 0.06$, and the probability that she chooses coffee and a rich tea biscuit is $0.2 \times 0.1 = 0.02$.

i Work out the other seven probabilities.

ii Add up all nine probabilities. Explain the result.

13.5 Experimental probability

This section will show you how to:

- calculate experimental probabilities and relative frequencies from experiments
- recognise different methods for estimating probabilities

Key words
bias
equally likely
experimental data
experimental probability
historical data
relative frequency
trials

ACTIVITY

Heads or tails?

Toss a coin 10 times and record the results like this.

H	T	H	H	T	T	H	T	H	H

Record how many heads you obtained.

Now repeat the above so that altogether you toss the coin 50 times. Record your results and count how many heads you obtained.

Now toss the coin another 50 times and once again record your results and count the heads.

It helps if you work with a partner. First, your partner records while you toss the coin. Then you swap over and record, while your partner tosses the coin. Add the number of heads you obtained to the number your partner obtained.

Now find three more people to do the same activity and add together the number of heads that all five of you obtained.

Now find five more people and add their results to the previous total.

Combine as many results together as possible.

You should now be able to fill in a table like the one on the next page. The first column is the number of times coins were tossed. The second column is the number of heads obtained. The third column is the number in the second column divided by the number in the first column.

The results below are from a group who did the same experiment.

Number of tosses	Number of heads	$\frac{\text{Number of heads}}{\text{Number of tosses}}$
10	6	0.6
50	24	0.48
100	47	0.47
200	92	0.46
500	237	0.474
1000	488	0.488
2000	960	0.48
5000	2482	0.4964

If you drew a graph of these results, plotting the first column against the last column, it would look like this.

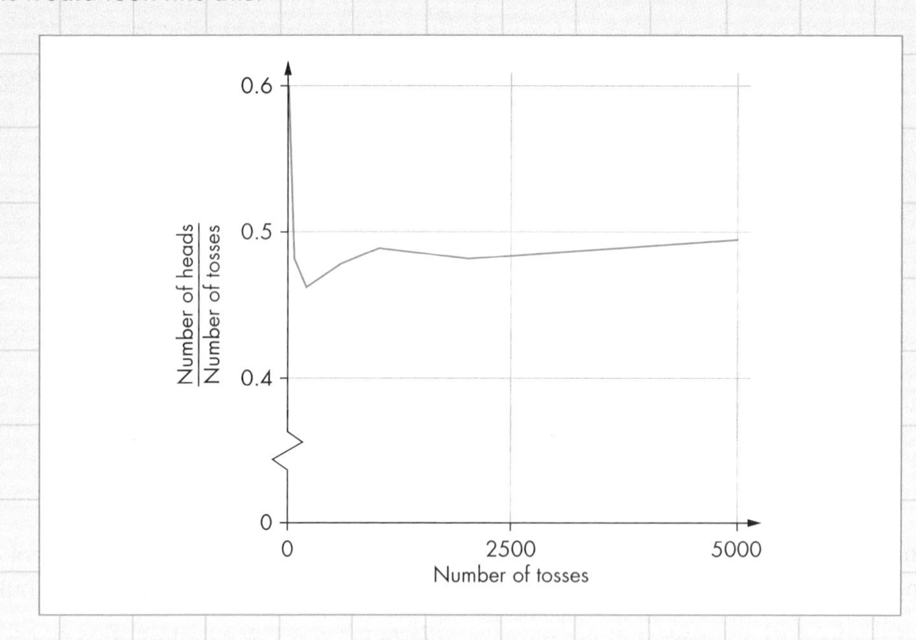

Your results should look very similar.

What happens to the value of $\frac{\text{number of heads}}{\text{number of tosses}}$ as the total number of tosses increases?

You should find that it gets closer and closer to 0.5.

The value of 'number of heads ÷ number of tosses' is called an **experimental probability**. As the number of **trials**, or experiments, increases, the value of the experimental probability gets closer to the true or theoretical probability.

Experimental probability is also known as the **relative frequency** of an event.
The relative frequency of an event is an estimate for the theoretical probability. It is given by:

$$\text{relative frequency of an outcome or event} = \frac{\text{frequency of the outcome or event}}{\text{total number of trials}}$$

EXAMPLE 7

The frequency table shows the speeds of 160 vehicles that pass a radar speed check on a dual carriageway.

Speed (mph)	20–29	30–39	40–49	50–59	60–69	70+
Frequency	14	23	28	35	52	8

a What is the experimental probability that a car is travelling faster than 70 mph?

b If 500 vehicles pass the speed check, estimate how many will be travelling faster than 70 mph.

a The experimental probability is the relative frequency, which is $\frac{8}{160} = \frac{1}{20}$.

b The number of vehicles travelling faster than 70 mph will be $\frac{1}{20}$ of 500.

That is:

$500 \div 20 = 25$

Finding probabilities

There are three ways in which the probability of an event can be found.

- **First method** If you can work out the theoretical probability of an outcome or event – for example, drawing a King from a pack of cards – this is called using **equally likely** outcomes.
- **Second method** Some probabilities, such as people buying a certain brand of dog food, cannot be calculated using equally likely outcomes. To find the probabilities for such an event, you can perform an experiment such as the one in the Activity on page 397, or conduct a survey. This is called collecting **experimental data**. The more data you collect, the better the estimate is.
- **Third method** The probabilities of some events, such as an earthquake occurring in Japan, cannot be found by either of the above methods. One of the things you can do is to look at data collected over a long period of time and make an estimate (sometimes called a 'best guess') at the chance of the event happening. This is called looking at **historical data**.

EXAMPLE 8

Which method (A, B or C) would you use to estimate the probabilities for the events **a** to **e**?

A: Use equally likely outcomes

B: Conduct a survey or collect data

C: Look at historical data

a Someone in your class will go abroad for a holiday this year.

b You will win the National Lottery.

c Your bus home will be late.

d It will snow on Christmas Day.

e You will pick a red seven from a pack of cards.

a You would have to ask all the members of your class what they intended to do for their holidays this year. You would therefore conduct a survey, Method B.

b The odds on winning are about 14 million to 1, so this is an equally likely outcome, Method A.

c If you catch the bus every day, you can collect data over several weeks. This would be Method C.

d If you check whether it snowed on Christmas Day for the last few years, you would be able to make a good estimate of the probability. This would be Method C.

e There are 2 red sevens out of 52 cards, so the probability of picking one can be calculated:

$$P(\text{red seven}) = \frac{2}{52} = \frac{1}{26}$$

This is Method A.

EXERCISE 13E

D

1 Which of these methods would you use to estimate or state the probabilities for each of the events **a** to **h**?

Method A: Use equally likely outcomes

Method B: Conduct a survey or experiment

Method C: Look at historical data

a How people will vote in the next election.

b A drawing pin dropped on a desk will land point up.

c A Premier League team will win the FA Cup.

d You will win a school raffle.

e The next car to drive down the road will be red.

f You will throw a 'double six' with two dice.

g Someone in your class likes classical music.

h A person picked at random from your school will be a vegetarian.

2 Naseer throws a fair, six-sided dice and records the number of sixes that he gets after various numbers of throws. The table shows his results.

Number of throws	10	50	100	200	500	1000	2000
Number of sixes	2	4	10	21	74	163	329

a Calculate the experimental probability of scoring a 6 at each stage that Naseer recorded his results.

b How many ways can a dice land?

c How many of these ways give a 6?

d What is the theoretical probability of throwing a 6 with a dice?

e If Naseer threw the dice a total of 6000 times, how many sixes would you expect him to get?

3 Marie made a five-sided spinner, like the one shown in the diagram. She used it to play a board game with her friend Sarah.

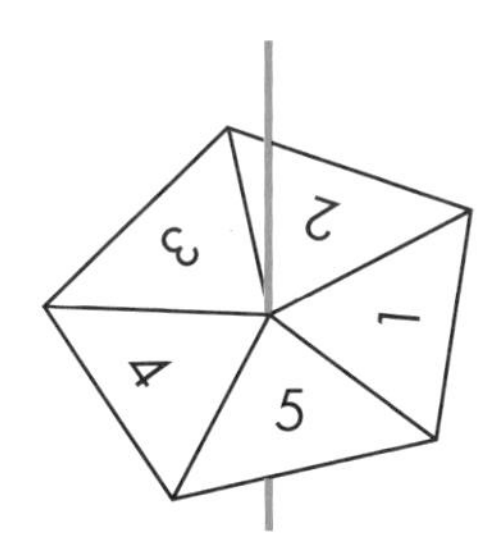

The girls thought that the spinner was not very fair as it seemed to land on some numbers more than others. They threw the spinner 200 times and recorded the results. The results are shown in the table.

Side spinner lands on	1	2	3	4	5
Number of times	19	27	32	53	69

a Work out the experimental probability of each number.

b How many times would you expect each number to occur if the spinner is fair?

c Do you think that the spinner is fair? Give a reason for your answer.

C

4 A sampling bottle contains 20 balls. The balls are either black or white. (A sampling bottle is a sealed bottle with a clear plastic tube at one end into which one of the balls can be tipped.) Kenny conducts an experiment to see how many black balls are in the bottle. He takes various numbers of samples and records how many of them showed a black ball.

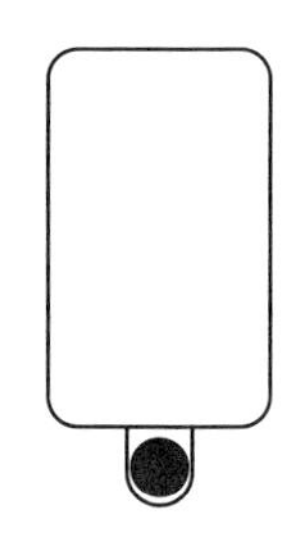

The results are shown in the table.

Number of samples	Number of black balls	Experimental probability
10	2	
100	25	
200	76	
500	210	
1000	385	
5000	1987	

a Copy the table and complete it by calculating the experimental probability of getting a black ball at each stage.

b Using this information, how many black balls do you think there are in the bottle?

5 Use a set of number cards from 1 to 10 (or make your own set) and work with a partner. Take turns to choose a card and keep a record each time of what card you get. Shuffle the cards each time and repeat the experiment 60 times. Put your results in a copy of this table.

Score	1	2	3	4	5	6	7	8	9	10
Total										

a How many times would you expect to get each number?

b Do you think you and your partner conducted this experiment fairly?

c Explain your answer to part **b**.

6 A four-sided dice has faces numbered 1, 2, 3 and 4. The 'score' is the face on which it lands. Five students throw the dice to see if it is biased. They each throw it a different number of times. Their results are shown in the table.

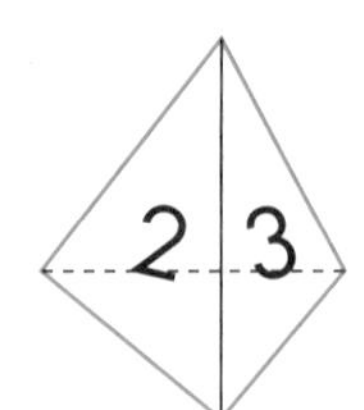

Student	Total number of throws	Score			
		1	2	3	4
Alfred	20	7	6	3	4
Brian	50	19	16	8	7
Caryl	250	102	76	42	30
Deema	80	25	25	12	18
Emma	150	61	46	26	17

a Which student will have the most reliable set of results? Why?

b Add up all the score columns and work out the relative frequency of each score. Give your answers to 2 decimal places.

c Is the dice biased? Explain your answer.

7 If you were about to choose a card from a pack of yellow cards numbered from 1 to 10, what would be the chance of each of the events **a** to **i** occurring? Read each of these statements and describe the probability with a word or phrase chosen from 'impossible', 'not likely', '50–50 chance', 'quite likely', or 'certain'.

a The next card chosen will be a 4.

b The next card chosen will be pink.

c The next card chosen will be a seven.

d The next card chosen will be a number less than 11.

e The next card chosen will be a number bigger than 11.

f The next card chosen will be an even number.

g The next card chosen will be a number more than 5.

h The next card chosen will be a multiple of 1.

i The next card chosen will be a prime number.

PS 8 Andrew made an eight-sided spinner.

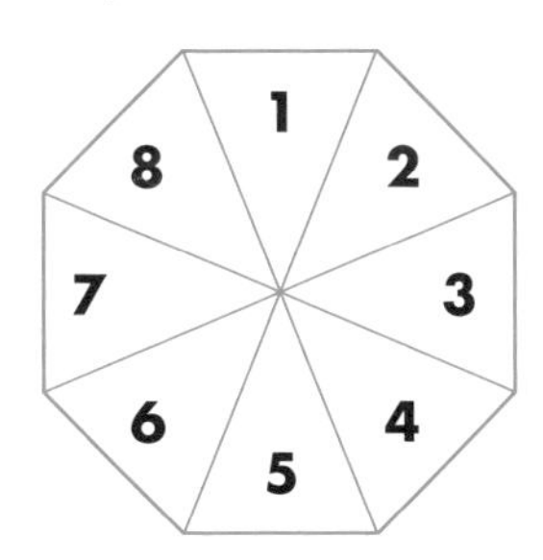

He tested it to see if it was fair.

He spun the spinner and recorded the results.

Unfortunately his little sister spilt something over his results table, so he could not see the middle part.

Number spinner lands on	1	2	3	6	7	8
Frequency	18	19	22	19	20	22

Assuming the spinner was a fair one, try to complete the missing parts of the table for Andrew.

C

FM **9** At a computer factory, tests were carried out to see how many faulty computer chips were produced in one week.

	Monday	Tuesday	Wednesday	Thursday	Friday
Sample	850	630	1055	896	450
Number faulty	10	7	12	11	4

On which day was it most likely that the highest number of faulty computer chips were produced?

AU **10** Steve tossed a coin 1000 times to see how many heads he got.

He said: "If this is a fair coin, I should get 500 heads."

Explain why he is wrong.

ACTIVITY

Biased spinner

You need a piece of stiff card, a cocktail stick and some sticky tack.

You may find that it is easier to work in pairs.

Make a copy of this hexagon on the card and push the cocktail stick through its centre to make a six-sided spinner. The size of the hexagon does not really matter, but it does need to be *accurately* drawn.

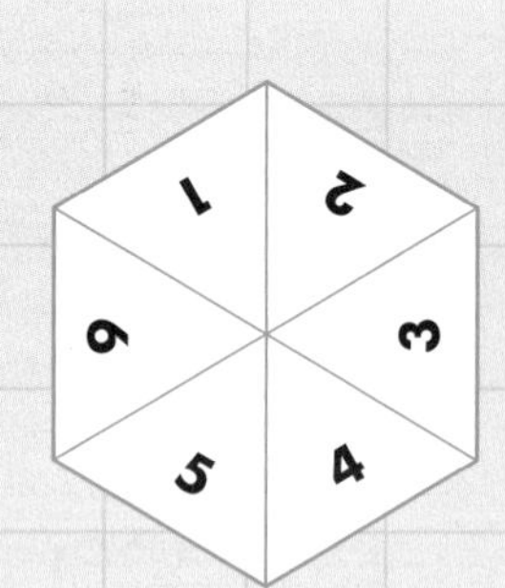

Stick a small piece of tack underneath one of the numbers. You now have a **biased** spinner.

Spin it 100 times and record your results in a frequency table.

Estimate the experimental probability of getting each number.

How can you tell that your spinner is biased?

Put some tack underneath a different number and see whether your partner can predict the number towards which the spinner is biased.

13.6 Combined events

This section will show you how to:

- work out the probabilities for two events occurring at the same time

Key words

probability space diagram
sample space diagram

There are many situations where two events occur together. Four examples are given below.

Throwing two dice

Imagine that two dice, one red and one blue, are thrown. The red dice can land with any one of six scores: 1, 2, 3, 4, 5 or 6. The blue dice can also land with any one of six scores. This gives a total of 36 possible combinations. These are shown in the left-hand diagram below, where combinations are given as (2, 3) and so on. The first number is the score on the blue dice and the second number is the score on the red dice.

The combination (2, 3) gives a total of 5. The total scores for all the combinations are shown in the diagram on the right-hand side. Diagrams that show all the outcomes of combined events are called **sample space diagrams** or **probability space diagrams**.

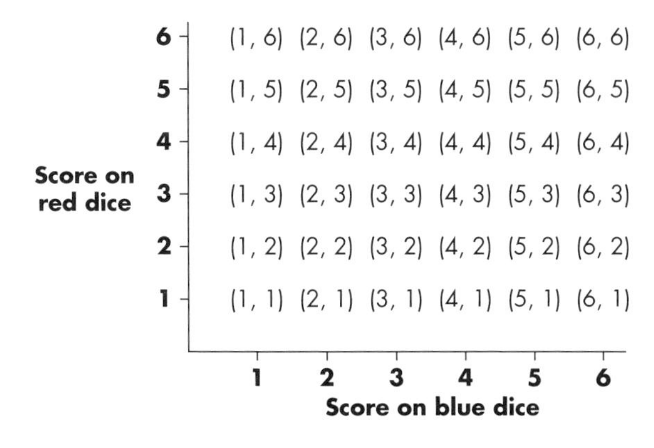

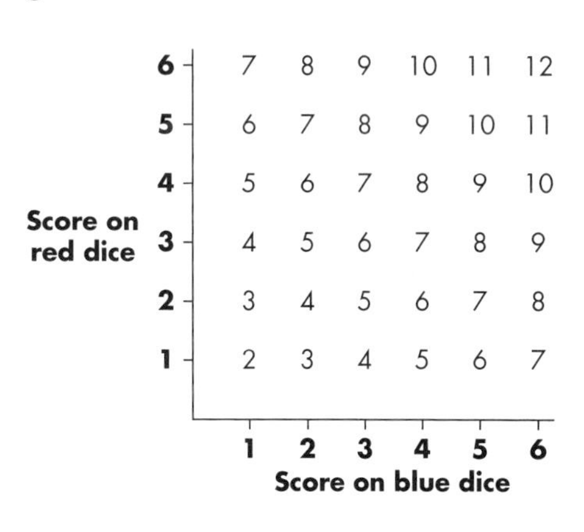

From the diagram on the right, you can see that there are two ways to get a score of 3. This gives a probability of scoring 3 as:

$$P(3) = \frac{2}{36} = \frac{1}{18}$$

From the diagram on the left, you can see that there are six ways to get a 'double'. This gives a probability of scoring a double as:

$$P(\text{double}) = \frac{6}{36} = \frac{1}{6}$$

Throwing coins

Throwing one coin

There are two equally likely outcomes, head or tail: 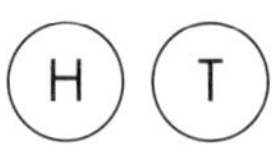

Throwing two coins together

There are four equally likely outcomes: 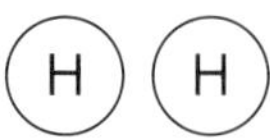 H T T H 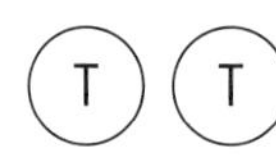

Hence:

$$P(2 \text{ heads}) = \frac{1}{4}$$

$$P(\text{head and tail}) = 2 \text{ ways out of } 4 = \frac{2}{4} = \frac{1}{2}$$

Dice and coins

Throwing a dice and a coin

Outcome on coin						
H	(1, H)	(2, H)	(3, H)	(4, H)	(5, H)	(6, H)
T	(1, T)	(2, T)	(3, T)	(4, T)	(5, T)	(6, T)
	1	2	3	4	5	6
	Score on dice					

Hence:

$$P(\text{head and an even number}) = 3 \text{ ways out of } 12 = \frac{3}{12} = \frac{1}{4}$$

EXERCISE 13F

1 To answer these questions, use the diagram on page 405 for all the possible scores when two fair dice are thrown together.

a What is the most likely score?

b Which two scores are least likely?

c Write down the probabilities of throwing all the scores from 2 to 12.

d What is the probability of a score that is:

i bigger than 10 **ii** between 3 and 7

iii even **iv** a square number

v a prime number **vi** a triangular number?

2 Use the diagram on page 405 that shows the outcomes when two fair, six-sided dice are thrown together as coordinates. What is the probability that:

a the score is an even 'double'

b at least one of the dice shows 2

c the score on one dice is twice the score on the other dice

d at least one of the dice shows a multiple of 3?

3 Use the diagram on page 405 that shows the outcomes when two fair, six-sided dice are thrown together as coordinates. What is the probability that:

a both dice show a 6

b at least one of the dice will show a 6

c exactly one dice shows a 6?

4 The diagram shows the scores for the event 'the difference between the scores when two fair, six-sided dice are thrown'. Copy and complete the diagram.

Score on second dice						
6	5	4			1	0
5	4	3				1
4	3					
3	2					
2	1					
1	0					
	1	**2**	**3**	**4**	**5**	**6**
	Score on first dice					

For the event described above, what is the probability of a difference of:

a 1 **b** 0

c 4 **d** 6

e an odd number?

5 When two fair coins are thrown together, what is the probability of:

a two heads **b** a head and a tail

c at least one tail **d** no tails?

Use the diagram of the outcomes when two coins are thrown together, on page 405.

6 Two five-sided spinners are spun together and the total score of the faces that they land on is worked out. Copy and complete the probability space diagram shown.

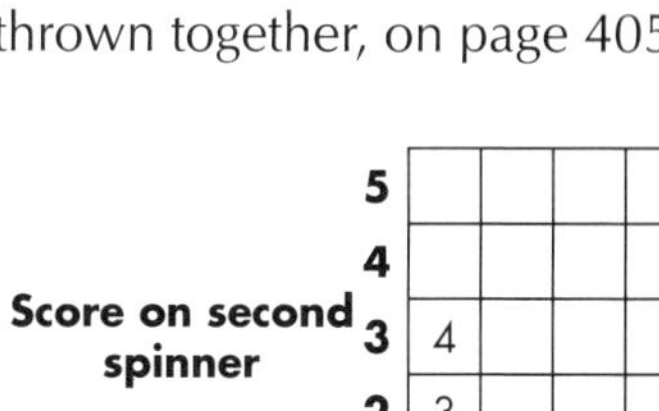

a What is the most likely score?

b When two five-sided spinners are spun together, what is the probability that:

i the total score is 5

ii the total score is an even number

iii the score is a 'double'

iv the total score is less than 7?

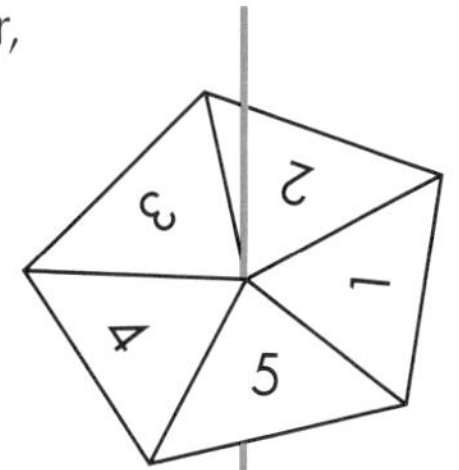

E

PS 7 Two eight-sided dice showing the numbers 1 to 8 were thrown at the same time.

What is the probability that the product of the two dice is an even square number?

AU 8 Isaac rolls two dice and multiplies both numbers to give their product. He wants to know the probability of rolling two dice that will give him a product between 19 and 35.

Explain why a probability space diagram will help him.

FM 9 Nic went to a garden centre to buy some roses.

She found they came in six different colours – white, red, orange, yellow, pink and copper.

She also found they came in five different sizes – dwarf, small, medium, large climbing and rambling.

a Draw a probability space diagram to show all the options.

b She buys a random rose for Auntie Janet. Auntie Janet only likes red and pink roses and she does not like climbing or rambling roses.

What is the probability that Nic has bought for Auntie Janet a rose:

i that she likes

ii that she does not like?

C

PS 10 Mrs Roberts asked: "What is the probability of rolling dice and getting a total of 10 or less. You must choose from impossible, very unlikely, unlikely, even chance, likely, very likely, or certain."

Evie replied: "It depends on how many dice I use."

For how many dice is each of the possible choices the answer?

13.7 Expectation

This section will show you how to:

- predict the likely number of successful events, given the number of trials and the probability of any one outcome

Key word

expect

When you know the probability of an outcome of an event, you can predict how many times you would expect that outcome to happen in a certain number of trials.

Note that this is what you **expect**. It is not necessarily what is going to happen. If what you expected always happened, life would be very dull and boring, and the National Lottery would be a waste of time!

EXAMPLE 9

A bag contains 20 balls, 9 of which are black, 6 white and 5 yellow. A ball is drawn at random from the bag, its colour is noted and then it is put back in the bag. This is repeated 500 times.

a How many times would you expect a black ball to be drawn?

b How many times would you expect a yellow ball to be drawn?

c How many times would you expect a black or a yellow ball to be drawn?

a P(black ball) $= \frac{9}{20}$

Expected number of black balls $= \frac{9}{20} \times 500 = 225$

b P(yellow ball) $= \frac{5}{20} = \frac{1}{4}$

Expected number of yellow balls $= \frac{1}{4} \times 500 = 125$

c Expected number of black or yellow balls $= 225 + 125 = 350$

EXERCISE 13G

1 **a** What is the probability of throwing a 6 with an ordinary dice?

b I throw an ordinary dice 150 times. How many times can I expect to get a score of 6?

2 **a** What is the probability of tossing a head with a coin?

b I toss a coin 2000 times. How many times can I expect to get a head?

3 **a** A card is taken at random from a pack of cards. What is the probability that it is:

i a black card **ii** a King **iii** a Heart **iv** the King of Hearts?

b I draw a card from a pack of cards and replace it. I do this 520 times. How many times would I expect to get:

i a black card **ii** a King **iii** a Heart **iv** the King of Hearts?

4 The ball in a roulette wheel can land in one of 37 spaces that are marked with numbers from 0 to 36 inclusive. I always bet on the same number, 13.

a What is the probability of the ball landing in 13?

b If I play all evening and there are exactly 185 spins of the wheel in that time, how many times could I expect to win?

C

C

5 In a bag there are 30 balls, 15 of which are red, 5 yellow, 5 green, and 5 blue. A ball is taken out at random and then replaced. This is done 300 times. How many times would I expect to get:

a a red ball

b a yellow or blue ball

c a ball that is not blue

d a pink ball?

6 The experiment described in question 5 is carried out 1000 times. Approximately how many times would you expect to get:

a a green ball

b a ball that is not blue?

7 A sampling bottle (as described in question 4 of Exercise 13E) contains red and white balls. It is known that the probability of getting a red ball is 0.3. If 1500 samples are taken, how many of them would you expect to give a white ball?

8 Josie said, "When I throw a dice, I expect to get a score of 3.5."

"Impossible," said Paul, "You can't score 3.5 with a dice."
"Do this and I'll prove it," said Josie.

a An ordinary dice is thrown 60 times. Fill in the table for the expected number of times each score will occur.

Score						
Expected occurrences						

b Now work out the average score that is expected over 60 throws.

c There is an easy way to get an answer of 3.5 for the expected average score. Can you see what it is?

FM **9** The probabilities of some cloud types being seen on any day are given below.

Cumulus	0.3
Stratocumulus	0.25
Stratus	0.15
Altocumulus	0.11
Cirrus	0.05
Cirrcocumulus	0.02
Nimbostratus	0.005
Cumulonimbus	0.004

a What is the probability of *not* seeing one of the above clouds in the sky?

b On how many days of the year would you expect to see altocumulus clouds in the sky?

C

PS 10 Every evening Anne and Chris cut a pack of cards to see who washes up.

If they cut a King or a Jack, Chris washes up.

If they cut a Queen, Anne washes up.

Otherwise, they wash up together.

In a year of 365 days, how many days would you expect them to wash up together?

AU 11 A market gardener is supplied with tomato plant seedlings and knows that the probability that any plant will develop a disease is 0.003.

How will she find out how many of the tomato plants she should expect to develop a disease?

13.8 Two-way tables

This section will show you how to:
- read two-way tables and use them to do probability and other mathematics

Key word
two-way table

A **two-way table** is a table that links together two variables. For example, the following table shows how many boys and girls there are in a form and whether they are left- or right-handed.

	Boys	Girls
Left-handed	2	4
Right-handed	10	13

This table shows the colour and make of cars in the school car park.

	Red	Blue	White
Ford	2	4	1
Vauxhall	0	1	2
Toyota	3	3	4
Peugeot	2	0	3

One variable is shown in the rows of the table and the other variable is shown in the columns of the table.

EXAMPLE 10

Use the first two-way table on the previous page to answer the following.

a How many left-handed boys are there in the form?

b How many girls are there in the form, in total?

c How many students are there in the form altogether?

d How many students altogether are right-handed?

e What is the probability that a student selected at random from the form is:

i a left-handed boy **ii** right-handed?

a 2 boys — Read this value from where the 'Boys' column and the 'Left-handed' row meet.

b 17 girls — Add up the 'Girls' column.

c 29 students — Add up all the numbers in the table.

d 23 — Add up the 'Right-handed' row.

e i P (left-handed boy) = $\frac{2}{29}$ — Use the answers to parts **a** and **c**.

ii P (right-handed) = $\frac{23}{29}$ — Use the answers to parts **c** and **d**.

EXAMPLE 11

Use the second two-way table on the previous page to answer the following.

a How many cars were in the car park altogether?

b How many red cars were in the car park?

c What percentage of the cars in the car park were red?

d How many cars in the car park were white?

e What percentage of the white cars were Vauxhalls?

a 25 — Add up all the numbers in the table.

b 7 — Add up the 'Red' column.

c 28% — 7 out of 25 is the same as 28 out of 100.

d 10 — Add up the 'White' column.

e 20% — 2 out of 10 is 20%.

EXERCISE 13H

G

1 The following table shows the top five clubs in the top division of the English Football League at the end of the season for the years 1965, 1975, 1985, 1995 and 2005.

		Year				
		1965	**1975**	**1985**	**1995**	**2005**
Position	**1st**	Man Utd	Derby	Everton	Blackburn	Chelsea
	2nd	Leeds	Liverpool	Liverpool	Man Utd	Arsenal
	3rd	Chelsea	Ipswich	Tottenham	Notts Forest	Man Utd
	4th	Everton	Everton	Man Utd	Liverpool	Everton
	5th	Notts Forest	Stoke	Southampton	Leeds	Liverpool

a Which team was in fourth place in 1975?

b Which three teams are in the top five for four of the five years?

c Which team finished three places lower in 1995 than in 1965?

E

2 Here is a display of 10 cards.

a Complete the two-way table.

		Shaded	Unshaded
Shape	**Circles**		
	Triangles		

b One of the cards is picked at random. What is the probability it shows either a shaded triangle or an unshaded circle?

3 The two-way table shows the number of doors and the number of windows in each room in a primary school.

		Number of doors		
		1	**2**	**3**
Number of windows	**1**	5	4	2
	2	4	5	4
	3	0	4	6
	4	1	3	2

a How many rooms are there in the school altogether?

b How many rooms have two doors?

c What percentage of the rooms in the school have two doors?

d What percentage of the rooms that have one door also have two windows?

e How many rooms have the same number of windows as doors?

E

4 Three cards are lettered A, B and C. Three discs are numbered 4, 5 and 6.

A B C (4) (5) (6)

One card and one disc are chosen at random.

If the card shows A, 1 is deducted from the score on the disc.
If the card shows B, the score on the disc stays the same.
If the card shows C, 1 is added to the score on the disc.

a Copy and complete the table to show all the possible scores.

		Number on disc		
		4	**5**	**6**
Letter on card	**A**	3		
	B	4		
	C	5		

b What is the probability of getting a score that is an even number?

c In a different game the probability of getting a total that is even is $\frac{2}{3}$.
What is the probability of getting a total that is an odd number?

D

5 The two-way table shows the ages and sexes of a sample of 50 students in a school.

		Age (years)					
		11	**12**	**13**	**14**	**15**	**16**
Sex	**Boys**	4	3	6	2	5	4
	Girls	2	5	3	6	4	6

a How many students are aged 13 years or less?

b What percentage of the students in the table are 16?

c A student from the table is selected at random. What is the probability that the student will be 14 years of age? Give your answer as a fraction in its lowest form.

d There are 1000 students in the school. Use the table to estimate how many boys are in the school altogether.

6 The two-way table shows the numbers of adults and the numbers of cars in 50 houses in one street.

		Number of adults			
		1	**2**	**3**	**4**
Number of cars	**0**	2	1	0	0
	1	3	13	3	1
	2	0	10	6	4
	3	0	1	4	2

a How many houses have exactly two adults and two cars?

b How many houses altogether have three cars?

c What percentage of the houses have three cars?

d What percentage of the houses with just one car have three adults living in the house?

7 Jane has two four-sided spinners.
Spinner A has the numbers 1 to 4 on it and
Spinner B has the numbers 5 to 8 on it.

Both spinners are spun together.

The two-way table shows all the ways the two spinners can land.

Some of the total scores are filled in.

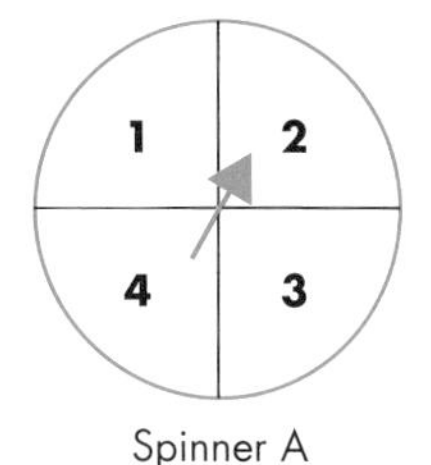

Spinner A

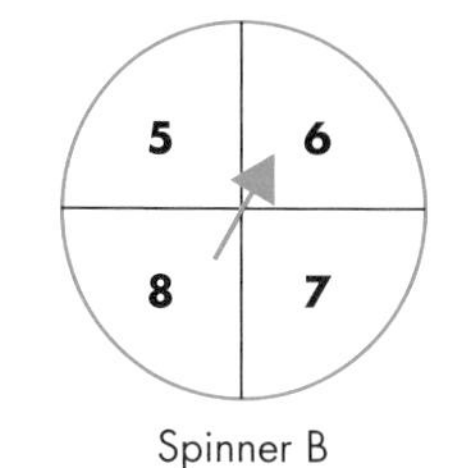

Spinner B

		Score on Spinner A			
		1	2	3	4
Score on Spinner B	5	6	7		
	6	7			
	7				
	8				

a Copy and complete the table to show all the possible total scores.

b How many of the total scores are 9?

c When the two spinners are spun together, what is the probability that the total score will be:

i 9

ii 8

iii a prime number?

8 The table shows information about the number of items in Zara's wardrobe.

		Type of item		
		Shoes (pairs)	Trousers	T-shirts
Colour	Blue	6	5	2
	Black	12	9	13
	Red	2	2	0

a How many pairs of blue shoes does Zara have?

b How many pairs of trousers does Zara have?

c How many black items does Zara have?

d If a T-shirt is chosen at random from all the T-shirts, what is the probability that it will be a black T-shirt?

D

9 Zoe throws a fair coin and rolls a fair dice.

If the coin shows a head, she records the score on the dice.
If the coin shows tails, she doubles the number on the dice.

a Copy and complete the two-way table to show Zoe's possible scores.

		Number on dice					
		1	2	3	4	5	6
Coin	Head	1	2				
	Tail	2	4				

b How many of the scores are square numbers?

c What is the probability of getting a score that is a square number?

C

AU 10 A gardener plants some sunflower seeds in a greenhouse and some in the garden. After they have fully grown, he measures the diameter of the sunflower heads. The table shows his results.

		Greenhouse	Garden
Diameter	Mean diameter	16.8 cm	14.5 cm
	Range of diameter	3.2 cm	1.8 cm

a The gardener, who wants to enter competitions, says, "The sunflowers from the greenhouse are better."

Using the data in the table, give a reason to justify this statement.

b The gardener's wife, who does flower arranging, says, "The sunflowers from the garden are better."

Using the data in the table, give a reason to justify this statement.

AU 11 Reyki plants some tomato plants in her greenhouse, while her husband Daniel plants some in the garden.

After the summer they compared their tomatoes.

		Garden	Greenhouse
Diameter	Mean diameter	1.8 cm	4.2 cm
	Mean number of tomatoes per plant	24.2	13.3

Use the data in the table to explain who had the better crop of tomatoes.

PS 12 Two hexagonal spinners are spun.

Spinner A is numbered 3, 5, 7, 9, 11 and 13.
Spinner B is numbered 4, 5, 6, 7, 8 and 9.

What is the probability that when the two spinners are spun, the result of multiplying two numbers together will give a product greater than 40?

13 Here are two fair spinners.

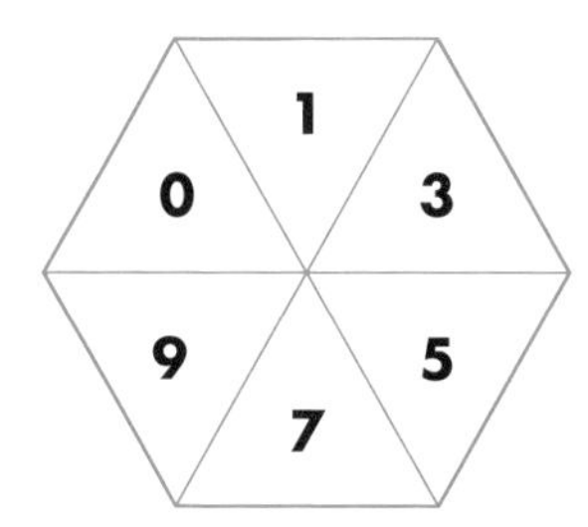

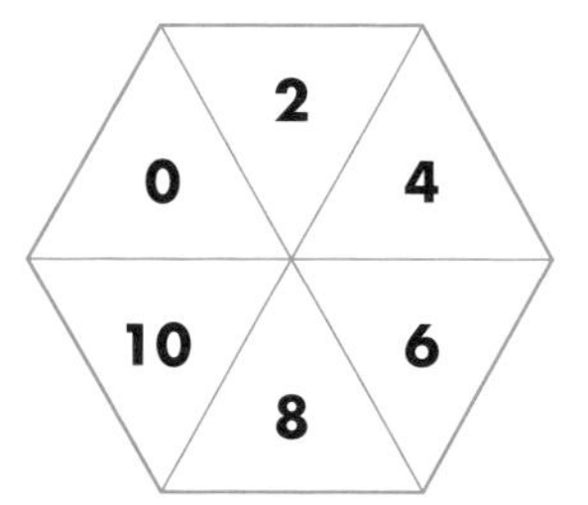

The spinners are spun.

The two numbers obtained are added together.

a Draw a probability sample space diagram.

b What is the most likely score?

c What is the probability of getting a total of 12?

d What is the probability of getting a total of 11 or more?

e What is the probability of getting a total that is an odd number?

GRADE BOOSTER

- **G** You can understand basic terms such as 'certain', 'impossible', 'likely'
- **F** You can understand that the probability scale runs from 0 to 1 and can calculate the probability of events
- **E** You can list all outcomes of two independent events such as tossing a coin and throwing a dice, and calculate probabilities from lists or tables
- **E** You can calculate the probability of an outcome not happening if you know the probability of it happening
- **D** You can understand that the total probability of all possible outcomes in a paticular situation is 1
- **C** You can predict the expected number of successes from a given number of trials if the probability of one success is known
- **C** You can calculate relative frequency from experimental evidence and compare this with the theoretical probability

What you should know now

- How to use the probability scale and estimate the likelihood of outcomes of events depending on their position on the scale
- How to calculate theoretical probabilities from different situations
- How to calculate relative frequency and understand that the reliability of experimental results depends on the number of experiments carried out

1 Here are some statements.

Draw an arrow from each statement to the word which best describes its likelihood.

One has been done for you. (3)

(Total 3 marks)

Edexcel, June 2008, Paper 5 Foundation, Unit 1 Test, Question 2

A head is obtained when a fair coin is thrown once.	Certain
A number less than 7 will be scored when an ordinary six-sided dice is rolled once.	Likely
It will rain every day for a week next July in London.	Even
A red disc is obtained when a disc is taken at random from a bag containing 9 red discs and 2 blue discs.	Unlikely
	Impossible

2 Some bulbs were planted in October. The ticks in the table shows the months in which each type of bulb grows into flowers.

		Month					
		Jan	Feb	March	April	May	June
Type of bulb	Allium					✓	✓
	Crocus	✓	✓				
	Daffodil		✓	✓	✓		
	Iris	✓	✓				
	Tulip				✓	✓	

a In which months do tulips flower?

b Which type of bulb flowers in March?

c In which month do most types of bulb flower?

d Which type of bulb flowers in the same months as the iris?

Ben puts one of each type of these bulbs in a bag. He takes a bulb from the bag without looking.

e **i** Write down the probability that he will take a crocus bulb.

ii Copy the probability scale and mark with a cross (X) the probability that he will take a bulb which flowers in February.

0 ———————— 1

3 There are 3 red pens, 4 blue pens and 5 black pens in a box.

Sameena takes a pen, at random, from the box.

Write down the probability that she takes a black pen. (2)

(Total 2 marks)

Edexcel, June 2008, Paper 2 Foundation, Question 18

4 **a** Alice has a spinner that has five equal sections. The numbers 1, 2 and 3 are written on the spinner.

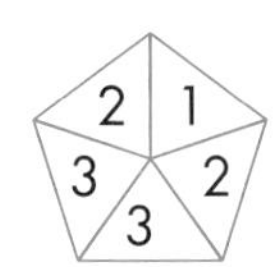

Alice spins the spinner once. On what number is the spinner least likely to land?

b

Alice thinks that the chance of getting a 2 is $\frac{2}{3}$. Explain why Alice is wrong.

5 A bag contains some beads which are red or green or blue or yellow.

The table shows the number of beads of each colour.

Colour	Red	Green	Blue	Yellow
Number of beads	3	2	5	2

Samire takes a bead at random from the bag. Write down the probability that she takes a blue bead.

Edexcel, March 2005, Paper 12A Intermediate, Question 1

6 Fifty people take a maths exam. The table shows the results.

	Pass	Fail
Male	12	16
Female	9	13

a A person is chosen at random from the group.

What is the probability that the person is male?

b A person is chosen at random from the group.

What is the probability that the person passed the test?

7 Doris has a bag in which there are nine counters, all green. Alex has a bag in which there are 15 counters, all red. Jade has a bag in which there are some blue counters.

a What is the probability of picking a red counter from Doris's bag?

b Doris and Alex put all their counters into a box.

What is the probability of choosing a green counter from the box?

c Jade now adds her blue counters to the box.

The probability of choosing a blue counter from the box is now $\frac{1}{3}$.

How many blue counters does Jade put in the box?

8

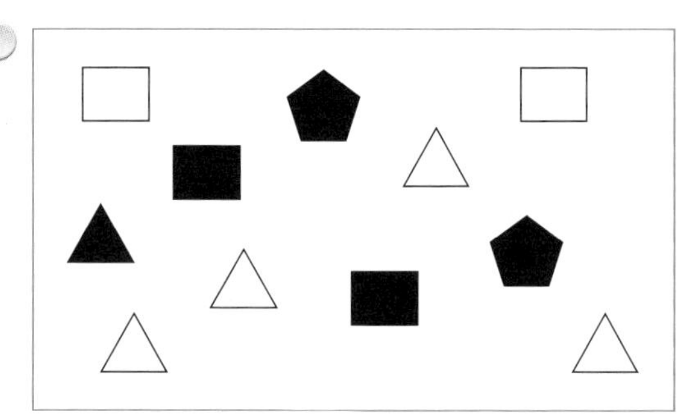

The diagram shows some 3-sided, 4-sided and 5-sided shaped.

The shapes are black or white.

a Complete the two-way table. (3)

	Black	White	Total
3-sided shape		4	5
4-sided shape	2		
5-sided shape		0	
Total			11

Ed takes a shape at random.

b Write down the probability the shape is white **and** 3-sided. (2)

(Total 5 marks)

Edexcel, March 2008, Paper 5 Foundation, Unit 1 Test, Question 3

9 Here are two fair spinners. They have numbers on each section.

Spinner A: 1, 4, 8, 5

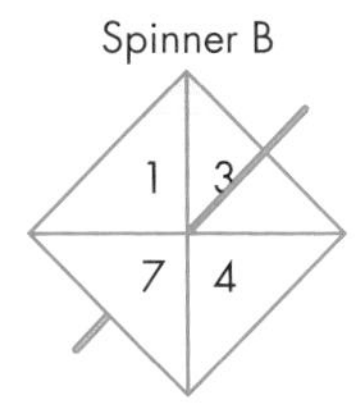

The spinners are spun. The two numbers are added together.

a Copy and complete the table to show all the possible total scores.

		Spinner B			
		1	3	4	7
Spinner A	1	2	4		
	4				
	5				
	8				

b What is the most likely score?

c What is the probability of getting a score of 5?

d What is the probability of getting a score of 11 or more?

e What is the probability of getting a score that is an odd number?

10 Here are the ages, in years, of 15 teachers.

35 52 42 27 36
23 31 41 50 34
44 28 45 45 53

a Draw an ordered stem-and-leaf diagram to show this information.
You must include a key. (3)

Key:

One of these teachers is picked at random.

b Work out the probability that this teacher is more than 40 years old. (2)

(Total 5 marks)

Edexcel, May 2008, Paper 1 Foundation, Question 22

Worked Examination Questions

1 Simon has a bag containing blue, green, red, yellow and white marbles.

a Complete the table to show the probability of each colour being chosen at random.

b Which colour of marble is most likely to be chosen at random?

c Calculate the probability that a marble chosen at random is blue or white.

Colour of marble	Probability
Blue	0.3
Green	0.2
Red	0.15
Yellow	
White	0.1

a 0.25

2 marks

The probabilities in the table should add up to 1 so,
$0.3 + 0.2 + 0.15 + 0.1 + \mathbf{?} = 1 \qquad 0.75 + \mathbf{?} = 1$
This calculation gets you 1 method mark. Using the calculation, the missing probability is 0.25. This answer gets you 1 mark.

b Blue marble

1 mark

The blue marble is most likely as it has the largest probability of being chosen. The correct answer is worth 1 mark.

c 0.4

2 marks

The word 'or' means you add probabilities of separate events.
The calculation P(blue or white) = P(blue) + P(white) = 0.3 + 0.1 = 0.4 is worth 1 method mark and the correct answer is worth 1 mark.

Total: 5 marks

FM **2** In a raffle 400 tickets have been sold. There is only one prize.
Mr Raza buys 5 tickets for himself and sells another 40.
Mrs Raza buys 10 tickets for herself and sells another 50.
Mrs Hewes buys 8 tickets for herself and sells just 12 others.

a What is the probability of:

i Mr Raza winning the raffle

ii Mr Raza selling the winning ticket

iii Mr Raza either winning the raffle or selling the winning ticket?

b What is the probability of either Mr or Mrs Raza selling the winning ticket?

c What is the probability of Mrs Hewes not winning the lottery?

a i $\frac{5}{400}$

You get 1 mark for the correct answer.

ii $\frac{40}{400}$

You get 1 mark for the correct answer.

iii $\frac{(5+40)}{400} = \frac{45}{400}$ 4 marks

You get 1 method mark for writing out the combined probability. You also get 1 accuracy mark for expressing this as a number out of 400.

b $\frac{(40+50)}{400} = \frac{90}{400}$ 2 marks

You get 1 method mark for writing out the combined probability. You also get 1 accuracy mark for expressing this as a number out of 400.

c $1 - \frac{8}{400} = \frac{392}{400}$ 2 marks

You get 1 method mark for subtracting from 1. You also get 1 mark for the final answer.

Total: 8 marks

Worked Examination Questions

PS **3** Here are two fair spinners.

The spinners are spun.

The two numbers are added together.

What is the probability of getting a score of an even number?

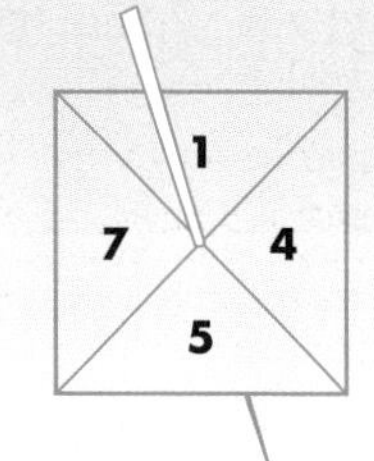

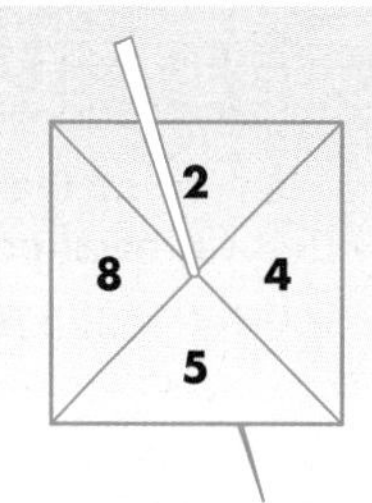

		Spinner A			
		1	4	5	7
Spinner B	2	3	6	7	9
	4	5	8	9	11
	5	6	9	10	12
	8	9	12	13	15

You get 1 mark for creating a table or, otherwise, for showing all the possible totals in order to solve the problem.

6

Counting how many even numbers are in the totals will get you 1 mark.

16

Counting how many numbers are in the table altogether will get you 1 mark.

$\frac{6}{16}$

Realising that the probability of getting an even number is $\frac{6}{16}$ will get you 1 mark.

Total: 4 marks

AU **4** Ellie cuts a pack of cards 100 times to see how many times she gets a heart.

She got 25 Hearts and her mother told her, "That's exactly what you would expect to get."

Explain why her mother is not quite correct.

Although the theoretical probability of cutting a Heart is $\frac{1}{4}$ and $\frac{1}{4}$ multiplied by 100 does give an expectation of 25, it would be unusual to cut exactly 25 Hearts. It is more likely that she would get a number close to 25, either just below it or just above it.

You get 1 mark for an explanation that clearly shows this understanding.

Total: 1 mark

13 Functional Maths
Fairground games

Joe had a stall at the local fair and wanted to make a reasonable profit from the game below.

In this game, the player rolls two balls down the sloping board and wins a prize if they land in slots that total more than seven.

Joe wants to know how much he should charge for each go and what the price should be. He would also like to know how much profit he is likely to make.

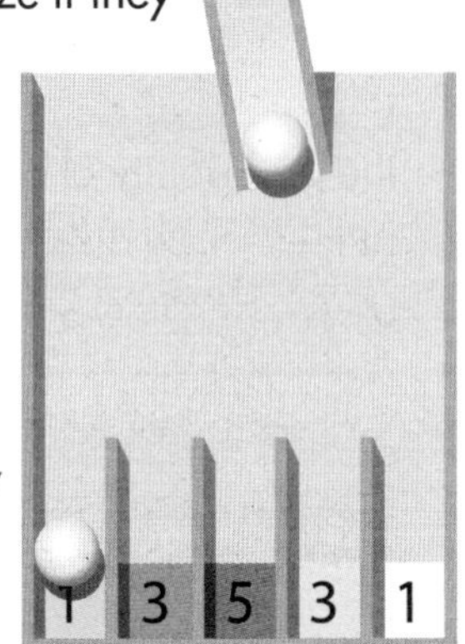

Getting started

Practise calculating probabilities using the spinner and questions below.

- What is the probability of spinning the spinner and getting:
  - a one
  - a five
  - a two
  - a number other than five?

  Give your answer as a fraction and as a decimal.
- If you spun the spinner 20 times, how many times would you expect to get:
  - a five
  - not five
  - an odd number?
- If you spun the spinner 100 times, how many times would you expect to get:
  - a two
  - not a two
  - a prime number?

Now, think about which probabilities you must calculate, in order to help Joe, and to design your own profitable game.

Your task

For this task you can work individually or in pairs.

1 Write a report for Joe which includes:

- a diagram to show the probability of each outcome
- the probability of at least three different outcomes
- the probability of winning
- at least three different ways of setting up the game, showing the cost of one go, the value of the prize and the expected profit for varying numbers of people playing the game.

Based on this information, advise Joe how he should set up the game. Justify your decision.

2 Design your own fairground game. Describe the rules for winning, show the probability of winning using a diagram, and give costings and the projected profit.

Explain why you think your game should be used at the next local fair.

Answers to Chapter 1

1.1 Adding with grids

Exercise 1A

1 a

1	3	7	11
9	2	8	19
6	5	4	15
16	10	19	45

b

0	6	7	13
8	1	4	13
9	5	3	17
17	12	14	43

c

0	8	7	15
1	6	2	9
9	3	4	16
10	17	13	40

d

2	4	6	12
3	5	7	15
8	9	1	18
13	18	14	45

e

5	9	3	17
6	1	8	15
2	7	4	13
13	17	15	45

f

0	8	3	11
7	2	4	13
1	6	5	12
8	16	12	36

g

9	4	8	21
7	0	5	12
1	6	3	10
17	10	16	43

h

0	8	6	14
7	1	4	12
5	9	2	16
12	18	12	42

i

1	8	7	16
6	2	5	13
0	9	3	12
7	19	15	41

2 a Yes (£33) with £1 left over **b** 23

3 a 7 and 11
b 31
c 7 + 11 + 12 = 30 (largest two odds and one even gives larger total than largest three evens)

4 Possible answer: 4, 6, 8

5 a

1	7	8	16
0	3	6	9
5	4	2	11
6	14	16	36

b

1	2	3	6
6	5	4	15
7	8	9	24
14	15	16	45

c

9	3	6	18
4	0	5	9
1	2	8	11
14	5	19	38

d

9	1	6	16
2	7	4	13
8	5	0	13
19	13	10	42

e

2	9	6	17
4	1	3	8
5	0	8	13
11	10	17	38

f

1	7	8	16
6	2	4	12
5	9	3	17
12	18	15	45

g

0	2	1	3
9	6	7	22
8	4	5	17
17	12	13	42

h

1	0	3	4
8	7	4	19
9	6	5	20
18	13	12	43

i

1	5	4	10
6	2	3	11
8	7	0	15
15	14	7	36

6 Possible answer: 9, 11, 16

7 Possible answer: 11 and 13

1.2 Multiplication tables check

Exercise 1B

1 a 20 **b** 21 **c** 24 **d** 15 **e** 16 **f** 12
g 10 **h** 42 **i** 24 **j** 18 **k** 30 **l** 28
m 18 **n** 56 **o** 25 **p** 45 **q** 27 **r** 30
s 49 **t** 24 **u** 36 **v** 35 **w** 32 **x** 36
y 48 **z** £48 × 2 = £96, yes

2 a 5 **b** 4 **c** 6 **d** 6 **e** 5 **f** 4
g 7 **h** 6 **i** 2 **j** 3 **k** 7 **l** 8
m 9 **n** 5 **o** 8 **p** 9 **q** 4 **r** 7
s 7 **t** 9 **u** 5 **v** 4 **w** 5 **x** 7
y 6 **z** £6 per hour, 10 hours

3 a 12 **b** 15 **c** 21 **d** 13 **e** 8 **f** 7
g 14 **h** 3 **i** 30 **j** 6 **k** 35 **l** 5
m 16 **n** 7 **o** 16 **p** 15 **q** 27 **r** 6
s 15 **t** 24 **u** 40 **v** 6 **w** 17 **x** 72
y 46 **z** Ahmed is paid more (£33) than Ben (£32)

4 a 86 **b** 56 **c** 358 + 6

5 a 30 **b** 50 **c** 80 **d** 100 **e** 120 **f** 180
g 240 **h** 400 **i** 700 **j** 900 **k** 1000 **l** 1400
m 2400 **n** 7200 **o** 10 000 **p** 2 **q** 7 **r** 9
s 17 **t** 30 **u** 3 **v** 8 **w** 12 **x** 29
y 50

6 a 8 (3 × 8) **b** 900 (900 ÷ 10)

7 For example: 4 × 5 = 20 and 5 × 6 = 30

8 For example: 3 × 4 × 5 = 60 = 10 × 6
and 5 × 6 × 7 = 210 = 35 × 6

1.3 Order of operations and BIDMAS/BODMAS

Exercise 1C

1 a 11 **b** 6 **c** 10 **d** 12 **e** 11 **f** 13
g 11 **h** 12 **i** 12 **j** 4 **k** 13 **l** 3

2 a 16 **b** 2 **c** 10 **d** 10 **e** 6 **f** 18
g 6 **h** 15 **i** 9 **j** 12 **k** 3 **l** 8

3 b 3 + 2 × 4 = 11 **c** 9 ÷ 3 − 2 = 1
d 9 − 4 ÷ 2 = 7 **e** 5 × 2 + 3 = 13
f 5 + 2 × 3 = 11 **g** 10 ÷ 5 − 2 = 0
h 10 − 4 ÷ 2 = 8 **i** 4 × 6 − 7 = 17
j 7 + 4 × 6 = 31 **k** 6 ÷ 3 ÷ 7 = 9
l 7 + 6 ÷ 2 = 10

4 a 38 **b** 48 **c** 3 **d** 2 **e** 5 **f** 14
g 10 **h** 2 **i** 5 **j** 19 **k** 15 **l** 2
m 20 **n** 19 **o** 54 **p** 7 **q** 2 **r** 7
s 7 **t** 38 **u** 42 **v** 10 **w** 2 **x** 10
y 10 **z** 24

5 a (4 + 1) **b** No brackets needed
c (2 + 1) **d** No brackets needed
e (4 + 4) **f** (16 − 4)
g No brackets needed **h** No brackets needed
i (20 − 10) **j** No brackets needed
k (5 + 5) **l** (4 + 2)
m (15 − 5) **n** (7 − 2)
o (3 + 3) **p** No brackets needed
q No brackets needed **r** (8 − 2)

6 a 8 **b** 6 **c** 6 **d** 13 **e** 11 **f** 9
g 12 **h** 8 **i** 15 **j** 16 **k** 1 **l** 7

7 No, correct answer is 5 + 42 = 47

8 **a** $2 + 3 \times 4 = 14$ **b** $8 - 4 \div 4 = 7$ (correct)
c $6 + 3 \times 2 = 12$ (correct) **d** $7 - 1 \times 5 = 2$
e $2 \times 7 + 2 = 16$ (correct) **f** $9 - 3 \times 3 = 0$

9 **a** $2 \times 3 + 5 = 11$ **b** $2 \times (3 + 5) = 16$
c $2 + 3 \times 5 = 17$ **d** $5 - (3 - 2) = 4$
e $5 \times 3 - 2 = 13$ **f** $5 \times 3 \times 2 = 30$

10 $4 + 5 \times 3 = 19$
$(4 + 5) \times 3 = 27$. So $4 + 5 \times 3$ is smaller

11 $(5 - 2) \times 6 = 18$

12 $10 \div (5 - 3) = 4$

13 $10 - 3 \times 1.5$ and $10 - 1.5 - 1.5 - 1.5$

1.4 Rounding

Exercise 1D

1 **a** 20 **b** 60 **c** 80 **d** 50 **e** 100 **f** 20
g 90 **h** 70 **i** 10 **j** 30 **k** 30 **l** 50
m 80 **n** 50 **o** 90 **p** 40 **q** 70 **r** 20
s 100 **t** 110

2 **a** 200 **b** 600 **c** 800 **d** 500 **e** 1000 **f** 100
g 600 **h** 400 **i** 1000 **j** 1100 **k** 300 **l** 500
m 800 **n** 500 **o** 900 **p** 400 **q** 700 **r** 800
s 1000 **t** 1100

3 **a** 1 **b** 2 **c** 1 **d** 1 **e** 3 **f** 2
g 3 **h** 2 **i** 1 **j** 1 **k** 3 **l** 2
m 74 **n** 126 **o** 184

4 **a** 2000 **b** 6000 **c** 8000 **d** 5000 **e** 10 000 **f** 1000
g 6000 **h** 3000 **i** 9000 **j** 2000 **k** 3000 **l** 5000
m 8000 **n** 5000 **o** 9000 **p** 4000 **q** 7000 **r** 8000
s 1000 **t** 2000

5 **a** 230 **b** 570 **c** 720 **d** 520 **e** 910 **f** 230
g 880 **h** 630 **i** 110 **j** 300 **k** 280 **l** 540
m 770 **n** 500 **o** 940 **p** 380 **q** 630 **r** 350
s 1010 **t** 1070

6 **a** True **b** False **c** True **d** True **e** True **f** False

7 Welcome to Swinton population 1400 (to the nearest 100)

8 **a** Man Utd v West Brom
b Blackburn v Fulham
c 40 000, 19 000, 42 000, 26 000, 40 000, 68 000, 35 000, 25 000, 20 000
d 39 600, 19 000, 42 100, 26 100,40 400, 67 800, 34 800, 25 500, 20 200

9 **a** 35 min **b** 55 min **c** 15 min
d 50 min **e** 10 min **f** 15 min
g 45 min **h** 35 min **i** 5 min
j 0 min

10 **a** 375
b 25 (350 to 374 inclusive)

11 A number between 75 and 84 inclusive added to a number between 45 and 54 inclusive with a total not equal to 130, for example $79 + 49 = 128$

1.5 Adding and subtracting numbers with up to four digits

Exercise 1E

1 **a** 713 **b** 151 **c** 6381
d 968 **e** 622 **f** 1315
g 8260 **h** 818 **i** 451
j 852

2 **a** 646 **b** 826 **c** 3818
d 755 **e** 2596 **f** 891
g 350 **h** 2766 **i** 8858
j 841 **k** 6831 **l** 7016
m 1003 **n** 4450 **o** 9944

3 **a** 450 **b** 563 **c** 482
d 414 **e** 285 **f** 486
g 244 **h** 284 **i** 333
j 216 **k** 2892 **l** 4417
m 3767 **n** 4087 **o** 1828

4 **a** 128 **b** 29 **c** 334
d 178 **e** 277 **f** 285
g 335 **h** 399 **i** 4032
j 4765 **k** 3795 **l** 5437

5 **a** 558 miles **b** 254 miles

6 252

7 **a** 6, 7 **b** 4, 7 **c** 4, 8
d 7, 4, 9 **e** 6, 9, 7 **f** 6, 2, 7
g 2, 6, 6 **h** 4, 5, 9 **i** 4, 8, 8
j 4, 4, 9, 8

8 Units digit should be 6 (from 14 – 8)

9 **a** 3, 5 **b** 8, 3 **c** 5, 8
d 8, 5, 4 **e** 6, 7, 5 **f** 1, 2, 1
g 2, 7, 7 **h** 5, 5, 6 **i** 8, 3, 8
j 1, 8, 8, 9

10 For example: $181 - 27 = 154$

1.6 Multiplying and dividing by single-digit numbers

Exercise 1F

1 **a** 56 **b** 65 **c** 51
d 38 **e** 108 **f** 115
g 204 **h** 294 **i** 212
j 425 **k** 150 **l** 800
m 960 **n** 1360 **o** 1518

2 **a** 294 **b** 370 **c** 288
d 832 **e** 2163 **f** 2520
g 1644 **h** 3215 **i** 3000
j 2652 **k** 3696 **l** 1880
m 54 387 **n** 21 935 **o** 48 888

3 **a** 219 **b** 317 **c** 315
d 106 **e** 99 **f** 121
g 252 **h** 141 **i** 144
j 86 **k** 63 **l** 2909
m 416 **n** 251 **o** 1284

4 **a** 705 miles **b** £3525

5 **a** 47 miles
b Three numbers with a total of 125. First number must be less than 50. second number less than first number, third number less than second number, for example 48, 42 and 35.

6 **a** 119 **b** 96 **c** 144
d 210 **e** 210

7 **a** 13 **b** 37 weeks **c** 43 m
d 36 **e** 45

8 **a** $152 + 190 = 324$
b $(10 \times 190) + 76 = 1976$
c $(100 \times 38) + 190 = 3990$

Examination questions

1 **i** 4 **ii** 2 **iii** 9

2 **a** **i** 28 000 000 **ii** 2800
b **i** 100 **ii** 10

3 7760 metres

4 **a** £1505
b 19 people

5 **a** 7750
b 7849

6 Murray did the addition first, Harry did the multiplication first.

7 **a** 96
b 32
c **i** 8 **ii** 2
d **i** 1080 **ii** 15

8 **a** **i** 105 **ii** 23
b **i** 72 **ii** 31

9 **a** He used BODMAS and did the power first, then the multiplication before the addition.
b **i** $(2 \times 3)^2 + 6 = 42$ **ii** $2 \times (3^2 + 6) = 30$

10 **a** Adam has calculated 5×2 instead of 5^2, Bekki has added 3 to 5 first, instead of working out the power first.
b 26

Answers to Chapter 2

2.1 Recognise a fraction of a shape

Exercise 2A

1 **a** $\frac{1}{4}$ **b** $\frac{1}{3}$ **c** $\frac{5}{8}$ **d** $\frac{7}{12}$ **e** $\frac{4}{9}$
f $\frac{3}{10}$ **g** $\frac{3}{8}$ **h** $\frac{15}{16}$ **i** $\frac{5}{12}$ **j** $\frac{7}{18}$
k $\frac{4}{8} = \frac{1}{2}$ **l** $\frac{4}{12} = \frac{1}{3}$ **m** $\frac{6}{9} = \frac{2}{3}$
n $\frac{6}{10} = \frac{3}{5}$ **o** $\frac{4}{8} = \frac{1}{2}$ **p** $\frac{5}{64}$

2 Check students' diagrams.

3

4 **a** g **b** i **c** neither **d** e
e neither **f** m **g** k **h** e

5 Fraction B is not $\frac{1}{3}$. Fraction C does not have a numerator of 4.

6 $\frac{1}{2}$

2.2 Recognise equivalent fractions, using diagrams

Exercise 2B

1 **a** $\frac{4}{24}$ **b** $\frac{8}{24}$ **c** $\frac{3}{24}$ **d** $\frac{16}{24}$ **e** $\frac{20}{24}$
f $\frac{18}{24}$ **g** $\frac{9}{24}$ **h** $\frac{15}{24}$ **i** $\frac{21}{24}$ **j** $\frac{12}{24}$

2 **a** $\frac{11}{24}$ **b** $\frac{9}{24}$ **c** $\frac{7}{24}$ **d** $\frac{19}{24}$ **e** $\frac{23}{24}$
f $\frac{23}{24}$ **g** $\frac{21}{24}$ **h** $\frac{22}{24}$ **i** $\frac{19}{24}$ **j** $\frac{23}{24}$

3 **a** $\frac{5}{20}$ **b** $\frac{4}{20}$ **c** $\frac{15}{20}$ **d** $\frac{16}{20}$ **e** $\frac{2}{20}$
f $\frac{10}{20}$ **g** $\frac{12}{20}$ **h** $\frac{8}{20}$ **i** $\frac{14}{20}$ **j** $\frac{6}{20}$

4 **a** $\frac{9}{20}$ **b** $\frac{14}{20}$ **c** $\frac{11}{20}$ **d** $\frac{19}{20}$ **e** $\frac{19}{20}$

5 **a** $\frac{2}{6}$ **b** $\frac{2}{3}$ **c** none **d** $\frac{6}{8}$

6 **a** $\frac{4}{10}, \frac{8}{20}, \frac{10}{25}$ **b** for example $\frac{12}{30}$

7 **a**

×	2	3	4	5
3	6	9	12	15
4	8	12	16	20

b $\frac{6}{8}, \frac{9}{12}, \frac{12}{16}, \frac{15}{20}$

8 **a** $\frac{1}{12}$
b **i** $\frac{7}{12}$ **ii** $\frac{9}{12}$ (or $\frac{3}{4}$) **iii** $\frac{11}{12}$
iv $\frac{9}{12}$ (or $\frac{3}{4}$) **v** $\frac{7}{12}$

2.3 Equivalent fractions and simplifying fractions by cancelling

Exercise 2C

1 **a** $\frac{8}{20}$ **b** $\frac{3}{12}$ **c** $\frac{15}{40}$ **d** $\frac{12}{15}$ **e** $\frac{15}{18}$
f $\frac{12}{28}$ **g** $\times 2, \frac{6}{20}$ **h** $\times 3, \frac{3}{9}$
i $\times 4, \frac{12}{20}$ **j** $\times 6, \frac{12}{18}$
k $\times 3, \frac{9}{12}$ **l** $\times 5, \frac{25}{40}$
m $\times 2, \frac{14}{20}$ **n** $\times 4, \frac{4}{24}$ **o** $\times 5, \frac{15}{40}$

2 **a** $\frac{1}{2} = \frac{2}{4} = \frac{3}{6} = \frac{4}{8} = \frac{5}{10} = \frac{6}{12}$

b $\frac{1}{3} = \frac{2}{6} = \frac{3}{9} = \frac{4}{12} = \frac{5}{15} = \frac{6}{18}$

c $\frac{3}{4} = \frac{6}{8} = \frac{9}{12} = \frac{12}{16} = \frac{15}{20} = \frac{18}{24}$

d $\frac{2}{5} = \frac{4}{10} = \frac{6}{15} = \frac{8}{20} = \frac{10}{25} = \frac{12}{30}$

e $\frac{3}{7} = \frac{6}{14} = \frac{9}{21} = \frac{12}{28} = \frac{15}{35} = \frac{18}{42}$

3 **a** $\frac{2}{3}$ **b** $\frac{4}{5}$ **c** $\frac{5}{7}$ **d** $\div 6, \frac{2}{3}$

e $25 \div 4, \frac{3}{5}$ **f** $\div 3, \frac{7}{10}$

4 **a** 32 **b** $\frac{1}{2}$ **c** $\frac{5}{6}$

5 **a** $\frac{2}{3}$ **b** $\frac{1}{3}$ **c** $\frac{2}{3}$ **d** $\frac{3}{4}$ **e** $\frac{1}{3}$

f $\frac{1}{2}$ **g** $\frac{7}{8}$ **h** $\frac{4}{5}$ **i** $\frac{1}{2}$ **j** $\frac{1}{4}$

k $\frac{4}{5}$ **l** $\frac{5}{7}$ **m** $\frac{5}{7}$ **n** $\frac{2}{3}$ **o** $\frac{2}{5}$

p $\frac{2}{5}$ **q** $\frac{1}{3}$ **r** $\frac{7}{10}$ **s** $\frac{1}{4}$

t $\frac{3}{2} = 1\frac{1}{2}$ **u** $\frac{2}{3}$ **v** $\frac{2}{3}$ **w** $\frac{3}{4}$

x $\frac{3}{2} = 1\frac{1}{2}$ **y** $\frac{7}{2} = 3\frac{1}{2}$

6 **a** $\frac{1}{2}, \frac{2}{3}, \frac{5}{6}$ **b** $\frac{1}{2}, \frac{5}{8}, \frac{3}{4}$ **c** $\frac{2}{5}, \frac{1}{2}, \frac{7}{10}$

d $\frac{7}{12}, \frac{2}{3}, \frac{3}{4}$ **e** $\frac{1}{6}, \frac{1}{4}, \frac{1}{3}$ **f** $\frac{3}{4}, \frac{4}{5}, \frac{9}{10}$

g $\frac{7}{10}, \frac{4}{5}, \frac{5}{6}$ **h** $\frac{3}{10}, \frac{1}{3}, \frac{2}{5}$

7 **a** $\frac{1}{3} + \frac{1}{4} = \frac{4}{12} + \frac{3}{12} = \frac{7}{12}$

Explanations may involve ruling out other combinations.

b $\frac{1}{2}$ as the smallest denominator is the biggest unit fraction.

Diagrams may be used but must be based on equal sized area.

8 **a** $\frac{1}{5}$ **b** $\frac{1}{77}$

9 **a** $\frac{3}{8}$ **b** $\frac{1}{2}$ **c** $\frac{5}{16}$

2.4 Finding a fraction of a quantity

Exercise 2D

1 **a** 18 **b** 10 **c** 18 **d** 28
e 15 **f** 18 **g** 48 **h** 45

2 **a** £1800 **b** 128 g **c** 160 kg
d £116 **e** 65 litres **f** 90 min
g 292 days **h** 21 h **i** 18 h
j 2370 miles

3 **a** $\frac{5}{8}$ of 40 = 25 **b** $\frac{3}{4}$ of 280 = 210 **c** $\frac{4}{5}$ of 70 = 56

d $\frac{5}{6}$ of 72 = 60 **e** $\frac{3}{5}$ of 95 = 57 **f** $\frac{3}{4}$ of 340 = 255

4 £6080

5 £31 500

6 13 080

7 52 kg

8 **a** 856 **b** 187 675

9 **a** £50 **b** £550

10 **a** 180 g **b** 900 g

11 **a** £120 **b** £240

12 Lion Autos

13 Offer B

2.5 One quantity as a fraction of another

Exercise 2E

1 **a** $\frac{1}{3}$ **b** $\frac{1}{5}$ **c** $\frac{2}{5}$ **d** $\frac{5}{24}$ **e** $\frac{2}{5}$ **f** $\frac{1}{6}$ **g** $\frac{2}{7}$ **h** $\frac{1}{3}$

2 $\frac{3}{5}$

3 $\frac{12}{31}$

4 $\frac{7}{12}$

5 Jon saves $\frac{30}{90} = \frac{1}{3}$

Matt saves $\frac{35}{100}$ which is greater than $\frac{1}{3}$, so Matt saves the greater proportion of his earnings.

6 $\frac{13}{20} = \frac{65}{100}$, $\frac{16}{25} = \frac{64}{100}$, so first mark is better.

2.6 Arithmetic with fractions

Exercise 2F

1 **a** $\frac{7}{10}$ **b** $\frac{2}{5}$ **c** $\frac{1}{2}$ **d** $\frac{3}{100}$ **e** $\frac{3}{50}$ **f** $\frac{13}{100}$

g $\frac{1}{4}$ **h** $\frac{19}{50}$ **i** $\frac{11}{20}$ **j** $\frac{16}{25}$

2 **a** 0.5 **b** 0.75 **c** 0.6 **d** 0.9 **e** 0.333 **f** 0.625
g 0.667 **h** 0.35 **i** 0.636 **j** 0.444

3 **a** 0.3, $\frac{1}{2}$, 0.6 **b** 0.3, $\frac{2}{5}$, 0.8 **c** 0.15, $\frac{1}{4}$, 0.35

d $\frac{7}{10}$, 0.71, 0.72 **e** 0.7, $\frac{3}{4}$, 0.8 **f** $\frac{1}{20}$, 0.08, 0.1

g 0.4, $\frac{1}{2}$, 0.55 **h** 1.2, 1.23, $1\frac{1}{4}$

4 Just shirts – $\frac{1}{3}$ (0.33) is greater than $\frac{1}{4}$ (0.25)

5 $\frac{7}{8}$ (= 0.875)

6 $\frac{2}{3}$ (= 0.67)

7 0.125

8 $\frac{1}{2}$

Exercise 2G

1 **a** $\frac{8}{15}$ **b** $\frac{7}{12}$ **c** $\frac{3}{10}$ **d** $\frac{11}{12}$ **e** $\frac{7}{8}$ **f** $\frac{1}{2}$
g $\frac{1}{6}$ **h** $\frac{1}{20}$ **i** $\frac{1}{10}$ **j** $\frac{1}{8}$ **k** $\frac{1}{12}$ **l** $\frac{1}{3}$
m $\frac{1}{6}$ **n** $\frac{7}{9}$ **o** $\frac{5}{8}$ **p** $\frac{3}{8}$ **q** $\frac{1}{15}$ **r** $1\frac{13}{24}$
s $\frac{59}{80}$ **t** $\frac{22}{63}$ **u** $\frac{37}{54}$

2 **a** $3\frac{5}{14}$ **b** $10\frac{3}{5}$ **c** $2\frac{1}{6}$ **d** $3\frac{31}{45}$ **e** $4\frac{47}{60}$ **f** $\frac{41}{72}$
g $\frac{29}{48}$ **h** $1\frac{43}{48}$ **i** $1\frac{109}{120}$ **j** $1\frac{23}{30}$ **k** $1\frac{31}{84}$

3 $\frac{1}{20}$

4 **a** $\frac{1}{6}$ **b** 30, must be divisible by 2 and 3

5 $\frac{13}{15}$

6 $\frac{1}{3}$

7 $\frac{3}{8}$

8 He has added the numerators and added the denominators instead of using a common denominator. Correct answer is $3\frac{7}{12}$.

Exercise 2H

1 **a** $\frac{1}{6}$ **b** $\frac{1}{10}$ **c** $\frac{3}{8}$ **d** $\frac{3}{14}$ **e** $\frac{8}{15}$ **f** $\frac{1}{5}$
g $\frac{2}{7}$ **h** $\frac{3}{10}$ **i** $\frac{1}{2}$ **j** $\frac{2}{5}$

2 **a** $\frac{3}{32}$ **b** $\frac{3}{8}$ **c** $\frac{7}{20}$ **d** $\frac{16}{45}$ **e** $\frac{3}{5}$ **f** $\frac{5}{8}$

3 3 km

4 $\frac{1}{12}$

5 $\frac{3}{8}$

6 21 tonnes

7 260 pupils

8 £51

9 18

10 **a** $\frac{5}{12}$ **b** $2\frac{1}{12}$ **c** $6\frac{1}{4}$ **d** $2\frac{11}{12}$ **e** $3\frac{9}{10}$ **f** $3\frac{1}{3}$
g $12\frac{1}{2}$ **h** 30

11 $\frac{2}{5}$ of $6\frac{1}{2} = 2\frac{3}{5}$

12 £5

13 Three-quarters of 68 = 51

14 7 min

15 £10.40

16 £30

Exercise 2I

1 **a** $\frac{3}{4}$ **b** $1\frac{2}{5}$ **c** $1\frac{1}{15}$ **d** $1\frac{1}{14}$ **e** 4 **f** 4
g 5 **h** $1\frac{5}{7}$ **i** $\frac{4}{9}$ **j** $1\frac{3}{5}$

2 18

3 40

4 15

5 16

6 **a** $2\frac{2}{15}$ **b** 38 **c** $1\frac{7}{8}$ **d** $\frac{9}{32}$ **e** $\frac{1}{16}$ **f** $\frac{256}{625}$

Examination questions

1 **a** $\frac{1}{2}$ **b** 9 squares shaded

2 $\frac{1}{3}, \frac{2}{5}, \frac{1}{2}, \frac{3}{4}$

3 $\frac{3}{10}$

4 **a** £72.96 **b** 87 miles

5 **a** $\frac{2}{5}$, because $\frac{2}{5} = 0.4$ **b** 15

6 $\frac{4}{5}$, because $\frac{4}{5} = 0.8$ and $\frac{3}{4} = 0.75$. One grid has 16 squares shaded from left to right to show 0.8 and the other has 15 squares shaded from top to bottom to show 0.75.

7 120 tissues

8 $\frac{3}{10}$

9 27 kg

10 **a** $\frac{1}{2}$ **b** $\frac{2}{5}$

11 **a** **i** 105 **ii** $\frac{1}{2}$ **b** $\frac{1}{4}$ **c** $\frac{3}{5}$

12 £4.80

13 172 g

14 **a** 0.625 **b** $\frac{3}{5}$

15 **a** 20 **b** $\frac{7}{10}$

16 Smaller, e.g. if the original packet was 500 g, the new packet size is $\frac{6}{5} \times 500 = 600$. If this is then reduced by $\frac{1}{5}$, $\frac{4}{5} \times 600 = 480$ g.

17 **a** 0.2 **b** $0.\dot{3}$

18 660 g

19 **a** $0.\dot{1}42\,85\dot{7}1$ **b** $0.\dot{3}84\,61\dot{5}3$

20 **a** $\frac{2}{3}$ **b** 0.6

21 **a** $\frac{5}{12}$ **b** $\frac{3}{20}$

22 $1\frac{11}{15}$

Answers to Chapter 3

3.1 Introduction to negative numbers

Activity

1 **a** 860 feet **b** 725 feet **c** 75 feet
d 475 feet **e** 1100 feet **f** 575 feet
g 425 feet **h** 310 feet **i** 700 feet
j 1010 feet

2 480 feet

3 1180 feet

4 Closed gate/Collapsed tunnel and Dead Man's seam/C seam

5 910 feet

6 Dead Man's seam/D seam and B seam/C seam

7 Collapsed tunnel/North gate

Exercise 3A

1 **a** 0 °C **b** 5 °C **c** −2 °C **d** −5 °C **e** −1 °C

2 **a** 11 degrees **b** 9 degrees

3 8 degrees

4

	Sump	River Cave	Lost Cave	Echo Cave	Angel Cavern	Bat Cave	Base	Camp 1	Camp 2	Camp 3	Camp 4	Summit
Summit	4030	4010	3930	3910	3795	3665	3420	2300	1070	670	220	0
Camp 4	3810	3790	3710	3690	3575	3445	3200	2080	850	450	0	
Camp 3	3360	3340	3260	3240	3125	2995	2750	1630	400	0		
Camp 2	2960	2940	2860	2840	2725	2595	2350	1230	0			
Camp 1	1730	1710	1630	1610	1495	1365	1120	0				
Base	610	590	510	490	375	245	0					
Bat Cave	365	345	265	245	130	0						
Angel Cavern	235	215	135	115	0							
Echo Cave	120	100	20	0								
Lost Cave	100	80	0									
River Cave	20	0										
Sump	0											

3.2 Everyday use of negative numbers

Exercise 3B

1 −£5

2 −£9

3 Profit

4 −200 m

5 −50 m

6 Above

7 −3 h

8 −5 h

9 After

10 −2 °C

11 −8 °C

12 Above

13 −70 km

14 −200 km

15 North

16 +5 m

17 −5 mph

18 −2

19 **a** You owe the bank £89.72.
b You are paying money out of the account.
c You are paying money into the account.

20 **a** −11 °C **b** 6 degrees

21 **a** −8 **b** 7 **c** 15

22 9 days

23 1.54am

3.3 The number line

Exercise 3C

1 Many different answers to each part

2 Many different answers to each part

3 **a** Is smaller than **b** is bigger than **c** Is smaller than
d Is smaller than **e** is bigger than **f** Is smaller than
g Is smaller than **h** is bigger than **i** is bigger than
j Is smaller than **k** Is smaller than **l** is bigger than

4 **a** Is smaller than **b** Is smaller than **c** Is smaller than
d is bigger than **e** Is smaller than **f** Is smaller than

5 **a** < **b** > **c** < **d** < **e** < **f** >
g < **h** > **i** > **j** > **k** < **l** <
m > **n** > **o** < **p** >

6 **a**
−5 −4 −3 −2 −1 0 1 2 3 4 5

b
−25 −20 −15 −10 −5 0 5 10 15 20 25

c −10, −8, −6, −4, −2, 0, 2, 4, 6, 8, 10

d −50, −40, −30, −20, −10, 0, 10, 20, 30, 40, 50

e −15, −12, −9, −6, −3, 0, 3, 6, 9, 12, 15

f −20, −16, −12, −8, −4, 0, 4, 8, 12, 16, 20

g $-2\frac{1}{2}$, −2, $-1\frac{1}{2}$, −1, $-\frac{1}{2}$, 0, $\frac{1}{2}$, 1, $1\frac{1}{2}$, 2, $2\frac{1}{2}$

h −100, −80, −60, −40, −20, 0, 20, 40, 60, 80, 100

i −250, −200, −150, −100, −50, 0, 50, 100, 150, 200, 250

7 6 °C −2 °C −4 °C 2 °C

8 $-4\frac{1}{2}$ $-\frac{1}{4}$ $+3\frac{3}{4}$

−6, −5, −4, −3, −2, −1, 0, 1, 2, 3, 4, 5, 6

3.4 Arithmetic with negative numbers

Exercise 3D

1 **a** −2° **b** −3° **c** −2° **d** −3° **e** −2° **f** −3°
g 3 **h** 3 **i** −1 **j** −1 **k** 2 **l** −3
m −4 **n** −6 **o** −6 **p** −1 **q** −5 **r** −4
s 4 **t** −1 **u** −5 **v** −4 **w** −5 **x** −5

2 **a** −4 **b** −4 **c** −10 **d** 2 **e** 8 **f** −5
g 2 **h** 5 **i** −7 **j** −12 **k** 13 **l** 25
m −32 **n** −30 **o** −5 **p** −8 **q** −12 **r** 10
s −36 **t** −14 **u** 41 **v** 12 **w** −40 **x** −101

3 **a** 6 **b** −5 **c** 6 **d** −1 **e** −2 **f** −6
g −6 **h** −2 **i** 3 **j** 0 **k** −7 **l** −6
m 8 **n** 1 **o** −9 **p** −9 **q** −5 **r** −80
s −7 **t** −1 **u** −47

4 Students' own check

5 **a** 7 degrees **b** −6 °C

6 **a** 2 − 8
b 2 + 5 − 8 or 2 + 4 − 7 or 8 − 4 − 5 or 8 − 2 − 7 or 5 − 4 − 2
c 2 − 5 − 7 − 8 **d** 2 + 5 − 4 − 7 − 8

7 500 ft

Exercise 3E

1 **a** 6 **b** 7 **c** 8 **d** 6 **e** 8 **f** 10
g 2 **h** −3 **i** 1 **j** 2 **k** −1 **l** −7
m 2 **n** −3 **o** 1 **p** −5 **q** 3 **r** −4
s −3 **t** −8 **u** −10 **v** −9 **w** −4 **x** −9

2 **a** −8 **b** −10 **c** −11 **d** −3 **e** 2 **f** −5
g 1 **h** 4 **i** 7 **j** −8 **k** −5 **l** −11
m 11 **n** 6 **o** 8 **p** 8 **q** −2 **r** −1
s −9 **t** −5 **u** 5 **v** −9 **w** 8 **x** 0

3 **a** 3 °C **b** 0 °C **c** −3 °C **d** −5 °C **e** −11 °C

4 **a** 10 degrees Celsius **b** 7 degrees Celsius
c 9 degrees Celsius

5 −9, −6, −5, −1, 1, 2, 3, 8

6 **a** −3 **b** −4 **c** −2 **d** −7 **e** −14 **f** −6
g −12 **h** −10 **i** 4 **j** −4 **k** 14 **l** 11
m −4 **n** −1 **o** −10 **p** −5 **q** −3 **r** 5
s −4 **t** −8

7 **a** 2 **b** −3 **c** −5 **d** −7 **e** −10 **f** −20

8 **a** 2 **b** 4 **c** −1 **d** −5 **e** −11 **f** 8

9 **a** 13 **b** 2 **c** 5 **d** 4 **e** 11 **f** −2

10 **a** −10 **b** −5 **c** −2 **d** 4 **e** 7 **f** −4

11 Check student's answers

12 Check student's answers

13 **a** −5 **b** 6 **c** 0 **d** 2 **e** 13 **f** 0
g −6 **h** −2 **i** 212 **j** 5 **k** 3 **l** 3
m −67 **n** 7 **o** 25

14 **a** −1, 0, 1, 2, 3 **b** −6, −5, −4, −3, −2
c −3, −2, −1, 0, 1 **d** −8, −7, −6, −5, −4
e −9, −8, −7, −6, −5 **f** 3, 4, 5, 6, 7
g −12, −11, −10, −9, −8 **h** −16, −15, −14, −13, −12
i −2, −1, 0, 1, 2, 3; −4, −3, −2, −1, 0, 1
j −12, −11, −10, −9, −8, −7; −14, −13, −12, −11, −10, −9
k −2, −1, 0, 1, 2, 3; 0, 1, 2, 3, 4, 5
l −8, −7, −6, −5, −4, −3, −2; −5, −4, −3, −2, −1, 0, 1
m −10, −9, −8, −7, −6, −5, −4; −1, 0, 1, 2, 3, 4, 5
n 3, 4, 5, 6, 7, 8, 9; −5, −4, −3, −2, −1, 0, 1

15 **a** −4 **b** 3 **c** 4 **d** −6 **e** 7 **f** 2
g 7 **h** −6 **i** −7 **j** 0 **k** 0 **l** −6
m −7 **n** −9 **o** 4 **p** 0 **q** 5 **r** 0
s 10 **t** −5 **u** 3 **v** −3 **w** −9 **x** 0
y −3 **z** −3

16 **a** +6 + +5 = 11 **b** +6 + −9 = −3
c +6 − −9 = 15 **d** +6 − +5 = 1

17 **a** +5 + +7 − −9 = +21 **b** +5 + −9 − +7 = −11
c +7 + −7, +4 + −4

18 It may not come on as the thermometer inaccuracy might be between 0° and 2° or 2° and 4°

19 −1 and 6

Exercise 3F

1 −12

−1	−9	−2
−5	−4	−3
−6	1	−7

2 0

1	−4	3
2	0	−2
−3	4	−1

3 −15

0	−14	−1
−6	−5	−4
−9	4	−10

4 −9

2	−12	1
−4	−3	−2
−7	6	−8

5 −18

−3	−6	−9
−12	−6	0
−3	−6	−9

6 −21

−2	−18	−1
−6	−7	−8
−13	4	−12

7

−4	−12	−5
−8	−7	−6
−9	−2	−10

−21

8

2	1	−3
−5	0	5
3	−1	−2

0

9

−2	−10	−3
−6	−5	−4
−7	0	−8

−15

10

−8	−1	−3	−14
−8	−9	−7	−2
−11	−6	−4	−5
1	−10	−12	−5

−26

11

−7	5	2	−16
−6	−8	−5	3
−11	−3	0	−2
8	−10	−13	−1

−16

Examination questions

1 **a** Tuesday
b Friday

2 **a** Oslo
b 13 °C

3 **a** Plymouth
b Edinburgh

4 −2 °C

5 **a** 3
b −8

6 **a** 9 °C
b 8 °C
c −1 °C

7 **a** 8
b −9
c −7

8 **a** 90 °C
b 540 °C
c Jupiter
d −230 °C

9 **a** −2 °C, −1 °C, 0 °C, 1 °C
b 3 °C

10 **a i** 8 **ii** 9 **iii** −7
b i −7 **ii** 8 **iii** 8

11 **i** Any number between −196 °C and −210 °C, e.g. −200 °C

ii Any number over −196 °C, e.g. −10 °C or any positive number
iii Any number below −210 °C, e.g. −250 °C

12 **a i** Liquid hydrogen **ii** 70 °C
b i 21 °C **ii** 91 °C

Answers to Chapter 4

4.1 Equivalent percentages, fractions and decimals

Exercise 4A

1 **a** $\frac{2}{25}$ **b** $\frac{1}{2}$ **c** $\frac{1}{4}$ **d** $\frac{7}{20}$ **e** $\frac{9}{10}$ **f** $\frac{3}{4}$

2 **a** 0.27 **b** 0.85 **c** 0.13 **d** 0.06 **e** 0.8 **f** 0.32

3 **a** $\frac{3}{25}$ **b** $\frac{2}{5}$ **c** $\frac{9}{20}$ **d** $\frac{17}{25}$ **e** $\frac{1}{4}$ **f** $\frac{5}{8}$

4 **a** 29% **b** 55% **c** 3% **d** 16% **e** 60% **f** 125%

5 **a** 28% **b** 30% **c** 95% **d** 34% **e** 27.5% **f** 87.5%

6 **a** 0.6 **b** 0.075 **c** 0.76 **d** 0.3125 **e** 0.05 **f** 0.125

7 150

8 none

9 20

10 **a** 77% **b** 39% **c** 63%

11 27%

12 61.5%

13 **a** 50% **b** 20% **c** 80%

14 12.5%, 25%, 37.5%, 50%, 75%

15 **a** 20% **b** 25% **c** 75% **d** 45% **e** 14% **f** 50%
g 60% **h** 17.5% **i** 55% **j** 130%

16 **a** 33.3% **b** 16.7% **c** 66.7% **d** 83.3% **e** 28.6% **f** 78.3%
g 68.9% **h** 88.9% **i** 81.1% **j** 20.9%

17 **a** 7% **b** 80% **c** 66% **d** 25% **e** 54.5% **f** 82%
g 30% **h** 89.1% **i** 120% **j** 278%

18 **a** $\frac{3}{5}$ **b** 0.6 **c** 60%

19 **a** 63%, 83%, 39%, 62%, 77% **b** English

20 6.7%

21 25.5%

22 34%, 0.34, $\frac{17}{50}$; 85%, 0.85, $\frac{17}{20}$; 7.5%, 0.075, $\frac{3}{40}$; 45%, 0.45, $\frac{9}{20}$; 30%, 0.3, $\frac{3}{10}$; 66%, 0.66, $\frac{2}{3}$; 84%, 0.84, $\frac{21}{25}$; 45%, 0.45, $\frac{9}{20}$; 37.5%, 0.375, $\frac{3}{8}$

23 $\frac{9}{10}$ or 90%

4.2 Calculating a percentage of a quantity

Exercise 4B

1 **a** 0.88 **b** 0.3 **c** 0.25 **d** 0.08 **e** 1.15

2 **a** 78% **b** 40% **c** 75% **d** 5% **e** 110%

3 **a** £45 **b** £6.30 **c** 128.8 kg **d** 1.125 kg **e** 1.08 h **f** 37.8 cm **g** £0.12 **h** 2.94 m **i** £7.60 **j** 33.88 min **k** 136 kg **l** £162

4 48

5 £2410

6 **a** 86% **b** 215

7 8520

8 287

9 Each team: 54 000, referees: 900, other teams: 9000 (100 to each, FA: 18 000, celebrities: 8100

10 990

11 Mon: 816, Tue: 833, Wed: 850, Thu: 799, Fri: 748

12 **a** £3.25 **b** 2.21 kg **c** £562.80 **d** £6.51 **e** 42.93 m **f** £24

13 480 cm^3 nitrogen, 120 cm^3 oxygen

14 13

15 £270

16 More this year as it was 3% of a higher amount than last year.

4.3 Ratio

Exercise 4C

1 **a** 1 : 3 **b** 3: 4 **c** 2 : 3 **d** 2 : 3 **e** 2 : 5 **f** 2 : 5 **g** 5 : 8 **h** 25 : 6 **i** 3 : 2 **j** 8 : 3 **k** 7 : 3 **l** 5 : 2 **m** 1 : 6 **n** 3 : 8 **o** 5 : 3 **p** 4 : 5

2 **a** 1 : 3 **b** 3 : 2 **c** 5 : 12 **d** 8 : 1 **e** 17 : 15 **f** 25 : 7 **g** 4 : 1 **h** 5 : 6 **i** 1 : 24 **j** 48 : 1 **k** 5 : 2 **l** 3 : 14 **m** 2 : 1 **n** 3 : 10 **o** 31 : 200 **p** 5 : 8

3 $\frac{7}{10}$

4 $\frac{10}{25} = \frac{2}{5}$

5 **a** $\frac{2}{5}$ **b** $\frac{3}{5}$

6 **a** $\frac{7}{10}$ **b** $\frac{3}{10}$

7 Amy $\frac{3}{5}$, Katie $\frac{2}{5}$

8 Fruit crush $\frac{5}{32}$, lemonade $\frac{27}{32}$

9 $13\frac{1}{2}$ litres

10 **a** $\frac{1}{2}$ **b** $\frac{7}{20}$ **c** $\frac{3}{20}$

11 James $\frac{1}{2}$, John $\frac{3}{10}$, Joseph $\frac{1}{5}$

12 sugar $\frac{5}{22}$, flour $\frac{3}{11}$, margarine $\frac{2}{11}$, fruit $\frac{7}{22}$

13 3 : 1

14 1 : 4

Exercise 4D

1 **a** 160 g, 240 g **b** 80 kg, 200 kg **c** 150, 350 **d** 950 m, 50 m **e** 175 min, 125 min **f** £20, £30, £50 **g** £36, £60, £144 **h** 50 g, 250 g, 300 g **i** £1.40, £2, £1.60 **j** 120 kg, 72 kg, 8 kg

2 **a** 175 **b** 30%

3 **a** 40% **b** 300 kg

4 21

5 **a** Mott: no, Wright: yes, Brennan: no, Smith: no, Kaye: yes
b For example: W26, H30; W31, H38; W33, H37

6 **a** 1 : 400 000 **b** 1: 125 000 **c** 1 : 250 000 **d** 1 : 25 000 **e** 1 : 20 000 **f** 1 : 40 000 **g** 1 : 62 500 **h** 1 : 10 000 **i** 1 : 60 000

7 **a** 1 : 1 000 000 **b** 47 km **c** 8 mm

8 **a** 1 : 250 000 **b** 2 km **c** 4.8 cm

9 **a** 1 : 20 000 **b** 0.54 km **c** 40 cm

10 **a** 1 : 1.6 **b** 1 : 3.25 **c** 1 : 1.125 **d** 1 : 1.44 **e** 1 : 5.4 **f** 1 : 1.5 **g** 1 : 4.8 **h** 1 : 42 **i** 1 : 1.25

Exercise 4E

1 **a** 3 : 2 **b** 32 **c** 80

2 **a** 100 **b** 160

3 0.4 litres

4 102

5 1000 g

6 10 125

7 5.5 litres

8 **a** 14 min **b** 75 min

9 **a** 11 pages **b** 32%

10 Kevin £2040, John £2720

11 **a** 160 cans **b** 48 cans

12 **a** lemonade 20 litres, ginger 0.5 litres
b This one, one-thirteenth is greater than one-fiftieth.

13 60

14 100

15 40 cc

4.4 Best buys

Exercise 4F

1 **a** £4.50 for a 10-pack **b** £1.08 for 6 **c** £2.45 for 1 litre **d** Same value **e** 29p for 250 grams **f** £1.39 for a pack of 6 **g** £4 for 3 cartons

2 **a** Large jar **b** 600 g tin **c** 5 kg bag **d** 75 ml tube **e** Large box **f** Large box **g** 400 ml bottle

3 **a** £5.11 **b** Large tin

4 **a** 95p **b** Family size

5 Bashir's

6 Mary

7 Kelly

4.5 Speed, time and distance

Exercise 4G

1 18 mph

2 280 miles

3 52.5 mph

4 11.50 am

5 500 s

6 **a** 75 mph **b** 6.5 h **c** 175 miles **d** 240 km **e** 64 km/h **f** 325 km **g** 4.3 h (4 h 18 min)

7 **a** 7.75 h **b** 52.9 mph

8 **a** 2.25 h **b** 99 miles

9 **a** 1.25 h **b** 1 h 15 min

10 **a** 48 mph **b** 6 h 40 min

11 **a** 120 km **b** 48 km/h

12 **a** 30 min **b** 6 mph

13 **a** 10 m/s **b** 3.3 m/s **c** 16.7 m/s **d** 41.7 m/s **e** 20.8 m/s

14 **a** 90 km/h **b** 43.2 km/h **c** 14.4 km/h **d** 108 km/h **e** 1.8 km/h

15 **a** 64.8 km/h **b** 28 s **c** 8.07 (37 min journey)

16 **a** 6.7 m/s **b** 66 km **c** 5 minutes **d** 133.3 metres

17 6.6 minutes

Examination questions

1 **a** 0.4375 **b** 0.27

2 History = 84%, geography = 80% so history is higher

3 20%

4 The small packet is better value as it costs 0.42p per gram and the large packet costs 0.44p per gram.

5 **a** £30 **b** £220

6 **a** 0.92 **b** $\frac{3}{100}$ **c** 20 grams

7 £23 800

8 £140

9 **a** 35% **b** $\frac{3}{10}$ **c** 280 **d** 2%

10 **a** $\frac{37}{100}$ **b** £153.55

11 50%

12 **a** £3.20 **b** 75% **c** £17

13 **a** 93.6 mph **b** 5 hours 12 minutes

14 40 mph

15 225 km

16 80 km/h

17 **a** 3.75 mph **b** 3.33 mph

18 **a** $\frac{3}{10}$ **b** 56

19 62.5 cm

20 **a** B and C **b** Jack, 24; Kenny, 28

21 **a** 3 : 7 **b** 2

Answers to Chapter 5

5.1 Long multiplication

Exercise 5A

1 **a** 12 138 **b** 45 612 **c** 29 988 **d** 20 654 **e** 51 732 **f** 25 012 **g** 19 359 **h** 12 673 **i** 19 943 **j** 26 235 **k** 31 535 **l** 78 399 **m** 17 238 **n** 43 740 **o** 66 065 **p** 103 320 **q** 140 224 **r** 92 851 **s** 520 585 **t** 78 660

2 3500

3 No, 62 pupils cannot get in

4 Yes, he walks 57.6 km

5 Yes, 1 204 000 letters

6 5819 litres

7 Yes, she raised £302.40

5.2 Long division

Exercise 5B

1 **a** 25 **b** 15 **c** 37
d 43 **e** 27 **f** 48
g 53 **h** 52 **i** 32
j 57 **k** 37 rem 15 **l** 25 rem 5
m 34 rem 11 **n** 54 rem 9 **o** 36 rem 11
p 17 rem 4 **q** 23 **r** 61 rem 14
s 42 **t** 27 rem 2

2 68

3 38

4 4 months

5 34 h

6 £1.75

5.3 Arithmetic with decimal numbers

Exercise 5C

1 **a** 4.8 **b** 3.8 **c** 2.2 **d** 8.3 **e** 3.7 **f** 46.9
g 23.9 **h** 9.5 **i** 11.1 **j** 33.5 **k** 7.1 **l** 46.8
m 0.1 **n** 0.1 **o** 0.6 **p** 65.0 **q** 213.9 **r** 76.1
s 455.2 **t** 51.0

2 **a** 5.78 **b** 2.36 **c** 0.98 **d** 33.09 **e** 6.01 **f** 23.57
g 91.79 **h** 8.00 **i** 2.31 **j** 23.92 **k** 6.00 **l** 1.01
m 3.51 **n** 96.51 **o** 0.01 **p** 0.07 **q** 7.81
r 569.90 **s** 300.00 **t** 0.01

3 **a** 4.6 **b** 0.08 **c** 45.716 **d** 94.85 **e** 602.1
f 671.76 **g** 7.1 **h** 6.904 **i** 13.78 **j** 0.1 **k** 4.002
l 60.0 **m** 11.99 **n** 899.996 **o** 0.1 **p** 0.01
q 6.1 **r** 78.393 **s** 200.00 **t** 5.1

4 **a** 9 **b** 9 **c** 3 **d** 7 **e** 3 **f** 8
g 3 **h** 8 **i** 6 **j** 4 **k** 7 **l** 2
m 47 **n** 23 **o** 96 **p** 33 **q** 154 **r** 343
s 704 **t** 910

5 £1 + £7 + £4 + £1 = £13

6 3, 3.46, 3.5

7 4.7275 or 4.7282

Exercise 5D

1 **a** 49.8 **b** 21.3 **c** 48.3 **d** 33.3 **e** 5.99 **f** 8.08
g 90.2 **h** 21.2 **i** 12.15 **j** 13.08 **k** 13.26 **l** 24.36

2 **a** 1.4 **b** 1.8 **c** 4.8 **d** 3.8 **e** 3.75 **f** 5.9
g 3.7 **h** 3.77 **i** 3.7 **j** 1.4 **k** 11.8 **l** 15.3

3 **a** 30.7 **b** 6.6 **c** 3.8 **d** 16.7 **e** 11.8 **f** 30.2
g 43.3 **h** 6.73 **i** 37.95 **j** 4.7 **k** 3.8 **l** 210.5

4 **a** £16.74 **b** £1.40

5 2.7 m + 2.5 m = 5.2 m (one pipe)
3.1 m + 1.7 m = 4.8 m (second pipe)
4.2 m (third pipe). Three pipes needed.

6 **a** 5.3 **b** 6.7 **c** 2.05 **d** 1.9 **e** 4.95 **f** 3.71

7 **a** 1.6 **b** 42.7 **c** 4.29 **d** 12.8 **e** 22.4 **f** 51.97

8 DVD, shirt and pen

Exercise 5E

1 **a** 7.2 **b** 7.6 **c** 18.8 **d** 37.1 **e** 32.5 **f** 28.8
g 10.0 **h** 55.2 **i** 61.5 **j** 170.8 **k** 81.6 **l** 96.5

2 **a** 9.36 **b** 10.35 **c** 25.85 **d** 12.78 **e** 1.82 **f** 3.28
g 2.80 **h** 5.52 **i** 42.21 **j** 56.16 **k** 7.65 **l** 48.96

3 **a** 1.8 **b** 1.4 **c** 1.4 **d** 1.2 **e** 2.13 **f** 0.69
g 2.79 **h** 1.21 **i** 1.89 **j** 1.81 **k** 0.33 **l** 1.9

4 **a** 1.75 **b** 1.28 **c** 1.85 **d** 3.65 **e** 1.66 **f** 1.45
g 1.42 **h** 1.15 **i** 3.35 **j** 0.98 **k** 2.3 **l** 1.46

5 **a** 1.89 **b** 1.51 **c** 0.264 **d** 4.265 **e** 1.224 **f** 0.182
g 0.093 **h** 2.042 **i** 1.908 **j** 2.8 **k** 4.25 **l** 18.5

6 Pack of 8 at £0.625 each

7 **a** £49.90 **b** Small, so that all four children can have an ice cream (cost £4.80)

8 Yes. She only needed 8 paving stones.

Exercise 5F

1 **a** 89.28 **b** 298.39 **c** 66.04
d 167.98 **e** 2352.0 **f** 322.4
g 1117.8 **h** 4471.5 **i** 464.94
j 25.55 **k** 1047.2 **l** 1890.5

2 **a** £224.10 **b** £223.75 **c** £29.90

3 $5 \times 7 = 35$

4 £54.20

5 **a** £120.75 **b** £17 − £3.45 = £13.55

Exercise 5G

1 **a** 0.48 **b** 2.92 **c** 1.12
d 0.12 **e** 0.028 **f** 0.09
g 0.192 **h** 3.0264 **i** 7.134
j 50.96 **k** 3.0625 **l** 46.512

2 Yes, with 25p left over.

3 a 35, 35.04, 0.04 b 16, 18.24, 2.24
c 60, 59.67, 0.33 d 180, 172.86, 7.14
e 12, 12.18, 0.18 f 24, 26.016, 2.016
g 40, 40.664, 0.664 h 140, 140.58, 0.58

4 a 572 b i 5.72 ii 1.43 iii 22.88

5.4 Multiplying and dividing with negative numbers

Exercise 5H

1 a −15 b −14 c −24 d 6 e 14 f 2
g −2 h −8 i −4 j 3 k −24 l −10
m −18 n 16 o 36 p −4 q −12 r −4
s 7 t 25 u 18 v −8 w −45 x 3
y −40

2 a −9 b 16 c −3 d −32 e 18 f 18
g 6 h −4 i 20 j 16 k 8 l −48
m 13 n −13 o −8 p 0 q 16 r −42
s 6 t 1 u −14 v 6 w −4 x 7
y 0

3 a −2 b 30 c 15 d −27 e −7

4 a 4 b −9 c −3 d 6 e −4

5 a −9 b 3 c 1

6 a 16 b −2 c −12

7 a 24 b 6 c −4 d −2

8 For example: 1 × (−12), −1 × 12, 2 × (−6), 6 × (−2), 3 × (−4), 4 × (−3)

9 For example: 4 ÷ (−1), 8 ÷ (−2), 12 ÷ (−3), 16 ÷ (−4), 20 ÷ (−5), 24 ÷ (−6)

10 a 21 b −4 c 2 d −16 e 2 f −5
g −35 h −17 i −12 j 6 k 45 l −2
m 0 n −1 o −7 p −36 q 9 r 32
s 0 t −65

11 a -12 b 12 degrees c 3 × −6

12 −5 × 4, 3 × −6, −20 ÷ 2, −16 ÷ −4

13 a 4 b 13 c 10 d 1

5.5 Approximation of calculations

Exercise 5I

1 a 50 000 b 60 000 c 30 000
d 90 000 e 90 000 f 50
g 90 h 30 i 100
j 200 k 0.5 l 0.3
m 0.006 n 0.05 o 0.0009
p 10 q 90 r 90
s 200 t 1000

2 a 60 000 b 30 000 c 80 000
d 30 000 e 10 000 f 6000
g 1000 h 800 i 100
j 600 k 2 l 4
m 3 n 8 o 40
p 0.8 q 0.5 r 0.07
s 1 t 0.01

3 a 65, 74 b 95, 149 c 950, 1499

4 Elsecar 750, 849; Hoyland 950, 1499; Barnsley 150 000, 249 999

5 15, 16 or 17

6 1, because there could be 450 then 449

Exercise 5J

1 a 35 000 b 15 000 c 960
d 12 000 e 1050 f 4000
g 4 h 20 i 1200

2 a £3000 b £2000 c £1500
d £700

3 a £15 000 b £18 000 c £18 000

4 £21 000

5 a 14 b 10 c 1.1 d 1 e 5 f $\frac{2}{3}$
g 3 or 4 h $\frac{1}{2}$ i 6 j 400 k 2 l 20

6 a 500 b 200 c 90 d 50 e 50 f 500

7 8

8 a 200 b 2800 c 10 d 1000

9 a 40 b 10 c £70

10 1000 or 1200

11 a 28 km b 120 km c 1440 km

12 400

13 a 3 kg b 200

Exercise 5K

1 a 1.7 m b 6 min c 240 g
d 80 °C e 16 miles f 14 m^2

2 82 °F, 5.3 km, 110 min, 43 000 people, 6.2 s, 67th, 1788, 15, 5 s

3 40 × £20 = £800

4 22.5 °C − 18.2 °C = 4.3 °C

Examination questions

1 **a** £1.80 **b** £1.25 **c** 40p

2 **a** £6.55 **b** £7.38 **c** £3.31

3 Answer in the region of £7.50 to £8

4 **a** 9 **b** 36

5 83p

6 **a** 450 **b** 2800

7 **a** 34 **b** 150

8 40.96

9 Chuck 30 + 14 ÷ 1.3 − 0.5 = 44 ÷ 0.8 = 55

10 800

11 600

Answers to Chapter 6

6.1 Reading scales

Exercise 6A

1 **a** **i** 4 **ii** 16 g **iii** 38 mph
b **i** 8 kg **ii** 66 mph **iii** 60 g
c **i** 13 oz **ii** 85 mph **iii** 76 kph
d **i** 26 **ii** 71.6 **iii** 64

2 **a**

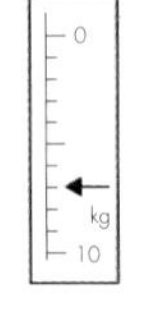

b 20 30 40 mph

c

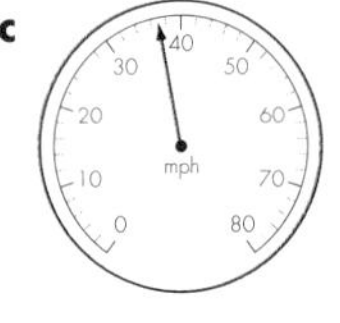

d

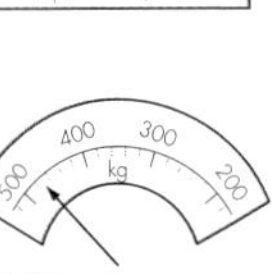

3 **a** 50 °C **b** 64 °C **c** −10 °C
d 82 °C **e** −16 °C

4 **a** 8 kg **b** 29 mph **c** 230 g
d 12.7 kg

5 360 g **b** weigh out 400 g, then weigh out 300 g

6 **a** 1.2 kg **b** 125 g

7 **a** 125 km/h
b 125 km/h shown on scale; the scale could go up in 10s, 20s or 25s, discuss this with students

8 No, it is pointing to 7.48 m; each division is 0.1 ÷ 5 = 0.02

6.2 Sensible estimates

Exercise 6B

1 Bicycle about 2 m, bus about 10 m, train about 17 m

2 Height about 4 m, length about 18 or 19 m

3 250 g

4 **a** About 4 m **b** About 5 m **c** About 5.5 m

5 About 9.5 g or 10 g

6 About 9 m; the ratio of Joel's height in the photograph to his real height must be the same as the ratio of the height on the statue in the photograph to its actual height

7 About 6 m

6.3 Systems of measurement

Exercise 6C

1 **a** metres **b** kilometres **c** millimetres
d kilograms or grams **e** litres **f** kilograms
g tonnes **h** millilitres **i** centilitres
j metres **k** kilograms **l** litres
m grams **n** centilitres **o** millimetres

2 Check individual answers.

3 The 5 metre since his height is about 175 cm, the lamp post will be about 525 cm

4 Inches, feet and yards are too small as units; this distance is an approximation and so needs to be a large unit as this is a large distance.

6.4 Metric units

Exercise 6D

1 **a** 1.25 m **b** 8.2 cm **c** 0.55 m
d 2.1 km **e** 2.08 cm **f** 1.24 m
g 4.2 kg **h** 5.75 t **i** 8.5 cl
j 2.58 l **k** 3.4 l **l** 0.6 t
m 0.755 kg **n** 0.8 l **o** 2 l
p 63 cl **q** 8.4 m^3 **r** 35 cm^3
s 1.035 m^3 **t** 0.53 m^3 **u** 34 000 m

2 **a** 3400 mm **b** 135 mm **c** 67 cm
d 7030 m **e** 7.2 mm **f** 25 cm
g 640 m **h** 2400 ml **i** 590 cl
j 84 ml **k** 5200 l **l** 580 g
m 3750 kg **n** 0.000 94 l **o** 2160 cl
p 15 200 g **q** 14 000 l **r** 0.19 ml

3 He should choose the 2000 mm × 15 mm × 20 mm

4 as 1 millilitre = 1 cm^3 and 1 litre = 1000 cm^3

5 1 000 000 000 000

6.5 Imperial units

Exercise 6E

1 **a** 24 in **b** 12 ft **c** 3520 yd
d 80 oz **e** 56 lb **f** 6720 lb
g 40 pt **h** 48 in **i** 36 in
j 30 ft **k** 64 oz **l** 5 ft
m 70 lb **n** 12 yd **o** 224 oz

2 **a** 5 miles **b** 120 pt **c** 5280 ft
d 8 ft **e** 7 st **f** 7 gal
g 2 lb **h** 5 yd **i** 5 tons
j 63 360 in **k** 8 lb **l** 9 gal
m 10 st **n** 3 miles **o** 35 840 oz

3 the 32-ounce bag

4 4 014 489 600

5 1 tonne = 1000 kilograms = 1000 × 2.2 pounds = 2200 pounds; 1 ton = 2240 pounds; 2240 is greater than 2200

6.6 Conversion factors

Exercise 6F

1 **a** 20 cm **b** 13.2 lb **c** 48 km
d 67.5 l **e** 2850 ml **f** 10 gal
g 12 in **h** 50 miles **i** 5 kg
j 3 pints **k** 160 km **l** 123.2 lb
m 180 l **n** 90.9 kg **o** 1100 yd
p 30 cm **q** 6.4 kg **r** 90 cm

2 ton

3 metre

4 **a** **i** 1000 g **ii** 1 kg
b **i** 4500 g **ii** 4.5 kg

5 **a** 135 miles **b** 50 mph **c** 2 h 42 min

6 4 hours 10 minutes

7 288

Examination questions

1 metres, kilograms, inches

2

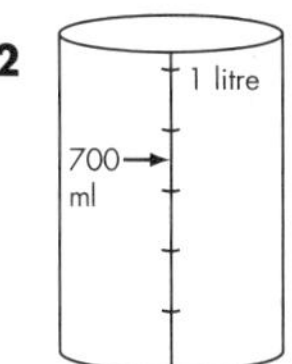

3 **a** 17.8 cm **b** −2 °C **c** 2.8 kg

4 **a** 40 000 m **b** 25 miles

5 No. 30 pints ≈ 17.1 litres (30 × 0.570)

6 **a** 1.8 m **b** about 7 m

7 **a** 250 mm **b** 324 m

8 **a** metres, grams, litres **b** 400 cm **c** 1.5 kg

9 **a** 4.6 kg
b **i** 2.2 lb **ii** 11 lb

10 81.25 mph

11 **a** 1 m **b** 2.82 m

12 Centimetres is too small a unit; this distance is an approximation of a large distance and so needs to be a large unit such as kilometres.

13 **a** $7\frac{1}{2}$ miles
b 2.3 kg of potatoes, 227 grams of butter, 2.25 litres of milk

Answers to Chapter 7

7.1 Measuring and drawing angles

Exercise 7A

1 **a** 40° **b** 30° **c** 35° **d** 43° **e** 100° **f** 125°
g 340° **h** 225°

2 Student's drawings of angles

3 Student's drawing and calculations

4 Yes, the angle is 75°

5 Any angle between 0° and 45°

6 (c) is an obtuse angle, the others are acute

7 **a** 80° **b** 50° **c** 25°

7.2 Angle facts

Exercise 7B

1 **a** 48° **b** 307° **c** 108° **d** 52° **e** 59° **f** 81°
g 139° **h** 51° **i** 138° **j** 128° **k** 47° **l** 117°
m 27° **n** 45° **o** 108° **p** 69° **q** 135° **r** 58°
s 74° **t** 23° **u** 55° **v** 56°

2 **a** 82° **b** 105° **c** 75°

3 45° + 125° = 170° and for a straight line it should be 180°.

4 **a** $x = 100°$ **b** $x = 110°$ **c** $x = 30°$

5 **a** $x = 55°$ **b** $x = 45°$ **c** $x = 12.5°$

6 **a** $x = 34°, y = 98°$ **b** $x = 70°, y = 120°$ **c** $x = 20°, y = 80°$

7 6 × 60° = 360°; imagine six of the triangles meeting at a point

8 $x = 35°, y = 75°$; $2x = 70°$ (opposite angles), so $x = 35°$ and $x + y = 110°$ (angles on a line), so $y = 75°$

7.3 Angles in a triangle

Exercise 7C

1 **a** 70° **b** 50° **c** 80° **d** 60° **e** 75° **f** 109°
g 38° **h** 63°

2 **a** No, total is 190° **b** Yes, total is 180° **c** No, total is 170°
d Yes, total is 180° **e** Yes, total is 180° **f** No, total is 170°

3 **a** 80° **b** 67° **c** 20° **d** 43° **e** 10° **f** 1°

4 **a** 60° **b** Equilateral triangle **c** Same length

5 **a** 70° each **b** Isosceles triangle **c** Same length

6 $x = 50°$, $y = 80°$

7 **a** 109° **b** 130° **c** 135°

8 65°

9 Isosceles triangle; angle DFE = 30° (opposite angles), angle DEF = 75° (angles on a line), angle FDE = 75° (angles in a triangle), so there are two equal angles in the triangle and hence it is an isosceles triangle

10 $a = 80°$ (opposite angles), $b = 65°$ (angles on a line), $c = 35°$ (angles in a triangle)

11 Missing angle $= y$, $x + y = 180°$ and $a + b + y = 180°$ so $x = a + b$

7.4 Parallel lines

Exercise 7D

1 **a** 40° **b** $b = c = 70°$ **c** $d = 75°$, $e = f = 105°$ **d** $g = 50°$, $h = i = 130°$ **e** $j = k = l = 70°$ **f** $n = m = 80°$

2 **a** $a = 50°$, $b = 130°$ **b** $c = d = 65°$, $e = f = 115°$ **c** $g = i = 65°$, $h = 115°$ **d** $j = k = 72°$, $l = 108°$ **e** $m = n = o = p = 105°$ **f** $q = r = s = 125°$

3 **a** $a = 95°$ **b** $b = 66°$, $c = 114°$

4 **a** $x = 30°$, $y = 120°$ **b** $x = 25°$, $y = 105°$ **c** $x = 30°$, $y = 100°$

5 **a** $x = 50°$, $y = 110°$ **b** $x = 25°$, $y = 55°$ **c** $x = 20°$, $y = 140°$

6 290°; x is double the angle allied to 35°, so is $2 \times 145°$

7 $a = 102°$ (angles on a line = 180°), $b = 78°$ (corresponding angle or allied angle), angle BDC = 66° (angles in a triangle = 180°) so $c = 114°$ (angles on a line = 180°), $d = 66°$ (corresponding angle or allied angle)

8 Angle PQD = 64° (alternate angles), so angle DQY = 116° (angles on a line = 180°)

9 Use alternate angles to see b, a and c are all angles on a straight line, and so total 180°

10 Third angle in triangle equals q (alternative angle), angle sum of triangle is 180°.

7.5 Special quadrilaterals

Exercise 7E

1 **a** $a = 110°$, $b = 55°$ **b** $c = 75°$, $d = 115°$ **c** $e = 87°$, $f = 48°$

2 **a** $a = c = 105°$, $b = 75°$ **b** $d = f = 70°$, $e = 110°$ **c** $g = i = 63°$, $h = 117°$

3 **a** $a = 135°$, $b = 25°$ **b** $c = d = 145°$ **c** $e = f = 94°$

4 **a** $a = c = 105°$, $b = 75°$ **b** $d = f = 93°$, $e = 87°$ **c** $g = i = 49°$, $h = 131°$

5 **a** $a = 58°$, $b = 47°$ **b** $c = 141°$, $d = 37°$ **c** $e = g = 65°$, $f = 115°$

6 both 129°

7 Marie, a rectangle must have right angles

8 **a** 65°
b Trapezium, angle A + angle D = 180° and angle B + angle C = 180°

Examination questions

1 D

2 50°

3 **a i** 30° **ii** opposite angles
b 3 angles do not add up to 360°

4 D 80°

5 60°

6 70°

7 Angle CDE = 35° alternate angles, angle CED = 105° angles in triangle, so angle AEC = 75° angles on straight line. There are also other methods.

8 **a i** 62° **ii** alternate angle
b 56°

9 67°

Answers to Chapter 8

8.1 The language of algebra

Exercise 8A

1 **a** $x + 2$ **b** $x - 6$ **c** $k + x$ **d** $x - t$ **e** $x + 3$ **f** $d + m$ **g** $b - y$ **h** $p + t + w$ **i** $8x$ **j** hj **k** $x \div 4$ or $\frac{x}{4}$ **l** $2 \div x$ or $\frac{2}{x}$ **m** $y \div t$ or $\frac{y}{t}$ **n** wt **o** a^2 **p** g^2

2 **a** **i** $P = 4$, $A = 1$ **ii** $P = 4x$, $A = x^2$ **iii** $P = 12$, $A = 9$ **iv** $P = 4t$, $A = t^2$
b **i** $P = 4s$ cm **ii** $A = s^2$ cm^2

3 **a** $x + 3$ yr **b** $x - 4$ yr

4 $F = 2C + 30$

5 Rule **c**

6 **a** $C = 100M$ **b** $N = 12F$ **c** $W = 4C$ **d** $H = P$

7 **a** $3n$ **b** $3n + 3$ **c** $n + 1$ **d** $n - 1$

8 Rob: $2n$, Tom: $n + 2$, Vic: $n - 3$, Will: $2n + 3$

9 **a** $P = 8n, A = 9n^2$ **b** $P = 24n, A = 36n^2$

10 **a** £4 **b** £$(10 - x)$ **c** £$(y - x)$ **d** £$2x$

11 **a** 75p **b** $15x$ p **c** $4A$ p **d** Ay p

12 £$(A - B)$

13 £$A \div 5$ or $\frac{£A}{5}$

14 **a** Dad: $(72 + x)$ yr, me: $(T + x)$ yr **b** 31

15 **a** $T \div 2$ or $\frac{T}{2}$ **b** $T \div 2 + 4$ or $\frac{T}{2} + 4$ **c** $T - x$

16 **a** $8x$ **b** $12m$ **c** $18t$

17 Andrea: $3n - 3$, Bert: $3n - 1$, Colin: $3n - 6$ or $3(n - 2)$, Davina: 0, Emma: $3n - n = 2n$. Florinda: $3n - 3m$

18 For example, $2 \times 6m$, $1 \times 12m$, $6m + 6m$, etc

19 Any values picked for l and w and substituted into the formulae to give the same answers

20 13

21 15p

22 **a** expression **b** formula **c** equation

8.2 Simplifying expressions

Exercise 8B

1 **a** $6t$ **b** $15y$ **c** $8w$ **d** $5b^2$ **e** $2w^2$ **f** $8p^2$ **g** $6t^2$ **h** $15t^2$ **i** $2mt$ **j** $5qt$ **k** $6mn$ **l** $6qt$ **m** $10hk$ **n** $21pr$

2 **a** All except $2m \times 6m$ **b** 2 and 0

3 $4x$ cm

4 **a** y^3 **b** $3m^3$ **c** $4t^3$ **d** $6n^3$ **e** t^4 **f** h^5 **g** $12n^5$ **h** $6a^7$ **i** $4k^7$ **j** t^3 **k** $12d^3$ **l** $15p^6$ **m** $3mp^2$ **n** $6m^2n$ **o** $8m^2p^2$

5 The number of people who get told the rumour doubles each day ie 2, 4, 8, 16, 32, 64, 128, 256, 512, 1024, 2048, but the number who know the rumour is 3, 7, 15, 31, 63, 127, 255, 511, 1023, 2047 so by the 10th day everyone in the school would know, plus 47 other people!

Exercise 8C

1 **a** £t **b** £$(4t + 3)$

2 **a** $10x + 2y$ **b** $7x + y$ **c** $6x + y$

3 **a** $5a$ **b** $6c$ **c** $9e$ **d** $6f$ **e** $4j$ **f** $3q$ **g** 0 **h** $-w$ **i** $6x^2$ **j** $5y^2$ **k** 0

4 **a** $7x$ **b** $3t$ **c** $-5x$ **d** $-5k$ **e** $2m^2$ **f** 0

5 **a** $7x + 5$ **b** $5x + 6$ **c** $5p$ **d** $5x + 6$ **e** $5p + t + 5$ **f** $8w - 5k$ **g** c **h** $8k - 6y + 10$

6 **a** $2c - 3d$ **b** $5d - 2e$ **c** $f + 3g + 4h$ **d** $6u - 3v$ **e** $7m - 7n$ **f** $3k + 2m + 5p$ **g** $2v$ **h** $2w - 3y$ **i** $11x^2 - 5y$ **j** $-y^2 - 2z$ **k** $x^2 - z^2$

7 **a** $8x + 6$ **b** $3x + 16$ **c** $2x + 2y + 8$

8 Any acceptable answers, e.g. $x + 4x + 2y + 2y$ or $6x - x + 6y - 2y$

9 **a** $2x$ and $2y$ **b** a and $7b$

10 **a** $3x - 1 - x$ **b** $10x$ **c** 25 cm

11 **a** $12p + 6s$ **b** 13 m and 50 cm

12 Maria is correct, as the two short horizontal lengths are equal to the bottom length and the two short vertical lengths are equal to the side length.

8.3 Expanding brackets

Exercise 8D

1 **a** $6 + 2m$ **b** $10 + 5l$ **c** $12 - 3y$ **d** $20 + 8k$ **e** $12d - 8n$ **f** $t^2 + 3t$ **g** $m^2 + 5m$ **h** $k^2 - 3k$ **i** $3g^2 + 2g$ **j** $5y^2 - y$ **k** $5p - 3p^2$ **l** $3m^2 + 12m$ **m** $15t - 12t^2$ **n** $6d^2 + 12de$ **o** $6y^2 + 8ky$ **p** $15m^2 - 10mp$

2 **a** $y^3 + 5y$ **b** $h^4 + 7h$ **c** $k^3 - 5k$ **d** $3t^3 + 12t$ **e** $15d^3 - 3d^4$ **f** $6w^3 + 3tw$ **g** $15a^3 - 10ab$ **h** $12p^4 - 15mp$ **i** $5m^2 + 4m^3$ **j** $t^4 + 2t^5$ **k** $5g^2t - 4g^4$ **l** $15t^3 + 3mt^2$

3 **a** $5(t - 1)$ and $5t - 5$
b Yes as $5(t - 1)$ when $t = 4.50$ is $5 \times 3.50 =$ £17.50

4 He has worked out 3×5 as 8 instead of 15 and he has not multiplied the second term by 3. Answer should be $15x - 12$.

5 **a** $3(2y + 3)$ **b** $2(6z + 4)$ or $4(3z + 2)$

Exercise 8E

1 **a** $7t$ **b** $3y$ **c** $9d$ **d** $3e$ **e** $3p$ **f** $2t$ **g** $5t^2$ **h** $5ab$ **i** $3a^2d$

2 **a** $22 + 5t$ **b** $21 + 19k$ **c** $10 + 16m$ **d** $16 + 17y$ **e** $22 + 2f$ **f** $14 + 3g$ **g** $10 + 11t$ **h** $22 + 4w$

3 **a** $2 + 2h$ **b** $9g + 5$ **c** $6y + 11$ **d** $7t - 4$ **e** $17k + 16$ **f** $6e + 20$ **g** $7m + 4$ **h** $3t + 10$

4 **a** $4m + 3p + 2mp$ **b** $3k + 4h + 5hk$ **c** $3n + 2t + 7nt$
d $3p + 7q + 6pq$ **e** $6h + 6j + 13hj$ **f** $6t + 8y + 21ty$
g $24p + 12r + 13pr$ **h** $20k - 6m + 19km$

5 **a** $5(f + 2s) + 2(2f + 3s) = 9f + 16s$
b £$(270f + 480s)$ **c** 42 450 − 30 000 = £12 450

6 for x-coefficients 3 and 1 or 1 and 4; For y-coefficients 5 and 1 or 3 and 4 or 1 and 7

7 $5(3x + 2) - 3(2x - 1) = 9x + 13$

8.4 Factorisation

Exercise 8F

1 **a** $6(m + 2t)$ **b** $3(3t + p)$ **c** $4(2m + 3k)$
d $4(r + 2t)$ **e** $m(n + 3)$ **f** $g(5g + 3)$
g $2(2w - 3t)$ **h** $2(4p - 3k)$ **i** $2(8h - 5k)$
j $2m(p + k)$ **k** $2b(2c + k)$ **l** $2a(3b + 2c)$
m $y(3y + 2)$ **n** $t(4t - 3)$ **o** $2d(2d - 1)$
p $3m(m - p)$

2 **a** $3p(2p + 3t)$ **b** $2p(4t + 3m)$ **c** $4b(2a - c)$
d $4a(3a - 2b)$ **e** $3t(3m - 2p)$ **f** $4at(4t + 3)$
g $5bc(b - 2)$ **h** $2b(4ac + 3ed)$ **i** $2(2a^2 + 3a + 4)$
j $3b(2a + 3c + d)$ **k** $t(5t + 4 + a)$
l $3mt(2t - 1 + 3m)$ **m** $2ab(4b + 1 - 2a)$
n $5pt(2t + 3 + p)$

3 **a** Not possible **b** $m(5 + 2p)$ **c** $t(t - 7)$
d Not possible **e** $2m(2m - 3p)$ **f** Not possible
g $a(4a - 5b)$ **h** Not possible **i** $b(5a - 3bc)$

4 **a** Mary has taken out a common factor
b Because the bracket adds up to £10 **c** £30

5 Bella. Aidan has not taken out the largest possible common factor. Craig has taken m out of both terms but there isn't an m in the second term.

6 There are no common factors.

8.5 Substitution

Exercise 8G

1 **a** 8 **b** 17 **c** 32

2 **a** 3 **b** 11 **c** 43

3 **a** 9 **b** 15 **c** 29

4 **a** 9 **b** 5 **c** −1

5 **a** 13 **b** 33 **c** 78

6 **a** 10 **b** 13 **c** 58

7 **a** £4 **b** 13 km
c No, 5 miles is 8 km so fare would be £6.50

8 **a** $2 \times 8 + 6 \times 11 - 3 \times 2 = 76$
b $5 \times 2 - 2 \times 11 + 3 \times 8 = 12$

9 Any values such that $lw = \frac{1}{2}bh$ or $bh = 2lw$

10 **a** 32 **b** 64 **c** 160

11 **a** 6.5 **b** 0.5 **c** −2.5

12 **a** 2 **b** 8 **c** −10

13 **a** 3 **b** 2.5 **c** −5

14 **a** 6 **b** 3 **c** 2

15 **a** 12 **b** 8 **c** $1\frac{1}{2}$

16 **a** $\frac{1050}{n}$ **b** £925

17 **a** **i** odd
ii odd
iii even
iv odd
b Any valid expression such as $xy + z$

18 **a** £20
b **i** −£40
ii Delivery cost will be zero.
c 40 miles

Examination questions

1 **a** $4p$ **b** $12qr$ **b** $-3t$

2 $5d$

3 $20x$

4 **a** $4a + 7b$ **b** $5p + 10q - 15r$

5 26.8

6 **a** $x + 8y$ **b** 1 **c** 25

7 **a** **i** £2.40 **ii** £7.60
b **i** $20c$ **ii** $20c + 15d$

8 **a** $y + 5$
b $\frac{y+5}{2}$ or $\frac{1}{2}y + 2\frac{1}{2}$

9 **a** $12ef$ **b** $8x + 20$ **c** $7r - 9t$

10 **a** $4a$ **b** $2pq$ **c** $4a + 2b$

11 **a** $3g$ **b** $5hk$

12 **a** $2a + 3b$

13 $3x + 2y$

14 **a** 64 **b** 2 **c** $5p + 2q$

15 19.1

16 **a** $2x + 6 + 5x + 10 = 7x + 16$
b $4x + y - 2x + y = 2x + 2y$

17 **a** $4x + y$ **b** $3(2c + 3)$ **c** $z(z + 6)$

18 **a** $5(x + 2)$ **b** $x(x - 8)$

19 **a i** $3x - 9 + 2x + 4 = 5x - 5$ **ii** $n^2 - 2n + 1$
b i $a(6a + 1)$ **ii** $2xy^2(3xy - 2)$

20 **a** $3x - 3 + 6x - 10 = 9x - 13$
b $x^2 - 3x - 2x + 6 = x^2 - 5x + 6$

Answers to Chapter 9

9.1 Conversion graphs

Exercise 9A

1 **a i** $8\frac{1}{4}$ kg **ii** $2\frac{1}{4}$ kg **iii** 9 lb **iv** 22 lb
b 2.2 lb
c Read off the value for12.6 (5.4 kg) and multiply this by 4 (21.6 kg)

2 **a i** 10 cm **ii** 23 cm **iii** 2 in **iv** $8\frac{3}{4}$ in
b $2\frac{1}{2}$ cm
c Read off the value for 9 in (23 cm) and multiply this by 2 (46 cm)

3 **a i** \$320 **ii** \$100 **iii** £45 **iv** £78
b \$3.20
c It would become less steep.

4 **a i** £120 **ii** £82
b i 32 **ii** 48

5 **a i** £100 **ii** £325
b i 500 **ii** 250

6 **a i** £70 **ii** £29
b i £85 **ii** £38

7 **a i** 95 °F **ii** 68 °F **iii** 10 °C **iv** 32 °C
b 32 °F

8 **a** Check student's graph **b** 2.15 pm

9 **a** Check student's graph **b** £50

10 **a** No trains on Christmas day and Boxing day
b Sudden drop in passenger numbers
c Increased shopping and last-minute present buying

11 No as 100 miles is about 160 km. He has travelled 75 km and has 125 km to go which is a total of 200 miles.

12 **a** Anya: CabCo £8.50, YellaCabs £8.40, so YellaCabs is best
Bettina: CabCo £11.50, YellaCabs £11.60, so CabCo is best
Calista: CabCo £10, YellaCabs £10, so either
b If they shared a cab, the shortest distance is 16 km, which would cost £14.50 with CabCo and £14.80 with Yellacabs.

9.2 Travel graphs

Exercise 9B

1 **a i** 2 h **ii** 3 h **iii** 5 h
b i 40 km/h **ii** 120 km/h **iii** 40 km/h **c** 6.30 am

2 **a** 30 km **b** 40 km **c** 100 km/h

3 **a i** 125 km **ii** 125 km/h
b i Between 2 and 3 pm **ii** 25 km/h

4 Jafar started the race quickly and covered the first 1500 metres in 5 minutes. He then took a break for a minute and then finished the race at a slower pace taking 10 minutes overall. Azam started the race at a slower pace, taking 7 minutes to run 1500 metres. He then sped up and finished the race in a total of $9\frac{1}{2}$ minutes, beating Jafar.

5 **a** Araf ran the race at a constant pace, taking 5 minutes to cover the 1000 metres. Sean started slowly, covering the first 500 metres in 4 minutes. He then went faster, covering the last 500 metres in $1\frac{1}{2}$ minutes, giving a total time of $5\frac{1}{2}$ minutes for the race.
b i 20 km/h **ii** 12 km/h **iii** 10.9 km/h

6 There are three methods for doing this question.
This table shows the first, which is writing down the distances covered each hour.

Time	9 am	9:30	10:00	10.30	11.00	11.30	12.00	12.30
Walker	0	3	6	9	12	15	18	21
Cyclist	0	0	0	0	7.5	15	22.5	30

The second method is algebra:
Walker takes T hours until overtaken,
so $T = \frac{D}{6}$; Cyclist takes $T - 1.5$ to overtake, so $T - 1.5 = \frac{D}{15}$.
Rearranging gives $15T - 22.5 = 6T$, $9T = 22.5$, $T = 2.5$

The third method is a graph:

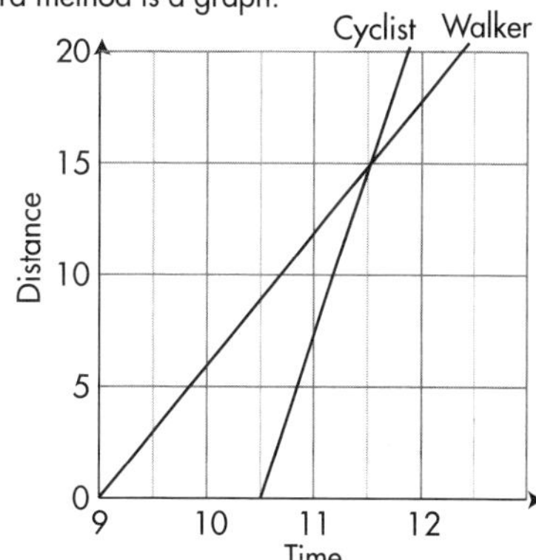

All methods give the same answer of 11:30 when the cyclist overtakes the walker.

7 **a i** Because it stopped several times **ii** Ravinder
b Ravinder at 3.55 pm or 3.58 pm, Sue at 4.20 pm, Michael at 4.35 pm
c i 24 km/h **ii** 20.6 km/h **iii** 5

9.3 Flow diagrams and graphs

Exercise 9C

1 **a** A(1, 2), B(3, 0), C(0, 1), D(−2, 4), E(−3, 2), F(−2, 0), G(−4, −1), H(−3, −3), I(1, −3), J(4, −2)
b i (2, 1) **ii** (−1, −3) **iii** (1, 1)
c $x = -3$, $x = 2$, $y = 3$, $y = -4$
d i $x = -\frac{1}{2}$ **ii** $y = -\frac{1}{2}$

2 Values of y: 2, 3, 4, 5, 6

3 Values of y: −2, 0, 2, 4, 6

4 Values of y: 1, 2, 3, 4, 5

5 Values of y: −4, −3, −2, −1, 0

6 **a** Values of y: 0, 4, 8, 12, 16 and 6, 8, 10, 12, 14
b (6, 12)

7 **a** Values of y: −3, −2, −1, 0, 1 and −6, −4, −2, 0, 2
b (3, 0)

8 Points could be (0, −1), (1, 4), (2, 9), (3, 14), (4, 19), (5, 24), etc

9 **a** £20.50
b £25.00, £28.50, £32.00, £35.50, £39.00, £42.50; (£20.00), £23.50, £27.00, £30.50, £34.00, £37.50; (£15.00), £18.50, £22.00, £25.50, £29.00, £32.50; (£10.00), (£13.50), £17.00, £20.50, £24.00, £27.50; (£5.00), (£8.50), (£12.00), £15.50, £19.00, £22.50; (£0.00), (£3.50), (£7.00), (£10.50), (£14.00), £17.50
c Yes, they had 3 cream teas and 4 high teas.

10 **a** 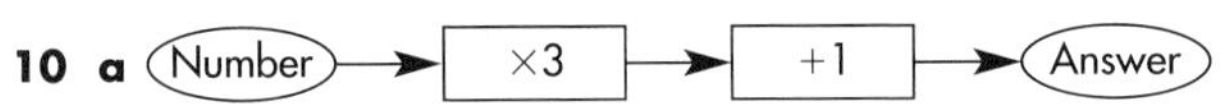

b $y = 3x + 1$
c Graph from (0, 1) to (5, 16)
d Read from 13 on the y-axis across to the graph and down to the x-axis. This should give a value of 4.

11 0, 8, 16, 24, 32, 40, 48

9.4 Linear graphs

Exercise 9D

1 Extreme points are (0, 4), (5, 19)

2 Extreme points are (0, −5), (5, 5)

3 Extreme points are (0, −3), (10, 2)

4 Extreme points are (−3, −4), (3, 14)

5 Extreme points are (−6, 2), (6, 6)

6 **a** Extreme points are (0, −2), (5, 13) and (0, 1), (5, 11)
b (3, 7)

7 **a** Extreme points are (0, −5), (5, 15) and (0, 3), (5, 13)
b (4, 11)

8 **a** Extreme points are (0, −1), (12, 3) and (0, −2), (12, 4)
b (6, 1)

9 **a** Extreme points are (0, 1), (4, 13) and (0, −2), (4, 10)
b Do not cross because they are parallel

10 **a** Values of y: 5, 4, 3, 2, 1, 0. Extreme points are (0, 5), (5, 0)
b Extreme points are (0, 7), (7, 0)

11 **a** Graph from (0, 25) to (8, 265) for Ian and graph from (0, 35) to (8, 255) for John.
b 4 hours

12 **a** Horizontal line through 4, vertical line through 1 and line from origin to (6, 6).
b 4.5 units squared

13 Graph passing through (x, z) (0, 2), (1, 3), (2, 4), (3, 5), (4, 6)

Exercise 9E

1 **a** 39.2 °C **b** Days 4 and 5, steepest line
c Days 8 and 9, steepest line **d** **i** Day 5 **ii** 2 **e** 37 °C

2 **a** 2 **b** $\frac{1}{3}$ **c** −3 **d** 1 **e** −2
f $-\frac{1}{3}$ **g** 5 **h** −5 **i** $\frac{1}{5}$ **j** $-\frac{3}{4}$

3 **a** 1 **b** −1
They are perpendicular and symmetrical about the axes.

4 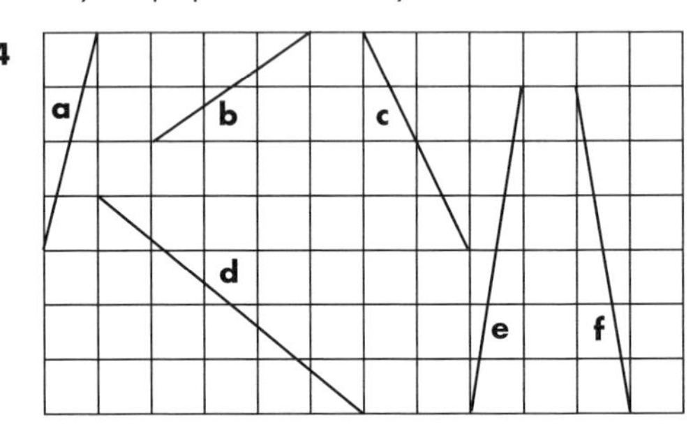

5 Rob has misread the scales. The gradient is actually 2. The line is $y = 2x + 2$. When $x = 10$, $y = 22$.

6 **a** 4 m **b** 1 m **c** **i** 1.375 m **ii** 3.2 m

Examination questions

1 **a** 19 – 21 feet
b 2.3 – 2.4 metres
c **i** Robert
ii The conversion graph shows that 4 metres is just over 13 feet, which is higher than 12 feet.

2 **a** (4, 3)
b cross drawn at point (−3, 1)
c (1, 0)

3 **a** **i** (1, 4) **ii** (4, 0)
b **i** Point P plotted at (3, 2)
ii Point Q plotted at (−4, 3)

4 **a**

x	−1	0	1	2	3
y	−5	−3	−1	1	3

b 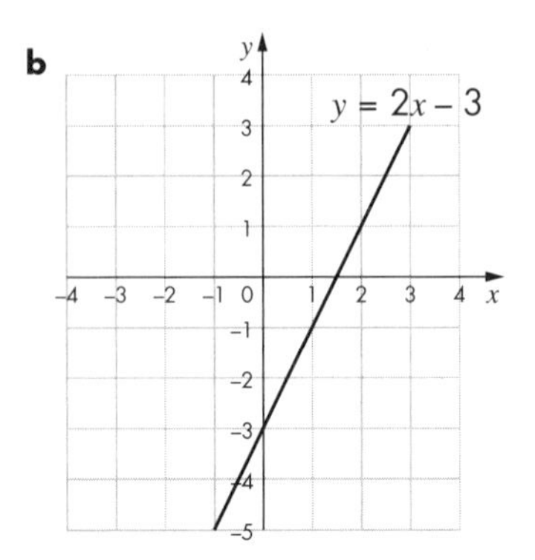

c (0.5, −2)

5 a

x	−2	−1	0	1	2
y	−7	−4	−1	2	5

b

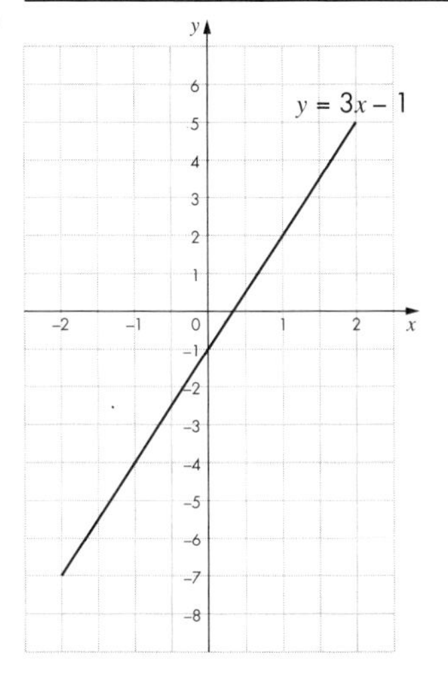

6 a

x	−3	−2	−1	0	1	2
y	−8	−5	−2	1	4	7

b

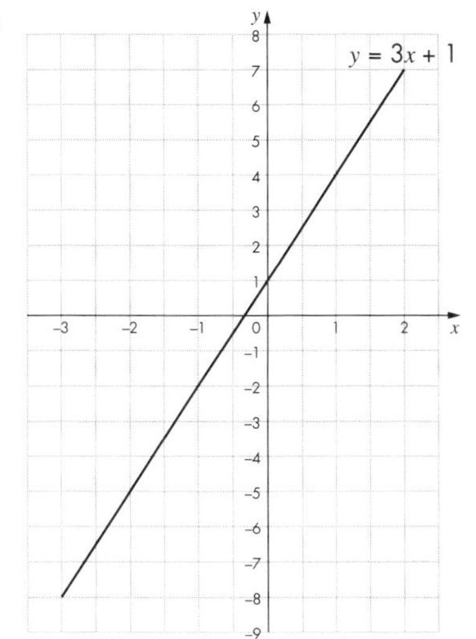

7 a

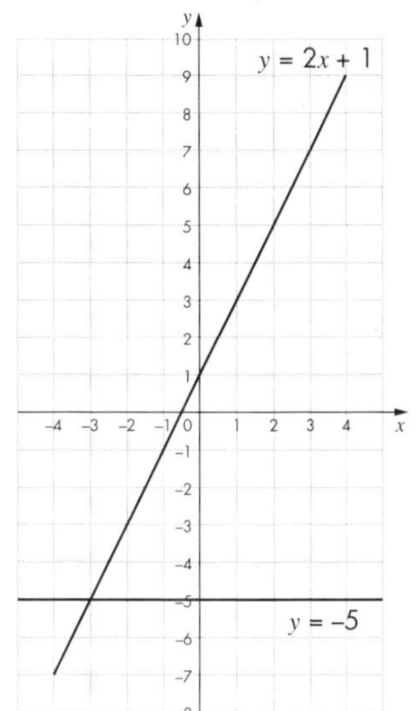

b (−3, −5)

8 a 10 km
b i 10:00 **ii** 30 minutes
c 11:20

9 a 1:30 pm
b i 10 km **ii** 30 minutes
c 36 km

10

The temperature starts at 0°c and keeps rising.	B
The temperature stays the same for a time and then falls.	D
The temperature rises and then falls quickly.	C
The temperature is always the same.	A
The temperature rises, stays the same for a time and then falls.	F
The temperature rises, stays the same for a time and then rises again.	E

Answers to Chapter 10

10.1 Frequency diagrams

Exercise 10A

1 a

Goals	0	1	2	3
Frequency	6	8	4	2

b 1 goal **c** 22

2 a

Temperature (°C)	14–16	17–19	20–22	23–25	26–28
Frequency	5	10	8	5	2

b 17–19 °C
c Getting warmer in the first half and then getting cooler towards the end.

3 a Observation **b** Sampling **c** Observation
d Sampling **e** Observation **f** Experiment

4 a

Score	1	2	3	4	5	6
Frequency	5	6	6	6	3	4

b 30 **c** Yes, frequencies are similar.

5 a

Height (cm)	151–155	156–160	161–165	166–170	171–175	176–180	181–185	186–190
Frequency	2	5	5	7	5	4	3	1

b 166–170 cm **c** Student's survey results.

6 various answers such as 1–10, 11–20, etc. or 1–20, 21–40, 41–60

7 The ages 20 and 25 are in two different groups.

8 Student's survey results and frequency tables.

10.2 Statistical diagrams

Exercise 10B

1

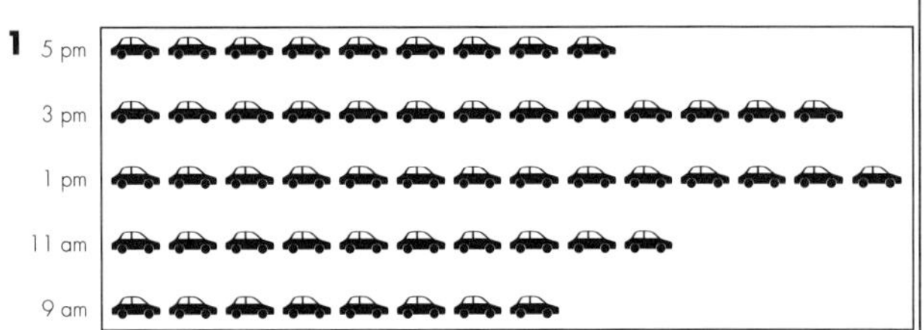

2

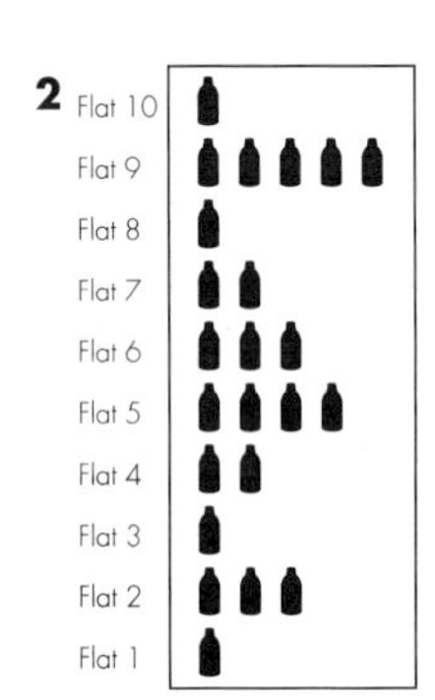

3 **a** May 9 h, Jun 11 h, Jul 12 h, Aug 11 h, Sep 10 h
b July **c** Visual impact, easy to understand.

4 **a** Simon **b** £165
c Difficult to show fractions of a symbol.

5 **a i** 12 **ii** 6 **iii** 13
b Check students' pictograms. **c** 63

6 Use a key of 17 students to one symbol.

7 There would be too many symbols to show.

8 **a–c** Student's pictograms.

9 The second teddy bear is much bigger than twice the size of the small teddy.

10.3 Bar charts

Exercise 10C

1 **a** Swimming
b 74
c For example: limited facilities
d No. It may not include people who are not fit.

2 **a**

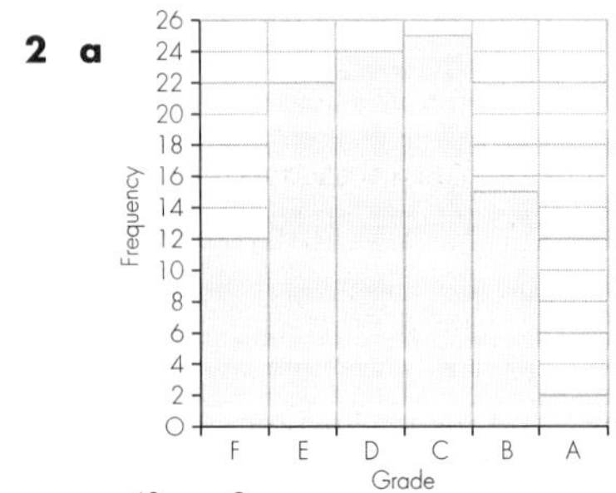

b $\frac{40}{100} = \frac{2}{5}$

c Easier to read the exact frequency.

3 **a**

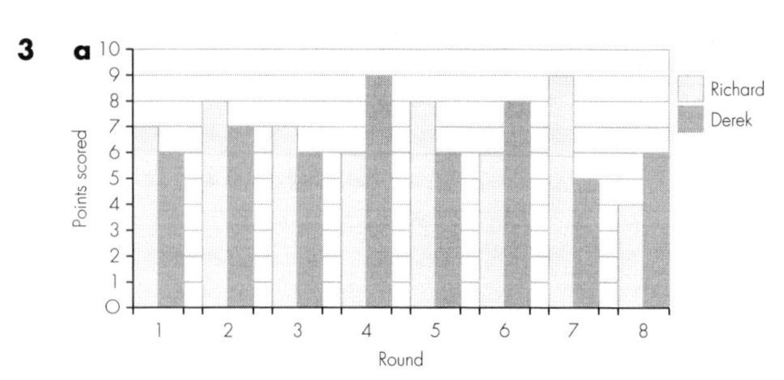

b Richard got more points overall, but Derek was more consistent.

4 **a**

Time (min)	1–10	11–20	21–30	31–40	41–50	51–60
Frequency	4	7	5	5	7	2

b

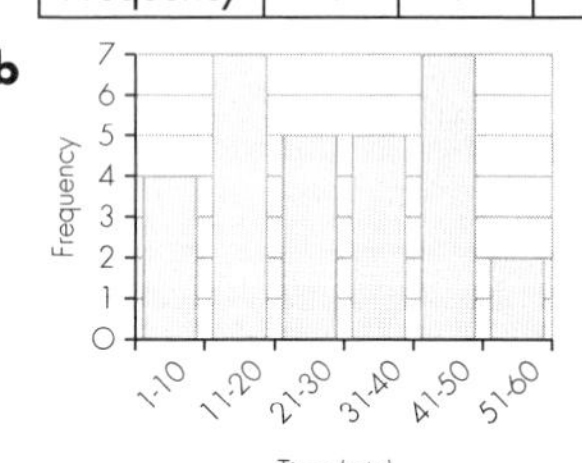

c Some live close to the school. Some live a good distance away and probably travel to school by bus.

5 **a**

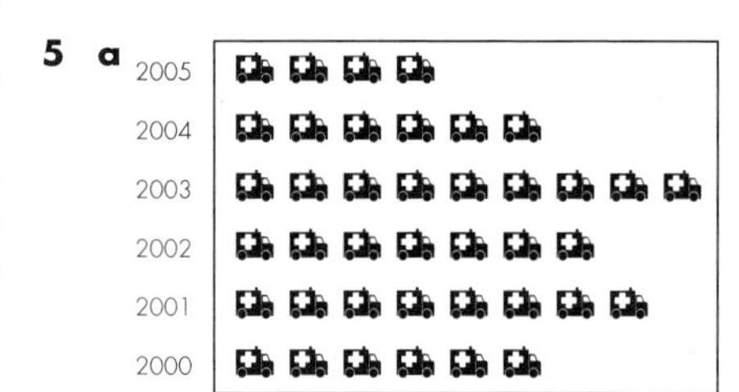

b

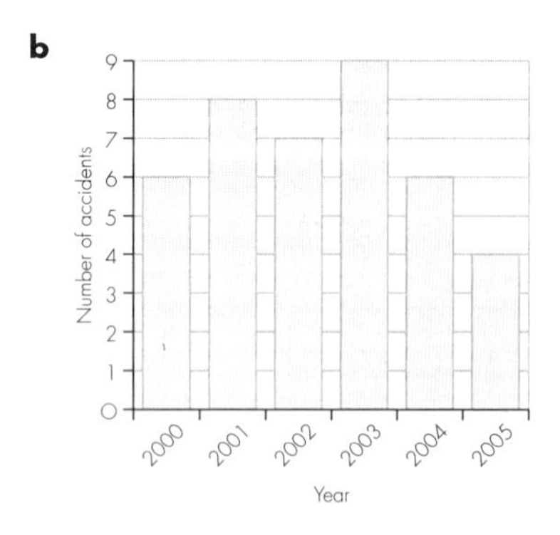

c Use the pictogram because an appropriate symbol makes more impact.

6 Yes. If you double the minimum temperature each time, it is very close to the maximum temperature.

7 The graphs do not start at zero.

8 Student's survey results and pictograms and bar charts.

9 Student's frequency tables and dual bar chart.

10 £44.40

10.4 Line graphs

Exercise 10D

1 **a** Tuesday, 52p **b** 2p **c** Friday **d** £90

2 **a**

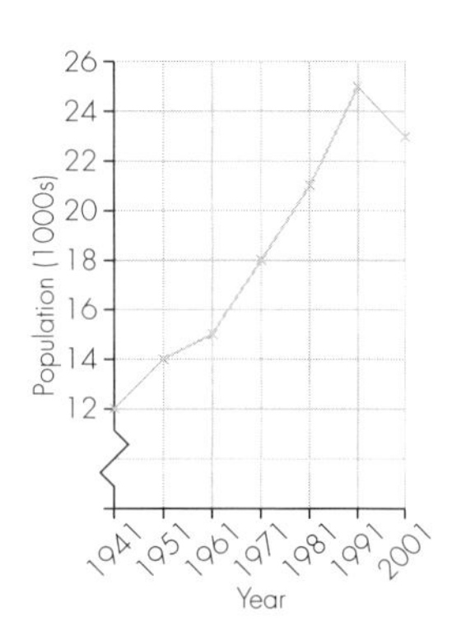

b about 16 500
c 1981 and 1991
d No; do not know the reason why the population started to decrease after 1991

3 **a**

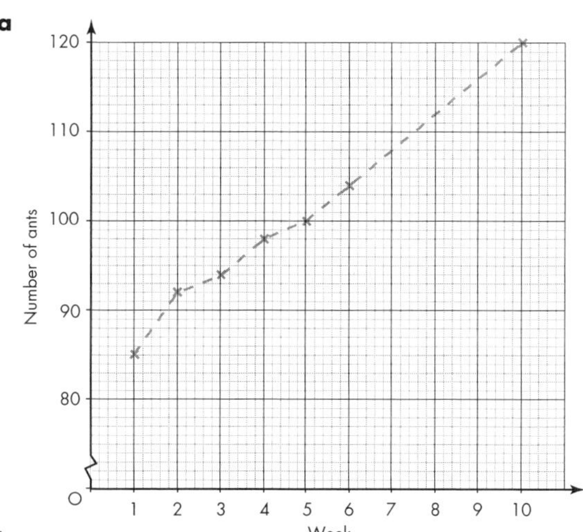

b 112

4 **a**

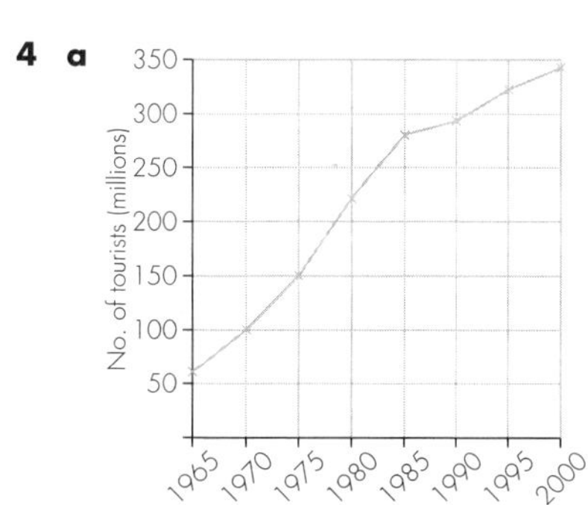

b About 410 million
c 1975 and 1980
d Student's explanation of trend.

5 **a**

Maximum temperature
Minimum temperature

b 7 °C and 10 °C

6 From a graph, about 1040 g

7 All the temperatures were presumably higher than 20 degrees.

10.5 Stem-and-leaf diagrams

Exercise 10E

1 **a** 17 s **b** 22 s **c** 21 s

2 **a** 57 **b** 55 **c** 56 **d** 48
e Boys did better, because their marks are higher.

3 **a**

2 | 8 9
3 | 4 5 6 8 8 9
4 | 1 1 3 3 3 8 8

Key 4 | 3 represents 43 cm

b 48 cm **c** 43 cm **d** 20 cm

4 **a**

0 | 2 8 9 9 9
1 | 2 3 7 7 8
2 | 0 1 2 3

Key 1 | 2 represents 12 messages

b 23 **c** 9

5 All the data start with a 5 and there are only two digits.

Examination questions

1 Tallies and frequencies: 3, 4, 7, 5, 4, 1

2 **a** 7
b Monday
c Tuesday and Wednesday

3 **a** 20
b 15
c 4 circles on Friday
$2\frac{1}{2}$ circles on Saturday

4 **a** chart filled in appropriately: Blue = 8 squares, Green = 5 squares, Yellow = 3 squares
b Blue

5 **a** **i** 8 **ii** 10
b Diana = 3 whole boxes
Erikas = 2 whole boxes and one small square

6 **a** Wednesday
Friday
b 320 minutes

7 The buses are all different sizes and there is no key.

8 1

9

```
2|3 5
3|1 2 7 8
4|0 6 8 9
5|6 6
```

Key: 2 | 5 = £25

10

```
0|7 8 9 9
1|0 0 0 2 5 5 6 7 8 9
2|1 1 3 8
3|0 2
```

Key: 2 | 5 = 25 years

11 **a** $150 \leqslant w < 200$
b Tom's tomatoes plotted at these points: (75, 21), (125, 28), (175, 26), (275, 9) and (325, 2) with a line connecting them
c Tom's were generally smaller, but more consistent

Answers to Chapter 11

11.1 The mode

Exercise 11A

1 **a** 4 **b** 48 **c** −1 **d** $\frac{1}{4}$ **e** no mode **f** 3.21

2 **a** red **b** sun **c** β **d** ★

3 **a** 32 **b** 6 **c** no
d no; boys generally take larger shoe sizes

4 **a** 5 **b** no; more than half the form got a higher mark

5 **a**

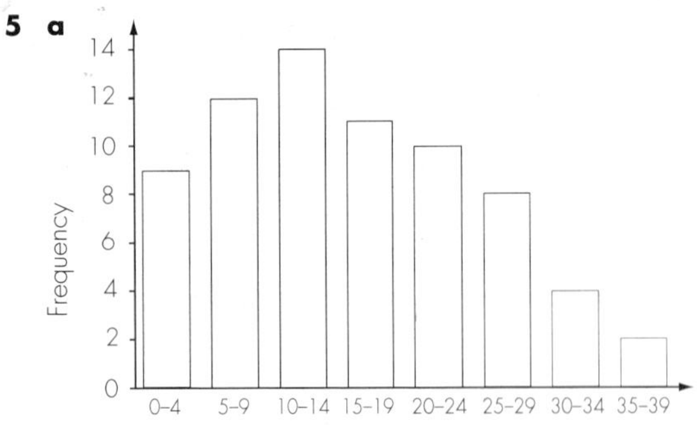

b 70 **c** 24
d cannot tell; know only that 9 households had between 0 and 4 e-mails
e 10–14

6 The mode will be the most popular item or brand sold in a shop.

7 **a** 28 **b** **i** brown **ii** blue **iii** brown
c Both students had blue eyes.

8 **a** May lose count.
b Put in a table, or arrange in order **c** 4

11.2 The median

Exercise 11B

1 **a** 5 **b** 33 **c** $7\frac{1}{2}$ **d** 24 **e** $8\frac{1}{2}$ **f** 0
g 5.25

2 **a** £2.20 **b** £2.25 **c** median, because it is the central value

3 **a** 5 **b** **i** 15 **ii** 215 **iii** 10 **iv** 10

4 **a** 13, Ella
b 162 cm, Pat
c 40 kg, Elisa
d Ella, because she is closest to the 3 medians

5 **a** 12 **b** 13

6 **a** 21 **b** 16
c

Mark	12	13	14	15	16	17	18
Frequency	1	3	4	3	6	3	1

d 15

7 Answers will vary

8 **a** 22 s **b** 25 s

9 **a** 56 **b** 48 **c** 49 **d** 45.5
e Boys have higher average and highest score.

10 12, 14, 14, 16, 20, 22, 24

11

```
2|4
3|8 8 9
4|1 2 5 7 7 7 8
5|0 3 5 8
6|0 3 4 4 8 9
7|1 3 5
8|9
```

key: 2 | 4 = 24 median = 53

12 **a** Possible answer: 11, 15, 21, 21 (one below or equal to 12 and three above or equal)
b Any four numbers higher than or equal to 12, and any two lower or equal
c Eight, all 4 or under

13 A median of £8 does not take into account the huge value of the £3000 so is in no way representative.

11.3 The mean

Exercise 11C

1 **a** 6 **b** 24 **c** 45 **d** 1.57 **e** 2

2 **a** 55.1 **b** 324.7 **c** 58.5 **d** 44.9 **e** 2.3

3 **a** 61 **b** 60 **c** 59 **d** Brian **e** 2

4 42 min

5 **a** £200 **b** £260 **c** £278
d Median, because the extreme value of £480 is not taken into account

6 **a** 35 **b** 36

7 **a** 6
b 16; all the numbers and the mean are 10 more than those in part a
c **i** 56 **ii** 106 **iii** 7

8 Possible answers: Speed – Kath, James, John, Joseph; Roberts – Frank, James, Helen, Evie. Other answers are possible.

9 36

10 24

11.4 The range

Exercise 11D

1 **a** 7 **b** 26 **c** 5 **d** 2.4 **e** 7

2 **a** 5°, 3°, 2°, 7°, 3° **b** Variable weather over England

3 **a** £31, £28, £33 **b** £8, £14, £4
c Not particularly consistent

4 **a** 82 and 83 **b** 20 and 12
c Fay, because her scores are more consistent

5 **a** 5 min and 4 min **b** 9 min and 13 min
c Number 50, because times are more consistent

6 **a** Issac, Oliver, Andrew, Chloe, Lilla, Billy and Isambard
b 70 cm to 92 cm

7 **a** Teachers because they have a high mean and students could not have a range of 20.
b Year 11 students as the mean is 15–16 and the range is 1.

11.5 Which average to use

Exercise 11E

1 **a** **i** 29 **ii** 28 **iii** 27.1 **b** 14

2 **a** **i** Mode 3, median 4, mean 5 **ii** 6, 7, $7\frac{1}{2}$ **iii** 4, 6, 8
b **i** Mean: balanced data **ii** Mode: 6 appears five times
iii Median: 28 is an extreme value

3 **a** Mode 73, median 76, mean 80
b The mean, because it is the highest average

4 **a** 150 **b** 20

5 **a** **i** 6 **ii** 16 **iii** 26 **iv** 56 **v** 96
b Units are the same
c **i** 136 **ii** 576 **iii** 435 **iv** 856
d **i** 5.6 **ii** 15.6 **iii** 25.6 **iv** 55.6 **v** 95.6

6 **a** Mean **b** Median **c** Mode
d Median **e** Mode **f** Mean

7 No. Mode is 31, median is 31, and mean is $31\frac{1}{2}$

8 **a** **i** £20 000 **ii** £28 000 **iii** £34 000
b **i** The 6% rise, because it gives a greater increase in salary for the higher paid employees.
ii 6% increase: £21 200, £29 680, £36 040;
+£1500: £21 500, £29 500, £35 500

9 **a** Median **b** Mode **c** Mean

10 Tom mean, David median, Mohaned mode

11 Possible answers: **a** 1, 6, 6, 6, 6 **b** 2, 5, 5, 6, 7

12 Boss chose the mean while worker chose the mode.

13 11.6

14 42.7 kg

11.6 Frequency tables

Exercise 11F

1 **a** **i** 7 **ii** 6 **iii** 6.4 **b** **i** 4 **ii** 4 **iii** 3.7
c **i** 8 **ii** 8.5 **iii** 8.2 **d** **i** 0 **ii** 0 **iii** 0.3

2 **a** 668 **b** 1.9 **c** 0 **d** 328

3 **a** 2.2, 1.7, 1.3 **b** Better dental care

4 **a** 0 **b** 0.96

5 **a** 7 **b** 6.5 **c**.6.5

6 **a** 1 **b** 1 **c** 0.98

7 **a** Roger 5, Brian 4 **b** Roger 3, Brian 8
c Roger 5, Brian 4 **d** Roger 5.4, Brian 4.5
e Roger, because he has the smaller range
f Brian, because he has the better mean

8 Possible answers: 3, 4, 15, 3 or 3, 4, 7, 9 …

9 Add up the weeks to see she travelled in 52 weeks of the year, the median is in the 26th and 27th week. Looking at the weeks in order, the 23rd entry is the end of 2 days in a week so the median must be in the 3 days in a week.

11.7 Grouped data

Exercise 11G

1 **a** **i** $30 < x \leqslant 40$ **ii** 29.5 **b** **i** $0 < y \leqslant 100$ **ii** 158.3
c **i** $5 < z \leqslant 10$ **ii** 9.43 **d** **i** 7–9 **ii** 8.4 weeks

2 **a** $100 < w \leqslant 120$ g **b** 10 860 g **c** 108.6 g

3 **a** 207 **b** 19–22 cm **c** 20.3 cm

4 **a** 160 **b** 52.6 min **c** modal group **d** 65%

5 **a** $175 < h \leqslant 200$ **b** 31% **c** 193.25 **d** No

6 Average price increases: Soundbuy 17.7p, Springfields 18.7p, Setco 18.2p

7 **a** Yes average distance is 11.7 miles per day.
b Because shorter runs will be completed faster, which will affect the average.
c Yes because the shortest could be 1 mile and the longest 25 miles.

8 The first 5 and the 10 are the wrong way round.

9 Find the midpoint of each group, multiply that by the frequency and add those products. Divide that total by the total frequency.

10 **a** As we do not know what numbers are in each group, we cannot say what the median is.
b Yes. The lowest number could be, for example, 28, and the highest could be 52, giving a range of 34.

11.8 Frequency polygons

Exercise 11H

1 a

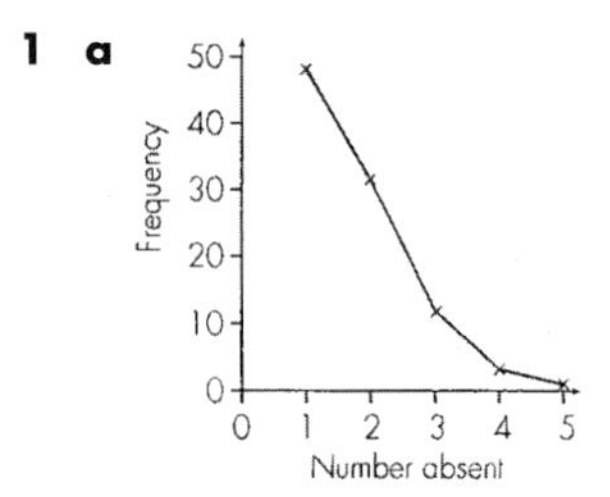

b 1.72

2 a

b 2.77

3 a

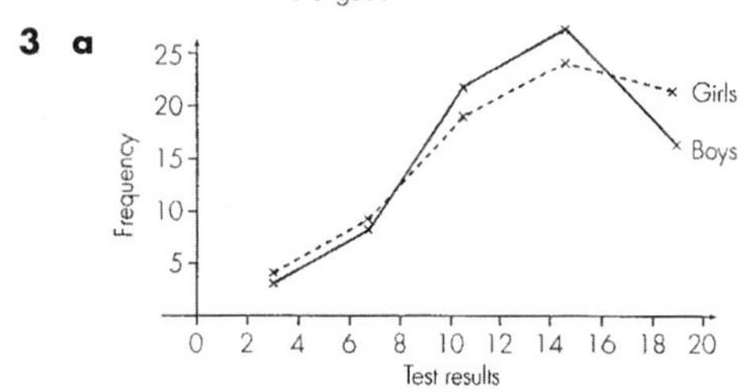

b boys 12.9, girls 13.1

4 a

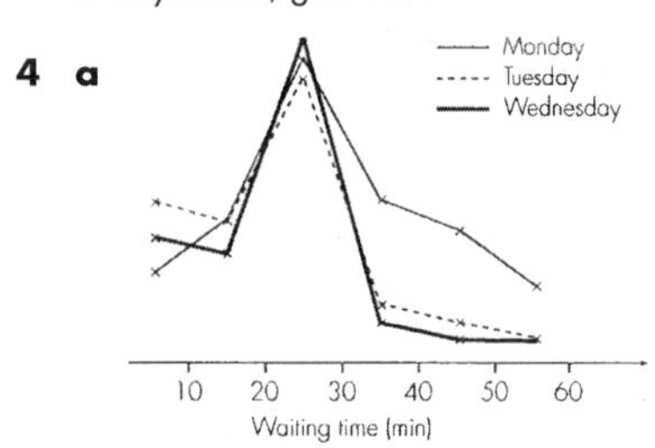

b Mon 28.4, Tue 20.9, Wed 21.3
c There are more people on a Monday as they became ill over the weekend.

5 a i 17, 13, 6, 3, 1 **ii** £1.45
b i

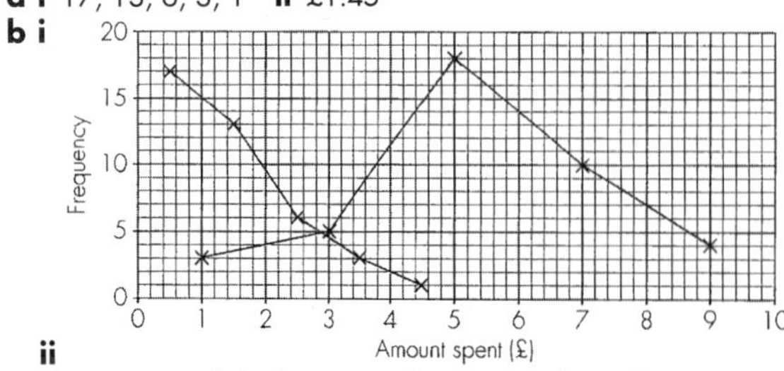

ii
c There is a much higher mean, first group of people just want a paper or a few sweets. Later people are buying food for the day.

6 a

Age, x	$20 < x \leqslant 30$	$30 < x \leqslant 40$	$40 < x \leqslant 50$	$50 < x \leqslant 60$	$60 < x \leqslant 70$
Frequency	14	16	14	12	4

b 41

7 2.17 hours

8 That is the middle value of the time group 0 to 1 minute, it would be very unusual for most of them to be exactly in the middle at 30 seconds.

Examination questions

1 a 6 **b** 9

2 a 6 **b** 6.2

3 a 65 k **b** 34 kg

4 a Tallies and frequencies 1, 4, 5, 3, 7
b 2 **c** USA

5 a Blond **b** Becky **c** 7

6 a They are quick growing
b They are of even height

7 a 37 **b** 52 **c** 48

8 a i 84 kg **ii** 53 kg **iii** 77.2 kg

9 a £210 **b** £360 **c** £459
d The mechanics get lower than the average pay.

10 a 39 **b** 49

11 a 5 **b** 41 kg

12 1

13 1.9

14 a 24 **b** 2 **c** 2 **d** 2.3

15 3.55 km

16 a 246 kg **b** 57 kg

17 6.08 hours

Answers to Chapter 12

12.1 Pie charts

Exercise 12A

1 a

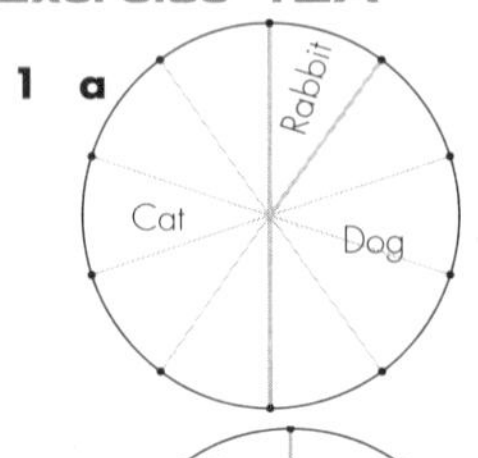

b

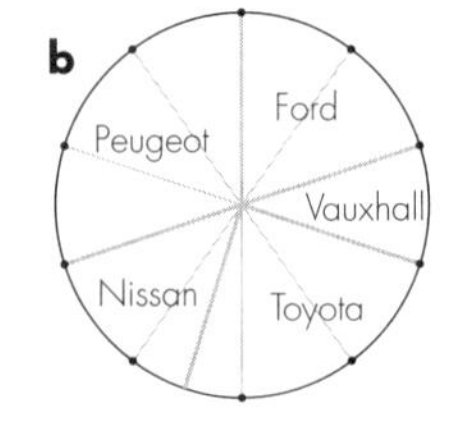

c

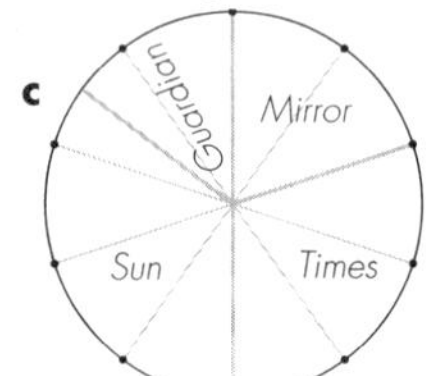

2 Pie charts with following angles:
a 36°, 90°, 126°, 81°, 27°
b 90°, 108°, 60°, 78°, 24°
c 168°, 52°, 100°, 40°

3 Pie charts with these angles: 60°, 165°, 45°, 15°, 75°

4 **a** 36
b Pie charts with these angles: 50°, 50°, 80°, 60°, 60°, 40°, 20°
c Student's bar chart.
d Bar chart, because easier to make comparisons.

5 **a** Pie charts with these angles: 124°, 132°, 76°, 28°
b Split of total data seen at a glance.

6 **a** 55° **b** 22 **c** $33\frac{1}{3}$%

7 **a** Pie charts with these angles:
Strings: 36°, 118°, 126°, 72°, 8°
Brass: 82°, 118°, 98°, 39°, 23°
Overall, the strings candidates did better, as a smaller proportion failed. A higher proportion of Brass candidates scored very good or excellent.

8 $\frac{1}{9}$

9 Choose a class in the school at random and ask each student how they get to school most mornings.

12.2 Scatter diagrams

Exercise 12B

1 **a** Positive correlation **b** Negative correlation
c No correlation **d** Positive correlation

2 **a** A person's reaction time increases as more alcohol is consumed.
b As people get older, they consume less alcohol.
c No relationship between temperature and speed of cars on M1.
d As people get older, they have more money in the bank.

3 **a**, **b** Student's scatter diagram and line of best fit.
c about 20 cm/s **d** about 35 cm

4 **a** Student's scatter diagram.
b Yes, usually (good correlation).

5 **a**, **b** Student's scatter diagram and line of best fit.
c Greta **d** about 70 **e** about 72

6 **a** Student's scatter diagram.
b no, because there is no correlation.

7 **a**, **b** Student's scatter diagram and line of best fit.
c about 2.4 km **d** 8 minutes

8 23 mph

9 Points showing a line of best fit sloping down from top left to bottom right.

12.3 Surveys

Exercise 12C

1–5 Student's own answers.

6 **a** The form should look something like this.
Question: Which of the following foods would you normally eat for your main meal of the day?

Name	Sex	Chips	Beef burgers	Vegetables	Pizza	Fish

b Yes, it looks correct as a greater proportion of girls ate healthy food.

7 The data-collection sheet should look something like this.
Question: What kind of tariff do you use on your mobile phone?

Name	Pay as you go		Contract	
	200 or more free texts	Under 200 free texts	200 or more free texts	Under 200 free texts

Various data-collection sheets would achieve the purpose; the student's answer will be accepted, provided he or she has offered some choices that can distinguish one tariff from another.

8 Examples are: shops names, year of student, tally space, frequency.

Exercise 12D

1 **a** Leading question, not enough responses.
b Simple 'yes' and 'no' response, with a follow-up question, responses cover all options and have a reasonable number of choices.

2 **a** Overlapping responses.
b ☐ £0–£2 ☐ over £2 up to £5
☐ over £5 up to £10 ☐ over £10

3–5 Students to provide own questionnaires.

6 **a** This is a leading question. There is no possibility of showing disagreement.
b This is a clear direct question that has an answer and good responses because only one selection can be made.
c Think of a question such as: Where do you usually buy your CDs?
Online ☐ Shopping centre ☐
Magazine offers ☐ Other ☐
Don't buy CDs ☐

7 A possible solution is:

Do you have a back problem? Yes ☐ No ☐

Tick the diagram that best illustrates how you sit.

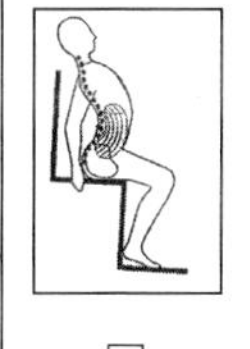
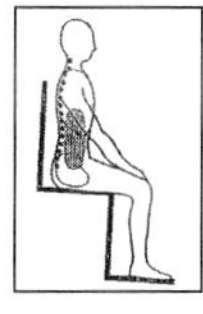
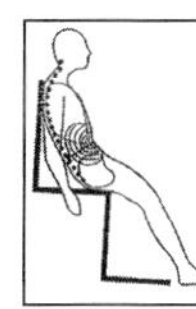
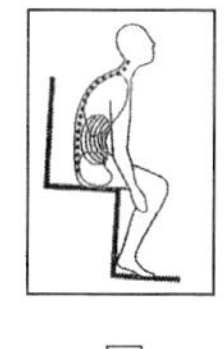

☐ ☐ ☐ ☐

8 The groups overlap each other. 'Less than £15' is included in the 'Less than £25'.

12.4 The data-handling cycle

Exercise 12E

1 Secondary data. Student to give own description of the data-handling cycle.

2 Primary data. Student to give own description of the data-handling cycle.

3 Primary or secondary. Student to give own description of the data-handling cycle.

4 Primary or secondary. Student to give own description of the data-handling cycle.

5 Primary. Student to give own description of the data-handling cycle.

6 Primary. Student to give own description of the data-handling cycle.

12.5 Other uses of statistics

Exercise 12F

1 Price: 78p, 80.3p, 84.2p, 85p, 87.4p, 93.6p

2 **a** 9.7 million **b** 4.5 years **c** 12 million
d 10 million

3 **a** £1 = $1.88
b Greatest drop was from June to July.
c There is no trend in the data so you cannot tell if it will go up or down.

4 £74.73

5 **a** Holiday month
b i 138–144 thousand **ii** 200–210 thousand

6 The general cost of living in 2009 dropped to 98% of that in 2008.

7 £51.50

Examination questions

1 The angles for the pie chart are: 60° gold, 80° silver, 220° bronze.

2 **a** Pie chart divided as follows: Understand – 6 sections; Understand some – 5 sections; Understand a little – 2 sections; Do not understand at all – 1 section
b Women had a better understanding. Around 85% of the women had some understanding or better, while of the men, the figure was only 75%.

3 **a** 27
b 195°

4 **a** Grade E
b $\frac{100}{360}$ or $\frac{5}{18}$
c i 16 **ii** 72
d Reason e.g. %, not actual numbers; do not know how many students, etc.

5 **a i** Diagram C **ii** Diagram A **iii** Diagram B
b Diagram A

6 **a**

	Eat vegetarian food	Do not eat vegetarian food	Total
Boys			
Girls			
Total			

b Yes, because 53% of the girls said they ate vegetarian compared with only 45% of the boys.

7 **a** Two of these statements: no time period, non-exhaustive response boxes, labels too vague
b Includes time period and proper response boxes

8 **a** There is no way to say the service is poor.
b How often do you visit the town centre?
Everyday ☐ 2, 3 or 4 times a week ☐ once a week ☐
2, 3 or 4 times a month ☐ once a month ☐
less than once a month ☐

9 **a** No time period and no option for 0
b One of too small a sample, not diverse enough as all from his class, or similar

10 **a** Graph accurately copied and points plotted at (50, 1.6) and (65, 1.75)
b Positive, i.e. the longer a mother's leg length, the heavier her baby's birth weight
c Line of best fit drawn with roughly 4 of the crosses on either side
d Around 1.64 kg

11 **a & b** Graph Time on horizontal axis from 0 to 20 and Distance km on vertical axis from 0 to 10 with the following points plotted: (3, 1.7) (17, 8.3) (11, 5.1) (13, 6.7) (9, 4.7) (15, 7.3) (8, 3.8) (11, 5.7) (16, 8.7) (10, 5.3) and with line of best fit drawn.
c & d Answers depend on student's plotting

12 **a** The longer the pike the more it weighs
b Cross at (78, 24) added to graph
c 15 kg

13 **a** Points (65, 100) and (80, 100) plotted on the graph.
b The greater the height, the longer the sheep.
c Roughly 109 cm

Answers to Chapter 13

13.1 Probability scale

Exercise 13A

1 **a** unlikely **b** certain **c** likely
d very unlikely **e** impossible **f** very likely
g evens

2 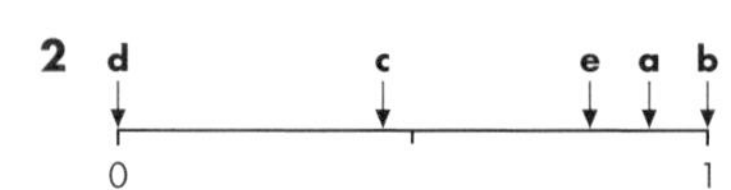

3

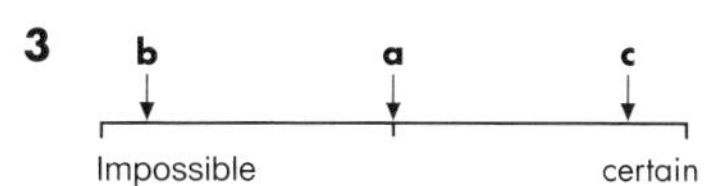

4 Student to provide own answers.

5 **a** very likely **b** very likely **c** very likely
d very likely **e** certain **f** very unlikely
g certain or very likely **h** very likely
i impossible **j** unlikely **k** certain
l very unlikely

6 As there is so little chance of winning the lottery with one ticket, even though having five tickets increases the chances fivefold, winning is still very unlikely.

13.2 Calculating probabilities

Exercise 13B

1 **a** $\frac{1}{6}$ **b** $\frac{1}{6}$ **c** $\frac{1}{2}$ **d** $\frac{1}{13}$ **e** $\frac{1}{4}$ **f** $\frac{1}{2}$
g $\frac{1}{3}$ **h** $\frac{1}{26}$ **i** $\frac{1}{13}$ **j** 0

2 **a** $\frac{1}{2}$ **b** $\frac{1}{2}$ **c** $\frac{1}{2}$ **d** $\frac{1}{52}$ **e** $\frac{4}{13}$ **f** $\frac{1}{52}$

3 **a** 0 **b** 1

4 **a** $\frac{1}{10}$ **b** $\frac{1}{2}$ **c** $\frac{2}{5}$ **d** $\frac{1}{5}$ **e** $\frac{2}{5}$

5 **a** $\frac{1}{3}$ **b** $\frac{1}{3}$ **c** $\frac{2}{3}$

6 **a** $\frac{6}{11}$ **b** $\frac{5}{11}$ **c** $\frac{6}{11}$

7 **a** $\frac{1}{5}$ **b** $\frac{1}{2}$ **c** $\frac{1}{2}$ **d** $\frac{7}{10}$

8 **a** $\frac{7}{15}$ **b** $\frac{2}{15}$ **c** $\frac{8}{15}$ **d** 0 **e** $\frac{8}{15}$

9 $\frac{1}{50}$

10 **a** AB, AC, AD, AE, BC, BD, BE, CD, CE, DE
b 1 **c** $\frac{1}{10}$ **d** 6 **e** $\frac{3}{5}$ **f** $\frac{3}{10}$

11 **a** 2 **b** 7 **c i** $\frac{5}{9}$ **ii** $\frac{4}{9}$

12 **a** $\frac{1}{13}$ **b** $\frac{1}{13}$ **c** $\frac{1}{2}$ **d** $\frac{2}{13}$ **e** $\frac{7}{13}$ **f** $\frac{1}{26}$

13 **a i** $\frac{12}{25}$ **ii** $\frac{7}{25}$ **iii** $\frac{6}{25}$
b They add up to 1. **c** All possible outcomes are mentioned.

14 35%

15 0.5

16 Class U

17 There might not be the same number of boys as girls in the class.

13.3 Probability that an outcome of an event will not happen

Exercise 13C

1 **a** $\frac{19}{20}$ **b** 55% **c** 0.2

2 **a i** $\frac{3}{13}$ **ii** $\frac{10}{13}$ **b i** $\frac{1}{4}$ **ii** $\frac{3}{4}$
c i $\frac{2}{13}$ **ii** $\frac{11}{13}$

3 **a i** $\frac{1}{4}$ **ii** $\frac{3}{4}$ **b i** $\frac{3}{11}$ **ii** $\frac{8}{11}$

4 **a** $\frac{1}{2}$ **b** 1 **c** $\frac{1}{3}$

5 Taryn

6 Because it might be possible for the game to end in a draw.

13.4 Addition rule for events

Exercise 13D

1 **a** $\frac{1}{6}$ **b** $\frac{1}{6}$ **c** $\frac{1}{3}$

2 **a** $\frac{1}{4}$ **b** $\frac{1}{4}$ **c** $\frac{1}{2}$

3 **a** $\frac{2}{11}$ **b** $\frac{4}{11}$ **c** $\frac{6}{11}$

4 **a** $\frac{1}{3}$ **b** $\frac{2}{5}$ **c** $\frac{11}{15}$ **d** $\frac{11}{15}$ **e** $\frac{1}{3}$

5 **a** 0.6 **b** 120

6 **a** 0.8 **b** 0.2

7 **a** $\frac{17}{20}$ **b** $\frac{2}{5}$ **c** $\frac{3}{4}$

8 Because these are three separate events. Also, probability cannot exceed 1.

9 $\frac{3}{4}$

10 $\frac{8}{45}$

11 The probability for each day stays the same, at $\frac{1}{4}$.

12 **a** The choices of drink and snack are not connected.
b (C, D), (T, G), (T, R), (T, D), (H, G), (H, R), (H, D)
c A possibility is not the same as a probability. Nine possibilities do not mean a probability of $\frac{1}{9}$.
d i (C, D) = 0.12, (T, G) = 0.15, (T, R) = 0.05, (T, D) = 0.3, (H, G) = 0.09, (H, R) = 0.03, (H, D) = 0.18
ii The total is 1 as this covers all the possibilities.

13.5 Experimental probability

Exercise 13E

1 **a** B **b** B **c** C **d** A **e** B **f** A
g B **h** B

2 **a** 0.2, 0.08, 0.1, 0.105, 0.148, 0.163, 0.1645
b 6 **c** 1 **d** $\frac{1}{6}$ **e** 1000

3 **a** 0.095, 0.135, 0.16, 0.265, 0.345
b 40 **c** No; all numbers should be close to 40.

4 **a** 0.2, 0.25, 0.38, 0.42, 0.385, 0.3974 **b** 8

5 **a** 6 **b, c** Student to provide own answers.

6 **a** Caryl, threw the greatest number of times.
b 0.39, 0.31, 0.17, 0.14
c Yes; all answers should be close to 0.25.

7 **a** not likely **b** impossible **c** not likely
d certain **e** impossible **f** 50–50 chance
g 50–50 chance **h** certain **i** quite likely

8 The missing top numbers are 4 and 5, the bottom two numbers are both likely to be close to 20.

9 Thursday

10 Although he might expect the probability to be close to $\frac{1}{2}$, giving 500 heads, the actual number of heads is unlikely to be exactly 500, but should be close to it.

13.6 Combined events

Exercise 13F

1 **a** 7 **b** 2 and 12

c $\frac{1}{36}, \frac{1}{18}, \frac{1}{12}, \frac{1}{9}, \frac{5}{36}, \frac{1}{6}, \frac{5}{36}, \frac{1}{9}, \frac{1}{12}, \frac{1}{18}, \frac{1}{36}$

d **i** $\frac{1}{12}$ **ii** $\frac{1}{3}$ **iii** $\frac{1}{2}$ **iv** $\frac{7}{36}$ **v** $\frac{5}{12}$ **vi** $\frac{5}{18}$

2 **a** $\frac{1}{12}$ **b** $\frac{11}{36}$ **c** $\frac{1}{6}$ **d** $\frac{5}{9}$

3 **a** $\frac{1}{36}$ **b** $\frac{11}{36}$ **c** $\frac{5}{18}$

4

Score on second dice						
6	5	4	3	2	1	0
5	4	3	2	1	0	1
4	3	2	1	0	1	2
3	2	1	0	1	2	3
2	1	0	1	2	3	4
1	0	1	2	3	4	5
	1	**2**	**3**	**4**	**5**	**6**

Score on first dice

a $\frac{5}{18}$ **b** $\frac{1}{6}$ **c** $\frac{1}{9}$ **d** 0 **e** $\frac{1}{2}$

5 **a** $\frac{1}{4}$ **b** $\frac{1}{2}$ **c** $\frac{3}{4}$ **d** $\frac{1}{4}$

6

Score on second spinner					
5	6	7	8	9	10
4	5	6	7	8	9
3	4	5	6	7	8
2	3	4	5	6	7
1	2	3	4	5	6
	1	**2**	**3**	**4**	**5**

Score on first spinner

a 6 **b** **i** $\frac{4}{25}$ **ii** $\frac{13}{25}$ **iii** $\frac{1}{5}$ **iv** $\frac{3}{5}$

7 $\frac{8}{64} = \frac{1}{8}$

8 It will show all the possible products.

9 **a**

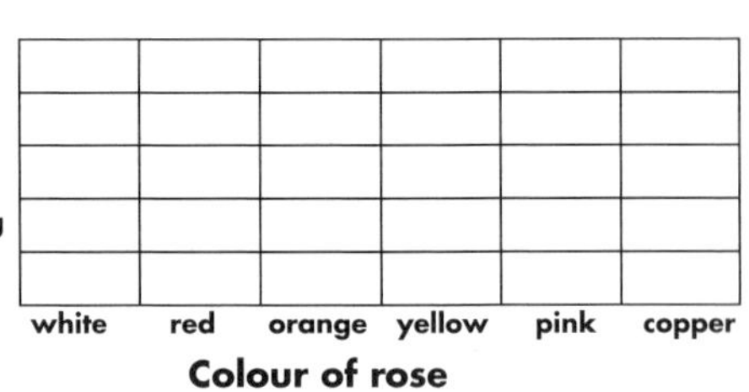

b **i** $\frac{6}{30} = \frac{1}{5}$ **ii** $\frac{4}{5}$

10 impossible: no dice; very unlikely: 5 or more dice; unlikely: 4 dice; evens: not possible; likely: 3 dice; very likely: 2 dice; certain: 1 dice

13.7 Expectation

Exercise 13G

1 **a** $\frac{1}{6}$ **b** 25

2 **a** $\frac{1}{2}$ **b** 1000

3 **a** **i** $\frac{1}{2}$ **ii** $\frac{1}{13}$ **iii** $\frac{1}{4}$ **iv** $\frac{1}{52}$

b **i** 260 **ii** 40 **iii** 130 **iv** 10

4 **a** $\frac{1}{37}$ **b** 5

5 **a** 150 **b** 100 **c** 250 **d** 0

6 **a** 167 **b** 833

7 1050

8 **a** 10, 10, 10, 10, 10, 10 **b** 3.5
c Find the average of the scores (21 ÷ 6)

9 **a** 0.111 **b** 40

10 281 days

11 Multiply the number of plants by 0.003.

13.8 Two-way tables

Exercise 13H

1 **a** Everton **b** Man Utd, Everton, Liverpool **c** Leeds

2 **a**

		Shaded	Unshaded
Shape	Circle	3	3
	Triangle	2	2

b $\frac{1}{2}$

3 **a** 40 **b** 16 **c** 40% **d** 40% **e** 16

4 **a**

		No. on disc		
		4	5	6
Letter on card	A	3	4	5
	B	4	5	6
	C	5	6	7

b $\frac{4}{9}$ **c** $\frac{1}{3}$

5 **a** 23 **b** 20% **c** $\frac{4}{25}$ **d** 480

6 **a** 10 **b** 7 **c** 14% **d** 15%

7 **a**

		Spinner A			
		1	2	3	4
Spinner B	5	6	7	8	9
	6	7	8	9	10
	7	8	9	10	11
	8	9	10	11	12

b 4 **c** **i** $\frac{1}{4}$ **ii** $\frac{3}{16}$ **iii** $\frac{1}{4}$

8 **a** 6 **b** 16 **c** 34 **d** $\frac{13}{15}$

9 **a**

		Number on dice					
		1	2	3	4	5	6
Coin	H	1	2	3	4	5	6
	T	2	4	6	8	10	12

b 2 (1 and 4) **c** $\frac{1}{4}$

10 **a** Those from the greenhouse have a larger mean diameter.
b Those from the garden have a smaller range, so are more consistent.

11 Either Reyki, because she had bigger tomatoes, or Daniel, because he had more tomatoes.

12 $\frac{22}{36} = \frac{11}{18}$

13 **a**

Score on second spinner						
10	10	11	13	15	17	19
8	8	9	11	13	15	17
6	6	7	9	11	13	15
4	4	5	7	9	11	13
2	2	3	5	7	9	11
0	0	1	3	5	7	9
Score on first spinner	0	1	3	5	7	9

b 9 **c** 0 **d** $\frac{15}{36} = \frac{5}{12}$ **e** $\frac{30}{36} = \frac{5}{6}$

Examination questions

1 A head is obtained when a fair coin is thrown once – Even
A number less than 7 will be scored when an ordinary six-sided dice is rolled once – Certain
A red disc is obtained when a disc is taken at random from a bag containing 9 red discs and 2 blue discs – Likely

2 **a** April and May
b Daffodil
c February
d Crocus
e **i** $\frac{1}{5}$ **ii**

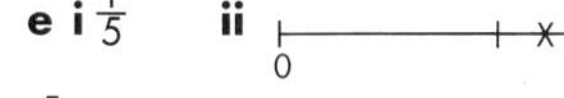

3 $\frac{5}{12}$

4 **a** 1
b There are five possible outcomes and only two of them are 2. The probability is $\frac{2}{5}$.

5 $\frac{5}{12}$

6 **a** $\frac{14}{25}$
b $\frac{21}{50}$

7 **a** 0
b $\frac{5}{8}$
c 12

8 **a**

	Black	White	Total
3-sided shape	1	4	5
4-sided shape	2	2	4
5-sided shape	2	0	2
Total	5	6	11

b $\frac{4}{11}$

9 **a**

	Spinner B			
Spinner A	1	3	4	7
1	2	4	5	8
4	5	7	8	11
5	6	8	9	12
8	9	11	12	15

b 8
c $\frac{1}{8}$
d $\frac{5}{16}$
e $\frac{1}{2}$

10 **a**

```
2 | 3 7 8
3 | 1 4 5 6
4 | 1 2 4 5 5
5 | 0 2 3
```

Key: 3 | 7 = 37 years

b $\frac{8}{15}$

GLOSSARY

above/below 1. A number greater than/less than another number. For example, above average, below freezing (point of water). 2. An object over/under another object.
acute angle An angle with a value between 0° and 90°. (See also **reflex angle**.)
add See **addition**
addition One of the basic operations of arithmetic. The process of combining two or more values to find their total value. Addition is the inverse operation to subtraction.
after/before 1. A number greater than/less than another number. For example 9 is the number after 8, 20 comes before 40. 2. An event occurring later/earlier than another event.
allied angles When two parallel lines are cut by a third line (transversal), two pairs of allied angles are formed between the lines, each pair on one side of the transversal. Each pair of allied angles add up to 180°.
alternate angles When two parallel lines are cut by a third line (transversal), two pairs of alternate angles are formed between the lines. The alternate angles lie one on each side of the transversal. Alternate angles are of equal size.
angle The space (usually measured in degrees [°]) between two intersecting lines or surfaces (planes). The amount of turn needed to move from one line or plane to the other.
angles around a point The angles around a point add up to 360°.
angles on a straight line The angles on a straight line add up to 180°.
approximate An inexact value that is accurate enough for the current situation.
approximation See **approximate**
average A single number that represents or typifies a collection of values. The three commonly used averages are mode, mean and median.
axis A fixed line used for reference, along or from which distances or angles are measured. A pair of coordinate axes is shown.
bar chart A diagram where quantities are represented by rectangles of the same width but different, appropriate heights.
below See **above/below**
best buy A purchase that gives best value for money spent.
better value A purchase that costs less or buys a greater quantity than another.
bias A die, coin, or spinner has a bias if it is more likely to land on one number/side than another.
brackets The symbols '(' and ')' which are used to separate part of an expression. This may be for clarity or to indicate a part to be worked out individually. When a number and/or value is placed immediately before an expression or value inside a pair of brackets, the two are to be multiplied together. For example, $6a(5b + c) = 30ab + 6ac$.
calculator An electronic device for working out mathematical operations. It is used by pressing keys and the results are shown on the screen.
cancel A fraction can be simplified to an equivalent fraction by dividing the numerator and denominator by a common factor. This is called cancelling.
capacity The volume of a liquid or gas.
centilitre (cl) A metric unit of volume or capacity. One hundredth of a litre. 100 cl = 1 litre.
centimetre (cm) A metric unit of length. One hundredth of a metre. 100 cm = 1 m.
certain Definite. An event is definitely going to occur. In this case the probability that the event will occur = 1.
chance The likelihood, or probability, of an event occurring.
class A collection of values grouped under one category or range.
class interval The size or spread of the measurement defining a class. For example, heights could be grouped in 1 cm or 10 cm class intervals.
coefficient The number in front of an unknown quantity (the letter) in an algebraic term. For example, in $8x$, 8 is the coefficient of x.
column 1. A vertical list of numbers or values. 2. The vertical parts of a table. 3. A way of arranging numbers to be added or subtracted.
column method (or traditional method) A method of calculating a 'long multiplication' by multiplying the number by the value of each digit of the multiplier and displaying the results in columns before adding them together to find the result.
common unit Express two or more quantities in the same unit, for the purpose of comparison or use in calculations. For example, if one time is given in hours and another in days, they need to be converted to a common unit before they can be compared or used in a calculation.
consistency A set of data is consistent if all the values are within a small or common range. If a few of the results fall outside this range, the data is not consistent and these 'rogue' results will affect or skew the conclusions drawn from the data.
continuous data Data that can be measured rather than counted, such as weight and height. It can have an infinite number of values within a range.
conversion factor A number that is used to convert a measurement in one unit to another unit. For example, $\times \frac{5}{8}$ converts kilometres to miles.
conversion graph A graph that can be used to convert from one unit to another. It is drawn by joining two or more points where the equivalence is known. Sometimes, but not always, it will pass through the origin.
correlation One measurement is affected by or affects another. For example, weight and height may correlate, but weight and hair colour do not.
corresponding angles When two parallel lines are cut by a third line (transversal), four pairs of corresponding angles are formed: a and e, b and f, c and g, and d and h. Corresponding angles are equal.
credit When you credit money to an account, you pay money in. If you pay 'using credit' (or with a credit card), you are paying with money you don't have. (Your bank is crediting you – lending you – money to use. This usually has to be paid back with additional interest.) If your account is 'in credit', you have money in the account. You have a positive amount of money. (See also **debit**.)

data collection sheet A form or table which is used for recording data collected during a survey.
debit When money is taken from an account, the account is debited with that amount. A debit card can only be used to spend money that you have – the cost is debited from the account. If your account is 'in debit', you have spent more than you had. You now have a negative amount – you owe money to the bank.
decimal Any number using base 10 for the number system. It usually refers to a number written with one or more decimal places.
decimal equivalents A number can be written as decimal, a fraction or a percentage. A fraction or percentage can be converted to a decimal equivalent – a decimal of the same value.
decimal fraction Usually refers to the part of a decimal number after (to the right of) the decimal point, that is, the part less than 1.
decimal place Every digit in a number has a place value (hundreds, tens, ones, etc.). The places after (to the right of) the decimal point have place values of tenths, hundredths, etc. These place values are known as decimal places.
decimal point The dot used to separate the integer (whole number) place values from the fraction place values (tenths, etc.)
denominator The number below the line in a fraction. It tells you the denomination, name, or family of the fraction. For example, a denominator of 3 tells you are thinking about thirds; a single unit has been divided into three parts. (See also **denominator** and **numerator**.)
difference The result of the subtraction of two numbers; the amount by which one number is greater than another number.
digit A number symbol. Our (decimal or denary) number system uses the digits 0, 1, 2, 3, 4, 5 , 6, 7, 8, and 9.
discrete data Data that is counted, rather than measured, such as favourite colour or a measurement that occurs in integer values, such as the number of days.
distance The separation (usually along a straight line) of two points.
division One of the basic operations of arithmetic. Division shows the result of sharing. For example, share 12 books among 3 people, they get 4 books each ($12 \div 3 = 4$). It is also used to calculate associated factors. For example, How many threes in twelve? There are four threes in twelve ($12 \div 3 = 4$). It is the inverse operation to multiplication. It can be written using $A \div B$, A/B or $\frac{A}{B}$.
divisions The marks or partitions on a scale that break it into sections.
dual bar chart This shows two bar charts on one set of axes. It might show the heights of boys and the heights of girls, rather than the heights of all children.
equally likely Two events are described as equally likely if the probabilities of the occurrence of each of the events are equal. For example, when a die is thrown, the outcomes 6 and 2 are equally likely. (They both have a probability of $\frac{1}{6}$)
equation A number sentence where one side is equal to the other. An equation always contains an equals sign (=).
equation of a line An equation, usually containing two variables (such as x and y), and from which you can plot a line on graph paper. For example, $y = 2x + 3$.
equivalent The same, equal in value. For example, equivalent fractions, equivalent expressions.
equivalent fractions Equivalent fractions are fractions which can be cancelled down to the same value, such as $\frac{10}{20} = \frac{5}{10} = \frac{1}{2}$.
estimate (as a verb) to state or calculate a value close to the actual value by using experience to judge a distance, weight, etc. or by rounding numbers to make the calculation easier; (as a noun) the same as estimation.
evaluate Work out the value of something.
event Something that happens. An event could be the toss of a coin, the throw of a die, or a football match.
expand Make bigger. Expanding brackets means you must multiply the terms inside a bracket by the number or letters outside. This will take more room to write, so you have 'expanded' the expression.
expect See **expectation**
expectation Something you expect to happen.
experiment A controlled test carried out to validate a hypothesis or make a discovery.
experimental A result from an experiment.
experimental data Data gathered from an experiment.
experimental probability The probability found by a series of trials or experiments. It will be an estimate of the true probability.
expression Collection of symbols representing a number. These can include numbers, variables (x, y, etc.), operations (+, ×, etc.), functions (squaring etc.), but there will be no equals sign (=).
extreme values Extreme values in a collection of data are very high and very low values. They are often noticeable because they are much greater or smaller than the rest of the data.
factor A whole number that divides exactly into a given number.
factorisation Finding one or more factors of a given number.
flow diagram A diagram showing the progression of a calculation or a logical path.
foot (ft) An imperial measurement of length, about 15 cm long. 12 inches = 1 foot, 3 feet = 1 yard.
formula (Plural: formulae) an equation that enables you to convert or find a measurement from another known measurement or measurements. For example, the conversion formula from the Fahrenheit scale of temperature to the more common Celsius scale is: $C = \frac{5}{9}(F - 32)$ where C is the temperature on the Celsius scale and F is the temperature on the Fahrenheit scale.
fraction A fraction means 'part of something'. To give a fraction a name, such as 'fifths', we divide the whole amount into equal parts (in this case, five equal parts). A 'proper' fraction represents an amount less than one (the numerator is smaller than the denominator.) Any two numbers or expressions can be written as a fraction, i.e. they are written as a numerator and denominator. (See also **numerator** and **denominator**.)

frequency How often something occurs.
frequency polygon A line graph drawn from the information given in a frequency table.
frequency table A table showing values (or classes of values) of a variable alongside the number of times each one has occurred.
function A function of x is any algebraic expression in which x is the only variable. This is often represented by the function notation $f(x)$ or function of x.
gallon (gal) An imperial measurement of volume and capacity. An average-sized bucket holds about 2 gallons. 8 pints = 1 gallon.
gradient How steep a hill or the line of a graph is. The steeper the slope, the larger the value of the gradient. A horizontal line has a gradient of zero.
gram (g) A metric unit of mass. 1000 grams = 1 kilogram.
greater than (>) Comparing the value of two numbers or quantities. For example, $8 > 4$ states that 8 has a higher value than 4.
grid A table of columns and rows such as those used to add up numbers. Squared or graph paper should be used for drawing charts and graphs.
grid method (or box method) A method of calculating a 'long multiplication' by arranging the value of each digit in a grid and multiplying them all separately before adding them together to find the result.
grouped data Data from a survey that is grouped into classes.
grouped frequency table A method of tabulating data by grouping it into classes. The frequency of data values that occur within a class is recorded as the frequency of that class.
historical data Data cannot always be found by experiment or from a survey. Data collected by other people, sometimes over a long period of time, is called historical data. Weather records would provide historical data.
hypothesis A theory or idea.
imperial The description of measurements used in the UK before metric units were introduced. They often have a long history (for example originating in Roman times) and are commonly based on units of twelve or 16 rather than ten used by the metric system.
impossible If something cannot happen, it is said to be impossible. The probability of it happening = 0.
improper fraction An improper fraction is a fraction whose numerator is greater than the denominator. The fraction could be re-written as a mixed number. For example, $\frac{7}{2} = 3\frac{1}{2}$
inch (in) An imperial unit of length. One inch is about $2\frac{1}{2}$ cm long. 12 inches = 1 foot.
inequality An equation shows two numbers or expressions that are equal to each other. A inequality shows two numbers or expressions that are not equal. The symbols $<, \leq, \geq, >$ are used.
input value The number or value that goes into a flow diagram. The value at the start of the flow diagram.
interior angle An angle between the sides inside a polygon. An internal angle.
key A key is shown on a pictogram and stem and leaf diagram to explain what the symbols and numbers mean. A key may also be found on a dual bar chart to explain what the bars represent.
kilogram (kg) A metric unit of mass. A bag of sugar has a mass of 1 kg. 1 kilogram = 1000 grams.
kilometre (km) A metric unit of distance. 1 kilometre = 1000 metres.
kite A quadrilateral with two pairs of adjacent sides that are equal. The diagonals on a kite are perpendicular, but only one of them bisects the kite.
length How long something is. We can talk about distances, such as the length of a table and also time, such as the length of a TV programme.
less than (<) Comparing the value of two numbers or quantities. Example: $2 < 7$ states that 2 is less than 7.
like terms Terms in algebra that are the same apart from their numerical coefficients. For example, $2ax^2$ and $5ax^2$ are a pair of like terms but $5x^2y$ and $7xy^2$ are not. Like terms can be combined by adding together their numerical coefficients so $2ax^2 + 5ax^2 = 7ax^2$.
likely If an event is likely to occur, there is a good chance that it will occur. There is no fixed number for its probability, but it will be between $\frac{1}{2}$ and 1.
line of best fit When data from an experiment or survey is plotted on graph paper, the points may not lie in an exact straight line or smooth curve. You can draw a line of best fit by looking at all the points and deciding where the line should go. Ideally, there should be as many points above the line as there are below it.
line segment A part of a line.
linear graphs A straight-line graph from an equation such as $y = 3x + 4$.
litre (l) A metric measure of volume or capacity. 1 litre = 1000 millilitres = 1000 cubic centimetres.
long division A method of division involving division by numbers with a large number of digits.
lowest terms A fraction which is in its simplest form is said to be written in its lowest terms. For example, the fraction $\frac{16}{24}$ may be written in its lowest terms as $\frac{2}{3}$.
margin of error Conclusions made from surveys and polls cannot be entirely accurate because, for example, the statistician can only interview a sample of people. The margin of error tells us how confident they are about the accuracy of their conclusion.
mean The mean value of a sample of values is the sum of all the values divided by the number of values in the sample. The mean is often called the average, although there are three different concepts associated with 'average': mean, mode and median.
median The middle value of a sample of data that is arranged in order. For example, the sample 3, 2, 6, 2, 2, 3, 7, 4 may be arranged in order as follows 2, 2, 2, 3, 3, 4, 6, 7. The median is the fourth value, which is 3. If there is an even number of values the median is the mean of the two middle values, for example, 2, 3, 6, 7, 8, 9, has a median of 6.5.
metre (m) A metric unit of length. 1 metre is approximately the arm span of a man. 1 metre = 100 centimetres.

metric A system of units of measurement where the sub-units are related by multiplying or dividing by ten. For example, for mass, 1 kilogram = 10 hectograms, 1 hectogram = 10 decagram, 1 decagram = 10 grams, 1 gram = 10 decigrams, 1 decigram = 10 centigrams, 1 centigram = 10 milligrams. The basic units of length and volume are metres and litres.
middle value The middle value of a sample of data is needed to find the median of that data. The values must be arranged in order first.
mile (mile) An imperial unit of length. One mile is almost 2 km. 1 mile = 1760 yards.
millilitre (ml) A metric unit of volume or capacity. One thousandth of a litre. 1000 ml = 1 litre.
millimetre (mm) A metric unit of length. One thousandth of a metre. 1000 mm = 1 metre.
mixed number A number written as a whole number and a fraction. For example, the improper fraction $\frac{5}{2}$ can be written as the mixed number $2\frac{1}{2}$
modal class If data is arranged in classes, the mode will be a class rather than a specific value. It is called the modal class.
modal value See **mode**.
mode The value that occurs most often in a sample. For example, the mode or modal value of the sample, 2, 2, 3, 3, 3, 3, 4, 5, 5, is 3.
more than (>) See **greater than**
multiplication A basic operation of arithmetic. Multiplication is associated with repeated addition. For example, 4 × 8 = 8 + 8 + 8 + 8 = 32. Multiplication is the inverse of division.
multiplication tables A grid or tables used to list the multiplication of numbers up to 12 × 12.
multiplier The number used to multiply by.
multiply out When brackets have a number or expression in front of them, it is a short way of writing, 'multiply everything inside the brackets by this'. Performing these multiplications is called 'multiplying out'.
mutually exclusive If the occurrence of a certain event means that another event cannot occur, the two events are mutually exclusive. For example, if you miss a bus you cannot also catch the bus.
national census The national census has been held every ten years since 1841. Everybody in the UK has to be registered and modern censuses ask for information about things such as housing, work, and certain possessions.
negative Something less than zero (in maths). The opposite of positive. (See also **negative number** and **positive**.)
negative coordinates To plot a point on a graph, two coordinates (x and y) are required. If either the x or y values are on the negative part of the number line, they are negative coordinates.
negative correlation If the effect of increasing one measurement is to decrease another, they are said to show negative correlation. For example, the time taken for a certain journey will have negative correlation with the speed of the vehicle. The slope of the line of best fit has a negative gradient.
negative number Describes a number whose value is less than zero. For example, –2, –4, –7.5, are all negative numbers. (See also **positive**.)
no correlation If the points on a scatter graph are random and do not appear to form a straight line, the two measurements show no correlation. One does not affect the other.
number line A continuous line on which all the numbers (whole numbers and fractions) can be shown by points at distances from zero.
numerator The number above the line in a fraction. It tells you the number of parts you have. For example, $\frac{3}{5}$ means you have three of the five parts. (See also **denominator**.)
observation Something that is seen. It can also be the result of a measurement during an experiment or something recorded during a survey.
obtuse angle An angle that is greater than 90° but less than 180°.
operation An action carried out on two or more numbers (in maths). It could be addition, subtraction, multiplication, or division.
opposite angles (or vertically opposite angles) When two straight lines cross, four angles are formed. The angles on the opposite side of the point of intersection are equal, so there are two pairs of equal opposite angles.
order 1. The sequence of carrying out arithmetic operations. 2. Arranged according to a rule. For example, ascending order.
ordered data Data or results arranged in ascending or descending order.
ounce (oz) An imperial unit of mass. 1 ounce is about 25 grams. 16 ounces = 1 pound.
outcome The result of an event or trial in a probability experiment, such as the score from a throw of a die.
output value The number or value that comes out of a flow diagram. The value at the end of the flow diagram.
parallelogram A four-sided polygon with two pairs of equal and parallel opposite sides.
partition method A method of multiplication in a grid where the results of the multiplication of each digit are written and read diagonally before addition.
percentage A number written as a fraction with 100 parts. writing $\frac{}{100}$, we use the symbol %. So $\frac{50}{100}$ is written as 50%.
pictograms A pictorial method of representing data on a graph. A very simple example of a pictogram is shown in which the profits of a shopkeeper over a three-month period are represented by 'bags of money', each bag being equivalent to £100.
pie chart A chart that represents data as slices of a whole 'pie' or circle. The circle is divided into sections. The number of degrees in the angle at the centre of each section represents the frequency.
pint (pt) An imperial unit of volume or capacity. 1 pint is about half a litre. Milk is usually sold in 1-pint or 2-pint cartons. 8 pints = 1 gallon.

poll A collection of data gathered from a survey of a group of people. An election is a poll because each person records their wish for the outcome of the election. Public opinion is gathered in polls where people state their preferences or ideas.
population All the members of a particular group. The population could be people or specific outcomes of an event.
positive Something greater than zero (in maths). The opposite of negative. (See also **positive number** and **negative**.)
positive correlation If the effect of increasing one measurement is to increase another, they are said to show positive correlation. For example, the time taken for a journey in a certain vehicle will have positive correlation with the distance covered. The slope of the line of best fit has a positive gradient. (See also **negative correlation**.)
positive number Describes a number whose value is greater than zero. The counting numbers are all positive numbers. (See also **negative**.)
pound (lb) An imperial unit of mass. 1 ounce is about half a kilogram. 1 pound = 16 ounces, 14 pounds = 1 stone.
primary data Data collected directly by a survey or experiment.
probability The measure of the possibility of an event occurring.
probability fraction All probabilities lie between 0 and 1. Any probability that is not 0 or 1 is given as a fraction – the probability fraction.
probability scale A line divided at regular intervals. It is usually labelled impossible, unlikely, even chance, etc., to show the likelihood of an event occurring. Possible outcomes can then be marked along the scale.
probability space diagram A diagram or table showing all the possible outcomes of an event.
profit/loss The gain or loss made from buying and selling.
protractor An instrument used for measuring angles.
quantity An measurable amount of something which can be written as a number or a number with appropriate units. For example, the capacity of a bottle.
random Haphazard. A random number is one chosen without following a rule. Choosing items from a bag without looking means they are chosen at random; every item has an equal chance of being chosen.
raw data Data in the form it was collected. It hasn't been ordered or arranged in any way.
reflex angle An angle that is greater than 180° and less than 360°.
relative frequency Also known as experimental probability. It is the ratio of the number of successful events to the number of trials. It is an estimate for the theoretical probability.
representative A number or quantity that is typical of a set of data. A person that is typical of a given group. An average is representative of all the given values.
retail price index (RPI) A measure of the variation in the prices of retail goods and other items.
rhombus A parallelogram that has sides of equal length. A rhombus has two lines of symmetry and a rotational symmetry of order two. The diagonals of a rhombus bisect each other at right angles and they bisect the figure.
rounding An approximation for a number that is accurate enough for some specific purpose. The rounded number, which can be rounded up or rounded down, may be used to make arithmetic easier or may be less precise than the unrounded number.
row A horizontal list of numbers or values. The horizontal parts of a table.
sample The part of a population that is considered for statistical analysis. The act of taking a sample within a population is called sampling. There are two factors that need to be considered when sampling from a population: 1. The size of the sample. The sample must be large enough for the results of a statistical analysis to have any significance. 2. The way in which the sampling is done. The sample should be representative of the population.
sample space diagram See **probability space diagram**
scales An instrument used to find the mass or weight of something.
scatter diagram A diagram of points plotted of pairs of values of two types of data. The points may fall randomly or they may show some kind of correlation.
secondary data Data collected by someone other than yourself.
sector A region of a circle, like a slice of a pie, bounded by an arc and two radii.
shape Could be a 2D or 3D shape. Any drawing or object. In mathematics we usually study simple shapes such as squares, prisms, etc., or compound shapes that can be formed by combining two or three simple shapes, such as an ice-cream cone made by joining a hemisphere to a cone.
sign A symbol used to represent something such as x, ÷ , =, and √. The sign of a number means whether it is a positive or a negative number.
significant figure The significance of a particular digit in a number is concerned with its relative size in the number. The first (or most) significant figure is the left-most, non-zero digit; its size and place value tell you the approximate value of the complete number. The least significant figure is the right-most digit; it tells you a small detail about the complete number. For example, if we write 78.09 to 3 significant figures we would use the rules of rounding and write 78.1. (See also **approximate**.)
simplest form A fraction cancelled down so it cannot be simplified any further. An expression where the arithmetic is completed so that it cannot be simplified any further.
simplify To make an equation or expression easier to work with or understand by combining like terms or cancelling down. For example: $4a - 2a + 5b + 2b = 2a + 7b$ or $\frac{12}{18} = \frac{2}{3}$
slope See **gradient**
social statistics Information about the condition and circumstances of people. For example, data about health and employment.
solve Finding the value or values of a variable (x) which satisfy the given equation.
speed How fast something moves.
spread See **range**

stone (st) An imperial unit of mass. 1 stone is about 6 kilograms. 1 stone = 14 pounds, 160 stone = 1 ton
substitution When a letter in an equation, expression, or formula is replaced by a number, we have substituted the number for the letter. For example, if $a = b + 2x$, and we know $b = 9$ and $x = 6$, we can write $a = 9 + 2 \times 6$. So $a = 9 + 12 = 17$.
subtraction One of the basic operations of arithmetic. It finds the difference between two numbers. Subtraction is the inverse operation to addition.
symbol A written mark that has a meaning. All digits are symbols of the numbers they represent. +, <, √, and ° are other mathematical symbols. A pictogram uses symbols to represent amounts.
tally chart A chart with marks made to record each object or event in a certain category or class. The marks are usually grouped in fives to make counting the total easier.
term 1. A part of an expression, equation, or formula. Terms are separated by + and – signs. 2. Each number in a sequence or arrangement in a pattern.
time How long something takes. Time is measured in days, hours, seconds, etc.
time series A sequence of measurements taken over a certain time.
times table See **multiplication table**.
ton (t) An imperial unit of mass. 1 ton is about 1 tonne. 1 ton = 160 stone.
tonne (t) A metric unit of mass. 1 tonne is about 1 ton. 1 tonne = 1000 kilograms.
trapezium A quadrilateral with one pair of parallel sides.
trial An experiment to discover an approximation to the probability of the outcome of an event will consist of many trials where the event takes place and the outcome is recorded.
two-way table These link two variables. One is listed in the column headers and the other in the row headers; the combination of the variables is shown in the body of the table.
units Ones, as in hundreds, tens, and units.
unlikely An outcome with a low chance of occurring.
unordered data See **raw data**.
value for money See **best buy**.
variable A quantity that can have many values. These values may be discrete or continuous. They are often represented by x and y in an expression.
volume The amount of space occupied by a substance or object or enclosed within a container.
weight How heavy something is. An object's weight is measured on scales or on a balance.
***x*-value** The value along the horizontal axis on a graph.
***y*-value** The value along the vertical axis on a graph.
yard (yd) An imperial unit of length. 3 feet = 1 yard. In metric units, 1 yard is about 91 cm.

INDEX

William Collins' dream of knowledge for all began with the publication of his first book in 1819. A self-educated mill worker, he not only enriched millions of lives, but also founded a flourishing publishing house. Today, staying true to this spirit, Collins books are packed with inspiration, innovation and practical expertise. They place you at the centre of a world of possibility and give you exactly what you need to explore it.

Collins. Freedom to teach.

Published by Collins
An imprint of HarperCollins*Publishers*
77–85 Fulham Palace Road
Hammersmith
London
W6 8JB

Browse the complete Collins catalogue at
www.collinseducation.com

10 9 8 7 6 5 4 3 2 1

ISBN-13 978-0-00-733986-0

Kevin Evans, Keith Gordon, Trevor Senior and Brian Speed assert their moral rights to be identified as the authors of this work

British Library Cataloguing in Publication Data
A Catalogue record for this publication is available from the British Library

Commissioned by Katie Sergeant
Project managed by Alexandra Riley
Edited and proofread by Brian Asbury, Joan Miller, Philippa Boxer, Margaret Shepherd and Karen Westall
Indexing by Michael Forder
Answer check by Amanda Dickson
Photo research by Jane Taylor
Cover design by Angela English
Content design by Nigel Jordan
Typesetting by Gray Publishing
Functional maths and problem-solving pages by EMC Design and Jerry Fowler
Production by Simon Moore
Printed and bound by L.E.G.O. S.p.A. Italy

Acknowledgements

The publishers have sought permission from Edexcel to reproduce questions from past GCSE Mathematics papers.

The publishers wish to thank the following for permission to reproduce photographs. Every effort has been made to trace copyright holders and to obtain their permission for the use of copyright material. The publishers will gladly receive any information enabling them to rectify any error or omission at the first opportunity.

Talking heads throughout © René Mansi/iStockphoto.com; p.6 © Karen Mower/iStockphoto.com, © D.Huss/iStockphoto.com; p.32–33 © Walesonview.com, © zwo5de/iStockphoto.com, © rydrych/iStockphoto.com, © BorisPamikov/iStockphoto.com; p.62–63 Cheshire Regiment/Wikipedia, © Xdrew, © reprorations.com; p.64 © ajb/iStockphoto.com, © Andrew Howe/iStockphoto.com, © fotoIE/iStockphoto.com, © Kiankhoon/iStockphoto.com, © theasis/iStockphoto.com, © Peter van Wagner/iStockphoto.com; p.88–89 © SDbT/iStockphoto.com, © JoeBiafore/iStockphoto.com, © laflor/iStockphoto.com, © dalton00/iStockphoto.com; p.90 © weareadventurers/iStockphoto.com, © Roland Frommknecht/iStockphoto.com, © Compassandcamera/iStockphoto.com, © Eugenio d'Orio/iStockphoto.com, © Steve Stone/iStockphoto.com; p.120–121 © Viktor Polyacov (Dr), © Jerry Fowler; p.122 © Bejan Fatur/iStockphoto.com, © Claudio Baba/iStockphoto.com, © Viktor Kitaykin/iStockphoto.com, © lisafx/iStockphoto.com, © nullplus/iStockphoto.com, © Joas Kotzch/iStockphoto.com, © Sean Locke/iStockphoto.com, © tjerophotography/iStockphoto.com; p.148–149 © Robert Stainforth/Alamy; p.150 © Anastazzo/Shutterstock Images LLC, © Kevin Gardner/Shutterstock Images LLC; p.178 © Henryk Sadura/iStockphoto.com, © MB Birdy/iStockphoto.com; p.206 Leonardo da Vinci/Wikimedia Commons, Chris73/Wikimedia Commons; p.234–235 © Peter Garbet/iStockphoto.com, © Iztok Grilic/iStockphoto.com, © Chritine Glade/iStockphoto.com, © Anastasia Pelikh/iStockphoto.com; p.236 William Playfair/Wikimedia Commons; p.272–273 © Catherine Karnow/Corbis; p.274 William Playfair/Wikimedia Commons, © HultonArchive/iStockphoto.com; p.304–305 © Roob/iStockphoto.com, © johnwoodcock/iStockphoto.com; p.306 © JenDen 2005/iStockphoto.com, © Domenico Pellegriti/iStockphoto.com; p.346–347 © SpatzPhoto/Alamy; p.348 William Playfair/Wikimedia Commons; p.380–381 © Robert Terry, © Jerry Fowler; p.382 © iLex/iStockphoto.com, © George Clerk/iStockphoto.com, © Owen Price/iStockphoto.com; p.424–425 © Olivier Meerson, © Jerry Fowler.

With thanks to Samantha Burns, Claire Beckett, Andy Edmonds, Anton Bush (Gloucester High School for Girls), Matthew Pennington (Wirral Grammar School for Girls), James Toyer (The Buckingham School), Gordon Starkey (Brockhill Park Performing Arts College), Laura Radford and Alan Rees (Wolfreton School) and Mark Foster (Sedgefield Community College).